Estuaries of the World

Series editor

Jean-Paul Ducrotoy, The University of Hull, Hull, United Kingdom

More information about this series at http://www.springer.com/series/11705

John W. Day • Jori A. Erdman
Editors

Mississippi Delta Restoration

Pathways to a sustainable future

Springer

Editors
John W. Day
Oceanography and Coastal Sciences
Louisiana State University
Baton Rouge, LA, USA

Jori A. Erdman
School of Architecture
Louisiana State University
Baton Rouge, LA, USA

ISSN 2214-1553 ISSN 2214-1561 (electronic)
Estuaries of the World
ISBN 978-3-319-65662-5 ISBN 978-3-319-65663-2 (eBook)
https://doi.org/10.1007/978-3-319-65663-2

Library of Congress Control Number: 2017957617

Printed on acid-free paper

This Springer imprint is published by Springer Nature
The registered company is Springer International Publishing AG
The registered company address is: Gewerbestrasse 11, 6330 Cham, Switzerland

To all those who live, work and recreate in both the Mississippi Delta and deltas worldwide.

Foreword

With this book coordinated by John Day and Jori Erdman – *Mississippi Delta Restoration: Pathways to a Sustainable Future* – the collection Estuaries of the World by Springer has taken a new step. This is the second volume in the series covering the Mississippi Delta. In the book published in 2014, John Day, Paul Kemp, Angelina Freeman and David Muth demonstrated that to secure the future of estuarine ecosystems, a multidisciplinary understanding of their complexity is needed. Taking the Mississippi Delta (United States) as a case study, they reviewed a number of questions about coastal Louisiana restoration. The occurrence of severe disruption, notably Hurricane Katrina, the fifth hurricane of the 2005 Atlantic hurricane season, made it necessary to consider the future of Louisiana coastal ecosystems through a multidisciplinary understanding of their complexity. In the new book, Day and Erdman propose a full set of tools (both conceptual and operational) for sustainable restoration of the delta, essential for policy makers and decision makers.

Katrina originated over the Bahamas on 23 August 2005 from a tropical atmospheric wave. It stood as the costliest natural disaster and one of the five deadliest hurricanes in the history of the United States. For decades, Louisiana has relied mainly on earthen levees to develop and protect reclaimed land. With sea level rise, and because of short-sighted policies costing billions of dollars to the state, one of the most ecologically important deltas is simply disappearing from the surface of planet Earth. Why then have ecological restoration approaches often been met with indifference or perplexity by local inhabitants? Some heritage advocates have even reacted in a counterproductive way that virtually rejects any landscape change. The conflict between development and preservation is a major factor and, sadly, a typical malaise in urban, peri-urban and rural environments that have been over-domesticated and industrialized. In this book, after Barbara Tuchman, John Day and Jori Erdman consider many current management strategies as "folly" because of mistaken and even counterproductive policies that have been and still are being promoted.

Feasible alternative courses of action must be proposed, but when these are ignored, we are in a state of "folly". Yet one should not fall into a romanticism according to which the present situation is so radically different from past situations that all past philosophies, rationalism above all, no longer suffice. They cannot be suddenly thrown out of the basket. Some writers go so far as to assert that the frontier between state and nature has been abolished as if the difference between the natural and the artificial had disappeared, the natural engulfed by the sphere of the artificial. Such an affirmation can only lead to inaction, suggesting that the humanization of nature has reached such a point that humans find only themselves and their own products outside of nature. This would lead to obscurantism. Such a position would hinder the integration of economic development (which of course does not necessarily mean growth) in the management of ecosystems. Is not the best framework for incorporating conservation into development strategies in the form of regional (and therefore decentralized) democratic governance, relying on reliable scientific knowledge?

Current strategies are a far cry from a governance perspective based on genuine decentralization, favouring the forging of links between scientists, sociologists, economists and local authorities. For their part, experts, whether ecologists, biologists or sociologists, should take better account in their work of the public and social context in which they are called upon to work. The political dimension requires alternatives to come from a group, not an individual leader, and should persist beyond one or few political lifetimes. Practical and theoretical alternatives to this impasse are needed. In this book, the authors are proposing an approach that encompasses both change and stability and unifies democratic and ecological values.

Landscapes are not passive objects, but are the result of human influence that involves the physical and conceptual organization of the environment in sustainable and coherent systems. Rehabilitated or re-natured coastal areas should not remain "uninhabited" and "not lived". There, the individual should be "reconstructed" and act. In this book, a fusion between disciplines has been achieved: from history to architecture through geology, hydrology, ecology, etc. The gap between so-called hard sciences and human sciences has vanished to benefit both economics and politics. A long way has been travelled. Scientific ecology has gone beyond the simplistic concepts of equilibrium systems, such as climax communities, and no longer uses simple deterministic models. In its place, as early as the 1970s, quantitative ecology began to consider ecosystems as complex hierarchical adaptive systems that possess unique structural qualities, such as resilience, multi-scalar assemblages, dissipative properties and descriptors of dynamics, such as nonlinearity, irreversibility, self-organization, emergence, development, directivity, history, co-evolution, indeterminism and chaos. This may explain why recently anthropology and other social sciences have considered ecology in an enthusiastic way. The time has come to consider relations between societies and their environments by promoting exchanges between these disciplines. How could one contribute to a general transdisciplinary theory of relations between humans and the environment – natural, social and cultural – without building bridges between science and sociology? The principles that guide such an opening of ecology require a systemic approach to contribute to a successful restoration, answering key questions: What do we want estuaries to do for us? How can society invest in sustainable restoration projects and not waste resources that will become scarce in the future? How much land are we ready to lose to the rising sea? Such a systemic approach will re-immerse humans in their environment as they are; it emphasizes the inseparability of humans with the ecological functions of the ecosystems to which they belong.

Hull, UK

Jean-Paul Ducrotoy

Preface to the Day and Erdman Text

Mississippi Delta Restoration: Pathways to a Sustainable Future

Perhaps I am predisposed to investigations on the nature and function of deltas. I grew up on the Kaministiquia Delta, the largest delta forming on the shores of Lake Superior. The city inhabiting the delta was once called Fort William but changed its name in 1967 to Thunder Bay. Like all deltas, the land was flat and easy to build on. Industries established themselves, employing the river as a transportation path to the world's largest complex of freshwater lakes, the Great Lakes. Grains from the prairie farmlands, ores from the Canadian north, and pulp and lumber from the more local boreal forest made the port an important cog of the Canadian economy. Ships used the extended "lakers" some 1000 ft long. Even ocean-going vessels visited our port. Rich farmlands that extended up river supported the ever-growing delta population.

I enjoyed biking along the flat delta streets and roads and boating and fishing along the rivers and delta shores. Lake swells, locally called rollers, would pour into the harbor, big enough to scare fishermen in their small vessels. I remember the rich biodiversity of the wetlands, or what we called swamplands. These delta areas were infested with the worst of our local insects, such as black flies, various mosquitoes, deer and horse flies, bees and wasps, and dragonflies that would fly around us, chewing up these pesky insects that swarmed us. Water snakes, turtles, and frogs scurried in the grasses. Many times we would see big black Norwegian rats that ships had brought to our shores; even cats were scared of them. Many other invasive species were brought to our shores, including the nefarious zebra mussels that often clogged our pipes and coated our piers.

Even as a youngster, I understood how an upstream flood wave would gather size and power, often to peak in height as the water passed through the city. I remember when upstream dams were emplaced, to provide the city with flood control but also to generate important electricity to power our homes and industry. Our soils were very fine grained – fine sands, silts, and clay – and very rich in organic matter. It was rather easy to grow vegetables, and most of our neighbors did. As with any major port, national rail and road services connected the city to the outside world.

If this all sounds similar to chapters within this book, it is. There is a bond between delta folks, and I certainly established a bond with my colleague and friend Professor John Day. John grew up on the Mississippi Delta, more than an order-of-magnitude larger than my "Kam" delta. But our experiences were similar, even though I froze in winter and he sweltered in summer. John and I are two of nearly 600 million citizens of the world who live on or very near deltas.

Perhaps the greatest challenge facing deltas is the human occupation of these sensitive environments. Many of the world's river deltas formed thousands of years ago, as global ocean volume stabilized. Other deltas such as the Po, Ebro, Rhone, and Yellow Deltas owe their current existence to increased flux of sediment that was liberated from uplands due to land use disturbance – deforestation, agriculture, mining, and urbanization. While deltas rapidly grew, and shorelines shifted seaward, delta dwellers focused on the ephemeral nature of water routes distributed across a delta, since distributary channels would shift every few years to decades. So while deltas offered lush vegetation, diverse forests, wildlife and fisheries, and rich organic muddy soils, movement through these ultra-flat environments remained difficult. Tropical and subtropical infections, particularly from disease-carrying insects, kept human populations low. While annual river floods nourished soils, floodwater inundation also destroyed habitats and livelihoods.

To control floods and reduce risk, human-fortified levees were built along the larger river branches. In the case of the Mississippi, embankments were made with the full knowledge that future land might be lost to the sea as a result of the embankments. Constraining floodwaters by levees forced sediment-laden water to enter and exit a delta largely intact, with most sediment being deposited at mouth bars or in offshore deposits. Where delta wetlands were drained for agriculture, soil peats oxidized and land surfaces subsided.

During the twentieth century, upstream dams sequestered billions of tons of sediment, and the sediment delivered to most deltas has become greatly reduced. In some rivers, such as the Colorado, Ebro, and Nile, little sediment is now delivered. Since soils are no longer enriched by floodwaters, farmlands have had to rely on commercial fertilizers. Civil engineering continues to transform world deltas. Windmills and other pumping mechanisms keep subsided soils from becoming waterlogged. Some societies continue to view engineering as our pathway to both safety and prosperity, with future risks being met with future solutions.

With modernity came the growth of delta megacities – Bangkok, Jakarta, Rangoon, Dhaka, Alexandria and Cairo, Seoul, Tokyo, Shanghai, Guangzhou, Tianjin, and arguably Mumbai, Kolkata, Karachi, Lagos, and Ho Chi Minh City. Each city owes their existence to the very flat land that river deltas provide, in combination with the rich ecological resources of the delta and their association with large river arteries for trade. As the delta cities skyrocketed – an order-of-magnitude growth in the last few decades – the need for fresh, potable water increased. The draw of groundwater became industrialized, particularly as many of the agricultural areas producing rice transformed into protein bowls through the growth of twenty-first-century shrimp farming and other aquaculture operations. Petroleum mining and peat mining further accelerated delta subsidence and degraded the wetlands in general.

Global mean sea level is now rising at rates exceeding 3 mm/year due to climate warming; delta land surfaces are sinking at rates of 10–100 s of mm/year when combining this sea level rise with land subsidence. Shorelines are rapidly receding, and great swaths of land are being given up to the marine environment. With the transformation away from wetland and mangrove forests to farmlands and aquaculture, coastal inundation from ocean storms more easily overwhelms these sensitive environments. Each year more than 20 million people are displaced from their coastal residence due to storm surges and river flooding.

As I read this fascinating book, my memories of growing up on a delta bubble up. But I also reflect on the trillions of US dollars being put at risk from the destructive force of extreme weather events, climate change, and human greed. But whether a particular delta ends its existence from a series of slowly moving injuries or from a few big events, civilization will have contributed to the destruction of one of nature's true biodiversity hotspots and former cradle of human existence. Most deltas now function under the influence of human governance. Little of the natural system remains. Unless delta cultures and inhabitants can develop approaches and infrastructure to survive past decisions and future weather events, then the advantages of world deltas (flat-lying food sources and transportation hubs) will become disadvantages.

I hope you enjoy this fact-rich treatise on the Mississippi Delta. The book covers the growth and formation of this world-class delta during the Holocene (since the last great Ice Age), its slow transformation through historical times by humans, and its rapid transition during the epoch we now refer to as the Anthropocene (since 1950). I particularly enjoyed the historical perspective and early photos of this unique delta. But the messages of delta sustainability are very clear. Americans must wake up to the value of their Mississippi Delta and the ecosystem services that it provides and the one and only gateway to the heartland of America with its legacy of history and culture. Humans have blindly reduced the many bounties of the Mississippi Delta, including its lush but delicate wetlands. Let's assess our human footprint and work to reduce it. Coastal resilience is the key, and this book provides the necessary path forward.

University of Colorado, USA
June 18, 2017

J.P.M. Syvitski

Acknowledgements

This book was supported by a grant from the Louisiana State University Coastal Sustainability Studio Small Projects Fund [award number 1512]. Additional support came from a grant from the Gulf Research Program of the National Academies of Sciences [award number 2000005991]. The principal investigators of the LSU project were Craig Colten, James Wilkins, Jori Erdman, Clint Willson and John Day, and each provided important contributions at meetings as the book developed. Jeff Carney and Mary Bergeron were very supportive and facilitated access to the Coastal Sustainability Studio for work sessions and charrettes. Hunter Odom provided assistance with managing footnotes and developing the table of laws in chapter "Developing Legal Strategies for a Sustainable Coast". Adrian Wiegman, Jeff Rutherford, Christopher James and Giovanni Coakley managed logistics during the book's completion.

Contents

Contributors

Christopher G. Brantley U.S. Army Corps of Engineers, Norco, LA, USA

Jeffrey A. Carney Coastal Sustainability Studio, and School of Architecture, Louisiana State University, Baton Rouge, LA, USA

Giovanni P. Coakley School of Architecture, Louisiana State University, Baton Rouge, LA, USA

Craig E. Colten Department of Geography and Anthropology, Louisiana State University, Baton Rouge, LA, USA

John W. Day Department of Oceanography and Coastal Sciences, Louisiana State University, Baton Rouge, LA, USA

Christopher F. D'Elia College of the Coast and Environment, Louisiana State University, Baton Rouge, LA, USA

Jori A. Erdman School of Architecture, Louisiana State University, Baton Rouge, LA, USA

Elaine Evers-Hebert Department of Oceanography and Coastal Sciences, College of the Coast and Environment, Louisiana State University, Baton Rouge, LA, USA

Scott A. Hemmerling The Water Institute of the Gulf, Baton Rouge, LA, USA

Guerry O. Holm CH2M, Baton Rouge, LA, USA

Christopher W. James School of Architecture, Louisiana State University, Baton Rouge, LA, USA

Robert R. Lane Department of Oceanography and Coastal Sciences, Louisiana State University, Baton Rouge, LA, USA

G. Paul Kemp Department of Oceanography and Coastal Sciences, Louisiana State University, Baton Rouge, LA, USA

Jeffrey S. Rutherford Department of Oceanography and Coastal Sciences, Louisiana State University, Baton Rouge, LA, USA

Charles E. Sasser Department of Oceanography and Coastal Sciences, College of the Coast and Environment, Louisiana State University, Baton Rouge, LA, USA

Gary P. Shaffer Department of Biological Sciences, Southeastern Louisiana University, Hammond, LA, USA

Adrian R. H. Wiegman Department of Oceanography and Coastal Sciences, Louisiana State University, Baton Rouge, LA, USA

James G. Wilkins Louisiana Sea Grant Law and Policy Program, Center for Natural Resource Economics and Policy, Louisiana State University, Baton Rouge, LA, USA

Elizabeth A. Williams Foundation for Louisiana, Baton Rouge, LA, USA

Introduction – Changing Conditions in the Mississippi Delta from 1700 to 2100 and Beyond: Avoiding Folly

John W. Day and Jeffrey A. Carney

Abstract

The Mississippi River delta developed over the past several 1000 years after sea level stabilized at the end of the last glaciation. Native Americans lived in the delta during this time but had almost no impact on its functioning. This all changed with the arrival of European colonialists at the beginning of the eighteenth century. Over the ensuing two centuries, human impact multiplied. This included levee construction, closing of distributaries, and reclamation. But it was not until the twentieth century that human intervention had a major impact, culminating in the loss of about a quarter of the wetlands of the Mississippi delta by the end of the twentieth century. Recent studies have documented how profoundly the delta has changed and what the future portends if nothing is done. Currently there is an ambitious plan to restore the sustainability of the Mississippi delta. This book contributes to this effort to define sustainable pathways for the restoration of the delta. The book grew out of two efforts. One is an earlier volume in the Estuaries of the World Series entitled Perspectives on the Restoration of the Mississippi delta. This effort was supported by environmental organizations to provide independent advice to the philanthropic community and non-governmental organizations on how best to support the restoration of one of North America's premier ecological assets, the Mississippi delta. The second effort, called Changing Course, was a design competition that sought to bring teams of some of the world's best delta engineers, scientists, and planners to address the question of the future of the Lower Mississippi River Delta and its dependent resources and communities. This Initiative specifically sought to understand how the Mississippi River's water and sediment can be used to maximize rebuilding the delta wetlands while meeting or even enhancing the needs of navigation, flood protection, industries and coastal communities. In the current book, we investigate possible future pathways for sustainable management of the Mississippi delta. We consider current conditions as well as future trajectories of climate and energy and resource scarcity. We conclude that without profound changes of how humans live in and manage the delta, sustainability of the delta will be profoundly compromised.

Keywords

Mississippi River history • New Orleans • Changing course • Coastal land loss • Wetland restoration

J.W. Day (✉)
Department of Oceanography and Coastal Sciences, Louisiana State University, Baton Rouge, LA 70803, USA
e-mail: johnday@lsu.edu

J.A. Carney
Coastal Sustainability Studio, and School of Architecture, Louisiana State University, Baton Rouge, LA, USA
e-mail: jcarney@lsu.edu

J.W. Day, J.A. Erdman (eds.), *Mississippi Delta Restoration*,
Estuaries of the World, https://doi.org/10.1007/978-3-319-65663-2_1

In 1699, in response to English territorial ambitions, the French government sent an expedition to the northern Gulf of Mexico coast. The expedition was under the command of Pierre Le Moyne, Sieur d'Iberville, who was accompanied by his younger brother Jean-Baptiste Le Moyne, Sieur de Bienville. Their orders were to locate "the mouth of the Mississippi River,"...and to "select a good site that can be defended with a few men, and block entry to the river by other nations." The expedition entered the river on March 2, 1699. On March 3, the day before Ash Wednesday and the beginning of Lent, they camped on the bank of the river a few leagues upriver from the mouth and named the location Point de Mardi Gras. This led to the first permanent settlement in 1718 on the Mississippi further upriver at what is now New Orleans. Many consider the encampment at Point de Mardi Gras to be the date that the French established a permanent presence along the lower Mississippi. This story is recounted by Lawrence Powell (2012) in his fascinating history of the first century of New Orleans, *The Accidental City*.

Let's conduct a thought experiment about south Louisiana beginning in 1700. Imagine that Bienville, standing on the bank of the Mississippi at Point de Mardi Gras in 1699, could have looked into the future until 1900. What would have happened over those two centuries? Sea level varied only a few cm and the mean annual temperature did not significantly increase, and carbon dioxide levels remained fairly constant around the preindustrial level of about 280 parts per million. The mean flow of the great river would remain more or less the same, albeit with considerable variability from year to year, and deliver large volumes of freshwater and sediments that flowed over and nourished almost all of the deltaic plain (Condrey et al. 2014), while the economy of south Louisiana grew and prospered. The City of New Orleans ended the nineteenth century with a population of about 290,000, nearly 25% of Louisiana's total population of 1.2 million, and practically all of the city was above sea level (Fig. 1). The world population did not reach 2 billion until 1927, the year of the great flood. So the human population, both in south Louisiana and globally, likely could have been supported by the ecosystem services by the biosphere. Fossil fuel use was less than 1% of what it is currently. The broad Mississippi delta ended the nineteenth century with a large area of marshes and freshwater forested wetlands that were much more extensive and extended much further south than at present. It is likely that had Bienville searched, he would have found the tropical black mangrove, *Avicennia geminans*, to be very rare. Hurricanes would pound the coast in the ensuing two centuries but their average strength and frequency did not grow and the extensive wetlands south of the city provided an effective buffer against the storms. Regardless of our thought experiment, it is clear that even if Bienville had been able to look into the future, he would not have understood what he saw because much of the scientific understanding we have now did not exist in 1700. But in the first quarter of the twenty-first century, we have extensive information about the past and can reconstruct much of what happened from 1700 to 1900. And there is now abundant information that allows us to have a fairly good general idea of what will have happen in the next two centuries.

All of this average stability and predictability changed in the twentieth century. CO_2 levels soared to over 400 ppm by 2015, a level not seen for over 3 million years. Global mean temperature rose by about one degree Centigrade in the twentieth century. At the end of the last ice age, it took about 1000 years for temperature to increase by a similar amount. Temperature is projected to increase by 3–5 °C by 2100. The rate of sea-level rise increased from about one mm per year in the late 1800s to between 3 and 4 mm per year in the early twenty-first century (Fitzgerald et al. 2008). By the end of the twentieth century, continuous levees separated the river from almost all of the deltaic plain and the great delta was collapsing (Day et al. 2007, 2014a, b).

With this in mind, we can imagine one of Bienville's descendants in 2016 looking out over the river and coast and thinking about what the ensuing two centuries would be like. Perhaps a dowager of the French Quarter looking out an upper story window, or a street person lounging on the levee in front of Jackson Square, or one of the culturally and racially diverse inhabitants of south Louisiana cutting through the coastal wetlands in a fishing boat. What might the vision of Bienville's descendant be if current environmental trajectories continue until 2217? Unlike Bienville, these people if they so choose can have a fairly in depth understanding of what the next two centuries will be like based on an extensive array of scientific information.

Temperature rise will be dramatic, probably several degrees Centigrade globally, and the entire Gulf coast will be tropical. Wetland loss in the delta is dramatic (Fig. 2). Where saline wetlands exist, if indeed they survive at all, they are dominated by mangroves. These include not just the most cold-tolerant black mangrove (*Avicennia germinans*) but all three species common to the southern Gulf of Mexico – red mangroves (*Rhizophora mangle*) with their graceful arching prop roots as well as white mangrove (*Laguncularia racemosa)* with finger like pneumataphores sticking up from the muddy sediment. Sea level is much higher, perhaps by as much as three to five meters. Katrina-strength hurricanes with wind speeds of 150 miles per hour or more are much more frequent and regularly ravage the coast. The number of large floods on the Mississippi River has increased significantly due to both climate change and land-use changes. Fossil fuels are largely non-existent and what is left is extremely expensive. The flood control system on the Mississippi River, the vaunted Mississippi River and Tributaries project that was developed after the 1927 flood, has largely failed. Humans have lost control of the river and it has broken through the levees and is restoring the coast in

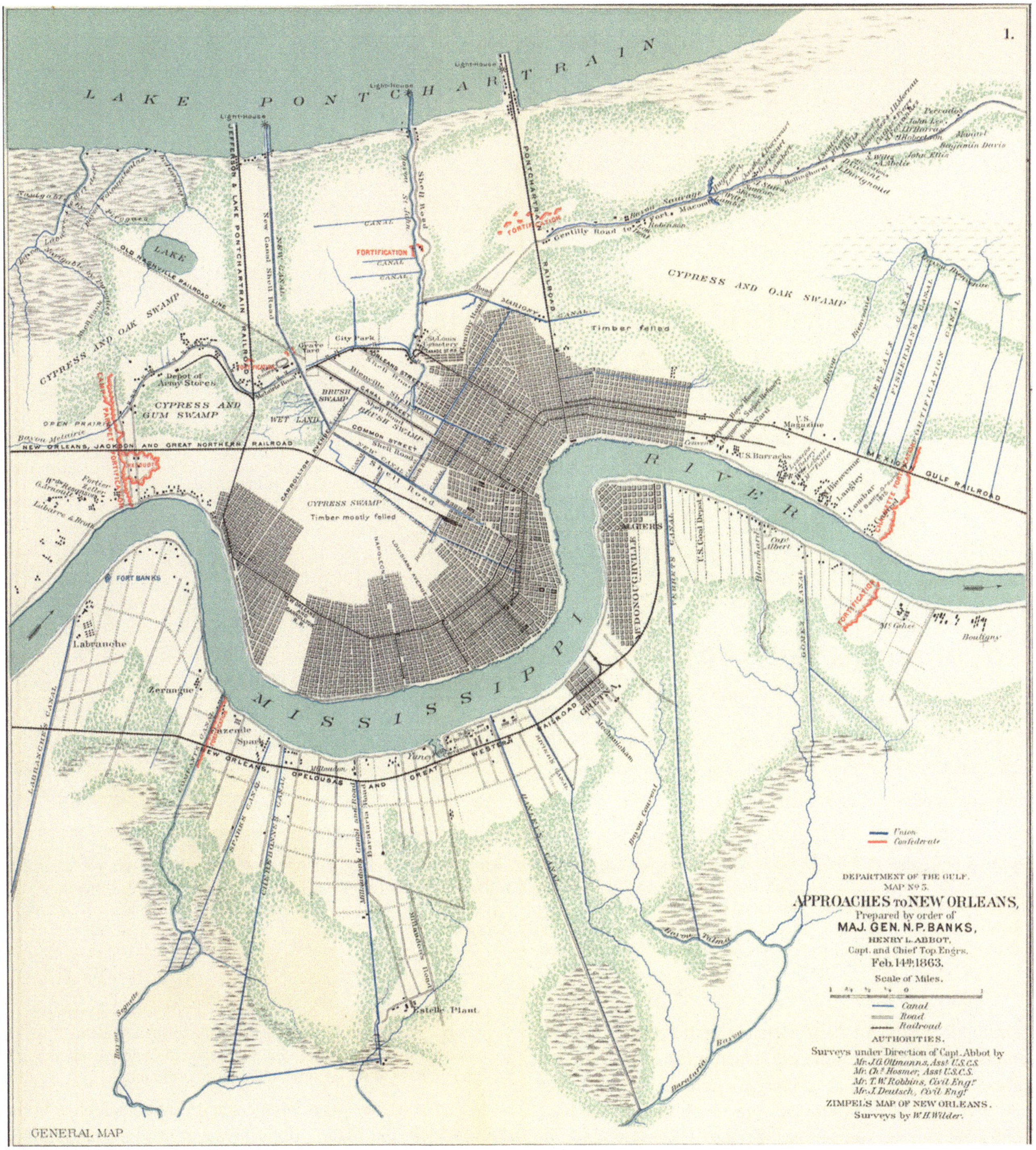

Fig. 1 Civil War era map of New Orleans dated 1863. The developed area is located on the natural levee of the river and to a lesser extent on the Metairie Ridge, an abandoned distributary of the St. Bernard delta. At this time, New Orleans was still almost completely above sea level, a situation that would soon change. There was also a broad band of cypress swamp between the city and Lake Pontchartrain that protected it against hurricane surge (Source: Department of the Gulf. Map No. 5. Approaches to New Orleans, Prepared by order of Maj. Gen. N.P. Banks, Henry L. Abbot, Capt. and Chief Top. Engrs. Feb. 14th, 1863. From the "War of the Rebellion Atlas", reproduced with permission: Courtesy of the Texas Collection, Baylor University http://digitalcollections.baylor.edu/)

places in a much more effective manner than the coastal master plan whose implementation sputtered by the middle of the twenty-first century. The river may well have changed channels. Below sea level areas like metropolitan New Orleans do not exist. Coastal communities have either retreated to high ground far from the coast or elevated structures to dizzying heights. These visions are also reflected in *The Not Yet* by Moira Crone (2012), an

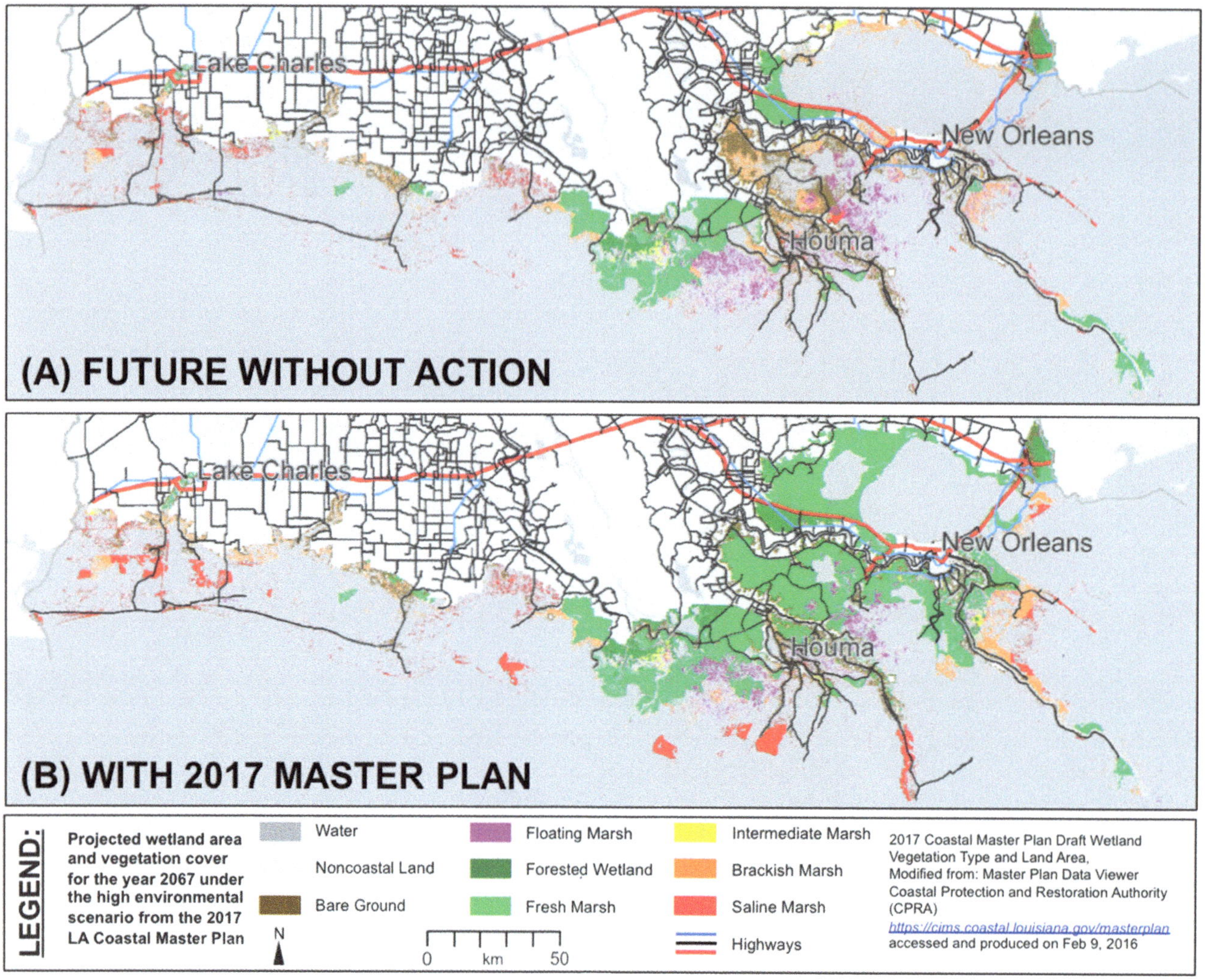

Fig. 2 A projection of the Louisiana coastline and vegetation types 50 years in the future under without action and with the 2017 Coastal Master Plan. The high scenario, shown here, considers the most severe possible sea-level rise, subsidence, storm intensity, and storm frequency. In this scenario, it is unlikely that the New Orleans metropolitan area or the natural levee south of Raceland would survive without protective wetlands (Modified from: Coastal Restoration and Protection Authority http://cims.coastal.louisiana.gov/masterplan/ Accessed 3/11/2017)

apocalyptic vision of New Orleans in the twenty-second century. She refers to Lake Pontchartrain as the open sea and has merged with Lake Maurepas to form a single water body and barrier beaches have disappeared. This is strikingly similar to current projections for the coast. The loss of control of the river are also reflected in the words of the poet and former LSU professor, William Hathaway:

> Our river will soon vault
> these flimsy levees and with one world's spin
> twist free in its wild thirst for salt.

Elizabeth Kolbert, the well known writer for the *New Yorker* and author of *The Sixth Extinction*, put these issues of sustainability and survivability into a broader context when she wrote:

> It may seem impossible to imagine that a technologically
> advanced society could choose, in essence to destroy itself.
> But that is what we are now in the process of doing.

In this book, we examine the science and societal actions that underlie the assumptions of the preceding paragraphs and offer some possible alternatives.

No Living Memory – Paradise Lost

When we speak of the current Mississippi delta and the services and benefits that it now provides to the human economy, we don't mean a pristine natural system. There are few ecosystems globally that have not experienced significant levels of human impact. And these impacted systems often don't provide the level of goods and services to humans as fully intact ones. For people living in the Mississippi delta and a broader audience concerned with the delta, there is no living memory of what a fully functioning natural deltaic ecosystem in its natural state was like.

The following passage from *Father and Son* by Edmund Gosse is a beautiful statement of such lost visions. Gosse was a writer popular in England around the turn of the twentieth century. His father, Philip, was a well-known naturalist, in the manner of nineteenth century gentlemen naturalists, who studied and published a number of books on sea life in tide pools along the Cornwall and Devon coasts of southern England. The elder Gosse was a Fellow of the Royal Society and an acquaintance of Charles Darwin. He was also an evangelical Christian and when Darwin published Origin of Species, Philip published a biblical interpretation of the diversity he saw in nature, for which he was much ridiculed. *Father and Son* is Edmund Gosse's (2004) personal account of growing up and coming of age while living with his father. Some of his fondest memories were of trips with his father to the Cornwall coast on collecting trips. The following is his remembrance of how the area had changed since he first visited it.

> It was down on the shore, tramping along the pebbled terraces of the beach, clambering over the great blocks of fallen conglomerate which broke the white curve with rufous promontories that jutted in the sea, or, finally, bending over those shallow tidal pools in the limestone rocks which were our proper hunting ground-it was in such circumstances as these that my father became most easy, most happy, most human. That hard look across his brows, which it wearied me to see, the look that came from sleepless anxiety of conscience, fades away and left the dark countenance still always stern indeed, but serene and unupbraiding. Those pools were our mirrors in which, reflected in the dark hyaline and framed by the sleek and shining fronds of oarweed, there used to appear the shapes of a middle-aged man and funny little boy, equally eager and, I almost find the presumption to say, equally well prepared for business.
>
> If anyone goes down to those shores now, if a man or boy seeks to follow in our traces, let him realize at once, before he takes the trouble to roll up his sleeves, that his zeal will end in labor lost. There is nothing now where in our days there was so much. Then, the rocks between tide and tide were submarine gardens of a beauty that seemed often to be fabulous, and was positively delusive, since, if we delicately lifted the weed curtains of a windless pool, though we might for a moment see its sides and floor paved with living blossoms, ivory-white, rosy-red, orange, and amethyst....
>
> Half a century ago, in many parts of the coast of Devonshire and Cornwall, where the limestone at the water's edge is wrought into crevices and hollows, the tideline was, like John Keats' Grecian vase, 'a still unravished bride of quietness.' These cups and basins were always full whether the tide was high or low, and the only way in which they were affected was that twice in twenty-four hours they were replenished by cold streams from the great sea, then twice were left brimming to be vivified by the temperate movement of the upper air. These were living flower beds, so exquisite in their perfection that my father used not seldom to pause before he began to rifle them, ejaculating that it was indeed a pity to disturb such congregated beauty. The antiquity of these rock pools and the infinite succession of the soft and radiant forms - sea anemones, seaweeds, shells, fish – which had inhabited them undisturbed since the creation of the world used to occupy my father's fancy. We burst in, he used to say, where no one had thought of intruding before, and if the Garden of Eden had been situated in Devonshire, Adam and Eve, stepping lightly down to bathe in the rainbow-colored spray, would have seen the identical sights that we now saw....
>
> All this is long over. The ring of living beauty drawn about our shores was a very thin and fragile one. It had existed all those centuries solely in consequence of the indifference, the blissful ignorance of man. These rock basins.... exist no longer, they are all profaned and emptied and vulgarized. An army of "collectors" has passed over them, and ravaged every corner of them. The fairy paradise has been violated; the exquisite product of centuries of natural selection has been crushed under the rough paw of well-meaning, idle-minded curiosity.... No one will see again on the shore of England what I saw in my early childhood, the submarine vision of dark rocks, speckled and starred with an infinite variety of color, and streamed over by silken flags of royal crimson and purple.

Similar stories and recollections could be told about thousands of places on earth from the tropics' disappearing rain forests to receding sea ice in polar regions. The Mississippi delta, compared to many places, has been ravished relatively recently. The delta was still almost completely natural when the European colonists arrived in the sixteenth century; we say almost because Native Americans had lived in the delta for millennia before the Europeans arrived. They prospered in the delta and built hundreds of mounds, called kitchen middens, composed of the shells of the oysters and clams they ate. Mixed in with the shells are pottery shards, some with beautiful markings, and items such as flint that suggests the active trade network with tribes in the southeast and Mississippi alluvial valley. But these early inhabitants had almost no negative impact on the dynamic delta.

Dr. Richard Condrey, an oceanography professor at Louisiana State University – now retired – and two colleagues studied old maps and journals from early French and Spanish mariners who explored the delta. From this information, they developed a description of the delta just before the beginning of the many interventions by the European invaders that led, by the end of the twentieth century, to the collapse of a large part of this enormously valuable ecosystem. They described what they called the last naturally active delta complexes of the Mississippi River (LNDM) as "a vast seaward-advancing arc that occupied, through four distributaries, all of the five most recent delta complexes of the Mississippi River and extended across all of coastal Louisiana..." It was characterized by plumes of freshwater that extended for more than 10 miles into the Gulf of Mexico during the spring flood of the Mississippi River. Overbank flooding of natural levees of the Mississippi and St. Bernard delta complexes occurred as far north as Bayou Baton Rouge (Condrey et al. 2014). This is in stark contrast to the delta today where almost all flow is confined to the leveed Mississippi River, where the offshore reef has completely disappeared, where more than 10,000 miles of canals have been cut through the wetlands, and where a quarter of these wetlands have disappeared over the last 75 years.

The portion of the LNDM west of the Birdsfoot delta received the outflows and overflows of the Mississippi's four distributaries. This coast was also low and subject to overflow; nearly continuous; and characterized by reefs, shoals, drift trees, and shallow inlets. The network of distributaries associated with the western portion of the LNDM nurtured a network of offshore oyster reefs which covered ~2000 km^2 in the Gulf of Mexico and extended along the coast for more than 150 km (Condrey et al. 2014). From 1500 to 1800, this offshore oyster reef restricted access to the Mississippi's western-most distributary called by the Spanish explorer Chaves the Río del Espíritu Santo. Here, in combination with Louisiana's eastern-most coastal cheniers (Chaves' Cabo de Cruz), the reef produced a natural harbor that was evidently of great importance to Spanish sailors caught in storms along the northern Gulf during the 1500s and 1600s.

Condrey et al. conclude that the State Coastal Master Plan does not take into consideration the functioning of the last naturally active complexes of the Mississippi River and is based upon an incomplete and incorrect consideration of the historical record. Their study indicates "much of Louisiana's coast was advancing into the sea at the onset of European colonization, that colonial and post-colonial modification of the Mississippi resulted in a cumulating loss of much of this potential, and that Louisiana's total land loss… peaked long before" the mid twentieth century. Condrey et al. also state that many of the restoration benchmarks are incompatible with a sustainable coast. These include diversions located too far down on the Mississippi and Atchafalaya Rivers, oyster reefs confined to estuarine environments, brackish-water dominated estuaries in the spring, and retention of most levees and deep shipping channels.

Condrey et al. conclude that the only way to return to a situation similar to the last naturally active delta complexes are "multiple, large-scale diversions of freshwater and sediment which begin at or above the headwaters of the Atchafalaya River, Bayou Plaquemines, and Bayou Lafourche and reconnect the Mississippi River to its Deltaic Plain…" Their analysis is consistent with Day et al.'s (2005, p. 1681) recommendation for "reconnecting the river to the deltaic plain via … the reopening of old distributaries", as well as the desirability of "a fully revised delta-lobe-scale chronostratigraphy" (Kulp et al. 2005, p. 282). This is also consistent with recent findings by Roberts et al. (2015), DeLaune et al. (2015), Twilley et al. (2016), and Xu et al. (2016). The vivid picture painted by Condrey et al. is of a wholly natural Mississippi delta where the river is fully connected to and spreads over the deltaic plain. Clearly, we can't go back to this but the clear message is that the resources of the river should be used to the fullest extent possible.

David Muth, an environmentalist and naturalist with the Gulf Restoration Program (and for many years with the National Park Service where he worked in the Jean Lafitte National Historical Park), echoes these sentiments (Muth 2014): "Louisiana's people are hampered by an inherent difficulty to comprehend how much the biophysical baseline has shifted. We lack an historic perspective, unaware of just how much more productive the system was and could be again. Many are engaged in a futile effort to hold onto what is doomed or put back what is already lost, rather than allow what could be: a vibrant new river management system that reignites the process that built the delta and its vast productivity in the first place."

Background for This Book

This book grew out of and was informed by two efforts that took place over the past several years. One is the book *Perspectives on the Restoration of the Mississippi Delta* and the second is the design competition for a new vision for the Mississippi delta called *Changing Course*.

Perspectives on the Restoration of the Mississippi Delta

In 2010, three environmental organizations, the National Audubon Society, Environmental Defense Fund and National Wildlife Federation, established the Science and Engineering Special Team (SEST). These national organizations, along with a number of local groups, were brought together by support from the Walton Family Foundation in 2008 to advocate for an urgent, scientifically credible campaign to restore the Mississippi River Delta (http://www.mississippiriverdelta.org). SEST was made up of carefully chosen group of experts on the Mississippi delta and river and chaired by Dr. John Day. SEST was initiated specifically to provide independent advice to the philanthropic community and non-governmental organizations on how best to support the restoration of one of North America's premier ecological assets, the Mississippi delta. The focus was on providing a clear-eyed, objective view of what was known about the delta and the tools available to resuscitate it. Initially, it was envisioned that the main product of the SEST was to provide a report to the NGO organizations that supported the effort. As the process evolved, it became clear that much new and exciting information was being produced that had a much broader value than just for the Mississippi delta. For this reason, the SEST team decided to publish a book on their findings. The work, *Perspectives on the Restoration of the Mississippi Delta* was published in 2014 (Day et al. 2014a, b) as part of the *Estuaries of the World* series produced by Springer. A portion of this current book, in the same series, is an update of information published in the SEST book.

Changing Course: A Bold and Pragmatic Future

A second important initiative that stimulated the present work was the *Changing Course* initiative also called the Lower Mississippi River Delta Design Initiative (LMRDDI). The design competition sought to bring teams of some of the world's best delta engineers, scientists, designers, and planners to address the question of the future of the Lower Mississippi River Delta and its dependent resources and communities. This Initiative specifically sought to understand how the Mississippi River's water and sediment can be used to maximize rebuilding the delta wetlands while meeting or even enhancing the needs of navigation, flood protection, industries and coastal communities. Solutions had to ensure the balanced interplay of ecosystem, society and economy well into the future. Several authors of the present book participated in the *Changing Course* competition either as members of advisory teams or design teams. The competition was based on the following understanding and thinking about the past, present and future of the delta.

The Missippippi River Delta is the center of a region in peril. For thousands of years a dynamic and tenuous balance existed between land and water. The formidable flow of the river provided a counter balance to the encroaching Gulf of Mexico. However, in recent generations humans tipped this balance, stifling its delta building energy. The stability provided through river control was a boon to a century of industrial expansion providing generations of people with relative comfort and safety. However, as the last decade has made clear, coastal Louisiana is no longer safe behind the veneer of protection provided by the levee system. In fact the protection based on controlling the Mississippi River is making us less safe in the long run. Following the past decade of hurricanes and floods, the region is now reacting to ever increasing risk from encroaching Gulf waters, increased storm intensity, and rising economic instability. Without direct, focused action to transform the relationship between society and environment, Louisiana will be too vulnerable and too unpredictable for sustained economic development. Something bold needs to be done.

Changing Course was an opportunity to think beyond the limits of our everyday decision making process and consider ways to overcome the tremendous risks that sea-level rise and land subsidence and other environmental, economic, and societal forces (e.g., Day et al. 2016) pose for New Orleans and South Louisiana. By convening a large multidisciplinary teams of scientists, engineers, planners, and designers the competition set out to see anew problems that have vexed the people of South Louisiana for generations. The Competition challenged teams to imagine new frameworks that would lead to better decision making, bolder directions, and effectively shared resources while maintaining the economic value of the Mississippi River.

This competition started from a field of 25 multidisciplinary teams formed by academics, NGOs, and large firms. The field was narrowed to eight and then finally to three teams representing the best in coastal engineering, ecology, planning, design, and other relevant fields. For 5 months the three teams competed to design a framework to transform the lower river, shedding new light on some of our most intractable problems. Teams were encouraged to come up with broad, conceptually new and bold solutions. At the same time, the competition expected designs to be implementable. For this, clear knowledge and understanding of the diverse needs of stakeholders was essential. The process of the competition was augmented with opportunities for engagement with stakeholders.

At the end of the process, the question remains, what is the value of a process like Changing Course? What does a design competition achieve that government and the private sector initiatives cannot? How has this particular competition changed the debate in Louisiana, and what is the value of this process to other places outside of the Gulf of Mexico? Will the necessity for change permeating these times provide the radical new ideas and the roadmap to achieve them? How will this thinking help change policy and lead to implementation?

Historically the design competition has been used precisely to shed light on and overcome our most intractable problems. Competitions have often provided public clients the opportunity to develop projects of importance in a context that exceeds the limitations of budget and politics, to provide transcendent, aspirational ideas – and true vision. The different designs suggested such things as realigning parts of the Mississippi delta, large new diversions, new building designs, and ecological engineering on a vast scale.

Context of the Competition

Design competitions have emerged to publicly engage issues that run far deeper than the typical design process can address. Maya Lin's Vietnam Memorial in Washington DC, and Michael Arad's September 11th memorial at the

World Trade Center site in Manhattan were projects that emerged from design competitions. Raw emotion and deep controversy surrounding both of these projects: a memorial to those who served and the nearly 60,000 US soldiers killed in the Vietnam War, and the nearly 3000 people killed by terrorist attacks on September 11th, 2001. The intense spotlight could easily have overly compromised projects designed through more traditional modes. But because these projects were conceived out of a competition, they were in a sense protected throughout the process. The singularity of these visions was not compromised, enabling them to become strong, hopeful symbols in extremely tumultuous times.

After Hurricane Katrina, the design competition emerged from the confines of architecture and public art to include a far broader range and scale of projects. Changing Course emerges out of this context; bridging engineering, ecological design, and urbanism to explore radical solutions to the tremendous challenges facing the River, the coast, and the New Orleans region.

The competition recognized that to overcome challenges rarely seen on such a massive scale, it would be essential to develop processes and solutions on the same scale. All of the three finalists' plans recognize that the Mississippi delta will shrink and it is within this context that planning must go forward (Fig. 3). *Changing Course* provides a vision for how we can bolster the physical environment against the threat of accelerated sea level rise and other factors while creating long-term economic and social opportunities that will enable people to remain and thrive in this environment. Coastal Louisiana has an opportunity most other coastlines do not: a dynamic ecological system that feeds and builds a rich deltaic zone. While other regional cities are exposed to rising seas on fixed coastal edges, Louisiana can leverage the Mississippi River to sustain a thriving wetland zone that protects and promotes its industrial economy, sustains ecosystem productivity, and nourishes human occupation.

The Future Is Now

In this book we will articulate a view that the current trajectory and manner of living in the Mississippi delta is unsustainable, likely profoundly so. Above all, as we move through this century, we must be aware of and avoid folly. But what is folly in the sense that we are speaking of it? The concept of folly comes from *The March of Folly* by Barbara Tuchman in her 1984 book. She defined folly as the pursuit of policies contrary to long-term public interest by a large group – governments or industry, or as far as the Mississippi delta is concerned, the coastal population of a state, a federal bureaucracy dealing with flood control and navigation on the Mississippi River, and a diverse group of industries involved in navigation and water borne trade, petrochemical production, and oil and gas exploration, production, and refining. For an event or series of events to be considered a folly, according to Tuchman, three criteria must be met: The policy must be perceived as counter-productive in its "own time." That is, a relatively large number of thoughtful people know that the policies are counter-productive and damaging.

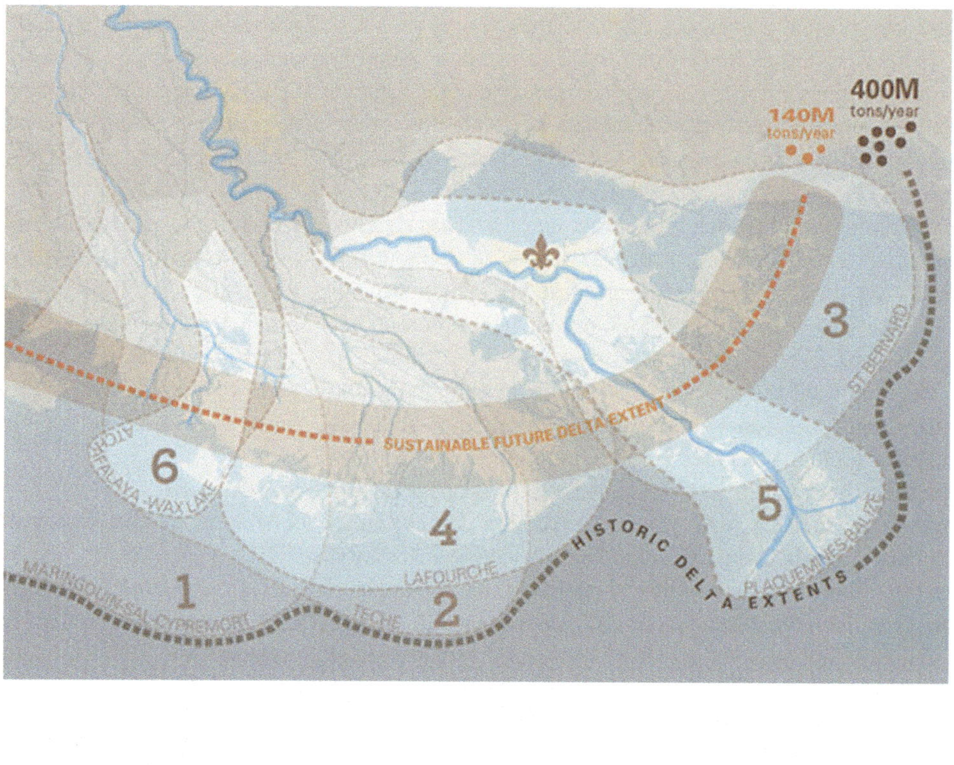

Fig. 3 Less land is sustainable today, compared to when the MRD was built, due to reduced sediment loads in the Mississippi River (The Baird Team 2015)

Second, feasible alternative courses of action must be proposed – but are ignored. And the suggested alternatives should come from a group, not an individual leader, and should persist beyond one political lifetime.

The human practices and policies that over a century and a half led to the disasters of the Mississippi delta and New Orleans are classic folly on a grand scale. There is the ongoing disaster of wetland loss as well as that of Hurricane Katrina (and other hurricanes) and the BP spill. If we are to restore the Mississippi to a functioning sustainable system we must avoid folly – don't make costly mistakes. Once used, energy and other resources are not available for a second chance. And time is of the essence. The great megatrends of the twenty-first century have non-linear trajectories so that the rate of change is accelerating. There is a need to understand what is possible and what is not. The current degree of river control as embodied in the Mississippi River and Tributaries project is likely not sustainable. The "Dutch" model will grow increasingly unworkable as the century proceeds and living below sea level will become increasingly untenable (Tessler et al. 2015; Day et al. 2016). The current trajectory is clearly unsustainable. Sustainability depends to a large degree on wise use of the resources of the River but these resources must be used wisely. There is a need to work with the river and delta is what is called ecological engineering.

In this book, we investigate possible future pathways for sustainable management of the Mississippi delta. We consider the history of the development of the Mississippi delta and the causes of its deterioration. In chapter "A Brief History of Delta Formation and Deterioration", Lane et al. provide a brief history of the formation of the Mississippi delta and causes for its deterioration and in chapter "Levees and the Making of a Dysfunctional Floodplain", Colten discusses the progressive isolation of the delta. In chapter "The Nutria in Louisiana: A Current and Historical Perspective", Sasser et al. discuss the role of the introduced rodent, Nutria, in wetland destruction. Shaffer et al. address the wise use of freshwater resources in maintaining fresh marshes and swamps in chapter "Optimum Use of Fresh Water to Restore Baldcypress – Water Tupelo Swamps and Freshwater Marshes and Protect Against Salt Water Intrusion: A Case Study of the Lake Pontchartrain Basin". In chapter "Energy and Climate – Global Trends and Their Implications for Delta Restoration", Rutherford et al. discuss the great megatrends of the twenty-first century, most importantly climate change and energy and resource scarcity and their implications for river and delta management. Wiegman et al. in chapter "The Costs and Sustainability of Ongoing Efforts to Restore and Protect Louisiana's Coast" present an overview of the State Master Plan for restoration of the delta. Diversions are a central element of the State Master Plan and in chapter "Large Infrequently Operated River Diversions for Mississippi Delta Restoration", Day et al. discuss the potential use of much larger diversions than are considered in the Master Plan. Large areas of the coast, especially the New Orleans area, are below sea level. Maintaining areas below sea level may become untenable in this century. In chapter "Raising Urban Land: Historical Perspectives on Adaptation", Colten provides a historical overview of elevating urban land with lessons for New Orleans. In chapter "Raising Buildings: The Resilience of Elevated Structures", Erdman and Williams review how architecture can be designed for a low-lying coastal region like New Orleans. Erdman et al. then present a bold plan (chapter "Raising New Orleans: The Marais Design Strategy") for elevating large portions of New Orleans to protect the city against sea-level rise and future hurricanes. In chapter "Eroding Communities and Diverting Populations: Historical Popula tion Dynamics in Coastal Louisiana", demographic trends for South Louisiana are discussed by Hammerling and alternatives for sustainable living are considered. A new legal framework to promote safe and sustainable development in coastal Louisiana is considered in chapter "Raising New Orleans: The Marais Design Strategy" by Wilkins. In chapter "Eroding Communities and Diverting Populations: Historical Population Dynamics in Coastal Louisiana", we bring together the elements of the book into a vision for a sustainable pathway forward to coastal Louisiana.

References

Condrey RE, Hoffman PE, Evers DE (2014) The last naturally active delta complexes of the Mississippi River (LNDM): discovery and implications. In: Perspectives on the restoration of the Mississippi Delta. Springer, Dordrecht, pp 33–50

Crone M (2012) The not yet. UNO Press, New Orleans

Day J, Barras J, Clairain E, Johnston J, Justic J, Kemp P, Ko J, Lane R, Mitsch W, Steyer G, Templet P, Yañez-Arancibia A (2005) Implications of global climatic change and energy cost and availability for the restoration of the Mississippi delta. Ecol Eng 24:251–263

Day JW, Boesch DF, Clairain EJ, Kemp GP, Laska SB, Mitsch WJ, Orth K, Mashriqui H, Reed DJ, Shabman L, Simenstad CA (2007) Restoration of the Mississippi Delta: lessons from hurricanes Katrina and Rita. Science 315(5819):1679–1684

Day JW, Kemp GP, Freeman AM (2014a) Summary and conclusions: the cost of inaction and the need for urgency. In: Perspectives on the restoration of the Mississippi Delta. Springer, Dordrecht, pp 185–192

Day J, Kemp GP, Freeman A, Muth DP (eds) (2014b) Perspectives on the restoration of the Mississippi Delta: the once and Future Delta. Springer, Dordrecht

Day JW, Agboola J, Chen Z, D'Elia C, Forbes DL, Giosan L, Kemp P, Kuenzer C, Lane RR, Ramachandran R, Syvitski J (2016) Approaches to defining deltaic sustainability in the 21st century. Estuar Coast Shelf Sci 183:275–291

DeLaune R, Sasser C, Evers E, White J, Roberts H (2015) Influence of the Wax Lake Delta sed- iment diversion on aboveground plant productivity and carbon storage in deltaic island and mainland coastal marshes. Estuar Coast Shelf Sci 177:83–89

FitzGerald D, Fenster M, Argow B, Buynevich I (2008) Coastal impacts due to sea- level rise. Annu Rev Earth Planet Sci 36:601–647
Gosse E (2004) Father and son. Oxford University Press, London
Kulp M, Fitzgerald D, Penland S (2005) Sand-rich lithosomes of the Holocene Mississippi River delta plain
Muth DP (2014) The once and future delta. In: Perspectives on the restoration of the Mississippi Delta. Springer, Dordrecht, pp 9–27
Powell L (2012) The accidental city: improvising New Orleans. The President and Fellows of Harvard College, USA
Roberts HH, DeLaune RD, White JR, Li C, Sasser CE, Braud D, Weeks E, Khalil S (2015) Floods and cold front passages: impacts on coastal marshes in a river diversion setting (Wax Lake Delta Area, Louisiana). J Coast Res 31:1057–1068
Tessler ZD, Vörösmarty CJ, Grossberg M, Gladkova I, Aizenman H, Syvitski JPM, Foufoula-Georgiou E (2015) Profiling risk and sustainability in coastal deltas of the world. Science 349 (6248):638–643
The Baird Team (2015) A Delta For All. Changing Course. Available online: http://deltaforall.com/our-approach/. Accessed 4/11/2017
Tuchman B (1984) The march of folly: from troy to Vietnam. Knopf, New York
Twilley R et al (2016) Co-evolution of wetland landscapes, flooding, and human settlement of the Mississippi River delta plain. Sustain, Sci, http://dx.doi.org/10.1007/s11625-016-0374-4
Xu K, Bentley S, Robichaux P, Sha X, Yang H (2016) Implicatoins of texture and erodibility for sedijment retention in receiving basins of coastal Louisiana diversions. Water. doi:10.3390/w8010026

A Brief History of Delta Formation and Deterioration

Robert R. Lane, G. Paul Kemp, and John W. Day

Abstract

The Mississippi River (MR) delta formed over the last several 1000 years as a series of overlapping deltaic lobes in various stages of progradation and deterioration that were sustained by overbank flooding and crevasses of the Mississippi River and active distributaries. To the west of the delta complex stretches the smooth shoreline of the Chenier Plain, which was created by the intermittent westward drift of sediments from the MR forming a series of beach ridges and mud flats. Beginning in the eighteenth century and greatly accelerating throughout the twentieth century, the delta has been impacted by a variety of human activities, most importantly by the separation of the Mississippi River from its deltaic plain, which has caused massive wetland loss during the last century. Other factors exacerbating wetland loss includes altered hydrology due to the proliferation of dredged canals and deep-well fluid withdrawal associated with the oil and gas industry, intentional impoundment for waterfowl management, and herbivory by nutria. The only place where wetland loss has not been high is at the Atchafalaya and Wax Lake delta complex, which are part of the beginning stages of a major new deltaic lobe development fed by the Atchafalaya River distributary that carries about a third of the combined lower MR discharge. One of the greatest threats to MR delta wetlands is accelerating sea-level rise due to a combination of subsidence and eustatic sea-level rise (ESLR), which ranges from 2–17 mm year^{-1} in the delta.

Keywords

Mississippi River delta • Atchafalaya River delta • Wetland loss • River water chemistry • Altered hydrology • Subsidence

Formation of the Mississippi Delta

The Mississippi River watershed has an area of 3.2 million km^2, encompassing about 40% of the area of the lower 48 U.S. states, and accounts for about 90% of river discharge to the Gulf of Mexico. Major tributaries to the lower Mississippi River include the Upper Mississippi, Missouri, Ohio, Arkansas and Red Rivers. As the Mississippi River (MR) approaches the ocean, water elevation rather than bed slope drives flow in the backwater zone that constitutes the MR delta, and for the last 724 km the bed of the river is actually below sea level; 4.6 m below sea level at Vicksburg, and over 52 m below sea level at New Orleans (Nittrouer et al. 2012). Before European settlement of the delta, functional MR distributaries branched off the mainstem of the river into Bayous Manchac, Plaquemines, Lafourche and Bayou La Loutre (Fig. 1). Now, river discharge is confined to two branches, the lower Mississippi and the Atchafalaya Rivers. The U.S. Army Corps of Engineers (USACE)

R.R. Lane (✉) • G. Paul Kemp • J.W. Day
Department of Oceanography and Coastal Sciences, Louisiana State University, Baton Rouge, LA 70803, USA
e-mail: rlane@lsu.edu; gpkemp@lsu.edu; johnday@lsu.edu

J.W. Day, J.A. Erdman (eds.), *Mississippi Delta Restoration*,
Estuaries of the World, https://doi.org/10.1007/978-3-319-65663-2_2

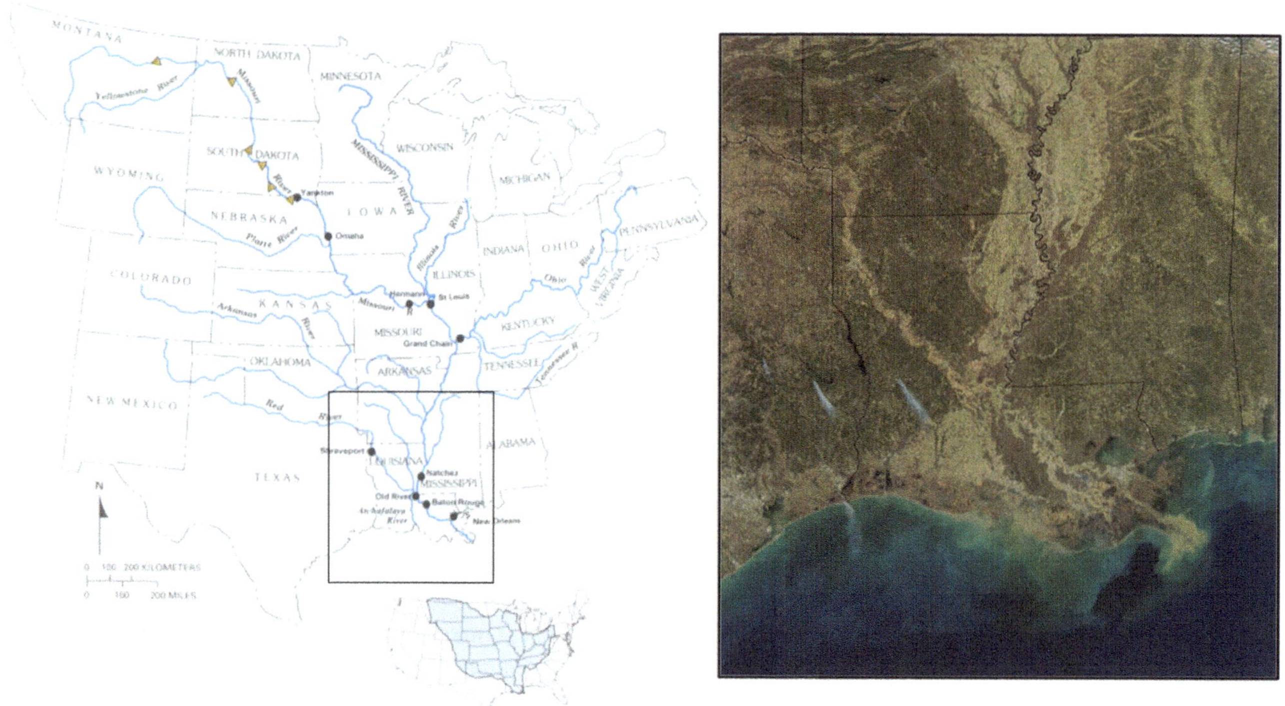

Fig. 1 The Mississippi River basin and its major tributaries on the left (Adapted from Meade and Moody 2010) and satellite image of the lower Misissippi River and delta complex on the right (from NASA Earth Observatory, https://earthobservatory.nasa.gov/)

manages water control structures upstream of Baton Rouge at the head of the delta to limit discharge from the MR to the Atchafalaya distributary to 30% of combined MR and Red River flow.

The MR mainstem carries about 70% of the total flow past New Orleans and discharges through a number of outlets that include three channels that diverge at Head-of-Passes in the Birdfoot (also known as Balize Delta (Kemp et al. 2014)). The Balize Delta has built out to the edge of the continental shelf so that its discharge of water, sediment and nutrients is into deep waters of the Gulf of Mexico. The Atchafalaya River, in contrast, discharges into the shallow waters of Atchafalaya Bay, which is fronted by a wide continental shelf nearly 200 km across (Roberts et al. 2003; Roberts et al. 2015). Over the past 30 years, two sand-dominant delta splays have grown to cover nearly 200 km^2 at the head of Atchafalaya bay. The new delta formations at the mouth of the Atchafalaya have formed and grown since 1973 at a combined rate of ~2.8 km^2 $year^{-1}$ (Rosen and Xu 2013).

The Mississippi delta is made up of several interdistributary estuarine basins separated by current and abandoned river distributary channels (Roberts 1997; Blum and Roberts 2012). To the west of the complex coast of the Deltaic Plain stretches the smooth shoreline of the Chenier Plain, which was created by the intermittent westward drift of sediments from the MR forming a series of beach ridges and mud flats over the past 3000 years (Fig. 2; Kemp and Wells 1987; Roberts 1997). The Deltaic and Chenier Plains are today characterized by a series of vegetation zones along the salinity gradient, with saline marshes at the coast that give way to brackish and freshwater marshes in the interior of the estuaries, and finally to cypress-tupelo swamp forests at the inland boundaries.

After sea level plateaued approximately 7000 years ago, but before massive human impacts during the twentieth century, the Mississippi River formed a vast deltaic wetland complex encompassing about 25,000 km^2 (Roberts 1997; Blum and Roberts 2012). Seasonal flooding of the river deposited large amounts of sediments and nutrients into the Mississippi River delta, allowing the delta to build regressively into the Gulf of Mexico in a series of offlapping deltaic lobes (Blum and Roberts 2012). Active and abandoned distributary channels formed a network of interconnected natural levee ridges. Reworked delta front sands became barrier islands that enhanced sediment trapping in estuarine lagoons and served to protect the delta from storm surge and salinity intrusion (e.g., Day et al. 2007; Xu et al. 2016). These skeletal elements sheltered wetlands from wave erosion and modulated the interaction of riverine and marine processes, limiting salinity intrusion. Wetland area in active deltaic lobes increased while decreasing in abandoned lobes, with a net increase in the area of deltaic wetlands during the past 3000 years.

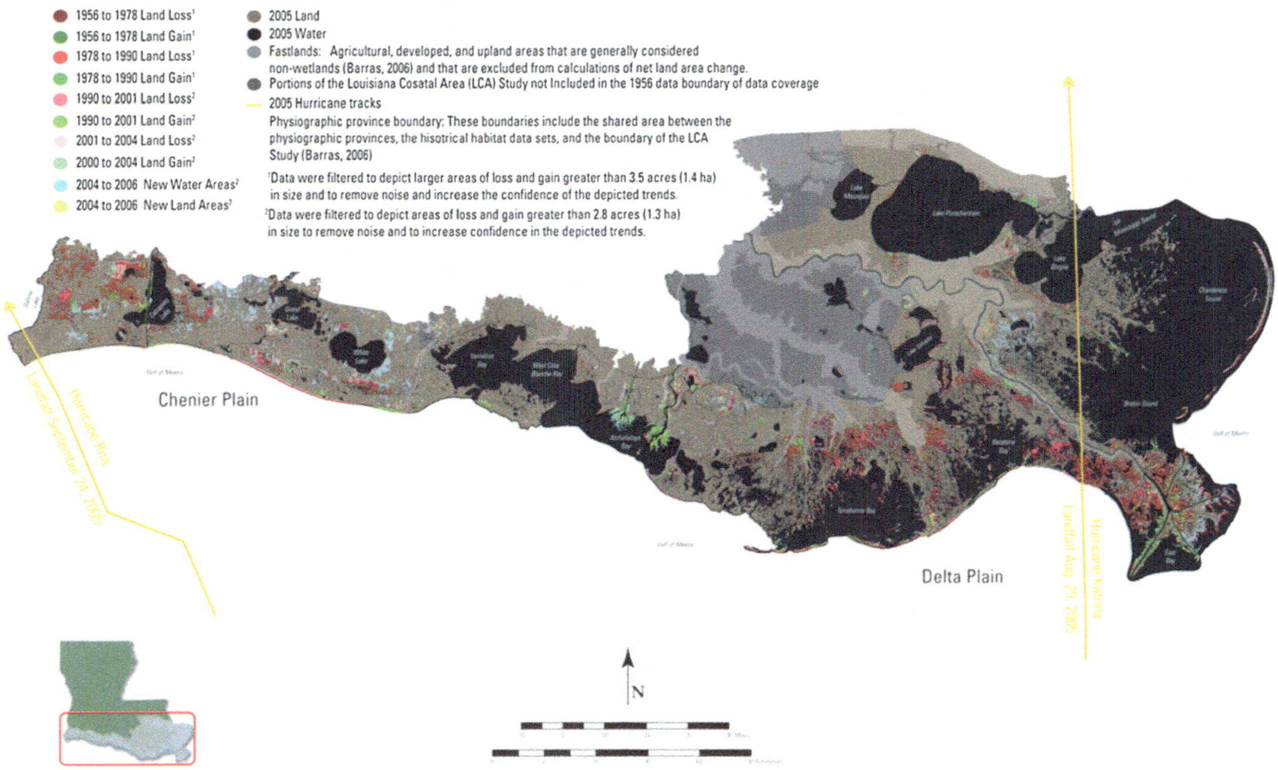

Fig. 2 Land loss/gain in the Louisiana coastal zone (Modified from Barras et al. 2008)

Until modified by human activity, some "abandoned" distributaries continued to deliver fresh water, sediments, and nutrients to large areas of the delta plain during floods. The fresh water inflow formed a buffer against saltwater intrusion, and provided sediments, rich in iron and other minerals that sustain healthier more productive wetlands. The distributary network, when combined with extensive wetland filled estuaries, was very efficient at capturing sediment, so that the delta retained virtually all of the sand, and 35–60% of the MR fine clays and silt (Kesel et al. 1992; Tornquist et al. 2008; Xu et al. 2016). Riverine input was important for coastal forested wetlands for several reasons, such as a buffer against saltwater intrusion and as a source of nutrients and mineral sediments that stimulated primary productivity and strengthened wetland soils.

Condrey et al. (2014) used maps and journals of early European explorers to describe what they called the last natural delta of the Mississippi that existed just prior to European settlement (Fig. 3). The delta was a vast seaward-advancing arc that occupied, through four distributaries, all of the five most recent delta complexes of the Mississippi River (Teche, St. Bernard, Lafourche, Modern, and Atchafalaya) and extended across the deltaic plain. It was characterized by plumes of fresh water that extended for more than 10 km into the Gulf of Mexico during the spring flood of the river and by a vast offshore oyster reef. Condrey et al. (2014) suggest that much of the Louisiana coast was advancing into the sea at the onset of European colonization, and that colonial and post-colonial modification of the Mississippi resulted in the loss of much of this potential. The natural delta described by Condrey et al. (2014) was formed and sustained by a hierarchical series of energetic forcings or events that occurred over a wide range of temporal and spatial scales. These energetic forcings included the shifting deltaic lobes, but also crevasse formation, great river floods, hurricanes, annual river floods, frontal passages, and tides (Roberts 1997; Roberts et al. 2015; Day et al. 1997, 2000, 2007; Vörösmarty et al. 2009).

Before river leveeing became widespread in many of the world's rivers, overbank flooding and crevasses were important and common mechanisms for replenishing floodplain and delta sediments and fertilizing the landscape (Condrey et al. 2014; Davis 1993; Saucier 1963; Syvitski et al. 2009; Vörösmarty et al. 2009). Crevasses function during high water via temporary channels through low points along the natural levee, forming crevasse splays, which have areas on the order of 10 to 100s of km^2 compared to 100 to 1000s of km^2 for full deltaic lobes (Saucier 1963; Welder 1959; Roberts 1997). These processes of floodplain and delta inundation and draining have built land along river corridors and in deltas the world over (Vörösmarty et al. 2009; Roberts 1997; Blum and Roberts 2009; Hensel et al. 1999; Stanley and Warne 1997). Where levees have not stopped river floods, overbank flooding occurs and leads to more

Fig. 3 Map of the Louisiana coast from 1764. Notice the three minor-distributaries connecting the Mississippi River to the Gulf of Mexico (Bellin (1764) taken from Condrey et al. (2014))

sustainable wetlands, such as in the Danube Delta on the border of Ukraine and Romania (Liviu Giosan, personal communication), several northwestern Mediterranean deltas (Day et al. 2011), and for the Atchafalaya delta of the Mississippi (Blum and Roberts 2012; Rosen and Jun Xu 2013; Delaune et al. 2016).

Davis (1993) documented hundreds of crevasses on the lower Mississippi after the arrival of Europeans. The footprints of these crevasses overlapped to form a more or less continuous band of sand-rich deposits essential to the formation and maintenance of both natural levees and coastal wetlands (Saucier 1963; Davis 2000; Allison and Meselhe 2010; Shen et al. 2015). Figure 4 shows the location of crevasses along the lower Mississippi River and the Bayou Lafourche distributaries as reported by Davis (1993, 2000). Day et al. (2016a, b) reported that an artificial crevasse opened at Caernarvon during the great flood of 1927 had a peak discharge of nearly 10,000 $m^3\ s^{-1}$ and deposited a crevasse splay of about 130 km^2 in about 3 months with sediment deposition as high as 42 cm. The artificial 1927 crevasse was similar in size and duration to naturally occurring historical crevasses. For example, the Davis Pond crevasse located on the west bank of the river upriver of New Orleans, which formed in 1884, was between 150 and 200 km^2 and is still clearly visible on photos of the area. And Shen et al. (2015) reported on a crevasse splay that developed off the Bayou Lafourche distributary just downstream of its juncture with the Mississippi River. They concluded that "discrete, episodically deposited sediment bodies dominate overbank stratigraphy" and that periodic aggradation rates of 1–4 cm $year^{-1}$ could persist for centuries. The total area of this splay was at least 60 km^2.

The Bonnet Carré crevasse, which was active from 1849 to 1882, created a large depositional sediment fan in wetlands and up to two meters of deposition in western Lake Pontchartrain (Fig. 5; Saucier 1963). The floodway has been opened 11 times since the 1930s (about once a decade and representing about 1% of the time since it was completed in 1933) with flows ranging from 3000 to 9000 $m^3\ s^{-1}$, and accretion rates in the spillway averaging about 25 mm $year^{-1}$ compared to about 4 mm $year^{-1}$ in adjacent wetlands without river input (Fig. 6; Nittrouer et al. 2012; Day et al. 2012). Total fine-grained sediment deposition in wetlands within the spillway near Lake Pontchartrain was as high as 2 m or an average of about 200 mm for each flood event.

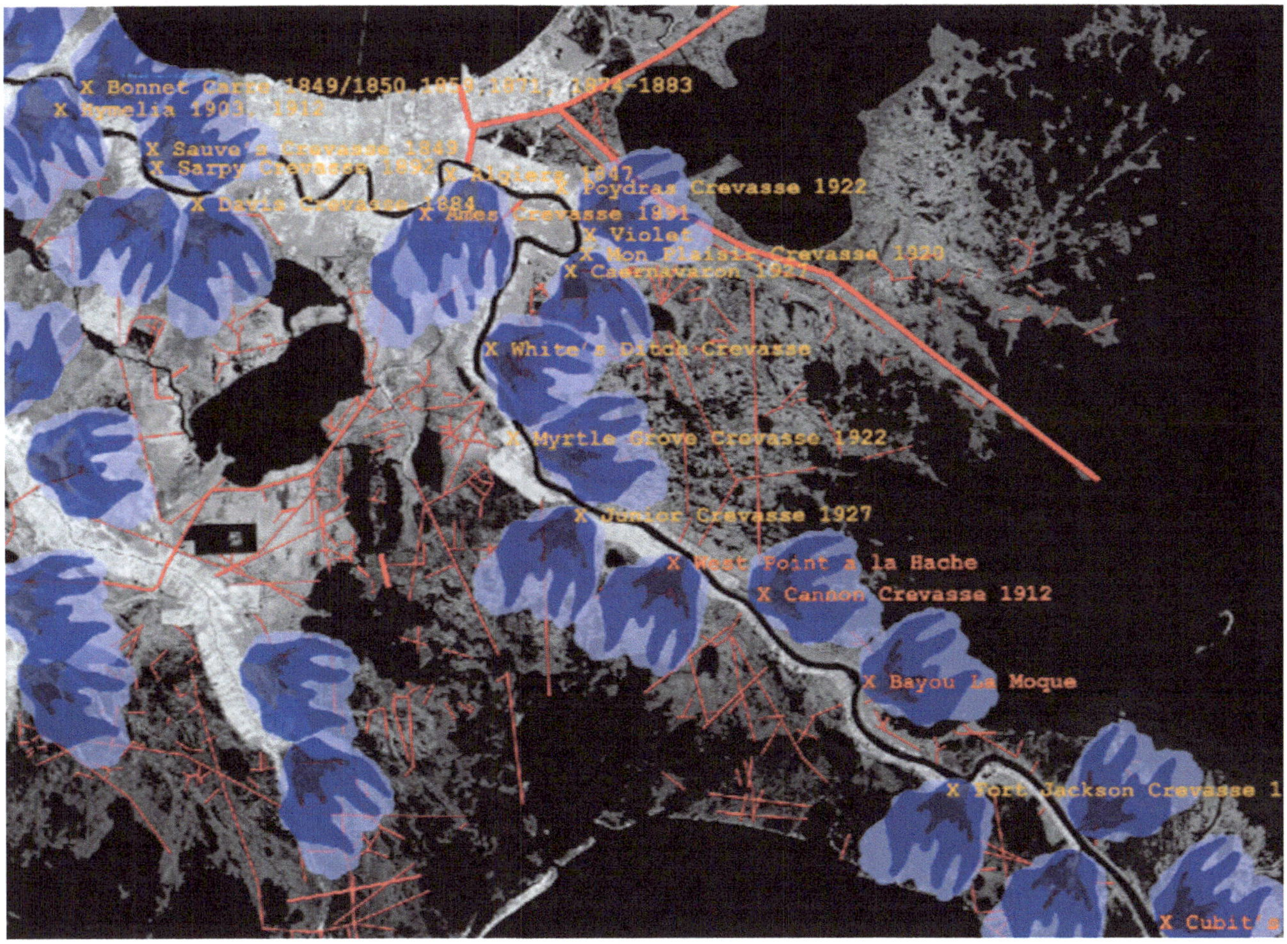

Fig. 4 Map of crevasses along the lower Mississippi and Bayou Lafourche distributaries as reported by Davis (1993, 2000). The *light and dark blue* coloring in the crevasse splays indicate the range in size of crevasses reported in the literature and summarized by (Day et al. 2016a, b)

Deterioration of the Mississippi Delta

Beginning in the eighteenth century and greatly accelerating throughout the twentieth century, the delta has been impacted by a variety of human activities such as levee construction and closure of distributaries, massive hydrological alternation, impoundments, and barrier island loss. Flood control levees built during the last two centuries have separated the Mississippi River from its deltaic plain, preventing seasonal flooding and inputs of freshwater, nutrients, and sediments to the surrounding deltaic plain (Kesel 1988, 1989; Mossa 1996; Roberts 1997; Day et al. 2000, 2007; Twilley and Rivera-Monroy 2009). Other factors exacerbating wetland loss include altered hydrology due to the proliferation of dredged canals and deep-well fluid withdrawal associated with the oil and gas industry (Turner et al. 1994; Day et al. 2000; Morton et al. 2002; Chan and Zoback 2007), intentional impoundment for waterfowl management (Boumans and Day 1994), and herbivory by nutria (Shaffer et al. 1992; Evers et al. 1998). This has led to salt water intrusion, deterioration of the skeletal framework, and massive wetland loss. These stresses are mostly the result of management in which navigation and flood control as well as tax revenues from the extraction of minerals and petroleum have prevailed over the ecological sustainability of the delta. The socioeconomic stakes include the nation's highest annual fisheries yield and massive economic investments and social amenities.

Mississippi River Water Quality Changes & Offshore Hypoxia

There has been a 70% decrease in suspended load in the Mississippi River since 1850 (Kesel 1988), mostly due to the trapping of sediments by dams constructed on the muddy part of the Missouri River during the 1950s, but also from other engineering activities on the river such as meander cutoffs, river-training structures, bank revetments, and soil erosion controls that have trapped sediment, eliminated sediment sources, or protected sediment that was once was

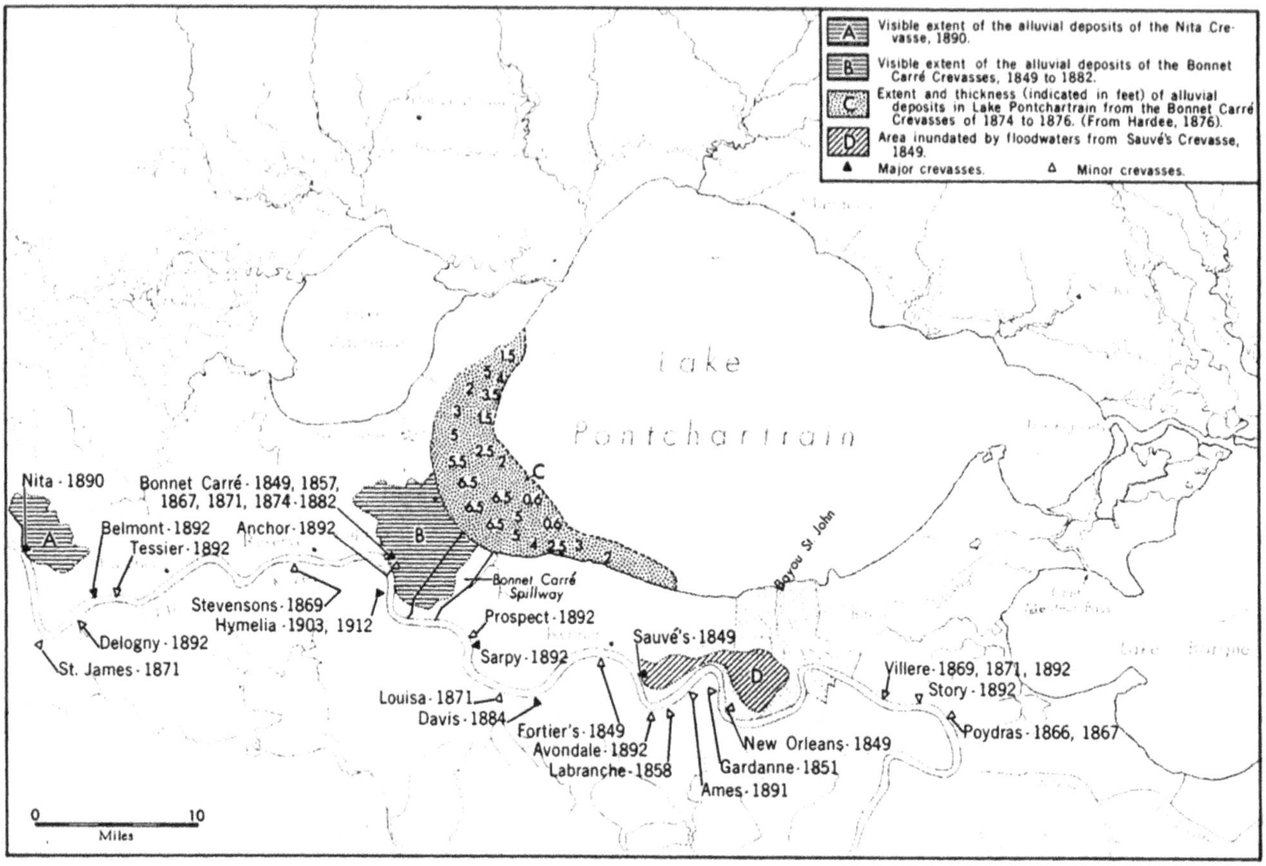

Fig. 5 Location of crevasses along the Mississippi River from 1849 to 1927 (from Saucier 1963)

available for transport episodically throughout the year (Meade and Moody 2010). Coinciding with decreased suspended sediment concentrations, intensive management practices associated with modern agriculture have significantly increased the export of nitrogen from farm fields into the watershed (Broussard and Turner 2009). The mean annual concentration of nitrate in the lower Mississippi River has doubled during the last half century, while the mean annual concentration of silicate has declined by 50% (Turner and Rabalais 1991). As a result, the DSi:DIN ratio of the lower Mississippi River has changed from about 4:1 to 1:1 (Rabalais et al. 1996). There has been a concurrent increase in the incidence of phytoplankton blooms and bottom water hypoxia in the coastal waters of the northern Gulf of Mexico, presumably in response to increased riverine DIN and DIP inputs, and more balanced nutrient ratios (Turner and Rabalais 1991; Justic et al. 1995; Rabalais et al. 1996). The northern Gulf of Mexico is presently the site of the largest (~20,000 km^2) and the most severe coastal hypoxic zone in the western Atlantic Ocean region (Fig. 7; Rabalais and Turner 2001). It is widely believed that decreasing DSi:DIN ratios in rivers exacerbate the eutrophication process in the adjacent coastal waters, primarily by reducing the potential for diatom growth in favor of noxious non-siliceous forms such as dinoflagellates (Officer and Ryther 1980; Cloern 2001). Diatoms are generally heavily grazed upon and are rarely a nuisance.

Wetland Loss in the Twentieth Century

Louisiana has more than 2 million ha of coastal wetlands, representing 40% of the country's total (Fig. 2). However, 80% of the coastal wetland loss in the entire continental United States has occurred in the Mississippi River Delta, with about 90% of the current coastal wetland loss in the continental United States occurring in Louisiana (LDNR 1998; Day et al. 2007; Couvillion et al. 2011). Coastal wetland loss occurs at a rate of one football field every hour (Couvillion et al. 2011). The most important factor leading to deterioration of the delta has been the near complete elimination of river water and sediment input to the deltaic plain due to the construction of flood control levees and closure of minor distributaries that connected the river to the deltaic plain (Day et al. 2000, 2007). Other factors include the reduction in sediment flux from the Mississippi

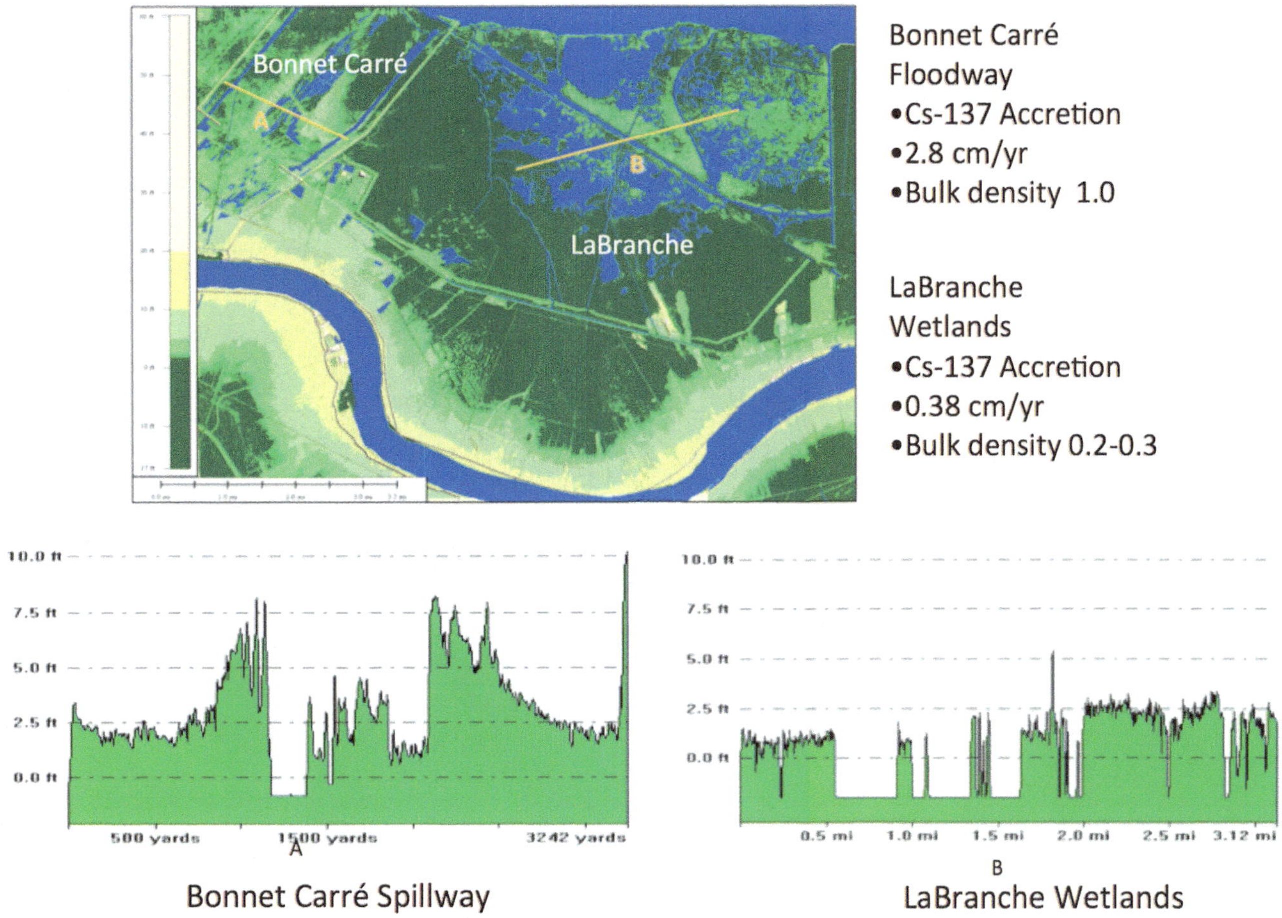

Fig. 6 Digital LIDAR elevation maps showing elevation (NGVD1984) in the Bonnet Carré Spillway (A–A′) and the LaBranche wetlands (B–B′; modified from Day et al. 2012)

River basin (Meade and Moody 2010), and the increased rate of sea-level rise experienced during the last several decades (Blum and Roberts 2012).

The construction of flood-control levees on the Mississippi River and closure of distributary channels began soon after the French colonization of New Orleans in 1719 (Boesch 1996; Welder 1959), and by the mid-twentieth century the Mississippi River delta was essentially separated from the river (Kesel 1988; Kesel 1989; Mossa 1996). Since that time there have been massive amounts of wetlands lost irrespective of any "no net loss" government policy, primarily due to levee and canal construction that caused impoundment, sediment and nutrient deprivation, land subsidence, and saltwater intrusion (Fig. 8; Barras et al. 2003; Boesch et al. 1994; Day et al. 2007; Turner et al. 1994). Impacts of wetland loss affect fisheries, oil and gas operations, navigation, hurricane protection, flood management, water quality, sediment availability, and hypoxia, all of which affect society. Wetland loss also threatens the economic value of the Mississippi delta to both Louisiana and to the Nation as a whole. The State of Louisiana predicted the public-use value of this loss by 2050 will be $37 billion (LDNR 1998).

From the 1930s until the end of the twentieth century, there was a dramatic loss of wetlands in the Mississippi Delta with loss rates as high as 101.7 km^2 per year and a total area of about 2500 km^2 of coastal wetlands has been lost since 1956 (Fig. 9; Barras et al. 2008). Over 95% of this loss was wetland, primarily marsh, conversion to open water. Wetland loss rates were highest in the 1960s and 1970s and have declined since, although rates remain high. These high rates of wetland loss are projected to continue for the next half century; from 2000 to 2050, it is estimated there will be an additional net wetland loss of ~2000 km^2 if no substantial restoration is undertaken (LDNR 1998).

The only place where wetland loss has not been high is at the outfall area of the Atchafalaya River. The Atchafalaya and Wax Lake delta complex, which are part of the beginning stages of a major new deltaic lobe development fed by the Atchafalaya River distributary that carries about a third of the combined lower Mississippi River discharge (Roberts

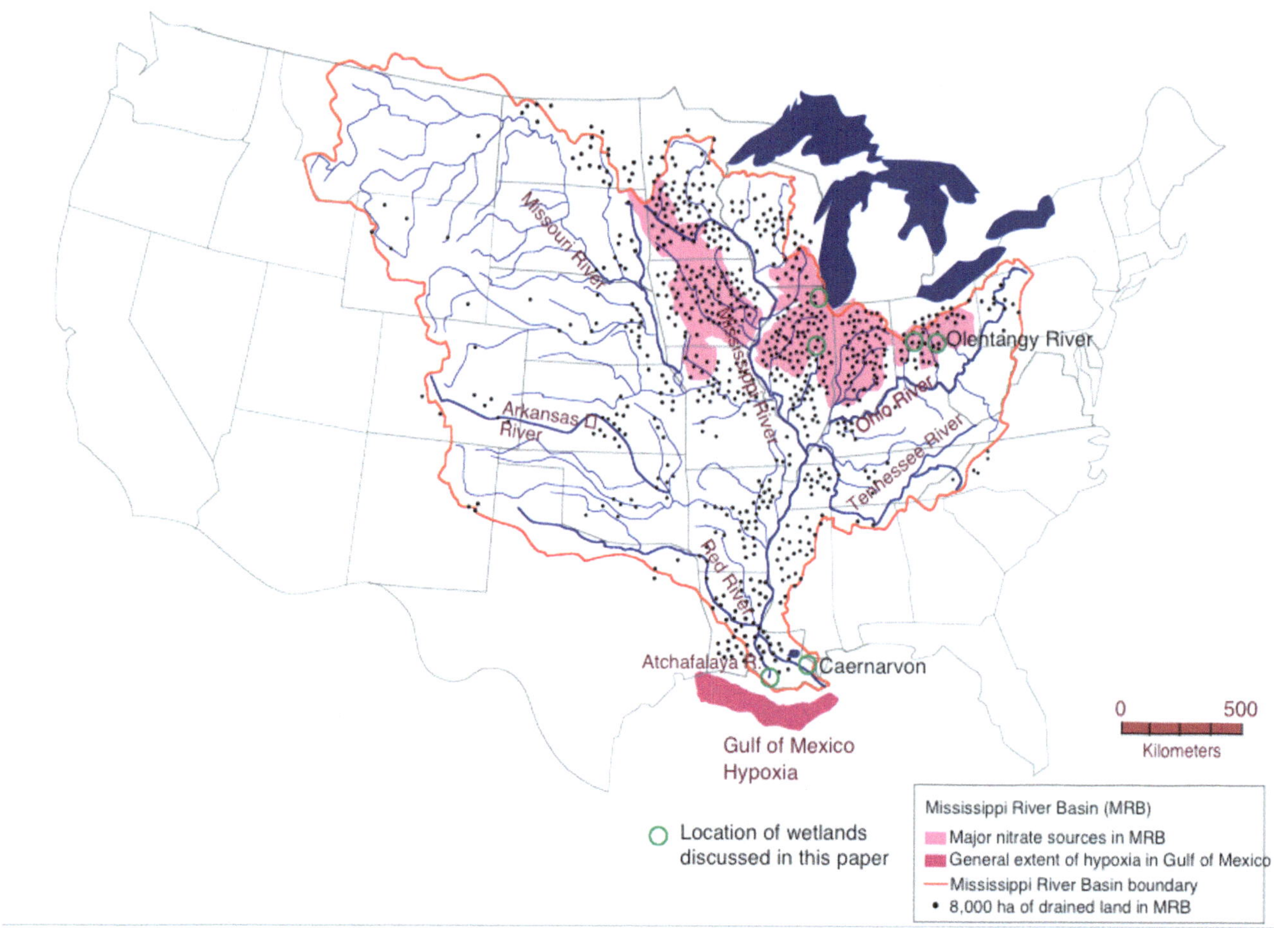

Fig. 7 Mississippi-Ohio-Missouri River Basin showing major sources of nutrients, major drainage areas, and general extent of Gulf of Mexico hypoxia (Modified from Mitsch et al. 2005)

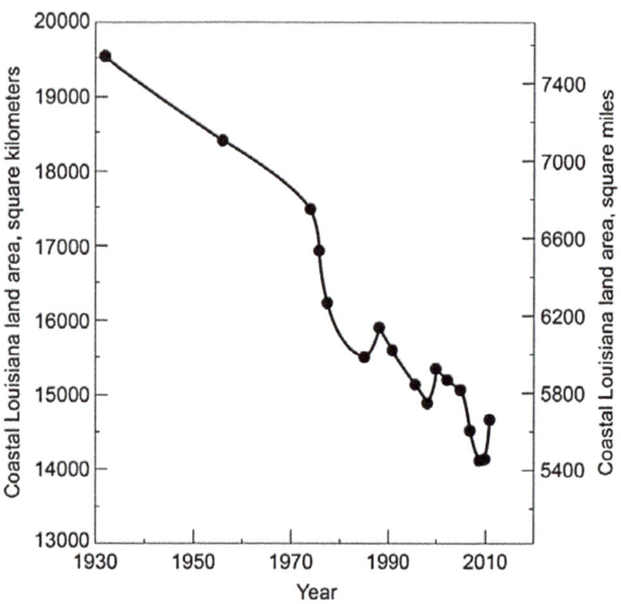

Fig. 8 Time series of change in coastal Louisiana land area from 1932 to the end of 2010 (from Olea and Coleman 2014, after Couvillion et al. 2011)

1997; Roberts et al. 2015; DeLaune et al. 2013). Subaqueous deposition occurred in Atchafalaya Bay for much of the twentieth century, first via the lower Atchafalaya River and then beginning in 1941 via the dredged Wax Lake Outlet. Both delta lobes first became subaerial during the large flood in 1973. The Atchafalaya delta lobe has a dredged navigation channel running through it but the Wax Lake lobe has developed naturally at the end of a 26 km long artificial conveyance channel. The delta lobes have grown at about 3 km^2/year to achieve total areas of 160 and 100 km^2, respectively, by 2015 (Roberts et al. 2015). Atchafalaya River discharge has also led to mineral sediment deposition in wetlands in a broad arc from Fourleague Bay to the east to Vermillion Bay to the west (e.g., Day et al. 2011) and these wetlands have the lowest rates of land loss in the Mississippi delta. Twilley et al. (2016) analyzed wetland loss in this area compared to the Terrebonne Bay complex to the east of Atchafalaya Bay that is isolated from Riverine input. They measured the change in the 50% land:water ratio in both basins from the 1930s to the present. The 50% land:water line retreated an average of 17,000 m in the Terrebonne

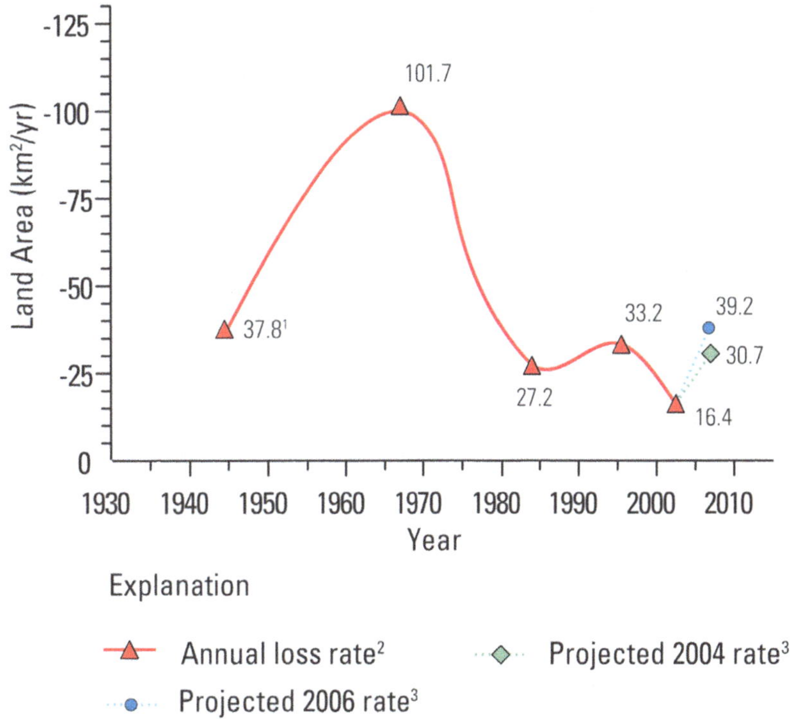

Fig. 9 Annual wetland loss rates in coastal Louisiana (from Barras et al. 2008)

region compared to only 35 m in the Atchafalaya region demonstrating the dramatic impact of river discharge.

Hydrologic Alteration

Natural hydrology of the wetlands in the delta is characterized by very low relief, slow flow, high surface friction, and complicated flow patterns due to a complex mosaic of vegetation and shallow ponds and channels (Blahnik and Day 2000; Mitsch and Gosselink 2015). Wetland hydrology is made even more complex by the effects of human activities such as dredging and channelization, barriers to flow such as embankments and roads, and increasing urbanization of the upland watershed that impacts incoming water quality (Cahoon et al. 2011; Conner et al. 2014; Lane et al. 2015a; Richardson 2005; Shaffer et al. 2009a, b; Wang et al. 2001). From a broader perspective, the hydrology and ecology of the whole delta has been dramatically changed due to the construction of flood control levees on the Mississippi River that resulted in the elimination of riverine input to most of the delta.

Construction of deep navigation and access canals combined with sea level rise, have exacerbated the frequency and intensity of saltwater intrusion events in surrounding wetlands. Any salinity for extended periods can result in high tree mortality rates and transitioning from freshwater forested to emergent wetlands (i.e., relic forests) and open water (Cahoon et al. 2011; Conner et al. 2014; Shaffer et al. 2009b). The construction of the Mississippi River Gulf Outlet (MRGO), for example, in the early 1960s led to dramatic increases in salinity in Lakes Borgne and Pontchartrain and to the death of over 6000 ha of forested wetlands in the Central Wetland Unit (Fig. 10; Shaffer et al. 2009b).

Several large canals were made during the construction of the interstate system, such as I-55, which connects I-12 to the north and I-10 to the south. The interstate was constructed in the 1960s and left a > 60 m wide by >5 m deep canal running between north and south bound lanes (Lane et al. 2015b; Shaffer et al. 2016). This canal now serves as the major drainage channel for the region, shunting upland runoff from north directly into Lake Maurepas, bypassing the Joyce wetlands in between. The canal also is a conduit for salt water during drought and storm surges, and has been a major contributor to the demise of the baldcypress-water tupelo forest in the southern half of the Joyce wetlands (Keddy et al. 2007). Practically all baldcypress and water tupelo on Jones Island and south of Pass Manchac have been killed due to high salinities (Shaffer et al. 2009b). Water tupelo has been eliminated from the lower two-thirds of the Joyce wetlands because of its lower salinity tolerance compared to baldcypress, which can tolerate salinities of 3–4 PSU and short-term salinity increases of >5 PSU (Allen et al. 1996; Conner et al. 2014; Shaffer et al. 2009a, b).

Oil access and pipeline canals are the most pervasive canal type in the delta, with over 16,000 km now dredged throughout

Fig. 10 Widening of the MRGO channel from 1959 to 2008 near Bayou Mercier. *Red dot* marks the same location in both photographs (from Shaffer et al. 2009b)

the Louisiana coastal zone. These canals are an important factor contributing to wetland loss (Day et al. 2000; Olea and Coleman 2014; Turner and Rao 1990). The first discovery of Louisiana oil in commercial quantities occurred at Jennings Salt Dome in the Acadia Parish in 1901, just 9 months after a similar discovery at Spindletop Dome in southeast Texas near the Louisiana border. Barge-mounted draglines were first used in 1938 to construct canals to enable barge-mounted drilling rigs to reach oil and natural gas drilling locations (Olea and Coleman 2014). Each drilling location required its own access canal because high-deviation directional drilling had not been developed and perfected before the peak of coastal Louisiana drilling activities in the 1970s. Spoil banks were created directly next to the dredged canals, which have been found to decrease the net flux of materials to and from nearby wetlands, making these areas prone to excessive inundation (Swenson and Turner 1987; Bryant and Chabreck 1998). Not only do spoil banks decrease the quantity of sediments and nutrients available to maintain wetland elevation (Boumans and Day 1994; Reed et al. 1997), but they also can increase flooding and lower soil Eh levels such that anoxic conditions and high sulfide concentrations cause dieback of vegetation (Cahoon and Turner 1989; Mendelssohn et al. 1981; Mendelssohn and Morris 2000). Because there is such a high density of canals and because most canals were dug in the twentieth century, some have argued that canals are responsible for most wetland loss (Turner 1997). However, wetland loss is a complex process that is caused by a number of interacting causative factors acting at different spatial and temporal scales including both isolation of river input from the deltaic plain and pervasive hydrologic alteration (e.g., Day et al. 2000, 2007; Olea and Coleman 2014).

A massive network of canals have also been dredged for the drainage of stormwater runoff along the northern reaches of the inter-distributary basins, where farming (mostly sugarcane) and urban development dominate the landscape (Peterson 2014; Yu et al. 2008). Stormwater is shunted away from developed areas as rapidly as possible by canals, often bypassing wetlands to drain directly into open water bodies such as lakes and major bayous where eutrophication is problematic (Southwick et al. 2002). These canals are dredged frequently, which can significantly affect hydrology. For example, Lane et al. (2015a) published a study of water quality and the potential of removing spoil banks along drainage canals in the Bayou Boeuf Basin, a sub-basin of Barataria Basin, to promote wetland overland flow. Hydrologic modeling of channel depths and surrounding wetland surface elevation indicated that the dredging of the channels for flow efficiency deepened them to such an extant that very little water would flow into surrounding wetlands even if the levees were removed. This effect most likely occurs at many areas where humans have dominated the deltaic environment.

Human Induced Subsidence

Subsidence is a natural geologic process due to the compaction, consolidation, and dewatering of sediments. Under natural conditions, sediment deposition from the river and *in situ* organic soil formation balanced subsidence in much of the delta. Now, however, these processes have been diminished and subsidence combined with sea-level rise, or relative sea-level rise (RSLR) is greater than accretion in much of the delta. In addition, recent evidence suggests that withdrawal of oil and natural gas can lower pressures in underlying geologic features enough to allow increased downfaulting, thereby increasing the rate of subsidence by two to three times in parts of the delta (Morton et al. 2002). Surface and subsurface data strongly indicate that the rapid subsidence and associated wetland loss in areas is largely induced by extraction of hydrocarbons and associated formation water (Morton et al. 2005). The proposed mechanism governing these fault-related manifestations of subsidence involves the venting of fluid (and gas) from geopressured shales vertically up fault planes (Kuecher et al. 2001). Saline fluids and gases exiting a basin via growth faults provide accommodation space at depth, resulting in active, fault-induced subsidence in the downthrown block (Fig. 11). Further studies indicate that increased downfaulting and reservoir compaction are important factors

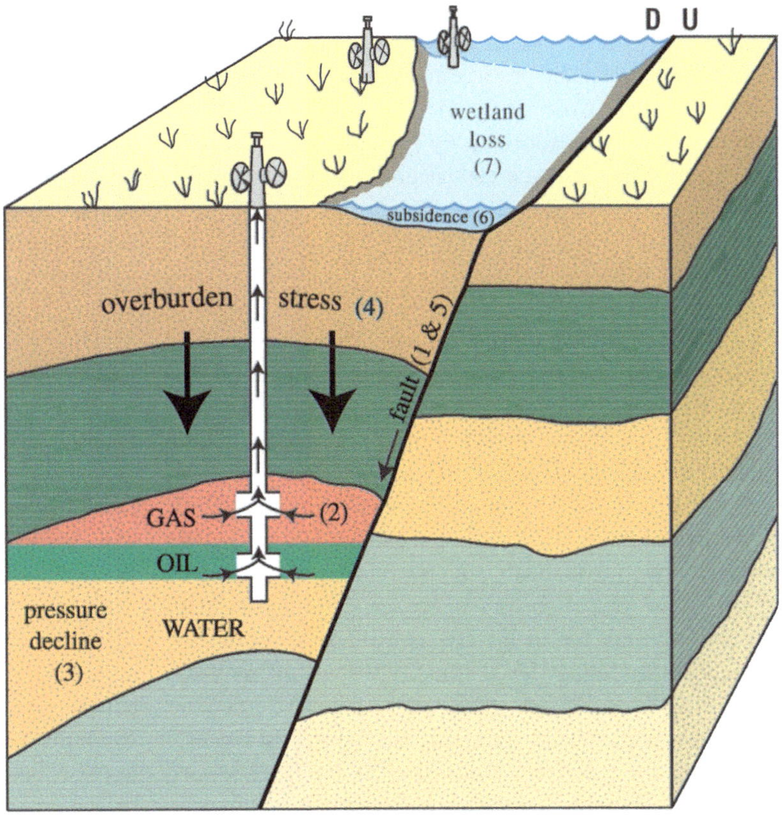

Fig. 11 Diagram of the possible effects of petroleum production in the coast. Prolonged or rapid production of oil, gas, and formation water (*2*) causes subsurface formation pressures to decline (*3*). The lowered pressures (*3*) increase the effective stress of the overburden (*4*), which causes compaction of the reservoir rocks and may cause formerly active faults (*1*) to be reactivated (*5*). Either compaction of the strata or downward displacement along faults can cause land-surface subsidence (*6*). Where subsidence and fault reactivation occur in wetland areas, the wetlands typically are submerged and changed to open water (*7*). Fig. is not to scale. *D* down, *U* up (from Morton and Purcell 2001)

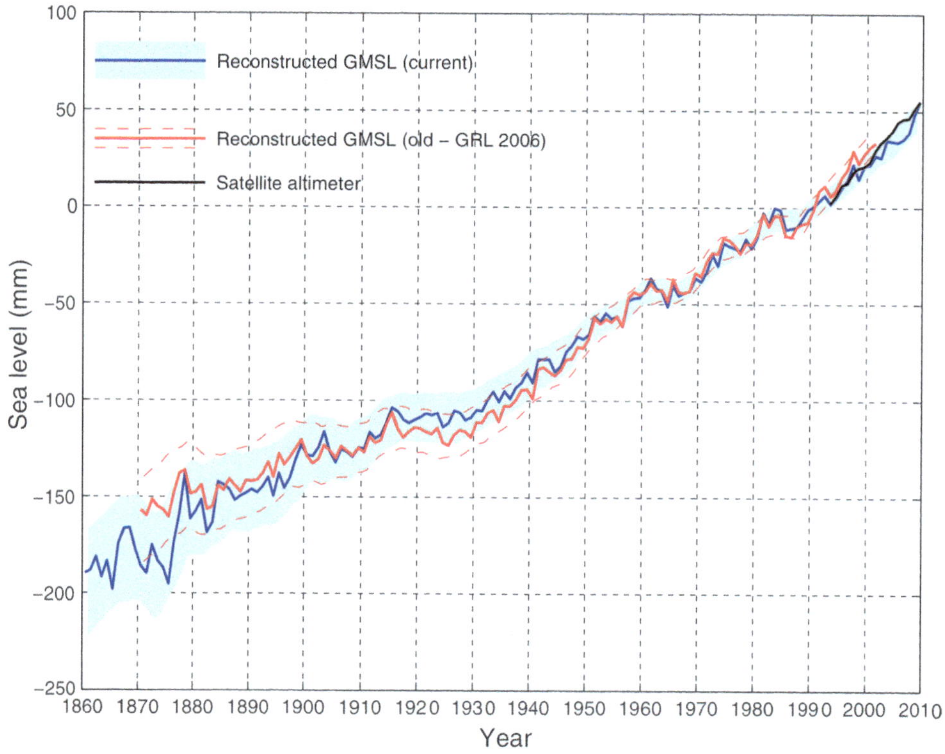

Fig. 12 Global average sea level from 1860 to 2009 as estimated from the coastal and island sea-level data (*blue*). The one standard deviation uncertainty estimates plotted about the low passed sea level are indicated by the shading. The Church and White (2006) estimates for 1870–2001 are shown by the *red solid line* and *dashed magenta lines* for the 1 standard deviation errors. The series are set to have the same average value over 1960–1990 and the new reconstruction is set to zero in 1990. The satellite altimeter data since 1993 is also shown in *black* (from Church and White 2011)

contributing to land loss in coastal Louisiana (Chan and Zoback 2007; Morton et al. 2003, 2005). Currently, a standard procedure in the long-term field management of high–flow rate reservoirs, especially oil reservoirs, is to inject fluids (usually co-produced brines) into the downdip portions of the reservoir to maintain pressure and hydrocarbon fluid composition (Olea and Coleman 2014). This activity substantially helps to reduce subsidence directly caused by reservoir pressure depletion and reservoir compaction.

Sea-Level Rise

One of the greatest threats to MR delta wetlands is accelerating sea-level rise due to a combination of subsidence and eustatic sea-level rise (ESLR). Current ESLR is between 3 and 4 mm year^{-1}, and there is a strong scientific consensus that the rate of ESLR will accelerate in association with global warming (Fig. 12; FitzGerald et al. 2008; Meehl et al. 2009; McCarthy and James 2009). Between 1993 and 2010, eustatic sea-level rise averaged 3.2 mm year^{-1} compared to <1.7 mm year^{-1} measured between 1901 and 2010, indicating an increase in the rate of eustatic sea-level rise during the last century (IPCC 2013). The Intergovernmental Panel on Climate Change (IPCC 2013) predicted sea-level rise of up to one meter by the end of the twenty-first century, with a range of uncertainty from 10 to 54 cm. Recent work based on semi-empirical methods suggests that ESLR may be more than one meter (Rahmstorf 2007; Pfeffer et al. 2008; Vermeer and Rahmstorf 2009; Horton et al. 2014). Deconto and Pollard (2016) reported that if CO_2 emissions continue unabated, Antarctic contribution to sea-level rise may add an additional meter resulting in a total sea-level rise of more >2 m by 2100. Sea level will continue rising over the next couple of centuries after 2100 by 2–3 m or higher depending on different climate scenarios. Increasing eustatic sea-level rise is especially critical in the Mississippi Delta, and other deltas, because it is augmented by high rates of subsidence. Relative sea level rise (RSLR), which is the combination of ESLR and subsidence, ranges from 2–17 mm year^{-1} in the delta (Shinkle and Dokka 2004).

Long-Term Water Level Cycles

Periodic drought is a major threat to the freshwater forested and emergent wetlands. For example, salinity at the LaBranche wetlands during the drought of 1999–2000 was over three times normal levels, reaching 10–12 PSU (Shaffer et al. 2009a, 2016). These high salinities led to extensive baldcypress mortality, not only in the LaBranche wetlands, but over a large area of western Pontchartrain Basin (Keddy et al. 2007). Droughts, such as the 1999–2000 drought, occur periodically, with others recorded during 1953, 1963, and 1969. Such droughts, combined with the altered hydrology, have caused much of the widespread mortality of baldcypress that is evident on the coast (Shaffer et al. 2009a).

Louisiana has experienced several episodes where there have been massive saltmarsh dieback events varying in size from 300 m^2 up to 5 km^2 in area (Fig. 13; McKee et al. 2004). For example, in 2000, Louisiana experienced a sudden and acute dieback event (termed "brown marsh") that affected over 100,000 ha of *Spartina alterniflora* dominated salt marsh throughout the Mississippi River deltaic plain (Lindstedt and Swenson 2006). Between May and October 2000, affected areas showed a progression from yellow to brown leaves to bare mud as *S. alterniflora* and, to a lesser extent, *Spartina patens*, died and decomposed, usually in interior portions of the marsh (Alber et al. 2008). However, other sympatric species such as mangroves (*Avicennia germinans*) were observed to remain unharmed by these conditions (Lindstedt and Swenson 2006). Possible triggers for brown marsh are extended periods of drought, high pore water salinities, heat, evaporation, and low river discharge (Lindstedt and Swenson 2006; McKee et al. 2004).

Species Conversion as a Measure of Climate Change

In coastal Louisiana, black mangroves (*Avicennia germinans*) were historically restricted to the southernmost barrier islands and beaches by winter freeze events (Penfound and Hathaway 1938). However, recent freeze-free winters have facilitated a noticeable expansion of Avicennia northward into Spartina marshes (Fig. 14; Giri et al. 2011; Perry and Mendelssohn 2009). This northward expansion is likely to continue if increases in temperature occur as predicted by climate change models (IPCC 2007). This has created a unique ecotone, where a species commonly found along neotropical coastlines (*Avicennia germinans*) coincides with a species that dominates temperate salt marshes (*Spartina alterniflora*; Visser et al. 1998). Zonation exists along an elevational gradient with *A. germinans* dominant at higher elevation creek banks and *S. alterniflora* at interior, lower elevation sites with greater

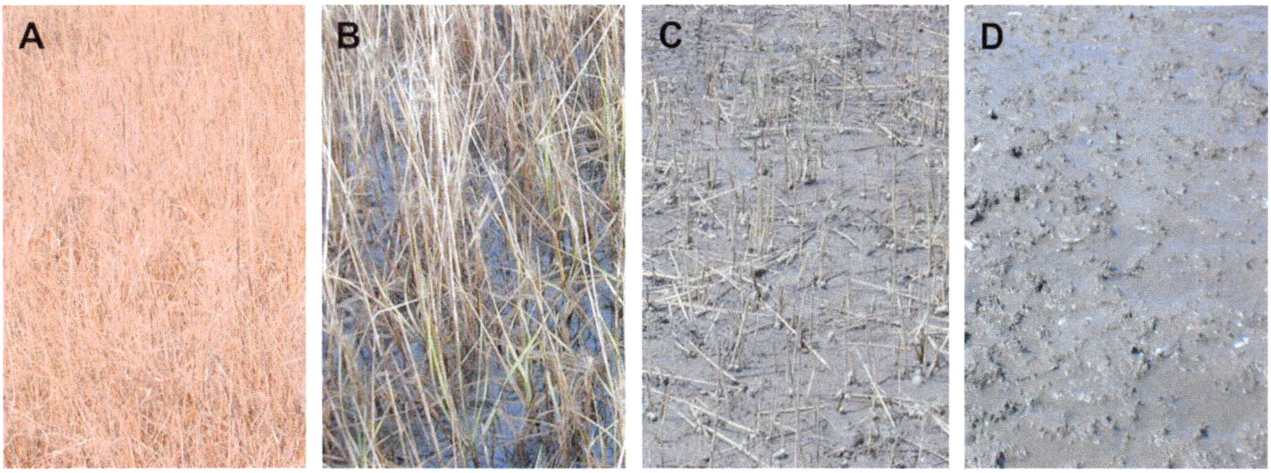

Fig. 13 Dieback in Louisiana: (**a**) standing dead *S. alterniflora* (June 2000); (**b**) initial thinning (Aug. 2000); (**c**) evidence of plant loss (Oct. 2000); (**d**) bare mud (April 2001; from Alber et al. 2008)

Fig. 14 Mangrove forest area of Louisiana from 1983 to 2010 (*left*, from Giri et al. 2011), and photos of expansion into *Spartina alterniflora* salt marsh (*middle* & *right*, from Perry and Mendelssohn 2009)

Fig. 15 Black mangrove propagules, seedlings, established forest, and flowers (from USDA NRCS 2009)

depth and duration of flooding (Patterson et al. 1997). Similar conversion from marsh to mangrove has been observed in other areas of the Gulf Coast (Raabe et al. 2012) and regions of the world (Saintilan and Williams 1999). Mangroves are halophytic and can be used to produce woody biomass in otherwise emergent marshes dominated by *spartina sp.*, providing important habitat and sustainability to Louisiana's coastal salt marshes via its woody structure, extensive root system, and ability to filter and trap sediments (Alleman and Hester 2011; Houck and Neill 2009).

The expansion of *Avicennia* in Louisiana may engender a greater degree of resilience in coastal salt marshes against climate change and sea-level rise via its extensive root system that filters and traps sediments (Fig. 15; Alleman and Hester 2011; Houck and Neill 2009) and raises wetland surface elevation (Comeaux et al. 2012). Black mangrove also serves as nesting habitat for many coastal birds, nursery habitat for crustaceans and fish, and as a food source for microorganisms that in turn are food for young marine life (Houck and Neill 2009). Mangroves have also been shown to attenuate storm surges (Danielsen et al. 2005), and most likely are more effective than marshes at storm surge reduction due to increased structure.

References

Alber M, Swenson EM, Adamowicz SC, Mendelssohn IA (2008) Salt marsh dieback: an overview of recent events in the US. Estuar Coast Shelf Sci 80:1–11

Alleman LK, Hester MW (2011a) Refinement of the fundamental niche of black mangrove (Avicennia Germinans) seedlings in Louisiana: applications for restoration. Wetl Ecol Manag 19:47–60

Allen JA, Pezeshiki SR, Chamberss JL (1996) Interaction of flooding and salinity stress on baldcypress (Taxodium Distichum). Tree Physiol 16:307–313

Allison MA, Meselhe EA (2010) The use of large water and sediment diversions in the lower Mississippi River (Louisiana) for coastal restoration. J Hydrol 387:346–360

Barras JA, Beville S, Britsch D, Hartley S, Hawes S, Johnston J, Kemp P, Kinler Q, Martucci A, Porthouse J, Reed D, Roy K, Sapkota S, Suhayda J (2003) Historical and projected coastal Louisiana land changes: 1978–2050. USGS open file report 03–334, 39 pp

Barras JA, Bernier JC, Morton RA (2008) Land area change in coastal Louisiana – a multidecadal perspective (from 1956 to 2006). U.S. Geological Survey Scientific Investigations Map 3019, scale 1:250,000, 14 p. pamphlet

Blahnik T, Day JW (2000) The effects of varied hydraulic and nutrient loading rates on water quality and hydrologic distributions in a natural forested treatment wetland. Wetlands 20:48–61

Blum MD, Roberts HH (2009) Drowning of the Mississippi Delta due to insufficient sediment supply and global sea-level rise. Nat Geosci 2:488–491

Blum MD, Roberts HH (2012) The Mississippi delta region: past, present, and future. Annu Rev Earth Planet Sci 40:655–683

Boesch DF (1996) Science and management in four U.S. coastal ecosystems dominated by land-ocean interactions. J Coast Conserv 2:103–114

Boesch DF, Josselyn MN, Mehta AJ, Morris JT, Nuttle WK, Simenstad CA, Swift DJP (1994) Scientific assessment of coastal wetland loss, restoration, and management in Louisiana. J Coast Res, Special Issue Number 20

Boumans RM, Day JW (1994) Effects of two Louisiana marsh management plans on water and materials flux and short-term sedimentation. Wetlands 14:247–261

Broussard W, Turner RE (2009) A century of changing land-use and water-quality relationships in the continental US. Front Ecol Environ. https://doi.org/10.1890/080085

Bryant JC, Chabreck RH (1998) Effects of impoundment on vertical accretion of coastal marsh. Estuaries 21:416–422

Cahoon DR, Turner RE (1989) Accretion and canal impacts in a rapidly subsiding wetland. Feldspar marker horizon technique. Estuaries 12:260–268

Cahoon DR, Perez BC, Segura BD, Lynch JC (2011) Elevation trends and shrinkeswell response of wetland soils to flooding and drying. Estuar Coast Shelf Sci 91:463–474

Chan AW, Zoback MD (2007) The role of hydrocarbon production on land subsidence and fault reactivation in the Louisiana coastal zone. J Coast Res 23:771–786

Church JA, White NJ (2011) Sea-level rise from the late 19th to the early 21st century. Surv Geophys 32(4–5):585–602

Cloern JE (2001) Our evolving conceptual model of the coastal eutrophication problem. Mar Ecol Prog Ser 210:223–253

Comeaux RS, Allison MA, Bianchi TS (2012) Mangrove expansion in the Gulf of Mexico with climate change: implications for wetland health and resistance to rising sea levels. Estuar Coast Shelf Sci 96:81–95

Condrey RE, Hoffman PE, Evers DE (2014) The last naturally active delta complexes of the Mississippi River (LNDM): discovery and implications. In: Day JW, Kemp GP, Freemen AM, Muth DP (eds) Perspectives on the restoration of the Mississippi Delta. Springer, Dordrecht, pp 33–50

Conner WH, Duberstein JA, Day JW, Hutchinson S (2014) Impacts of changing hydrology and hurricanes on forest structure and growth along a flooding/elevation gradient in a south Louisiana forested wetland from 1986 to 2009. Wetlands XX:1–12

Couvillion BR, Barras JA, Steyer GD, Sleavin W, Fischer M, Beck H, Trahan N, Griffin B, Heckman D (2011) Land area change in coastal Louisiana from 1932 to 2010: U.S. Geological Survey Scientific Investigations Map 3164, scale 1:265,000, 12 p. pamphlet

Danielsen F, Sørensen MK, Olwig MF, Selvam V, Parish F, Burgess ND, Hiraishi T, Karunagaran VM, Rasmussen MS, Hansen LB, Quarto A, Suryadiputra N (2005) The Asian tsunami: a protective role for coastal vegetation. Science 310:643

Davis DW (1993) Crevasses on the lower course of the Mississippi River. Coastal Zone'93, Proceedings of the eighth symposium on coastal and ocean management. American Society of Civil Engineers, pp 360–378

Davis DW (2000) Historical perspective on crevasses, levees, and the Mississippi River. In: Colten CE (ed) Transforming New Orleans and its environs. University of Pittsburgh Press, Pittsburgh, pp 84–106

Day JW Jr, Martin JF, Cardoch L, Templet PH (1997) System functioning as a basis for sustainable management of deltaic ecosystems. Coast Manag 25:115–153

Day JW Jr, Britsch LD, Hawes SR, Shaffer GP, Reed DJ, Cahoon D (2000) Pattern and process of land loss in the Mississippi Delta: a spatial and temporal analysis of wetland habitat change. Estuaries 23:425–438

Day JW, Boesch DF, Clairain EJ, Kemp GP, Laska SB, Mitsch WJ, Orth K, Mashriqui H, Reed DJ, Shabman L, Simenstad CA, Streever BJ, Twilley RR, Watson CC, Wells JT, Whigham DF (2007) Restoration of the Mississippi delta: lessons from hurricanes Katrina and Rita. Science 315:1679–1684

Day JW, Kemp GP, Reed DJ, Cahoon DR, Boumans RM, Suhayda JM, Gambrell R (2011) Vegetation death and rapid loss of surface elevation in two contrasting Mississippi delta salt marshes: the role of sedimentation, autocompaction and sea-level rise. Ecol Eng 37:229–240. https://doi.org/10.1016/j.ecoleng.2010.11.021

Day J, Hunter R, Keim RF, DeLaune R, Shaffer G, Evers E, Reed D, Brantley C, Kemp P, Day J, Hunter M (2012) Ecological response of forested wetlands with and without large-scale Mississippi River input: implications for management. Ecol Eng 46:57–67

Day JW, Cable JE, Lane RR, Kemp GP (2016a) Sediment deposition at the Caernarvon crevasse during the great Mississippi flood of 1927: implications for coastal restoration. Water 3(38):1–12

Day JW, Lane RR, D'Elia CF, Wiegman AR, Rutherford JS, Shaffer GP, Brantley CG, Kemp GP (2016b) Large infrequently operated river diversions for Mississippi delta restoration. Estuar Coast Shelf Sci 183:292–303

DeConto RM, Pollard D (2016) Contribution of Antarctica to past and future sea-level rise. Nature 531:591–597

DeLaune RD, Kongchum M, White JR, Jugsujinda A (2013) Freshwater diversions as an ecosystem management tool for maintaining soil organic matter accretion in coastal marshes. Catena 107:139–144

DeLaune RD, Sasser CE, Evers-Hebert E, White JR, Roberts HH (2016) Influence of the Wax Lake Delta sediment diversion on aboveground plant productivity and carbon storage in deltaic island and mainland coastal marshes. Estuarine Coast Shelf Sci 177:83–89

Evers DE, Sasser CE, Gosselink JG, Fuller DA, Visser JM (1998) The impact of vertebrate herbivores on wetland vegetation in Atchafalaya Bay, Louisiana. Estuaries 21:1–13

FitzGerald DM, Fenster MS, Argow BA, Buynevich IV (2008) Coastal impacts due to sea-level rise. Annu Rev Earth Planet Sci 36:601–647

Giri C, Long J, Tiezen L (2011) Mapping and monitoring Louisiana's mangroves in the aftermath of the 2010 Gulf of Mexico oil spill. J Coast Res 27:1059–1064

Hensel PF, Day JW, Pont D (1999) Wetland vertical accretion and soil elevation change in the Rhone River delta, France: the importance of riverine flooding. J Coast Res 15:668–681

Horton BP, Rahmstorf S, Engelhart SE, Kemp AC (2014) Expert assessment of sea-level rise by AD 2100 and AD 2300. Quat Sci Rev 84:1–6

Houck M, Neill R (2009) Plant fact sheet for black mangrove (*Avicennia germinans* (L.) L.). USDA-Natural Resources Conservation Service, Louisiana Plant Materials Center, Galliano, Louisiana 70354

IPCC (Intergovernmental Panel on Climate Change) (2007) Climate change 2007: the physical science basis. Contribution of Working Group I to the Fourth Assessment Report of the Intergovernmental Panel on Climate Change. Solomon S, Qin D, Manning M, Chen Z, Marquis M, Averyt KB, Tignor M, Miller HL (eds) Cambridge University Press, Cambridge/New York

IPCC (Intergovernmental Panel on Climate Change) (2013) Climate change 2013: the physical science basis. Contribution of Working Group 1 to the Fifth Assessment Report of the Intergovernmental Panel on Climate Change. Stocker TF, Qin D, Plattner GK, Tignor M, Allen SK, Boschung J, Nauels A, Xia Y, Bex V, Midgley PM (eds) Cambridge/New York, 1535 pp

Justic D, Rabalais NN, Turner RE (1995) Stoichiometric nutrient balance and origin of coastal eutrophication. Mar Pollut Bull 30:41–46

Keddy PA, Campbell D, McFalls T, Shaffer GP, Moreau R, Dranguet C, Heleniak R (2007) The wetlands of lakes Pontchartrain and Maurepas: past, present and future. Environ Rev 15:43–77

Kemp GP, Wells JT (1987) Observations of shallow-water waves over a fluid mud bottom: implications to sediment transport. In: Kraus NC (ed) Coastal sediments '87 Proc Spec Conf Adv. In Understanding coastal sediment processes, New Orleans, 12–14 May, 1987. Am Soc Civil Eng 1:363–378

Kemp GP, Willson CS, Rogers JD, Westphal KA (2014) Adapting to change in the lowermost Mississippi River: implications for navigation, flood control and restoration of the Delta ecosystem. In: Day JW Jr, Kemp GP, Freeman AM, Muth DP (eds) Perspectives on the restoration of the Mississippi Delta: the once and Future Delta. Springer, Dordrecht, pp 51–84

Kesel RH (1988) The decline in the suspended load of the lower Mississippi River and its influence on adjacent wetlands. Environ Geol Water Sci 11:271–281

Kesel RH (1989) The role of the lower Mississippi River in wetland loss in southeastern Louisiana, USA. Environ Geol Water Sci 13:183–193

Kesel RH, Yodis E, McCraw D (1992) An approximation of the sediment budget of the lower Mississippi river prior to major human modification. Earth Surf Process Landf 17:711–722

Kuecher GJ, Roberts HH, Thompson MD, Matthews I (2001) Evidence for active growth faulting in the Terrebonne delta plain, south Louisiana: implications for wetland loss and the vertical migration of petroleum. Environ Geosci 8:77–94

Lane RR, Day JW, Shaffer GP, Hunter RG, Day JN, Wood WB, Settoon P (2015a) Hydrology and water budget analysis of the East Joyce wetlands: past history and prospects for the future. Ecol Eng 87:34–44

Lane RR, Huang H, Day JW, Justic D, DeLaune RD (2015b) Water quality of a coastal Louisiana swamp and how dredging is undermining restoration efforts. Estuar Coast Shelf Sci 152:23–32

LDNR (Louisiana Department of Natural Resources. Baton Rouge) (1998) Coast 2050: Toward a Sustainable Coastal Louisiana. Louisiana Coastal Wetlands Conservation and Restoration Task

Force and the Wetlands Conservation and Restoration Authority. 161 pp
Lindstedt DM, Swenson EM (2006) The case of the dying marsh grass. Report submitted to Louisiana Department of Natural Resources, Baton Rouge
McCarthy, James J (2009) Reflections on: our planet and its life, origins, and futures. Science 326:1646–1655
McKee KL, Mendelssohn IA, Materne MD (2004) Acute salt marsh dieback in the Mississippi River deltaic plain: a drought-induced phenomenon? Glob Ecol Biogeogr 13(1):65–73
Meade RH, Moody JA (2010) Causes for the decline of suspended-sediment discharge in the Mississippi River system, 1940–2007. Hydrol Process 24:35–49
Meehl GA, Tebaldi C, Walton G, Easterling D, McDaniel L (2009) Relative increase of record high maximum temperatures compared to record low minimum temperatures in the US. Geophys Res Lett 36(23):L23701
Mendelssohn IA, Morris JT (2000) Eco-physiological controls on the productivity of Spartina Alterniflora loisel. In: Weinstein MP, Kreeger DA (eds) Concepts and controversies in tidal marsh ecology. Kuwer Acedemic Publishers, Boston, pp 59–80
Mendelssohn IA, McKee KL, Patrick WH (1981) Oxygen deficiency in Spartina Alterniflora roots: metabolic adaption to anoxia. Science 214:439–441
Mitsch WJ, Gosselink JG (2015) Wetlands. Wiley, Hoboken
Mitsch WJ, Day JW, Zhang L, Lane RR (2005) Nitrate-nitrogen retention in wetlands in the Mississippi River basin. Ecol Eng 24 (4):267–278
Morton RA, Purcell NA (2001) Wetland subsidence, fault reactivation, and hydrocarbon production in the U.S. Gulf Coast Region. USGS publication FS-091-01. pp 1–4
Morton RA, Bster NA, Krohn MD (2002) Subsurface controls on historical subsidence rates and associated wetland loss in southcentral Louisiana. Gulf Coast Assoc Geol Soc Trans 52:767–778
Morton RA, Tiling G, Ferina NF (2003) Causes of hot-spot wetland loss in the Mississippi delta plain. Environ Geosci 10:71–80
Morton RA, Bernier JC, Barras JA, Ferina NF (2005) Historical subsidence and wetland loss in the Mississippi Delta plain. Gulf Coast Assoc Geol Soc Trans 55:555–571
Mossa J (1996) Sediment dynamics in the lowermost Mississippi River. Eng Geol 45:457–479
Nittrouer JA, Best JL, Brantley C, Cash RW, Czapiga M, Kumar P, Parker G (2012) Mitigating land loss in coastal Louisiana by controlled diversion of Mississippi River sand. Nat Geosci 5:534–537
Officer CB, Ryther JH (1980) The possible importance of silicon in marine eutrophication. Mar Ecol Prog Ser 3(1):83–91
Olea RA, Coleman JL (2014) A synoptic examination of causes of land loss in southern Louisiana as they relate to the exploitation of subsurface geologic resources. J Coast Res 30:1025–1044
Patterson S, McKee KL, Mendelssohn IA (1997) Effects of tidal inundation and predation on Avicennia Germinans seedling establishment and survival in a sub-tropical mangal/salt marsh community. Mangrove Salt Marshes 1:103–111
Penfound WT, Hathaway ES (1938) Plant communities in the marshlands of southeastern Louisiana. Ecol Monogr 8(1):1–56
Perry CL, Mendelssohn IA (2009) Ecosystem effects of expanding populations of Avicennia Germinans in a Louisiana salt marsh. Wetlands 29(1):396–406
Peterson SN (2014) Failed agricultural impoundments: an interdisciplinary assessment of community structure and social resilience. M.S. thesis, Louisiana State University, Baton Rouge, LA. 69 pages
Pfeffer WT, Harper JT, O'Neel S (2008) Kinematic constraints on glacier contributions to 21st-century sea-level rise. Science 321:1340–1343
Raabe EA, Roy LC, McIvor CC (2012) Tampa Bay coastal wetlands: nineteenth to twentieth century tidal marsh-to-mangrove conversion. Estuar Coast 35:1–18
Rabalais NN, Turner RE, Justic D, Dortch Q, Wiseman WJ, Sen Gupta BK (1996) Nutrient changes in the Mississippi River and system responses on the adjacent continental shelf. Estuaries 17:850–861
Rabalais NN, Turner RE (2001) Hypoxia in the northern Gulf of Mexico: description, causes and change. In: Rabalais NN, Turner RE (eds) Coastal hypoxia: consequences for living resources and ecosystems. American Geophysical Union, Washington, DC, pp 1–36
Rahmstorf S (2007) A semi-empirical approach to projecting future sea-level rise. Science 315:368–370
Reed DJ, Luca ND, Foote AL (1997) Effect of hydrologic management on marsh surface sediment deposition in coastal Louisiana. Estuaries 20:301–311
Richardson CJ (2005) Highway construction and mitigation: on the road to increasing wetland function? Natl Wetl Newsl 27:17–20
Roberts HH (1997) Dynamic changes of the holocene Mississippi river delta plain: the delta cycle. J Coast Res 13:605–627
Roberts HH, Coleman JM, Bentley SJ, Walker N (2003) An embryonic major delta lobe: a new generation of delta studies in the Atchafalaya-wax Lake delta system. GCAGS/GCSSEPM Trans 53:690–703
Roberts HH, DeLaune RD, White JR, Li C, Sasser CE, Braud D, Weeks E, Khalil S (2015) Floods and cold front passages: impacts on coastal marshes in a river diversion setting (wax Lake Delta area, Louisiana). J Coast Res 31:1057–1068
Rosen T, Jun Xu Y (2013) Recent decadal growth of the Atchafalaya River Delta complex: effects of variable riverine sediment input and vegetation succession. Geomorphology 194:108–120
Saintilan N, Williams RJ (1999) Mangrove transgression into saltmarsh environments in south-east Australia. Glob Ecol Biogeogr 8 (2):117–124
Saucier RT (1963) Recent geomorphic history of the Pontchartrain basin. Louisiana State University Press, Baton Rouge
Shaffer GP, Sasser CE, Gosselink JG, Rejmanek M (1992) Vegetation dynamics in the emerging Atchafalaya Delta, Louisiana. USA J Ecol 80:677–687
Shaffer GP, Wood WB, Hoeppner SS, Perkins TE, Zoller JA, Kandalepas D (2009a) Degradation of Baldcypress – water tupelo swamp to marsh and open water in Southeastern Louisiana, USA: an irreversible trajectory? J Coast Res Spec Issue 54:152–165
Shaffer GP, Wood WB, Hoeppner SS, Perkins TE, Zoller J, Kandalepas D (2009b) Degradation of Baldcypress-water tupelo swamp to marsh and open water in southeastern Louisiana, USA: an irreversible trajectory? J Coast Res 54:152–165
Shaffer GP, Day JW, Kandalepas D, Wood WB, Hunter RG, Lane RR, Hillman ER (2016) Decline of the Maurepas swamp, Pontchartrain basin, Louisiana, and approaches to restoration. Water 8:1–28
Shen Z, Törnqvist TE, Mauz B, Chamberlain EL, Nijhuis AG, Sandoval L (2015) Episodic overbank deposition as a dominant mechanism of floodplain and delta-plain aggradation. Geology 43:875–878
Shinkle KD, Dokka RK (2004) Rates of vertical displacement at benchmarks in the lower Mississippi valley and the northern gulf coast. National Oceanic and Atmospheric Administration (NOAA) Technical Report NOS/NGS 50:135p
Southwick LM, Grigg BC, Kornecki TS, Fouss JL (2002) Potential influence of sugarcane cultivation on estuarine water quality of Louisiana's Gulf Coast. Agric Food Chem 50:4393–4399
Stanley DJ, Warne AG (1997) Holocene sea level change and early human utilization of deltas. GSA Today 7:1–7
Swenson EM, Turner RE (1987) Spoil banks: effects on a coastal marsh water-level regime. Estuar Coast Shelf Sci 24:599–609
Syvitski JPM, Kettner AJ, Overeem I, Hutton EWH, Hannon MT, Brankenridge GR, Day J, Vorosmarty C, Saito Y, Giosan L,

Nicholls RJ (2009) Sinking deltas due to human activities. Nat Geosci 2:682–686
Tornqvist TE, Wallace DJ, Storms JA, Wallinga J, van Dam RL, Blaauw M, Derksen MS, Klerks CJW, Meijneken C, Snijders EMA (2008) Mississippi delta subsidence primarily caused by compaction of Holocene strata. Nat Geosci 1:173–176
Turner RE (1997) Wetland loss in the northern Gulf of Mexico: multiple working hypotheses. Estuaries 20:1–13
Turner RE, Rabalais NN (1991) Changes in Mississippi River water quality this century. Bioscience 41:140–147
Turner RE, Rao YS (1990) Relationships between wetland fragmentation and recent hydrologic changes in a deltaic coast. Estuaries 13:272–281
Turner RE, Swenson EM, Lee JM (1994) A rationale for coastal wetland restoration through spoil bank management in Louisiana, USA. Environ Manag 18:271–282
Twilley RR, Rivera-Monroy V (2009) Sediment and nutrient tradeoffs in restoring Mississippi river delta: restoration vs. eutrophication. Contemp Water Res Educ 141:39–44
Twilley RR, Bentley SJ, Chen Q, Edmonds DA, Hagen SC, Lam NS-N, Willson CS, Xu K, Braud D, Peele RH, McCall A (2016) Co-evolution of wetland landscapes, flooding, and human settlement in the Mississippi River Delta Plain. Sustain Sci 11:711–731
Vermeer M, Rahmstorf S (2009) Global sea level linked to global temperature. Proc Natl Acad Sci 106:21527–21532
Visser JM, Sasser CE, Chabreck RH, Linscombe RG (1998) Marsh vegetation types of the Mississippi River deltaic plain. Estuaries 21:818–828
Vörösmarty C, Syvitski J, Day J, Sherbinin A, Giosan L, Paola C (2009) Battling to save the world's river deltas. Bull At Sci 65:31–43
Wang LZ, Lyons J, Kanehl P, Bannerman R (2001) Impacts of urbanization on stream habitat and fish across multiple spatial scales. Environ Manag 28:255–266
Welder FA (1959) Processes of deltaic sedimentation in the lower Mississippi River. Louisiana State University, Coastal Studies Institute Technical Report 12:1–90
Xu K, Bentley SJ, Robichaux P, Sha X, Yang H (2016) Implications of texture and erodibility for sediment retention in receiving basins of coastal Louisiana diversions. Water 8:26
Yu K, DeLaune RD, Tao R, Beine RL (2008) Nonpoint source of nutrients and herbicides associated with sugarcane production and its impact on Louisiana coastal water quality. Environmental Quality 37:2275–2283

Levees and the Making of a Dysfunctional Floodplain

Craig E. Colten

Abstract

Europeans colonists in Louisiana sought to detach their urban settlements from the invasive floodwaters of the Mississippi River. Early economic success of the French colony relied on cypress and rice – commodities that were reliant on annual inundations and wetland conditions – and thus regular flooding was beneficial across much of the landscape. With the turn toward sugar as the primary staple crop in the late eighteenth century, public policy and infrastructure investments fortified the levee system to reduce the risk of regular floods. With federal investment in the levee system in the late nineteenth century, structures increasingly severed the river from the floodplain wetlands. Even though a state scientist pointed out the value of interaction between the river and the floodplain, structural flood protection remained the dominant policy consideration through the twentieth century. Efforts are now underway to build sediment diversions to reactivate the connections between the river and the delta and will force those engaged in natural resource based to adapt to changing conditions.

Keywords

Levees • Floodplain • Mississippi River • Louisiana

Annual floods built much of the land adjacent to the lower Mississippi River – hence its designation as a floodplain. For millennia after the last glacial age, spring discharges flowed down the deeply incised channel, and overtopped the river banks, spilling across the deposits left in previous years and adding thin layers of fresh sediment. To cope with this cyclic pattern, Native Americans occupied settlements and built mounds on the upland terraces where they could take advantage of the floodplain ecology and yet avoid annual inundation. The early French settlers had different ideas about where to live, and although aware of the flood threat, they began separating the river from the floodplain.

Historian Chris Morris makes the argument that the compulsion to detach the river from its alluvial wetlands was driven by a desire to separate mud into dirt and water in order to dry out one of the world's great wetlands (Morris 2012). The underlying purpose behind this drive was to protect people and property from high water and to free floodplain agriculture from seasonal disruptions. Geographer Adam Mandelman argues that the French sought to erect barriers between settlements and the river to bring order to a disorderly landscape (Mandelman 2016). This impulse was a reflection of technological hubris, and also what Mandelman describes as part of a larger desire to create an impermeable society in the lower valley. This compulsion produced a dysfunctional floodplain. No longer does the annual inundation rejuvenate the alluvial soils and offset the slow process of deltaic subsidence, nor does it send water into the backswamps and contribute to a once-thriving wetland ecology.

C.E. Colten (✉)
Department of Geography and Anthropology, Louisiana State University, Baton Rouge, LA, USA
e-mail: ccolten@lsu.edu

J.W. Day, J.A. Erdman (eds.), *Mississippi Delta Restoration*,
Estuaries of the World, https://doi.org/10.1007/978-3-319-65663-2_3

To accomplish the separation, levees became the primary tool for flood protection by the mid-nineteenth century and dominated the human investments in landscape transformations. Nonetheless, there were also advocates for preserving connections between the river and the floodplain – both small-scale, local flumes and either natural waterways or massive engineered spillways. However, it was not until the flood of 1927 that reconnecting the river to segments of the floodplain via outlets became a principal part of the formal flood protection policy. At that same time, the first expressions of using outlets to restore the floodplain ecology emerge, albeit with little support (Viosca 1928). Over the next several decades, a series of small-scale diversions designed for specific purposes began to penetrate the impermeable levee lines. This chapter will explore the gradual separation of wet from dry with levees and then the eventual transition to a system of levees supplemented by large human-made outlets, followed by a series of efforts to restore the flow of water into the delta wetlands and thereby mimic natural processes in order to restore local ecologies. The flood protection impulse remains prominent and the parallel desire to restore wetlands creates tensions and challenges for those managing the floodplain and the river's delta.

The Levee Compulsion, 1718–1927

Seen through a twenty-first century lens, separating the river from its floodplain seems ecologically insensible. Yet, in the early eighteenth century European settlers to the lower Mississippi River valley had a different perspective. Strategic geopolitics impelled colonial governor Jean-Baptiste Le Moyne de Bienville to plant the colonial capital at the current flood-prone site of New Orleans. Rather than establishing the city well upstream on the bluffs at Baton Rouge, he opted for a site relatively close the mouth that enabled the French to fend off British intrusions and maintain hegemony over the entire basin. The site's superb portage potential to Lake Pontchartrain provided a second major impetus for its selection. The linkage between lake and city, via Bayou St. John, enabled maritime vessels to sail close to the city without fighting the river's current or facing the challenges of sailing up the meandering waterway. Additionally, furs and other products from the upriver colony could arrive at the city with relative ease and then enter the export commodity stream (Lewis 2003; Colten 2003; Morris 2012).

But then there was the issue of regular flooding. Iberville noted flood marks on trees during his initial voyage up the river in 1699. When digging wells and foundations, colonists encountered large tree trunks in the alluvial soil and realized they had been buried by flood-delivered sediments. After a flood in 1719, only a year after work began on the settlement, Bienville ordered levees built and drainage canals dug. Little more than incomplete berms along the river front, they offered minimal protection in 1722 when the river, driven by hurricane winds, once again flooded the incipient capital. Again in 1724, the river entered the city and high water remained on the muddy streets for several months (Morris 2012, 70–71). The initial levees stood about 3 ft high and ran for about a mile along the river by 1727 (Harrison 1951, 3). Yet, these barriers were an inadequate defense. The river could still wash across the floodplain and enter New Orleans from the backswamps. Leaders realized flood protection was a continuing concern.

Despite the seasonal threat, New Orleans remained the colonial capital and entrepôt to the Mississippi valley. The desire to inoculate it from regular high water prompted a series of ongoing flood-proofing steps. Colonial laws enacted in 1728 and 1743 required individual land owners to erect levees to protect their properties. This produced a longer line of levees and by 1735 levees extended about 30 miles above New Orleans and for about 12 miles downstream (Elliot 1932, 159). While not built to a consistent standard, these levees offered a bit more protection, although they did not truly sever the river from the floodplain. Regular flooding and especially crevasses, or levee breaches, allowed water to spread across the alluvial plain. Given the imperfect quality of the embankments, builders raised houses in New Orleans and along the river to elevate living quarters above the regular inundations. By the late French period, it was common for structures to stand more than 6 ft above the ground (Morris 2012, 71).

The rural landscape also endured these floods and interest in protecting food crops and export commodities contributed to levee building and the adoption of crops that benefitted from the environmental realities. The requirement for landowners to build their own levees was part of this strategy. However, after unsuccessful attempts the French abandoned wheat cultivation and turned to rice as a staple crop. It was much better suited to the annual flood cycle. Planters quickly adopted the crop, and it became a fundamental ingredient in local cuisine. Indeed, rice was among the first major adaptations to the chronically wet environment (Morris 2012; Mandelman 2016). Floods aided in the success of "providence" rice becoming the colony's key starch, and given its ecological needs planters sought to maintain connections to the river.

Despite early efforts to grow tobacco and indigo, cypress, harvested from the backswamps, proved to be the first truly successful export commodity. It was a native and abundant source of timber ideally suited to the local setting. Seasonal high water aided in the harvesting and transport of the massive cypress trunks. Trees could be felled during the low water season for drying and then floated out for transfer to saw mills during the high water season. Clearing the forests close to New Orleans provided the raw materials for shingles, sugar boxes, and lumber for the colony and markets in the Caribbean. Exploitation of this wetland resource was another adaptation to local conditions (Norgress 1947; Mancil 1972).

These human adaptations to floodplain conditions did not eliminate the levee building urge. By the end of the colonial period in 1803, levees extended as far up river as Baton Rouge and ran along the banks of the distributary Bayou Lafourche. Additionally, crevasses and regular levee failures periodically re-connected the river to its backswamps. Before the Louisiana Purchase, serious floods produced noteworthy crevasses in 1775, 1780, 1782, 1785, 1791, and 1796 (Elliott 1932, 105–6; Monette 1903). These events exposed the weakness in the levee system, but for rice planters such lapses in protection were tolerable, even desirable. However, the 1796 flood presented a serious problem to an emergent agricultural commodity – sugarcane. By 1795 New Orleans area planters successfully raised and processed sugar and this launched a land-use revolution along the lower river and its distributaries. Sugarcane cultivation spread quickly up the river, displacing rice, and spread down Bayous Lafourche and Teche in the early 1800s (Rehder 1999). In contrast to rice, sugar demands well-drained soils and hence effective flood protection, and this need reinforced the desire to build effective levees.

After Louisiana became a U.S. territory in 1803, regular floods continued and with them crevasses and damage to both the city of New Orleans and riverside agriculture. As a territory and then a state, Louisiana initially assigned land owners with the responsibility for maintaining their levees. In the 1840s, the state created the office of the state engineer. Among his responsibilities was overseeing the creation of a more consistent levee system. P.O. Herbert, the state engineer in 1846, questioned the efficiency of levees. He pointed out that confining the river between levees and closing outlets, elevated the flood risk. Indeed, he even noted that levees gradually raised the river bed and thereby elevated flood stages (Hebert 1846). This situation demanded a perpetual cycle of adding to the height of levees to fend off higher water – a costly proposition for the state engineer. Despite his observations, levees continued to dominate flood protection activities, and by the 1850s the 3 to 4 foot high berms of the colonial era where buried within an impressive line of earthen mounds that stood 15 ft high in some locations (Colten 2005, 29).

State authorities realized they were dealing with a problem that arose well beyond Louisiana's borders. As upstream farmers cleared land for agriculture and states built levees, runoff increased and flood risks rose in Louisiana. Following a series of particularly devastating floods in the 1840s and effective lobbying by elected officials from the lower river valley, congress enacted a pair of Swamp Land Acts in 1849 and 1850. These acts transferred 9 million acres of swamp lands to Louisiana, and considerable acreage to other states. Recipient states could sell these lands and apply the proceeds to levee construction. While the program produced little in the way of effective levees, Karen O'Neill argues that it accelerated commercial development of the wetlands and created a new contingent of land owners in the floodplain who joined the ranks of levee advocates demanding protection from high water (O'Neill 2006, 49–50).

The acts also stimulated funding for surveys of the river and its dynamics. These reports strengthened support for levees, although they were not completely in alignment on other aspects of flood protection. The prominent engineer Charles Ellet Jr. prepared a pair of reports in 1851 and 1853. He concluded that the expansion of upstream cultivated land and the extension of the lower river's levee system were major contributors to increasing flood risks (Pabis 1998, 432). To contend with this problem he advocated the construction of stronger levees, a combination of improved natural outlets and artificial outlets, and upstream reservoirs to regulate the discharge of annual floodwaters (Pabis 1998, 434). His advocacy for outlets corresponded to another State Engineer A.D. Woodridge's recommendation of outlets, particularly down the Atchafalaya basin (Woodridge 1850, 12–14). Such views by prominent river experts underscores the recognition that maintaining connections between the river and its floodplain remained part of the flood management discussion on the eve of the Civil War even though levees remained the most common landscape features associated with flood management. Ecological health of the floodplain and delta were absent from these discussions.

A second major study was begun by the Army engineer Andrew A. Humphreys and his assistant Henry L. Abbot at about the same time. Although not completed until 1861, Humphreys and Abbot produced a massive and thorough account of the Mississippi's hydrodynamics. Based on careful analysis of the floods of 1851 and 1858, the Humphrey report challenged Ellet's argument that conversion of swamplands into farm land aggravated flooding. Humphrey claimed that swamps did not serve as effective reservoirs in a river without levees and only served that function when levees broke. He also contended that conversion of wetlands to agriculture benefitted the nation and thereby served a higher purpose than flood control (Pabis 1998, 444). He further argued against the notion that outlets would cause sediment accumulation in the main channel – thus weakening the case for outlets (Pabis 1998, 447). In addition, he asserted that enlarging the channels of natural outlets like Bayou Lafourche and constructing artificial outlets would be prohibitively expensive. After analyzing the possibility of an outlet at Bonnet Carré he argued that it would be devastating to navigation on Lake Pontchartrain (Pabis 1998, 448).

Ultimately, he concluded that levees were the only feasible flood protection option for the lower Mississippi valley.

Humphrey also concluded that the bed of the Mississippi River consisted of hard clay and was not subject to regular incising by annual flood stages. This conclusion ran counter the theories of James Eads who advocated for the use of jetties to scour the river channel at its mouth. Humphrey and Eads engaged in a public feud over the feasibility of jetties to maintain a deep-water channel. Eads convinced congress in 1875 that it should allow him to prove his theory at his own expense, despite Humphrey's rejection of the idea. After several years of work on the jetties, Eads and his crews completed the project, and in 1879 the Army Corps of Engineers certified that they met the contract's requirements to create a 30-foot channel. Historian Martin Reuss reports that Humphrey's steadfast objection to a plan that Eads successfully accomplished undermined the credibility of the Corps of Engineers (Reuss 1985, 23). And indeed, Eads' success contributed to his influence in arguing that confining a river would enable it to scour a sufficiently deep channel and that closing all outlets would aid in this process (Reuss 1985, 26). This concept became cornerstone of the federal government's levees-only policy.

Even before the resolution of the Humphrey-Eads feud other factors began to align stronger public support for levees. During the Civil War, riparian land owners had neglected their levees which left flood defenses in a deteriorated condition by the late 1860s. This situation and the impoverished condition of most floodplain planters led to the expansion of rice cultivation in the immediate post-war period. Planters, seeking to quickly restore income from their lands and without enslaved labor to shore up the levees, turned to a crop that could thrive under annual floods and required less investment in terms of capital and labor (Mandelman 2016). To deliver water to their crops, they commonly installed wooden flumes in levees. Thus in the immediate post-war period, there was an expansion of rice cultivation and with it a tolerance for river-floodplain connections.

Following the Civil War, urban flood-protection advocates, influenced by Eads' success with the jetties and his arguments that a confined channel scours a deeper channel, tried to persuade congress to shore up the levee system. One step to aid scouring was to permanently close the natural outlet Bayou Plaquemine in 1868. The broader issue of investing federal dollars in protecting the land of recently rebellious Southerners met with limited support in congress. In order to consolidate support, congressmen from the lower valley turned the debate toward the value of levees for improving navigation. A massive flood in 1874 prompted still another study of the Mississippi and levees. It helped convince President Hays to create the Mississippi River Commission (MRC) in 1879 to act as an advisory body on flood control and navigation issues although it was not able to launch any immediate actions (Kemper 1929; Frank 1930; O'Neill 2006).

Crevasses returned as a prominent concern by the 1880s, and in 1884 a major flood produced some 284 crevasses. This catastrophic flood occurred during a significant rebound of sugar cultivation and a corresponding heightened expectation that the floodplain be divorced from the river. Levee advocates argued that many of the levee failures were due to faulty flumes inserted in the levees for rice cultivation. They convinced the state legislature to outlaw flumes and thus discontinue the limited river-floodplain connections (Mandelman 2016). The flood of 1882 had prompted planters to appeal to the Mississippi River Commission for more effective levees. The commission gradually adopted its levees-only policy and took steps to prohibit flumes as well (O'Neill 2006, 65; Mandelman 2016). Thus, by the 1890s, federal policy and flood-protection practices sought to fortify the riverfront with levees and largely dismissed the outlet option. This approach contributed to the closing of one of the last major distributaries, Bayou Lafourche, in 1904, although the Atchafalaya River continued to serve as a natural outlet. Improved levees and a significant decline in the number of crevasses from the high in 1884 to 12 in 1912 and 8 in 1913 seemed to support the levees-only policy (Frank 1930, 147). A 1913 MRC study concluded outlets would not significantly alter flood stages, thus fortifying the levee option (Camillo and Pearcy 2004, 100).

Serious floods continued, however, and a particularly damaging event in 1916, compelled congress to alter the stated purpose of levees along the lower river from serving navigation to flood control. A major crevasse below New Orleans in 1922 reinvigorated a lingering interest in outlets. Water broke through at Poydras and flowed into Breton Sound. Business and shipping interests in New Orleans feared that closure of a floodplain outlet far upstream in Arkansas had contributed to the high river stage on the lower river that year. They were apprehensive about the challenges of continually raising levees and docks and advocated for additional outlets. New Orleans interests championed a diversion into Lake Pontchartain but met with resistance from oyster fishermen and port officials in Mississippi. Ultimately, the Louisiana legislature approved the acquisition of property downstream from at Pointe a la Hache and in 1926 removed 11 miles of levee at this location downstream from New Orleans to create the Bohemia Spillway (Lopez et al. 2013). This outlet was to serve as a second wide mouth for the river during flood stages and thereby lower the crests at New Orleans (Kemper 1929; Elliot 1932, 319; Camillo and Pearcy 2004). Reports in the aftermath of the 1927 flood indicate the Bohemia Spillway effectively served that purpose (Lopez et al. 2013). And it has continued to serve this function, with modifications, to the present. In this way, deliberate steps to reconnect the river to its delta commenced. New Orleans interests continued to make a case for outlets to protect the city especially after the devastating 1927 flood.

Controlled Connections for Flood Control

The devastating flood of 1927 enabled those calling for additional outlets to reorient the basic policy of the Mississippi River Commission. Massive crevasses occurred during the extended high water in the spring of 1927, and they resulted in unprecedented damage. Floodwaters covered some 25,000 square miles, damaged over 160,000 homes, and caused an estimated $100 million in losses to crops and livestock. Some 900,000 floodplain residents had to flee their homes and at least 240 lost their lives (American Red Cross 1929; Cowdrey 1977; Kelman 2003). This tragic event represented a massive failure of the levees-only policy, and as a consequence congress launched hearings to determine how to reduce the risk of a repeat event.

At public hearings held in the aftermath of the flood, New Orleans interests pointed to the deliberate breach of the levees at Caernarvon in 1927 as proof that outlets added protection to the port city (U.S. Congress 1927a). The idea of outlets, long advocated by New Orleans interests (Kemper 1929), gained greater support, contrary to the recommendations made in previous engineering studies. Following the 1927 flood and in response to a request from congress, the Corps of Engineers prepared a report on the possibility of building controlled spillways on the lower Mississippi. The report examined the possibility of diversions at Bonnet Carré, Caernarvon, and through the Atchafalaya basin. The huge costs of the 1927 flood thoroughly undermined the previous argument that spillways were too costly, and the Corps reversed its previous opposition and recommended two of the spillways – Bonnet Carré and the Atchafalaya – along with the enlargement of levees upstream (U.S. Congress 1927b).

Agricultural and urban interests in the Atchafalaya voiced opposition to the outlets option. They declared the basin was the heart of the Louisiana "sugar bowl" and not undeveloped swamp land. Officials in Morgan City claimed that recent upstream levees along the Atchafalaya had increased flooding and use of the basin as a spillway would only exacerbate an already worsening situation. Fishermen in Mississippi charged that oyster beds would be destroyed if fresh water were diverted into Lake Pontchartrain, and thereby into Mississippi Sound (Mississippi River Commission 1927). Such were the arguments that swirled around the decision-making process.

Congress passed the 1928 Flood Control Act which included plans for both the Atchafalaya and the Bonnet Carré spillways and massive investments in levees along the lower river. The Atchafalaya River and floodways on either side would carry some 1.5 million cfs during a "super flood" and the Bonnet Carré spillway would redirect 250,000 cfs into Lake Pontchartrain. These outlets were to be used only in extreme flood years and not as regular passages for modest overflows. Nonetheless, this plan as adopted ensured continued use of the Atchafalaya as a permanent outlet. The Corps argued these outlets would eliminate the need to raise the levees and overhaul the port facilities at New Orleans. Levees and outlets emerged as the new policy for flood protection on the Mississippi, but restoring ecological functions was not part of the justification for outlets at the time (Camillo and Pearcy 2004; Reuss 1998).

Percy Viosca, a biologist and Director of Fisheries for the Louisiana Department of Conservation, addressed the issue of biological impacts of flood control shortly after the 1927 flood. He pointed out that when floods passed unimpeded across the floodplains they fertilized the alluvial soils and created "an inconceivably large supply of living plant and animal organisms, the fundamental food supply of our fresh and salt water food fishes, frogs, turtles, alligators, shrimp, oysters, fur bearing animals, and our ducks and other water birds." He noted that the 1927 flood impacted some 6 million acres and created a "gigantic fish hatchery and brood pond" (Viosca 1927, 51). He also inserted public health as a rationale by pointing out that the regularly flooded swamps were free of malaria (Viosca 1927, 53). Viosca was not advocating removing levees, but he observes, presciently, "A proper system of levees and back levees, designed for the purpose of protecting the more valuable front lands along the rims of the great lowland basins, combined with relief outlets into those basins which should be dedicated to the floods, would restore much of our formerly wet areas to their primitive condition" (Viosca 1927, 57). In a subsequent paper in *Ecology* he argues that by interfering with natural processes, humans contributed to the declining function of the wetlands. This condition would have serious repercussions in the freshwater fisheries, marine fisheries, oyster production, and also other wetland livelihoods such as trapping. All told, he reported that the wetlands sustained $20.5 in economic activity at the time, to say nothing of the recreational benefits (Viosca 1928). While powerful arguments, they did not influence the flood protection planning of the time.

After passage of the 1928 Flood Control Act, work commenced on the Bonnet Carré Spillway, the smaller of the two projects, and it was completed in 1931. A flood in 1937 prompted the Corps to open it for the first time to 81% of its capacity. There were few repercussions to this event. In 1945, the Corps opened all the bays and allowed the full 250,000 cfs to flow toward Lake Pontchartrain, and the predictions of the Mississippi oystermen came true. The rush of fresh water into Mississippi Sound caused a near total destruction of the oyster beds (Hopkins [U.S. Fish and Wildlife Service] in U.S. Congress 1946, 26–27). Biologist Percy Viosca countered that "the effect of the spillway was, on the whole, very beneficial because of its fertilizing effect on the waters of Lakes Pontchartrain and Borgne, and Mississippi Sound." He claimed that the fresh waters stimulated the organisms at the lower end of the food chain and this contributed to increased populations of mullet, menhaden, and shad. "The commercial shrimp crop taken in Lake Borgne and Mississippi Sound was the greatest since the shrimp trawl was introduced in 1917" (Viosca [U.S. Biological Survey] quoted in U.S. Congress 1946, 96). In addition to the oyster mortalities near Lake Pontchartrain, additional water moving through the Atchafalaya wiped out oyster production in Terrebonne Parish (O'Neal [Louisiana Department of Wildlife and Fisheries] in U.S. Congress1946, 79–81).

Damage to oysters did not dissuade the Corps from proceeding with the second major outlet. The Atchafalaya floodway was a much more ambitious environmental management undertaking and encountered political, engineering, and funding challenges along the way. Although there were already levees lining the Atchafalya River by 1927, new guide levees set back from the main channel had to be built for the new floodway system. By 1941 most of the guide levees were in place. Another challenge emerged in assembling the land and acquiring flowage rights form residents in the basin and this process proceeded as levee construction was underway through the 1930s (Reuss 1998, 188–90). By the mid-1940s, the Corps had acquired offers to sell easements on 80% of the property within the Morganza portion of the floodway. Reaching this level enabled construction work on other components to begin. World War II intervened and re-prioritization of national resources delayed construction funding. Nonetheless, in 1953 the Corps finally completed the structures that would divert water through the Morganza floodway. This provided a gateway to divert a specified volume of water through the Atchafalaya during floods. The primary purpose of the project was to control flooding on the Mississippi and not to sustain floodplain or coastal wetlands. These engineering priorities contributed to several environmental issues involving the Atchafalaya in the following decades (Reuss 1998).

During construction of the Atchafalaya floodway, environmental changes took place that dramatically altered the basin's ecology. Most significantly, the Corps worked diligently to guide the direction of the Atchafalaya's flow by dredging and trying to maintain one main channel where braided channels and large open lakes had existed. Dredging on the upper Atchafalaya, accelerated sedimentation on the lower river. Between 1930 and 1952, sediment dropped by the river created 60 square miles of land in Grand Lake. By the early 1950s, it hardly resembled either a river or lake (Reuss 1998, 154). This proved highly disruptive to local commercial fishermen who had worked the lake for decades (Delahoussaye 2010, 2014). According to Corps historian Martin Reuss, none of the Corps' documents dealing with dredging discussed the environmental effects of its activity (Reuss 1998).

As early as 1945, the Corps recognized the distinct possibility that the steeper gradient and shorter route of the Atchafalaya threatened to capture the majority of the Mississippi River flow (Reuss 1998, 214). To contend with this situation, the Corps proposed and received funding to construct the Old River Control Structure that would regulate the flow of river water down the Atchafalaya and preserve the status quo in terms of discharge via the Mississippi. In 1963, the engineers completed the structures that would send 30% of the combined flow of the Mississippi and Red rivers down the Atchafalaya. (Reuss 1998, 240–2).

In the 1960s and 1970s changing environmental concerns, new federal legislation and the institution of environmental impact studies for federal projects, debates over priorities in managing the Atchafalaya basin, and emerging concern with coastal land loss prompted rethinking of floodplain management policies and practices. These changes revealed the conflicted perspectives on how best to manage the basin. A local writer argued the Corps' regulation of flow through the basin would disrupt sport and commercial fishing (Gresham 1963). Despite these concerns, Louisiana's congressional delegation pressed for completion of the flood-control works. When federal officials proposed establishing a large wildlife refuge as part of a national program to preserve wetlands, local political officials pushed back against the effort. Eventually, a compromise led to the creation of several small state and federal wildlife areas, along with hydrologic management that prioritized flood management but also addressed wildlife and habitat concerns while also protecting the interests of land owners (Reuss 1998; Colten 2014).

Despite sometimes contentious debates between the many stakeholders, flood control remained the primary concern and use of the Morganza floodway in 1973 and again in 2011. Opening both the Morganza and the Bonnet Carré spillways clearly demonstrated the priority assigned to managing high water. During the first of those events terrestrial wildlife was either displaced or drowned as the flood waters coursed through the floodway. The exceptional discharges of freshwater damaged oyster beds near the mouth of the Mississippi and Atchafalaya rivers and also in Mississippi Sound (GSRI 1973, 44). Follow-up investigations reported that some 70% of the state's oyster leases suffered damages with some reporting mortalities of 45% (St. Amant 1973, 16–20). The 2011 opening produced oyster mortalities in the range of 85% in Mississippi and Breton sounds (USACE 2012, v.22). In both instances when the dual outlets were operated simultaneously, flood protection trumped other concerns and indeed the city of New Orleans, other smaller towns, riverfront industries, and the shipping facilities along the lower river remained dry.

Restoring Connections to Wetlands

Re-connections between the river and its alluvial plain and delta that were not driven primarily by flood control concerns became more prominent in the 1950s. Although state wildlife officials and local residents had advocated for freshwater diversions in the early 1900s to aid oyster cultivation and improve habitat for other wildlife, little was done on that front. The general thinking in the first half of the century was that huge freshwater discharges damaged oyster beds, at least in the short term. Supporting this position, the Louisiana Department of Wild Life and Fisheries (LDWLF) documented heavy oyster losses due to large freshwater releases through the Bohemia Spillway in 1944 and the Bonnet Carré in 1945 (LDWLF 1946, 185). Oyster productivity, however, proved to recover quickly after these events. By the 1950s, officials began reporting a more pervasive problem – land loss and its associated impacts. In its biennial report of 1954–55, the LDWLF noted increased salinity in oyster producing regions of the delta. Caused by flood control diversions, gulf encroachment (land loss), and canal dredging, mean salinities had increased over the preceding 20 years. This condition contributed to increased damage to oyster by "drills," rising mortalities due to a harmful bacteria (*D. marinum*), and other factors related to reduced concentrations of freshwater (LWLFC 1956, 145–46). To address this problem, the state launched construction of the Bayou LaMoque fresh water diversion in 1955. It consisted of four 10 × 10 foot concrete boxes in the levee with controllable flood gates that could be opened during high river stages to deliver fresh water to oyster reefs in St. Bernard and Plaquemines parishes (LWLFC 1956, 137). Completed in 1956, the diversion was unable to function at full capacity due to heavy scouring by the waters flowing into the coastal bays. Nonetheless, the department reported even at one third capacity, the effects were beneficial and that oyster yields were up 100% in American Bay (LWLFC 1957, 66). Thus, freshwater diversions gained credibility among oyster producers and state wildlife managers.

Louisiana officials met with the Corps of Engineers in 1957 to discuss additional freshwater diversion into the Mississippi River delta marshes below New Orleans. The discussions led to the consideration of four areas for diversions and the state preference for several small diversions, rather than large ones – with one exception being a larger diversion in upper Barataria Bay (LWLFC 1957, 66). A follow-up federal report provided additional details on these proposed outlets. It noted that in additional to the factors reported by the state, drought in the early 1950s had exacerbated low river stages and amplified the impacts associated with increasing salinity (USFWS 1959, 9).

During the next several decades, the state in collaboration with the Corps of Engineers, and initially with strong support of the oyster industry, installed a series of freshwater diversions. Congress authorized the first large-scale diversion project in 1965 and the Corps constructed the Caernarvon freshwater diversion between 1988 and 1991. At is inception, it had support among oyster producers, but presented serious liability issues to the state. Opened in 1991 with a capacity of 8000 cfs, the structure has contributed to diminishing salinity levels in Breton Sound, although extensive litigation by oyster fishermen ensued (Keithly and Wilkins 2006). Some leases have been harmed, but production in the state's oyster seed grounds has improved. Analysis also indicates other environmental impacts have been favorable. Marsh conditions have improved since the diversion's opening and shrimping has not been impacted (Penland and Beall 2002). Steps have been taken since 2013 to modify the diversion to enhance its wetland building capabilities (USACE 2013).

Building on the promising results of the Caernarvon diversion, several smaller diversions also opened in the 1990s. The West Point a la Hache siphons, completed in 1992, can deliver some 2144 cfs to brackish marsh on the west bank of the Mississippi River to stabilize salinities and promote vegetation (LDNR 2002). The Violet Diversion can deliver some 10,000 to 15,000 cfs into the Biloxi Marsh area to the east of the Mississippi. Opened in 1992 it has lowered salinity in this wetland area (Georgiou et al. 2007). The Naomi Diversion consists of a set of siphons and was completed in 1993. It diverts water in Plaquemines Parish to the wetland west of the river and has a capacity of just over 2100 cfs. It has reduced salinity in the discharge area since its opening. A second major freshwater diversion, Davis Pond, opened in 2002. It is situated above New Orleans and consists of a sizable concrete conduit through the levee to deliver over 10,000 cfs to an area of over 10,000 acres in the upper reaches of Barataria Bay. As with other projects, its operation has been erratic, but it has restored some of the historic wetland functions in its project area. These diversions, designed to deliver freshwater and not sediment, have reconnected the river with its delta in limited and highly controlled ways.

Planning and design work for two major sediment diversions gained approval in 2015 and the state Coastal Protection and Restoration Agency is moving forward on the Mid-Barataria and Mid-Breton diversions with capacities of 75,000 and 35,000 cfs respectively. They will differ from the freshwater diversion by using designs that will optimize sediment delivery to wetlands with the intent to reverse the rapid land loss occurring along the lower delta (Schleifstein 2015). Despite a CPRA supported report that underscored the importance of a social impact assessment before making a decision, the agency moved forward without taking that step (Colten and Hemmerling 2014). This could undermine public acceptance and support for these projects.

Conclusions

Levees along the lower Mississippi River are a dominant element of the human landscape. In towns and cities they loom over the riverfront neighborhoods and during the spring flood, ocean-going ships behind the earthen barriers appear to ride above the urban traffic. There have been no major breaches in recent decades, and the Corps proudly proclaims the levees have averted billions in flood damage since 1927. In both 1973 and 2011, the two principal outlets served their functions and rerouted excess flow. Thus, the highly engineered waterway has been contained for the time being.

Louisiana engineers voiced arguments for outlets in the mid-nineteenth century and a biologist proclaimed the benefits of re-connecting the river to the floodplain in the 1920s. In each instance, it took at least half a century or so to make strides toward those goals. Rivers and outlets became the official policy in 1928 and deliberate steps toward building freshwater diversions gained momentum in the 1990s. More ambitious sediment diversions are under design in the 2010s. Yet, the question remains, will these expensive projects build land fast enough to protect New Orleans and other coastal communities in the face of sea-level rise and climate change? Are these projects just another form of experimental engineering like the cutoffs of previous generations? To rely only on these experiments and not take steps to protect life and property within the other flood protection structures is insufficient. To diminish public participation in shaping the plans undermines democracy in a situation where lives, livelihoods, and property are at risk.

References

American National Red Cross (1929) The Mississippi Valley flood disaster of 1927. American National Red Cross, Washington, DC

Camillo CA, Pearcy MT (2004) Upon their shoulders: a history of the Mississippi River commission. Mississippi River Commission, Vicksburg

Colten CE (2003) Bayou St. John: strategic waterway of the Louisiana purchase. Hist Geogr 31:21–30

Colten CE (2005) Unnatural metropolis: wresting New Orleans from nature. Louisiana State University Press, Baton Rouge

Colten CE (2014) Southern waters: the limits to abundance. Louisiana State University Press, Baton Rouge

Colten CE, Hemmerling SA (2014) Social impact assessment methodology for diversions and other Louisiana coastal master plan projects. The Water Institute for the Gulf, Baton Rouge

Cowdrey AE (1977) Land's end: a history of the New Orleans District corp of engineers. U.S. Army Corps of Engineers, New Orleans District

Delahoussaye J (2010) James Delahoussaye collection of Atchafalaya River basin recordings. Library of Congress, Washington, DC

Delahoussaye J (2014) Personal interview. December 22

Elliott DO (1932) The improvement of the lower Mississippi River for flood control and navigation. U.S. Waterways Experiment Station, Vicksburg

Frank AD (1930) The development of the federal program of flood control on the Mississippi River. Columbia University Press, New York

Georgiou IJ, McCorquodale A, Retana AG (2007) Hydrodynamic and salinity modelling in the Pontchartrain Basin: final report. University of New Orleans, Pontchartrain Institute for Environmental Studies, New Orleans. www.pbc.uno.edu/reports/Georgiou/Hydrodynamic/Violet_UNO_Final_Report_AUG_2007.pdf

Gresham G (1963) Atchafalaya basin crisis. Louisiana Wild Life and Fisheries Commission, New Orleans

Gulf States Research Institute (1973) Flood of '73: post flood report: fish and wildlife supplement. Gulf South Research Institute, Baton Rouge

Harrison RW (1951) Swamp land reclamation in Louisiana, 1849–1879. Louisiana Bureau of Agricultural Economics, Baton Rouge

Herbet PO (1846) Annual report of the state engineer to the legislature of the state of Louisiana. The Jeffersonian, New Orleans, pp 10–11

Keithly WR, Wilkins JG (2006) Compensable property interests and takings: a case study of the Louisiana oyster industry. Am Fish Soc Symp 587–597

Kelman A (2003) A river and its city: the nature of landscape in New Orleans. University of California Press, Berkeley

Kemper JP (1929) Floods in the valley of the Mississippi a National Calamity. National Flood Commission, New Orleans

Lewis PF (2003) New Orleans: the making of an urban landscape. Center for American Places, Santa Fe

Lopez J, Henkel T, Boyd E, Conner P, Milliken M, Baker A, Gustavson K, Martinez L (2013) Bohemia spillway in southeastern Louisiana: history, general description, and 2011 hydrologic surveys. Lake Pontchartrain Basin Foundation, New Orleans

Louisiana Department of Natural Resources (LDNR) (2002) Annual Monitoring Report: Pointe a al Hache. Baton Rouge. http://lacoast.gov/reports/project/BA04Amr.pdf

Louisiana Department of Wild Life and Fisheries (LDWLF) (1946) First biennial report 1944–1945. Louisiana Department of Wild Life and Fisheries, New Orleans

Louisiana Wild Life and Fisheries Commission (1956) Sixth biennial report. Louisiana Wild Life and Fisheries Commission, New Orleans

Louisiana Wild Life and Fisheries Commission (LWLFC) (1957) Seventh biennial report, 1956–1957. Louisiana Wild Life and Fisheries Commission, New Orleans

Mancil E (1972) A historical geography of industrial cypress lumbering in Louisiana. Ph.D. dissertation, Louisiana State University

Mandelman A (2016) Porous boundaries and permeable places: ordering nature in the Mississippi River delta. Ph.D. dissertation, University of Wisconsin

Mississippi River Commission (1927) Public hearings of the Mississippi River Commission (July 7, 1927–August 20, 1927), Appendix A. Vicksburg: Miss.: Mississippi River Commission
Monette JW (1903) The Mississippi floods. Publications of the Mississippi Historical Society 7:428–478
Morris C (2012) The big muddy: an environmental history of the Mississippi River and its peoples from Hernando de Soto to Hurricane Katrina. Oxford University Press, New York
Norgress RE (1947) The history of the cypress lumber industry in Louisiana. Louisiana Historical Quarterly 30(3):1–83
O'Neill KM (2006) Rivers by design: state power and the origins of U.S. flood control. Duke University Press, Durham
Pabis GS (1998) Delaying the deluge: the engineering debate over the flood control on the lower Mississippi River. J South Hist 64:421–454
Penland S, Beall A (2002) Environmental atlas of the Lake Pontchartrain Basin. U.S. Geological Survey, Open File Report 02–206. http://pubs.usgs.gov/of/2002/of02-206/restoration/other-restoration.htm
Rehder JB (1999) Delta sugar: Louisiana's vanishing plantation landscape. Johns Hopkins University Press, Baltimore
Reuss M (1985) Andrew A. Humphreys and the development of hydraulic engineering: politics and technology in the Army Corps of Engineers, 1850–1950. Technol Cult 26(1):1–33
Reuss M (1998) Designing the bayous: the control of water in the Atchafalaya Basin, 1800–1995. U.S. Army Corps of Engineers, Office of History, Alexandria
Schleifstein M (2015) Move forward with Mid-Barataria, Mid-Breton sediment diversions, state says. New Orleans Times Picayune, 21 October. www.nola.com/environment/index.ssf/2015/10/move_forward_with_mid-baratari.html
St. Amant L (1973) Testimony to congress. In: U. S. Congress, Senate. Damage to oyster beds due to federal action, hearing before the subcommittee on water resources. 93rd Cong., 1st sess., 16–23
U.S. Army Corps of Engineers (2012) 2011 Post-flood report. U.S. Army Corps of Engineers, Mississippi Valley Division, Vicksburg
U.S. Army Corps of Engineers (2013) Modification of the Caernarvon diversion. www.mvn.usace.army.mil/Portals/56/docs/environmental/LCA/Near-Term%20Projects/ModofCaernarvonFactSheetJanuary2013PAO.pdf
U.S. Congress, House of Representatives (1927a) Flood control: hearings before the committee on flood control, 6 parts. 70th Cong., 1st sess. November 7 to November 22
U.S. Congress, House of Representatives (1927b) Spillways on the Lower Mississippi. H. Doc. 95. 70th Cong., 1st sess
U.S. Congress, House of Representatives (1946) Rehabilitation of oyster beds: hearings before the committee on flood control. 79th Cong., 2nd sess
U.S. Fish and Wildlife Service (1959) A plan for freshwater introduction from the Mississippi River into sub-delta marshes below New Orleans. U.S. Department of the Interior, Fish and Wildlife Service, Atlanta
Viosca P Jr (1927) Flood control in the Mississippi Valley in its relation to Louisiana fisheries. Trans Am Fish Soc 57:49–64
Viosca P Jr (1928) Louisiana wet lands and the value of their wild life and fishery resources. Ecology 9(2):216–229
Woodridge AD (1850) Report on the internal improvements of Louisiana for 1849. Documents of the first session of the third legislature of the State of Louisiana. (Baton Rouge: 1850)

The Nutria in Louisiana: A Current and Historical Perspective

Charles E. Sasser, Guerry O. Holm, Elaine Evers-Hebert, and Gary P. Shaffer

Abstract

Nutria is an exotic, aquatic rodent that was introduced to Louisiana wetlands during the early 1930s and can make coastal restoration more challenging. From 1960 to 1990, greater than 36-million nutria were taken when the fur market was lucrative. By 2000, the fur market had collapsed and nutria populations increased. Nutria reach sexual maturity at 4–8 months, may have 2–3 litters per year, and average 13.1 young per female per year. Nutria generally have a small home range in marsh habitats occurring primarily in fresh and intermediate marsh. The organic nature of these soils makes them particularly susceptible to destruction with grazing. Nutria are opportunistic feeders, with a broad diet comprising more than 60 plant species in Louisiana and they are attracted to wetlands that contain a reliable source of nutrient-rich fresh water, such as river diversions and assimilation wetlands. It is imperative that restoration projects that increase input of nutrient-rich fresh water into wetlands have a nutria control program in place. Nutria can consume large amounts of marsh biomass and in certain cases can cause the collapse of marsh locally. Scientific studies investigating effects of nutria on marsh habitats consistently conclude that nutria grazing is damaging to marsh and young forest vegetation. It is generally accepted that nutria damage – in addition to larger scale subsidence, sea level rise, and salinity intrusion – can create an accelerated deterioration of wetlands. Nutria grazing on baldcypress and water tupelo seedlings is extensive and remains a major factor in the inability of baldcypress-water tupelo forests to regenerate. Projects designed to restore coastal swamp forests should include a nutria control component and suitable protection of transplants should be used to minimize mortality from grazing. Eruptions of populations of nutria can cause severe wetland damage and loss. Some areas of the coast have persistent populations creating "Hot Spots" of severe damage, especially in freshwater-intermediate salinity areas of Terrebonne, Barataria, and Breton Sound basins. Nutria densities are relatively low in the Chenier Plain currently compared to historic observations and harvest records. The Coastwide Nutria Control Program (CNCP) was implemented in 2002–2003 by the LDWF, and since then there has been a reduction in 70,000 acres of marsh damage from 2003 to 2010. Approximately 446,000 nutria were

C.E. Sasser (✉) • E. Evers-Hebert
Department of Oceanography and Coastal Sciences, College of the Coast and Environment, Louisiana State University, Baton Rouge, LA 70803, USA
e-mail: csasser@lsu.edu; eevers@lsu.edu

G.O. Holm
CH2M, Baton Rouge, LA, USA
e-mail: guerry.holm@ch2m.com

G.P. Shaffer
Department of Biological Sciences, Southeastern Louisiana University, Hammond, LA, USA
e-mail: shafe@selu.edu

J.W. Day, J.A. Erdman (eds.), *Mississippi Delta Restoration*, Estuaries of the World, https://doi.org/10.1007/978-3-319-65663-2_4

harvested in 2010 in the CNCP, primarily in the deltaic plain. When considering the costs of creating new wetlands (approximately $50,000–$70,000 per acre), the CNCP is a successful wetland conservation program that has produced measureable reduction in marsh damage. Since 2002–2003, 2,571,480 nutria have been harvested. This program is a success and, from a resource management perspective, should be continued with improvement and expansion if possible. Nutrient enrichment of coastal landscapes may cause nutria population growth and habitat damage. Coastal restoration projects with areas receiving nutrient enrichment should include nutria control to ensure plant productivity, establishment, and expansion.

Keywords
Invasive species • Nutria • Herbivory • Wetland loss • Population dynamics

Nutria in Louisiana Wetlands: A Current and Historical Perspective

> The influence of nutria alone is sufficient to cause the marshes to continually decline, jeopardizing their existence. Biologist, Allan Ensminger (Addison 2000)

Introduction

More than 70 years since its introduction, the exotic nutria (*Myocastor coypus*) has changed the landscape ecology of Louisiana. Here we review some of the evidence for the capacity of this animal to affect ecological succession and the long-term stability of vegetated wetlands and look at current nutria distribution and relative abundance across the major geographic provinces. We also examine trends in the extent and severity of nutria grazing damage to wetlands under the Coastwide Nutria Control Program (CNCP) developed by the Louisiana Department of Wildlife and Fisheries (LDWF). We highlight the essential life history characteristics of nutria as documented by a number of studies (Kinler et al. 1987; Bounds et al. 2003; Baroch et al. 2002; Willner 1982). This chapter is primarily based on a report to the Lake Pontchartrain Basin Foundation (Holm et al. 2011).

A once thriving global fur market fueled the development of a successful fur industry, based primarily on muskrat, in Louisiana dating back to the early 1900s. After the nutria's introduction to Louisiana in the late 1930s and its range expansion in the 1940s, the nutria was at first thought a threat to the coastal resident's livelihood, because muskrat populations declined as nutria were becoming more prolific. The State facilitated a market transition from muskrat to nutria by transplanting and protecting the nutria, until it was clear the nutria population had formed a stronghold in Louisiana's coastal marshes. Eventually nutria pelts became an increasingly valuable component of coastal Louisiana's economy, climaxing in the 1970s, when European processors offered luxury nutria fur products.

Fashions changed, economies foundered, and animal rights activists challenged the ethics of trapping and fur ranching. The crash of the fur market in the 1980s left nutria population growth unchecked on over 1 million ha of ideal marsh habitat. The nutria has since emerged as a significant threat to coastal wetland sustainability. William Conner ranked nutria herbivory – along with saltwater intrusion and flood inundation – as the top three problems facing baldcypress-water tupelo forest regeneration and sustainability in Louisiana. While it was recognized early that nutria caused acute damage to sugar cane and rice crops, it was only during the 1980s as trapping relaxed that land managers complained of chronic damage to wetlands.

Quite common were the early accounts of the capacity of nutria to severely damage wetlands:

> I can show you places where nutrias and muskrats grew so thick 40 to 50 years ago, trappers could not catch enough of them. They ate up all the grass, and those areas are now ponds (Ignace Collins, in Felsher 2000).
>
> Two of the most intensively trapped and managed properties, on a tract of 150,000 acres in Vermilion Parish and a tract of 155,000 acres in Cameron Parish, were completely leveled by nutria, with peak takes of marketable pelts in the 60,000 to 70,000 figure, after which came sharp drops in production accompanied by poor pelt quality due to the lack of food supply. The active Mississippi River Delta, comprising about 350,000 acres. . .also went to pieces by 1956–57, as a result of the 250 nutria transplanted there in 1951. There was a gradual rise [of the population] from 1951 to the peak in the 1955–56 season, when nutria were everywhere with vegetative cover still standing. By the following season, only the pass banks appeared to be holding the delta together. (O'Neil , b)

There is ample scientific evidence that suggests that nutria have the capacity to alter wetland stability. Our current understanding emphasizes that nutria herbivory alone as a stress on plant productivity rarely explains what we could term 'irreversible marsh collapse', except at the highest densities. Instead, nutria grazing, interacting with other stresses (inundation or salt stress), may force a tipping point where the wetland is forced to another 'stable' state, perhaps with reduced productivity, altered plant composition, or ultimately a reversion to open water. These effects are real and have consequences for wetland loss and coastal restoration that can take the form of: (1) inhibition of the reproduction potential for herbaceous and forested (baldcypress- water tupelo) wetlands in the coastal zone; (2) marsh elevation loss with reduced organic matter accumulation, which further exacerbates inundation/salt stress to plants already experiencing broad-scale soil subsidence and rising sea levels; and (3) soil destabilization and susceptibility to accelerated erosion.

Coastal managers and scientists have wrestled with understanding just how severe the nutria damage problem is in Louisiana. The perceived impact of an exotic marsh rodent may seem negligible compared to the scale of the geological land subsidence, sediment deficits, and hydrologic impoundment afflicting coastal wetlands. Decades of cumulative deleterious effects of nutria grazing on the coastal landscape have been difficult to quantify and separate from other factors causing wetland loss. The extreme cases of population eruption and wetland damage that arise at local scales are a testament to the capacity of nutria to cause marsh collapse or inhibit forest regeneration, but the more subtle, chronic effects of grazing have been harder to document.

Most coastal managers and wildlife biologists advocate nutria control rather than eradication as the only realistic option, given the breadth of Louisiana's wetlands. Eradication has been successful in the Blackwater Refuge in Maryland, USA and Great Britain where the wetlands comprising suitable nutria habitat are relatively discrete and accessible. Several points are worth considering about nutria population control, past and present in Louisiana:

1. The fur market once provided enough incentive to control the nutria population; this industry declined and collapsed but has recovered somewhat in recent years.
2. Of the density independent mortality factors, severe freezes and tropical storms regulate the population. In Louisiana, no documented widespread disease outbreak has ever caused a significant reduction in the nutria population.
3. It is hypothesized that conserving large alligators will control nutria density. This is intuitively reasonable but there is a lack of strong evidence to support this at present.
4. Poisoning on a large-scale is unacceptable because of possible mortality of non-target species.
5. With adequate incentive, human predation is an effective control on the nutria population.

A Short History of the Nutria in Louisiana

Before the nutria was introduced into Louisiana, the native muskrat (*Ondatra zibethicus*) was the most numerous furbearer. Muskrat populations increased in Louisiana's marshes as marsh burning became more common in the late 1800s and early 1900s (O'Neil 1949). This helped increase access for trapping mink, otter, and raccoons, in addition to alligator hunting. Burning improved the muskrat's ideal forage, Olney's three-square grass (*Schoenoplectus americanus*). Eventually, muskrats moved from the southeastern marshes of the Delta plain to western marshes of the Chenier Plain. Muskrats became so numerous that in the early 1900s they were being hunted with 'dogs and pitchforks' to keep them from devouring cattle range. Although the earliest record of muskrat pelt trading from Louisiana to the northern markets occurred in New Orleans during 1878, it was not until the 1920s that Louisiana muskrat harvests became substantial. The good trapping years during the 1930s and 1940s yielded in excess of 4 million muskrat pelts statewide. During the 'great muskrat eat out' from 1945 to 1947, each year more than 7 million muskrats were harvested. The muskrat harvest largely collapsed thereafter (Fig. 1). This collapse coincided with the introduction and expansion of the nutria.

Wild populations of nutria became established in Louisiana in the early 1940s after escaping from fur-ranches and with intentional introductions across the coast (O'Neil 1968; Evans 1970; Bernard 2002).

By 1950, nutria became abundant in western Louisiana in the Chenier Plain region in St. Mary, Iberia, Vermilion, and Cameron Parishes. By the last half of the 1950s, the overall population of nutria peaked at about 20 million. Big events such as the passage of the very large Hurricane Audrey in 1957 caused mortality to the nutria population across southwest Louisiana, but also dispersed significant numbers inland leading to damage to agricultural crops, especially sugar cane. This impact to agriculture interests was followed in 1958 by the Louisiana legislature declaring nutria an "outlaw quadruped" and establishing a 25 cent bounty.

A boost for the nutria fur industry came about with the development of a nutria fur market with Germany in 1960. The nutria trapping industry continued to grow, with more nutria than muskrat taken during the 1961–1962 season, with muskrat harvests never again exceeding that of nutria. During this period in the 1960s the nutria population was so large that nutria caused extensive agricultural damage to sugar cane crops in coastal parishes, leading to disputes between the fur and agricultural (sugar) industries. In 1963 the nutria was considered an outlaw animal in 17 coastal parishes; however in 1965 nutria were given state-wide recognition as a protected furbearer except on agricultural lands. After nutria harvests surpassed that of muskrat in Louisiana in the 1960s,

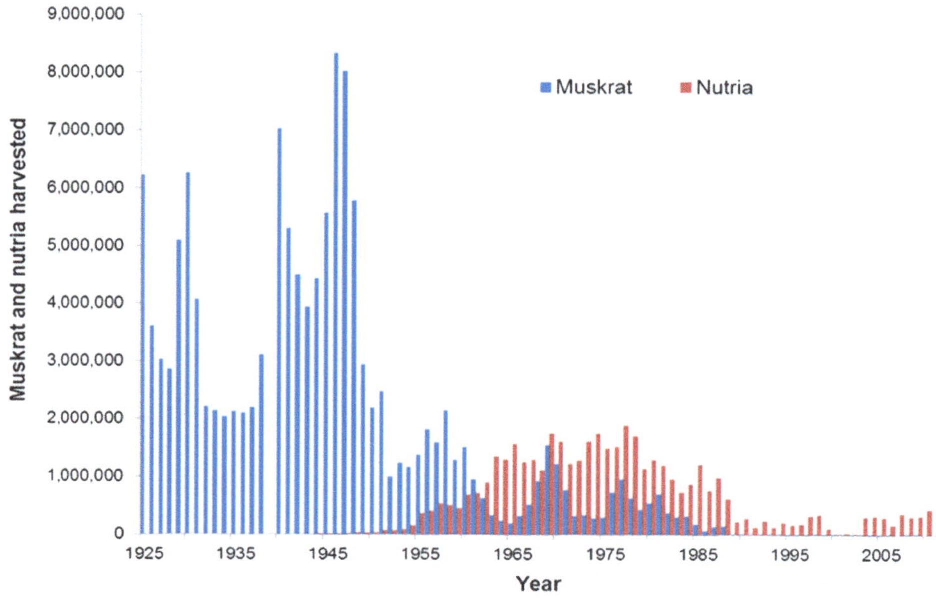

Fig. 1 Historical harvest of muskrat and nutria from Louisiana during 1924–2010 (See Table 1 for a summary of important events that affected harvest numbers)

the Louisiana Department of Wildlife and Fisheries started a practice of capturing nutria from one area of the state and releasing in another to share the fur industry in Louisiana (Edmond Mouton, LDWF, pers. comm.).

There were large harvests and sales of pelts in the 1970s and 1980s. By the end of the 1980s, however, changes were occurring, with reports of severe damage to marsh and agricultural lands, a foreign and domestic fur market in decline, and fur ranching and trapping ethics that were challenged. In 1988–1989, nutria harvest fell below 500,000. With hindsight, it is not surprising that increased reports of severe marsh damage were coming in as fur markets, and ultimately the numbers of nutria trapped, declined. The decadal harvest of nutria decreased in the 1990s to about 2.17 million compared to 15.26 million in the 1970s and 11.99 million in the 1960s. By 2000 there was virtually no demand for nutria fur (Wiebe and Mouton 2008). Not surprisingly, LDWF reported nutria damage to coastal marshes at about 105,000 acres for 1999. The Coastwide Nutria Control Program (CNCP) began in 2002 with 300,000 nutria harvested in 2003, up to 445,000 taken in 2009–2010.

Reproductive Potential of Nutria

Nutria have a high reproductive potential. Based on a population of 8000–11,000 nutria in Great Britain, Gosling (1974) calculated that a doubling could occur (15,000–18,000) within a 12-month period if not controlled. In Louisiana, the number of nutria harvested during the 1943–44 season was 436 and by the 1976–77 season, the peak of nutria harvested exceeded 1.8 million. This exceptional reproductive potential is attributable to extensive favorable habitat and climate, in addition to the nutria's natural history traits, such as an early sexual maturity (occurring at 6 months age for male and female) and a high fecundity of 13 or more young produced per female per year. Fecundity also increases with age (Willner et al. 1979). Contributing to this high rate of production is the nutria's continuous breeding cycle in Louisiana and potentially localized lack of predators.

Movements, Home Range, Dispersal

Nutria home range and dispersal distances are influenced by habitat type; long distances were traveled in agricultural settings compared to herbaceous marshes. For example, in an agricultural environment (sugar cane fields), nutria moved over 50 miles, but they averaged approximately three miles (Evans 1970). Most studies have shown that marsh nutria generally do not move much more than a few kilometers. Linscombe et al. (1981) reported marsh nutria moving up to 3.2 km; however, 279 out of 310 recaptures were within a 324 ha study area and most of the nutria were within 400 m of the study area. Nolfo-Clements (2009) calculated 77 m as the mean distance traveled by nutria with radio transmitters in a freshwater floating marsh. In ponds near Belle Chasse, La., Warkentin (1968) found that marked nutria remained within 300 m of their capture site, with one nutria moving 600 m. In southwestern Louisiana marshes, daily cruising ranges of nutria were mostly less than 200 m (Adams 1956).

Nutria in marsh habitats have restricted home ranges. A home-range of 13 ha was estimated by LeBlanc (1994). Using two different calculations, nutria home-range estimates of 29 and 33 ha were reported for freshwater floating marshes (Nolfo-Clements 2009), and there were no differences between sexes. Tropical storm and hurricane surge is responsible for inadvertently dispersing nutria to otherwise isolated areas, where they may establish new colonies (E. Mouton LDWF, pers. comm.).

Habitat Preferences

Nutria prefer freshwater wetlands compared to more saline environments in Louisiana. Ramsey et al. (1981) proposed that freshwater habitats provide more available calcium and digestible matter than more saline wetlands. Protein can be twice as high in freshwater versus brackish marshes (Wilsey et al. 1991). Fresh and intermediate marshes and swamp forest cover over one million ha in Louisiana. Ialeggio and Nyman (2014) reported that nutria grazing was higher in fertilized marsh plants than unfertilized controls.

Freshwater floating marshes are optimal habitat for nutria (O'Neil 1949; Palimsano 1972 in McNease 2005). There are likely several reasons for this, including the quality and diversity of forage, reduced exposure to flooding, and choice of escape routes, whether running aboveground on the marsh or diving beneath the surface and into holes or rivulets in the marsh mat. Given the susceptibility of coastal freshwater marshes to seasonal flooding from tropical storms and frontal set-up, floating or buoyant patches of marsh accommodate feeding, grooming, and resting areas that remain dry. In Barataria and Terrebonne basins, there is an estimated 140,000 ha of floating/buoyant marsh (Sasser et al. 1994). Widespread deterioration of floating marshes has been documented in northern Terrebonne basin, and it is unknown to what degree historical nutria grazing can explain this decline in marsh quality and quantity (McNease 2005; Sasser et al. 1995; Sasser et al. 2004). This area has experienced persistent and severe nutria damage (Marx et al. 2004).

Food Selection

Nutria are opportunistic foragers that exploit a variety of emergent, floating aquatic, submersed, and woody species. Seasonal availability in forage quality and quantity influences their diet. More than 60 species occurring in Louisiana are eaten by nutria (see Holm et al. 2011). Nutria are considered wasteful feeders (Harris and Webert 1962), often destroying ten times more than they consume (G. Linscombe pers. comm., in Bounds et al. 2003). They will consume leaves, roots, and reproductive organs and also consume bark when other food is scarce.

Given the prevalence of fast growing herbaceous species in Louisiana marshes, there is a broad selection of species that have foliage or organs that are palatable to nutria, at least during some parts of the growing or dormant seasons. There seem to be only a few plant species that nutria clearly avoid, such as *Peltandra virginica*, *Zizaniopsis mileacea* and *Justicia ovata* (Llewellyn and Shaffer 1993), and there are some species that have low preference. Nutria are capable of digesting a wide variety of herbaceous species to gain some nutritional value.

Some species of plants that nutria feed on can persist even under very intense grazing, such as *Eleocharis*, *Hydrocotyle*, and *Bacopa* (Sasser et al. 2008; Izdepski et al. 2009). That is, these plants have a high resiliency or plasticity to grazing and may exhibit compensatory growth with grazing pressure. We found only one reference of nutria feeding on *Ludwigia* spp., which can be prolific in nutria damaged areas (Sasser et al. 2008). Some species are largely avoided given a high content of digestibility reducers, such as silica in *Justicia ovata*, or the presence of a noxious compound, like calcium oxalate in *Colocasia esculenta*. Based on our personal observations, there are species within a genus that may have different palatability or higher nutritional value, such as *Typha latifolia* (more preferred) and *Typha domingensis* (less preferred; but see Geho et al. 2007). Within the genus *Sagittaria*, the species *S. platyphylla*, *S. graminea*, and *S. latifolia* are highly preferred by nutria especially on the early successional delta islands of the Mississippi and Atchafalaya Rivers (Fuller et al. 1985; Shaffer et al. 1992; Evers et al. 1998). On the other hand, *S. lancifolia* may be less palatable, but still constitutes a significant portion of the nutria diet when other species are scarce (Linscombe et al. 1981). Several studies have documented the effect of nutria grazing on the tender roots of *Taxodium distichum* seedlings (e.g., Myers et al. 1995). A survey of survival rates of baldcypress plantings in Louisiana show that a very small percentage of plantings survive without protection (Conner 1995; Myers et al. 1995).

Nutria Grazing Impacts to Coastal Louisiana Wetlands

> Nutria only had an impact upon biomass if another disturbance was present, and tended to amplify effects of disturbance. (Keddy et al. 2007)

The use of cage-like exclosures has been the most popular method of assessing the impacts of vertebrate herbivores on marsh damage. In the Louisiana wetland studies reviewed here, nutria have been associated with most of the grazing impacts. Other herbivores such as boar (*Sus scrofa*), waterfowl (typically ducks and geese), and muskrat (*Ondatra zibethicus*) may have been present, but their contribution was not measured or in some cases was not considered important. In most cases, nutria herbivory has been shown to markedly decrease plant biomass or produce changes in species composition. Significant reductions to aboveground biomass (50% to 100% loss) commonly have been observed in exclosure studies (Table 1). The increase in dead aboveground biomass noted by Ford and Grace (1998) can be attributed to the destructive nature of nutria feeding where an order of magnitude more marsh is destroyed than is consumed. Although few studies have measured belowground impacts, those that have found where grazing pressure is extreme, it is possible for nutria to keep live root stocks depleted. For example, Shaffer et al. (2015) reported that belowground biomass was significantly higher in cages compared to uncaged controls. Marshes protected from grazing for 4 years and transplanted with maidencane resulted in 141% increase in belowground biomass compared to grazed areas (Sasser et al. 2008). Reduced soil organic matter accumulation and detectable losses in soil elevation have been documented with moderate nutria grazing (Grace and Ford 1996).

Some of the most dramatic visual nutria damage to wetlands has occurred in the young delta marshes near the mouth of the Atchafalaya River (Figs. 2 and 3; Fuller et al. 1985; Shaffer et al. 1992; Evers et al. 1998; Sasser et al. 2009). Fuller and Sasser (1988) found that significant damage in the Atchafalaya Delta was caused by nutria and waterfowl. Severe nutria grazing also occurs in more mature, freshwater peat-based marshes. With persistent grazing, the peat soil can become destabilized as organic matter accumulation decreases and the soil weakens. This condition has been observed in thin-mat floating marshes in both Barataria and Terrebonne basins (Sasser et al. 1994) as well as in the Hammond Assimilation Wetland (Shaffer et al. 2015). While there are multiple causes for freshwater marsh deterioration, herbivory in the floating freshwater habitat can keep these habitats from recovering (Sasser et al. 2008, Sasser et al. 2009; Shaffer et al. 2015).

Nutrient additions can cause preferential nutria grazing to wetland plants (Howell 2003; Shaffer et al. 2009). Nutrient enrichment of plant tissue, particularly with nitrogen, can increase the protein content of herbaceous plants. Wetland restoration projects often add fertilizer to recently-transplanted marsh plants. In a test to evaluate attractants for nutria in Louisiana, Jojola and Witmer (2006) found that nutria spent more time at fertilized plots of both *Panicum hemitomon* and *Spartina* plants. They concluded that nutria identified olfactory cues of fertilized plants. Ialeggio and Nyman (2014) reported that nutria grazing was higher in fertilized marsh plants.

The use of assimilation wetlands can elevate wetland plant nutrients and cause increased grazing by nutria. Shaffer et al. (2015) concluded that wetland deterioration at the Hammond Assimilation Wetland was caused primarily by grazing due to large numbers of nutria. Ten exclosures all had high biomass levels compared to paired unprotected controls that were 100% defoliated. The degree to which assimilation wetlands or river diversions increase plant nutrition and, with time, enhance nutria population growth could hypothetically act as a positive feedback loop. Much of the potential coastal areas to be restored will be in fresher marsh areas and these areas will likely experience nutria grazing. Where fresh water and nutrient enriched waters are employed as restoration tools, nutria control measures will contribute to wetland establishment and maintenance.

The direct and indirect impacts of nutria grazing have the capability of causing significant wetland loss in Louisiana, perhaps not as a stress alone, but combined with other stresses of salt, inundation and physical energy. From their field studies in the oligohaline marshes of Lake Pontchartrain, Slocum and Mendelssohn (2008) concluded: "these effects of nutria in the marshes of the Mississippi River Delta suggest that the exotic rodent may create a cascade of multiple stressors that leads to wetland loss." Gough and Grace (1998) summarized the susceptibility of wetlands to grazing: "High levels of herbivory will likely intensify or alter the effects of sea level rise on coastal wetland communities. Increased environmental stresses in the presence of intense herbivore activities may affect community biomass production, the relative abundance of dominant plant species, and diversity." They found evidence for all these processes.

Nutria grazing directly influences the regeneration of baldcypress trees in Louisiana's forested wetlands. In efforts to re-establish baldcypress forests, transplanted seedlings were heavily grazed by nutria. For example, Blair and Langlinais (1960) reported that 90% of 2100 test seedlings were uprooted within 4 months after planting. Myers et al. (1995) and Brantley and Platt (1992) reported nearly 100% mortality of transplanted baldcypress seedlings due to nutria herbivory. Areas planted with baldcypress saplings do fair better possibly because their root masses are larger and can withstand attempts at uprooting. An expert on baldcypress forests, William Conner eventually ranked nutria herbivory – along with saltwater intrusion and flood inundation stress (land subsidence) – in the top three problems facing baldcypress-water tupelo forest regeneration and sustainability in Louisiana.

Table 1 Summary of controlled field studies that report grazing impacts to above- and belowground biomass, soil processes and rates of change in Louisiana coastal wetlands

Study	Wetland type	Plant/soil component	Outcomes and differences
Shaffer et al. (1992)	Deltaic, early-successional freshwater herbaceous	Species richness	47% decline 4.5 (G) vs. 8.5 (UG)
		Plant coverage (cover values)	74% reduction 63 (G) vs. 240 (UG)
		Species changes	In the presence of intense herbivory, these species significantly increased coverage over the 6-year study: *Justicia ovata, Colocasia esculenta, Polygonum punctatum, Leersia oryzoides*
Evers et al. (1998)	Deltaic, early-successional freshwater herbaceous	Total aboveground biomass	50% reduction, 597 (G) vs. 1235 (UG) g m^{-2}
		Total belowground biomass	25% reduction, 248 (G) vs. 346 (UG) g m^{-2}
		Species changes	With grazing:
			1. Decline in species richness
			2. Decrease in *Sagittaria* spp.
			3. Increase in *Polygonum punctatum, Justicia ovata*, and *Leersia oryzoides*
Fuller et al. (1985)	Deltaic, early-successional freshwater herbaceous	Total aboveground biomass	48–62% reduction,
Ford and Grace (1998)	River mouth, late-successional, brackish herbaceous	Live aboveground biomass	75% reduction, 112 (G) vs. 460 (UG) g m^{-2}
		Dead aboveground biomass	58% increase, 1063 (G) vs. 614 (UG) g m^{-2}
		Belowground biomass	50% decrease, 0.75 (G) vs. 1.35 (UG) g cm^{-3}
		Species richness	NSD
		Soil elevation increase	−0.50 cm difference, +0.22 (G) vs. +0.75 (UG) cm
		Root zone thickness	0.48 cm decrease 0.19 (G) vs. 0.67 (UG) cm
		Soil bulk properties	NSD
Gough and Grace (1998)	River mouth, late- successional, intermediate herbaceous; *Sagittaria lancifolia*	Total aboveground biomass[a]	28% reduction 1000 (G) vs. 1400 (UG) g m^{-2}
	River mouth, late- successional, intermediate herbaceous; *Spartina patens*	Total aboveground biomass[a]	NSD
		Species biomass; *Schoenoplectus americanus*[a]	40% reduction 200 (G) vs. 350 (UG) g m^{-2} *S. americanus* is selected over *Spartina patens*
Randall and Foote (2005)	Habitat type	Species biomass; *Spartina patens*	12% reduction 1870 (G) vs. 2132 (UG) g m^{-2}
		Species biomass; *Schoenoplectus americanus*	70% reduction 388 (G) vs. 1343 (UG) g m^{-2}
Sasser et al. (2004)	Restoration of floating freshwater marsh degraded by nutria herbivory with transplanting and protection	Transplant mortality of *Panicum hemitomon* within 6 months, without protection	Almost complete mortality 90–100%
		Coverage of *Panicum hemitomon* after 4-year with transplanting and protection	70% areal expansion
		Total aboveground biomass with transplanting	187.5% increase 400 (G) vs. 1150 (UG-with transplanting) g m^{-2}
		Live belowground biomass with transplanting	141% increase 567 (G) vs. 1367 (UG- with transplanting) g m^{-2}
		Root mat thickness	80% increase 10 (G) vs. 16 (UG- with transplanting) cm
		Soil strength (0–10 cm)	80% increase 0.25 (G) vs. 0.40 (UG- with transplanting) kg cm^{-2}

(continued)

Table 1 (continued)

Study	Wetland type	Plant/soil component	Outcomes and differences
Geho et al. (2007)	Oligohaline marsh, Lakes Maurepas and Pontchartrain; examined the effects of herbivory, neighbors, and sediment addition	16 species transplanted	Overall significant reduction in biomass was largely controlled by herbivory, less so for neighbors, and no effect of sediment addition; only two species had significant biomass reduction with grazing (*Taxodium distichum*, *Typha domingensis*)
Slocum and Mendelssohn (2008)	Oligohaline marsh, Lake Pontchartrain	Species composition	NSD
		Total aboveground biomass [a]	After 14 months, 800 (G) vs. 1700 (UG) g m^{-2} after 26 months, NSD
Day et al. (2011)	Fresh marsh, between Lakes Pontchartrain and Maurepas	Live belowground biomass[a]	157% increase 2167 (G) vs. 5573 (UG) g m^{-2}
Blair and Langlinais (1960)	Swamp, near Cypremort, LA	Bald cypress seedling planting	100% mortality, (G, 2100 of 2100)
Brantley and Platt (1992)	Former swamp, Manchac WMA	Bald cypress seedling planting	95.8% mortality (G, 308 of 321) vs negligible mortality (UG)
		Bald cypress sapling planting	28.6% mortality (G, 46 of 161)
Myers et al. (1995)	Former swamp Manchac WMA	Bald cypress seedling planting	(G) nearly 100% mortality
		Bald cypress seedling planting	50% mortality (G)
		Bald cypress seedling planting after 2 seasons	100% mortality (G)

[a]Interpolated from graph
G grazed, *UG* un-grazed, *NSD* no significant difference

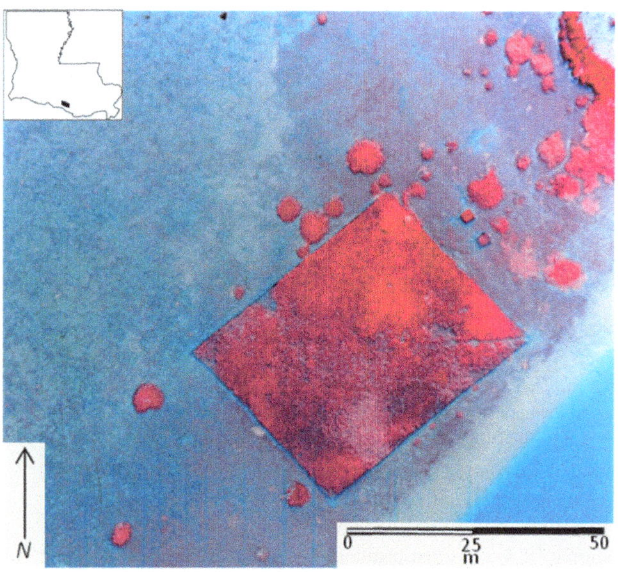

Fig. 2 A color-infrared image illustrating the effects of nutria herbivory in a recently- developing delta in the Atchafalaya Bay in St. Mary Parish, Louisiana (located as shown in *inset*). *Red* indicates vegetated areas. The 40 m × 50 m rectangular feature excluded nutria. The area outside of the rectangle was open to herbivory, with only Justicia ovata (*circular red areas*) and de-vegetated mudflat (*blue-gray*) remaining

Nutria Biomass Consumption Relative to Marsh Production

We calculated a budget of organic matter consumed and respired by nutria to develop a first order estimate of how much organic matter could be lost from marsh production (see Table 2). The amount of carbon lost from the wetland system with nutria grazing can be significant. In one year, an average-sized nutria can consume 72.4 kg of dry organic matter. While 64% of material consumed will be recycled back to the marsh, the other 36% is ingested. Thus, the entire amount of biomass that is ingested is equivalent to the biomass produced and available for consumption each year on approximately 24 m^2 of marsh. The biomass completely lost is equivalent to 7.8 m^2 of marsh, and this number is conservative since nutria may cause living plant destruction.

Historical Nutria Abundance Estimates

An early estimate of Louisiana's nutria population was approximately 20 million during the period of 1955–59 (O'Neil and Linscombe 1977); however, it is unclear

Fig. 3 Aerial views of a thin-mat floating marsh site where nutria exclosures were constructed and planted with *Panicum hemitomon* in Terrebonne Parish, depicting the differences inside and outside of the exclosures. Image shows the study site with four rectangular exclosures

whether this included the entire state or just coastal parishes. Ascertaining nutria densities on different wetland habitats has been a complex issue in Louisiana. These estimates have been derived by mark-recapture studies and harvest statistics on tracts of known area. Mark-recapture techniques may over-bias density (pers. comm. G. Linscombe 2010). The published historical estimates that we gathered vary considerably (Table 3).

It is generally accepted that nutria carrying capacity is highest in freshwater marshes, and decreases with increasing salinity of the habitat. The actual density of any given wetland may fluctuate with localized conditions (weather, marsh management) that operate across seasons and years. The highest reported densities (43.7 nutria ha^{-1}) of nutria can be found on their optimal habitat of freshwater floating marsh (Kinler et al. 1987). Studies in brackish marshes show that at a density of approximately 24 nutria ha^{-1} may exceed the carrying capacity of this habitat type and can result in devegetation (Linscombe et al. 1981).

Contemporary Harvest of Nutria as an Indication of Abundance

Based on the take of nutria from different habitat types in the dataset of the CNCP, we provide a synthesis of the range in harvest (trapping and shooting) from 2003 to 2010, a period when the bounty was at or near $5.00 per tail. The nutria control program started with the 2002–2003 season and has been one of the best success stories in protecting and restoring wetlands of the ongoing coastal restoration program. Seasonal variability is determined by both the population of nutria and the ability to get to nutria. The ability to reach hunting locations is better with higher water levels, which concentrates nutria on areas of high ground where they are more easily hunted. The greatest number of nutria harvested was 445,963 in the 2009–2010 (Table 4), which was a year with very high water in the marshes. The smallest number of animals harvested was 170,407 in 2005–2006 season, during the period of hurricanes Katrina and Rita. The totals for other years were between 297,500

Table 2 A budget of nutria metabolism and plant biomass consumption on an herbaceous marsh over one year

Nutria metabolism	Explanation	Quantity
Average nutria mass	A conservative estimate; 15% lower than Kays (1956) population average of 5.4 kg	4.72 kg
Nutria daily fresh tissue consumption	25% of body mass consumed daily[a]	1.18 kg (FW)
Nutria daily dry tissue consumption	Leaf tissue water content = 85%[b]	0.177 kg (DW)
Daily loss from marsh (36% of biomass goes to respiration + maintenance)	60% and 4% of mass consumed is excreted in feces and urine, respectively; this leaves 36% for respiration and maintenance to hold a constant body mass[c]	0.037 kg (DW)
Yearly biomass consumed from marsh	Respiration + maintenance + excretion + growth/reproduction; 12% increase in mass for growth and reproduction over one year [c]; (365 *0.177)*(1.12)	72.358 kg (DW)
Yearly biomass lost from marsh	Respiration + maintenance + growth/reproduction; 12% increase in mass for growth and reproduction over one year; (365 d*0.037)*(1.12)	26.049 kg (DW)
Marsh production		
Yearly marsh production		
Aboveground		2.0 kg m^{-2} $year^{-1}$
Belowground		3.0 kg m^{-2} $year^{-1}$
Total		5.0 kg m^{-2} $year^{-1}$
Standing stock available for consumption		
Aboveground	Assumes that 50% of annual production is available for consumption (unavailable: Litter production, decomposition)	1.0 kg
Belowground	Assumes that 70% of annual root/rhizome production is available for consumption (unavailable: decomposition)	2.0 kg
Total		3.0 kg m^{-2} $year^{-1}$ or 30,000 kg ha^{-1}
Nutria-marsh relations		
Density of nutria	40 nutria ha^{-1}[d]	
Biomass consumed from marsh	(72.358 kg nutria–^{1} $year^{-1}$)*(40 nutria ha^{-1}) 2894 kg ha^{-1} $year^{-1}$	
Biomass lost from marsh	(26.049 kg nutria–^{1} $year^{-1}$)*(40 nutria ha^{-1}) 1042 kg ha^{-1} $year^{-1}$	
% available biomass consumed		9.6% $year^{-1}$
% available biomass lost		3.5% $year^{-1}$

These calculations result in a 3.5% $year^{-1}$ permanent removal of plant production over a given area, with a density of 40 nutria ha^{-1}. At this density, approximately 9.6% $year^{-1}$ of all the biomass produced would be consumed (6.1% returns to the marsh as fecal matter)
[a]Gosling (1974); [b]Touchette et al. (2010); [c]K. Nagy et al. (1999); [d]Kinler et al. (1987)

Table 3 Nutria population density estimates in Louisiana

Study	Density (individuals ha^{-1})	Wetland type
O'Neil (1949)	19.8–24.7	Freshwater marsh, delta plain
Harris and Webert (1962)	7.4	Brackish marsh, Chenier plain
Valentine et al. (1972)	0.1–1.29	Brackish marsh, Chenier plain
Robicheaux (1978)	Summer 1.3; winter 6.5	Brackish marsh, Chenier plain
Linscombe et al. (1981)	24	Brackish marsh, marginal delta plain
Kinler et al. (1987)	43.7	Freshwater floating marsh, delta plain

and 380,000. Based on harvest of nutria, Edmond Mouton (LDWF, pers. comm.) estimates that there are currently several million nutria in the Louisiana coastal zone.

Habitat Type

Fresh and intermediate marsh types are considered optimum habitat for nutria, and as would be expected, most of the nutria harvested in each of the nutria control program years were in the fresh and intermediate marshes. In 2008, 47% of the Louisiana wetlands were classified as fresh and intermediate. About 60% of the nutria are harvested in fresh and intermediate marshes. Although salt marsh is not considered good habitat for nutria, some nutria do use salt marshes. The greatest number of nutria harvested in salt marsh occurred in 2003–2004 season (8336), with less than 1000 taken in the 2005–2006 season.

Table 4 Summary of Coastwide Nutria Control Program (CNCP) nutria harvest by marsh types, 2003–2010 (data from LDWF reports)

	2002–2003 (%)	2003–2004 (%)	2004–2005 (%)	2005–2006 (%)	2006–2007 (%)	2007–2008 (%)	2008–2009 (%)	2009–2010 (%)	Summary (%)
Freshwater	157,305 (51%)	158,568 (47%)	153,134 (51%)	96,916 (57%)	157,104 (41%)	130,284 (42%)	150,114 (45%)	206,109 (46%)	1,052,232 (46%)
Intermediate	67,008 (22%)	93,064 (28%)	44,571 (15%)	9093 (5%)	21,087 (6%)	52,230 (17%)	54,118 (16%)	72,819 (16%)	346,983 (15%)
Brackish	20,158 (7%)	25,311 (8%)	17,694 (6%)	8767 (5%)	10,201 (3%)	16,968 (6%)	22,217 (7%)	11,724 (3%)	112,882 (5%)
Salt	4088 (1%)	8336 (2%)	4384 (1%)	660 (0%)	3615 (1%)	3228 (1%)	1274 (0%)	4140 (1%)	25,637 (1%)
Other[a]	37,700 (12%)	49,366 (15%)	77,852 (26%)	54,371 (32%)	186,413 (49%)	102,923 (33%)	105,957 (32%)	151,171 (34%)	728,055 (32%)
Open water	7489 (2%)	307 (0%)		600 (0%)	1970 (1%)	2579 (1%)	358 (0%)		5814 (0%)
Unknown			200 (0%)						200
Grand Total	**308,160**	**334,952**	**297,835**	**170,407**	**380,390**	**308,212**	**334,038**	**445,963**	**2,271,797**

[a]"Other" includes swamp, mixed forest, open water and agriculture land types

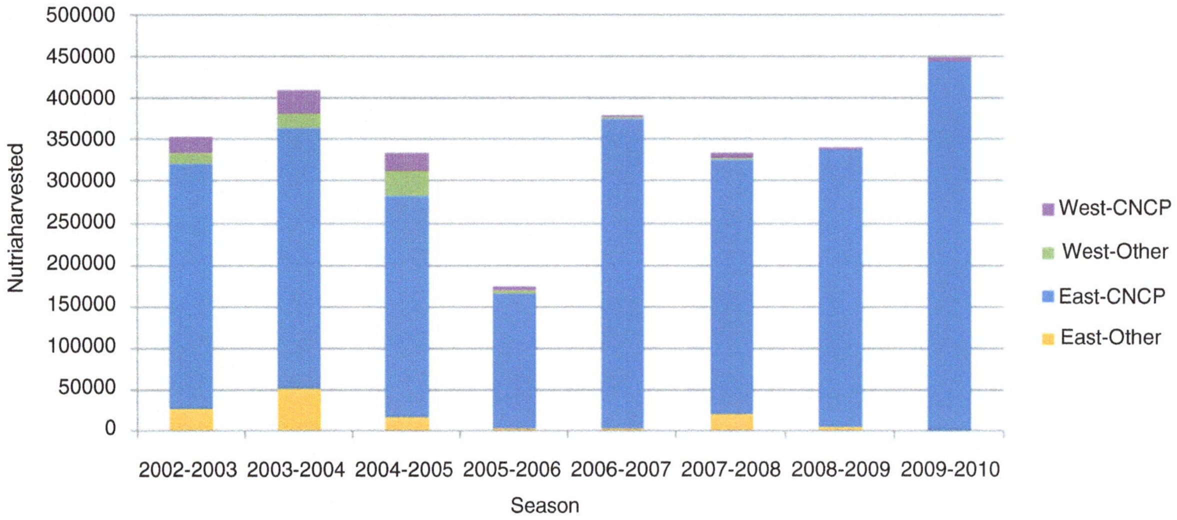

Fig. 4 Total nutria harvested in western and eastern Louisiana, including that within CNCP and others harvested for pelts as reported to Louisiana Department of Wildlife and Fisheries for 2002–2010

Region

The eastern, or Delta Plain, wetlands have consistently accounted for the majority of nutria harvest since the CNCP began. The poor harvests of the Chenier Plain are indicative of low numbers of nutria in the region, which has not been true of this region historically. Figure 4 represents both CNCP harvests and nutria harvested for pelts outside of the CNCP in Louisiana from 2002 to 2010. It illustrates that many fewer nutria were harvested in the western Parishes (Chenier Plain) than in eastern Louisiana in the Delta Plain. It also shows the effect of the CNCP, with a much greater number of nutria harvested as part of the CNCP.

Nutria are pervasive in coastal Louisiana, with damage and harvest reported for 12 coastal parishes, along with Jefferson Davis Parish (Fig. 5). The highest harvest rates were consistently in Terrebonne Parish, which each year comprised over 21% of the nutria harvested. In the last two seasons, less than 2450 nutria were harvested in Cameron Parish, with none harvested in Vermilion. In Vermilion and Calcasieu parishes – the two main parishes in the western portion of Louisiana – nutria harvesting associated with CNCP netted 74,506 nutria between 2003 and 2010. More than 100,000 nutria were harvested between 2002 and 2010 in each of seven parishes in the Delta Plain. The greatest nutria harvest after the 2005–2006 season occurred in St. Martin Parish.

Hurricanes Katrina and Rita affected the nutria harvest in 2005–2006, when the fewest nutria were harvested. LDWF attributed the reductions in harvest to Hurricanes Katrina and Rita due to destruction and/or displacement of nutria populations, as well as the displacement of trappers and hunters from their homes prior to the trapping season.

Methods to Control Nutria Populations

The control of nutria in Louisiana historically has been related to the global demand for furs (Linscombe 1992), with the method of harvesting being trapping. In fact, so many nutria were harvested in the 1960s that by 1965 the Louisiana state legislature returned the nutria to the protected list. Fur demands from Germany (1960s) and Russia (1996–1998) kept the nutria market flourishing. The market for nutria crashed in 1998 when the Russian economy collapsed and demand for the fur fell. During the late 1980s and 1990s reports from land managers of marsh vegetative damage due to nutria became common, and a search began for new ways of controlling the nutria population. A comprehensive review of possible methods to control nutria was conducted for the LDWF including information available through 2001and is included in Task III of Baroch et al. (2002). The outcome of the review was the recommendation of implementing an incentive payment program to provide control on the Louisiana nutria population. Table 5 presents a summary of those methods. The Coastwide Nutria Control Program (CNCP) initiated in 2002.

Previously, the incentive/bonus plan was used in Great Britain (Gosling and Baker 1987) and Maryland successfully. This plan differed from Louisiana's plan in that the hunters were salaried, and received an incentive/bonus to make sure that all of the nutria were taken. With this type of plan in Maryland's (USA) Blackwater Refuge, nutria had successfully been eradicated by 2004. The effort took

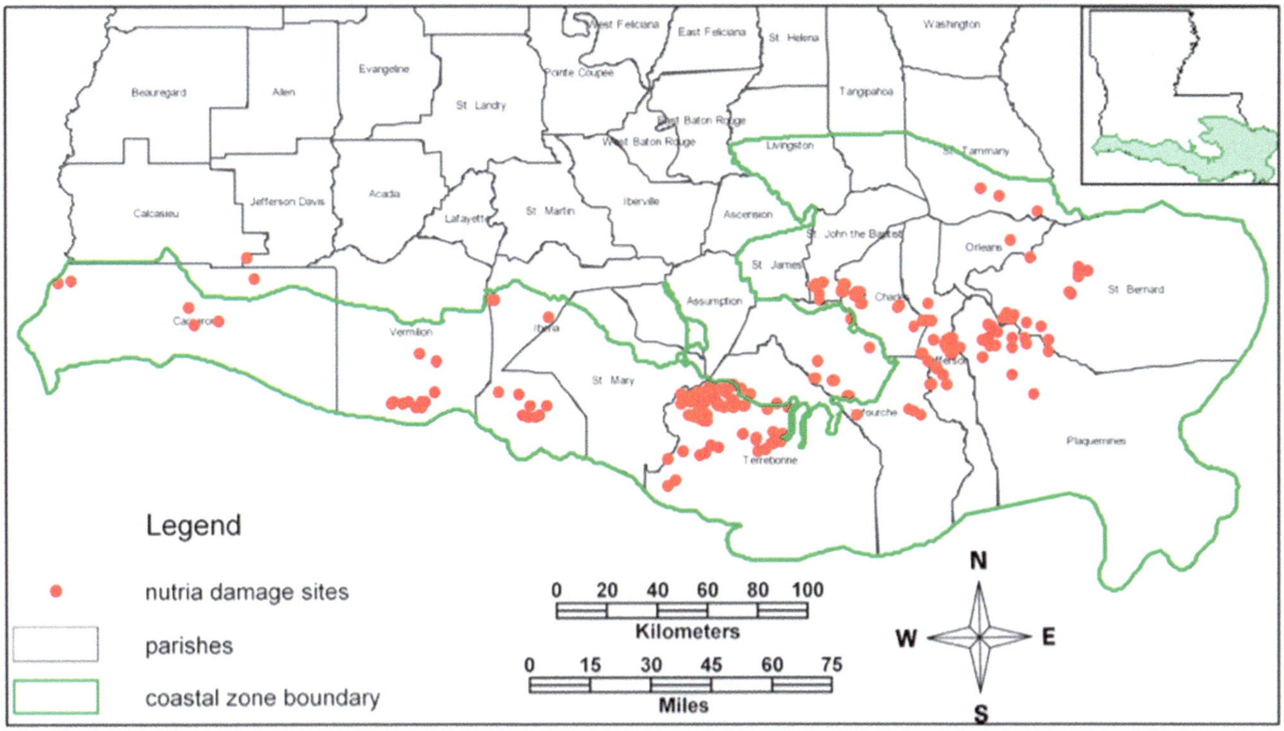

Fig. 5 Nutria damage as reported by LDWF for 2003–2010

Table 5 Methods used to control nutria populations based on Baroch et al. (2002)

Method	Purpose	Implementation	Louisiana: successful/ unsuccessful?	Reason	Ranking[a]
Trapping: Fur market	Control	Nutria trapped and sold for fur with no other compensation/reward	Successful and Unsuccessful	Success depends on fur market; only a few trappers remain	IV
Controlled hunting		Open season by licensed hunters	Unsuccessful	With little pelt value, no incentive	V
Bounty	Control	Used in Louisiana in early 1960s – turn in tail for price	Unsuccessful	Lack of appropriate funds	
Incentive/ pay	Control	CNCP	Successful		I
Incentive/ bonus	Eradicate	Incentive salary; Bonus after eradication to stop husbandry	Not useful	Too many sources of new nutria (ineffective spatially)	III
Chemical control:					
Toxicants	Control	Bait an area, then add toxicants		Labor intensive; Effects on other wildlife need to be determined	II
Fumigants	Eradicate	Agricultural areas – into burrows	Probably not	Must be contained to be effective; none registered nor found to be effective	NA
		Marsh areas – no burrows			
Chemical repellants			Doesn't kill; just moves	None registered nor found to be effective	NA
Induced fertility:	Control				
Bait		Must be baited approximately every 3 months; not species specific		Would be hard to stop invasion from untreated animals	NA
Injected		Must be injected; expensive but long-lasting results		Would be hard to stop invasion from untreated animals	NA

[a]Cost-effectiveness ranking by Baroch et al. (2002), with "I" being most cost-effective

2 years, $2 million dollars and 15 trappers to eradicate the nutria by killing all 8300 nutria on 95,000 acres, which is a small number compared to the 450,000 harvested over a much larger area in 2009–2010 in Louisiana. A similar program was implemented in Great Britain where eradication of nutria was achieved in 1989.

It generally is not considered feasible to eradicate nutria in Louisiana due to climate (mild winters), landscape features (many islands vs. mainland), difficult accessibility to nutria habitat. The coastal area of Louisiana also is surrounded on three sides with sources of nutria immigration. Texas, Mississippi, north Louisiana and Arkansas all have nutria populations that could conceivably immigrate into coastal Louisiana.

Local actions have been taken in some coastal parishes to control nutria populations. For example, to try and stop nutria from destroying canals in Jefferson Parish, LA, sheriff's deputies began shooting nutria at night. An interesting side to this control effort is that nutria seem to have learned to react to sounds associated with hunting, including hunting by truck and airboat. When airboats are used in the daily harvest in dense areas of nutria, a similar learned reaction cause the nutria to dive into deep water or hide under vegetation when an airboat is near.

Chemical controls are an option for nutria control, but presently only one chemotoxicant is licensed for the extermination of nutria. It is usually applied to a root vegetable like sweet potatoes on a floating mat. A major problem with the use of chemotoxicants is that they are not species-specific. Baroch et al. (2002) determined that the only really feasible method of control was the incentive payment program. In 2008, over 6600 nutria were shot in Big Branch National Wildlife Refuge and over 2000 in the Hammond Assimilation Wetland.

Table 6 Nutria removed by shooting during 2008 at the Big Branch Marsh National Wildlife Refuge

Jan 26	115
Jan 27	115
Jan 31	115
Feb 5	214
Feb 22	354
Feb 29	308
Mar 1	139
Apr 1	259
Apr 2	189
Apr 7	243
Apr 22	274
May 11	450
May 27	243
May 28	275
Jun 24	208
Jun 30	155
Aug 5	266
Aug 10	138
Aug 13	225
Aug 20	454
Aug 25	49
Sep 5	268
Sep 11	535
Sep 13	180
TOTAL	6369

About 1000 additional nutria were culled by trapping
Source: Daniel Breaux, Big Branch Marsh National Wildlife Refuge, U.S. Fish and Wildlife Service

Nutria Population Control at Big Branch Marsh National Wildlife Refuge

We have documented that extensive damage to marshes due to excessive nutria grazing has occurred in coastal marshes. However, careful quantification of the number of nutria killed in an attempt to control nutria is rare. An exception is at the Big Branch Marsh National Wildlife Refuge on the north shore of Lake Pontchartrain. Big Branch Marsh is a 7288 ha refuge with forested wetlands and fresh to saline marshes. It is one of the last large undeveloped areas on the north shore of Lake Pontchartrain. Nutria have been a continuing problem on the refuge and refuge staff have permitted trapping to control grazing. But in 2007, there was dramatic increase in nutria numbers and their grazing destroyed several hundred hectares of marsh. To control nutria, a culling program was initiated in 2008. Refuge staff went out regularly in airboats to shoot nutria and carefully documented the number shot (Table 6). From January to September over 6300 nutria were killed. In addition, during this same period, over 1000 nutria were trapped. A marsh creation project supported by the CWPPRA program and plantings restored the marsh.

A similar culling of nutria occurred at the Hammond Assimilation Wetland during the winter of 2007–2008. Nutria grazing damaged over 130 ha of marsh. Shooting of nutria took place in winter and spring. A running tally indicated that about 2000 nutria were killed (based on observations and reports by Chris Carroll and Jason Day). These results show that aggressive culling of nutria in addition to the statewide trapping program can reduce nutria populations during eat outs.

Coastwide Nutria Control Program

If we can reduce the nutria population by at least 500,000 each year, the vegetative damage caused by these animals will be dramatically reduced. Biologist, Greg Linscombe (Addison 2000)

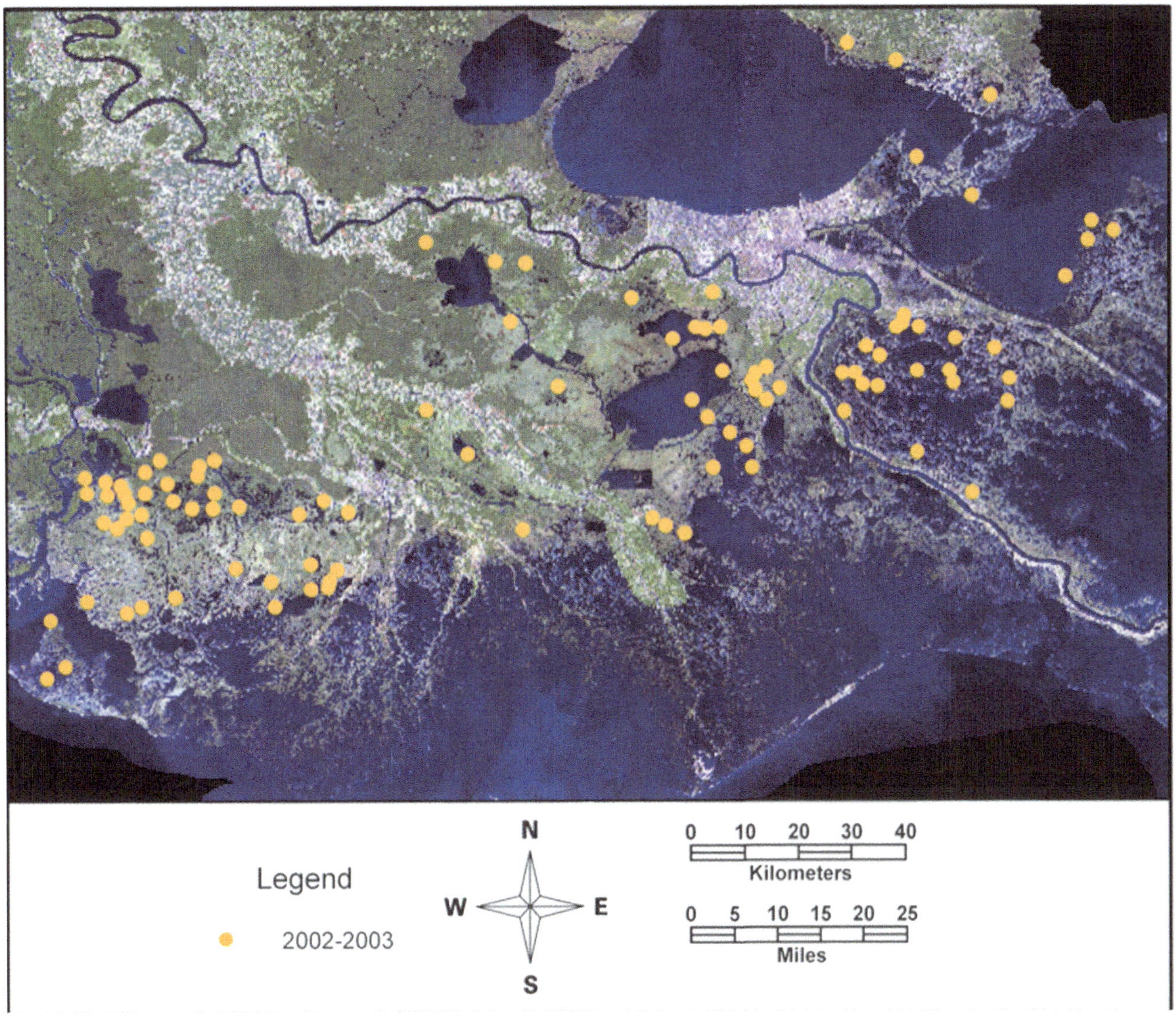

Fig. 6 Aerial image of the Louisiana Delta Plain showing the concentration of nutria damage sites in 2003. Data on damage were collected by LDWF during spring/summer of each year following the aerial survey techniques used in Kinler and Linscombe (1998) for the Coastwide Nutria Control Program funded by CWPPRA. Background image is Landsat5 TM data from 2005, courtesy of LOSCO

The Coastwide Nutria Control Program (CNCP) was first funded by CWPPRA to implement the nutria harvesting program in 2002–2003. A pilot Demonstration Program was initiated in 2000 with harvest occurring in 2001–2002. CNCP has offered $5.00 for each nutria tail since the 2006–2007 season. Coastal zone land can be leased through the CNCP by landowners or their responsible parties, but it is not mandatory. The control program is an incentive plan that pays individuals for nutria tails turned in, meaning that participants may decide how much they participate in the program. It may not be their main source of income. For instance, a harvester may only harvest enough to cover expenses for another business.

CNCP Nutria-Damaged Marsh Distribution

The distribution of nutria-damaged marsh is concentrated in the upper coastal basins, primarily in freshwater and intermediate marshes. Figure 6 presents a view of the Louisiana Delta Plain region of the coast with dots representing damage areas determined by aerial surveys during spring/summer of 2003, which had the highest concentration of damage of all years.

"Hot Spots" of nutria activity occurred in fresh and intermediate marshes in the Penchant area of the upper Terrebonne Basin, the "Pen" area of Barataria Basin, the Caernarvon area of the Breton Sound Basin, and east of Lac des Allemands (Fig. 7).

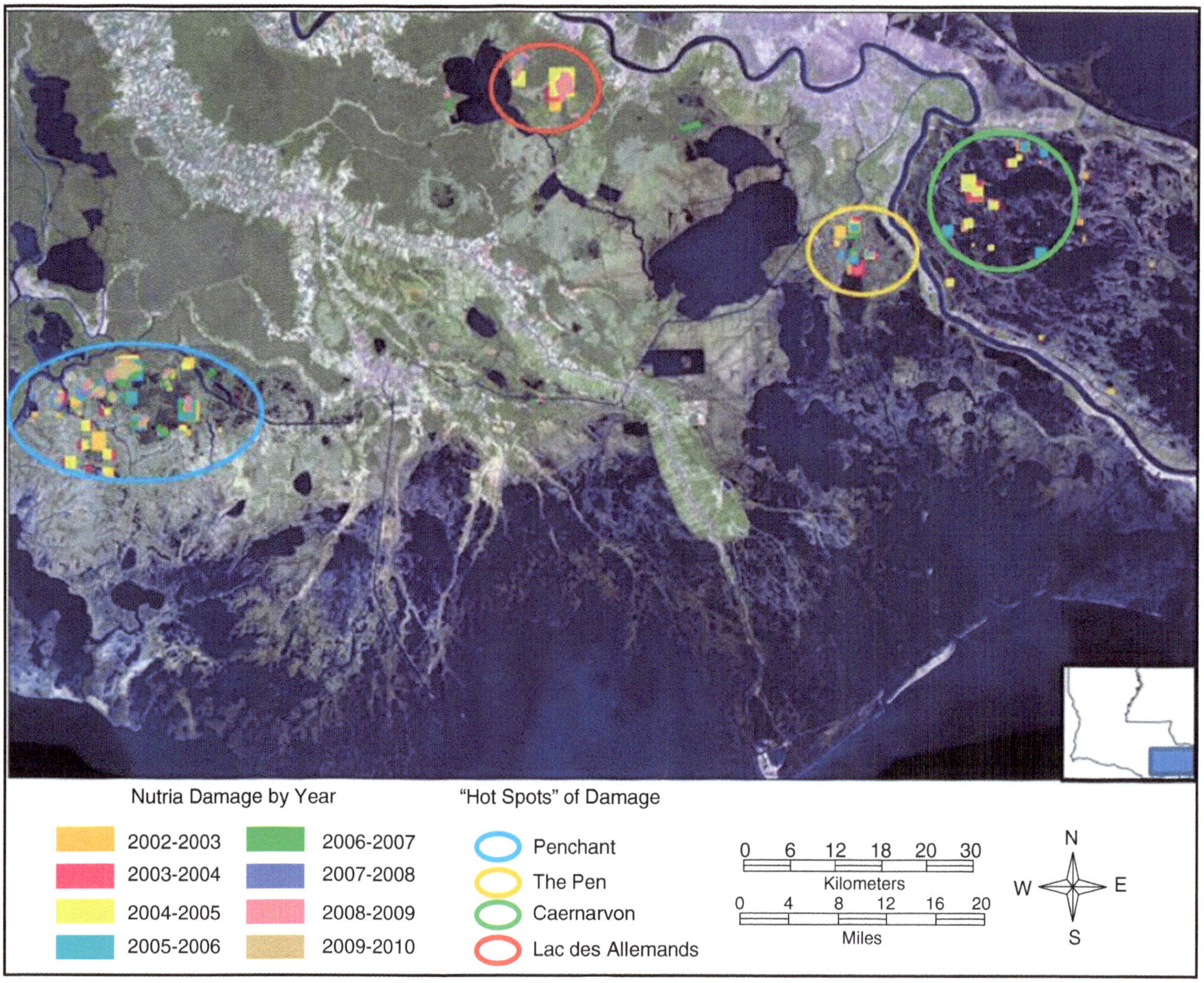

Fig. 7 Aerial image of the Louisiana Delta Plain highlighting the "Hot Spots" of severe nutria damage within the circled areas. Data on damage were collected by LDWF during spring/summer of each year following the aerial survey techniques used in Kinler and Linscombe (1998) for the Coastwide Nutria Control Program funded by CWPPRA. Background image is Landsat5 TM data from 2005, courtesy of LOSCO

The most damaged areas are consistently in the fresh/intermediate marsh, however some damage is reported in the brackish marsh. This is evident from Fig. 6 that shows the distribution of damaged marsh, but also from Fig. 8 that indicates the acreage damaged by marsh type. The upper Terrebonne Basin is particularly heavily used by nutria. This is a region of mostly freshwater floating marshes that provide excellent habitat for nutria. The floating mats rise and fall with changes in water level, always providing a floating marsh mat for nutria to access for feeding, resting, and nesting. The damage is more or less continuous in these areas, although the intensity does seem to vary interannually – probably as marsh vegetative material is removed (e.g. quality is degraded) and the population moves to a more desirable location. It also is clear from the nutria control program data set that as more animals are removed, there is a trend toward vegetation recovery.

CNCP Program Success

The extrapolated acres of marsh damage in coastal Louisiana with the numbers of nutria harvested for the years 1999–2010 is indicated in Fig. 9. This graph indicates a clear trend of decreasing acres of damaged marsh from 2003 to 2010. The area of nutria-damaged marsh decreased from 80,000 acres in 2003 to less than 10,000 acres in 2010. At the same time the harvest of nutria was steady in the range of approximately 300,000 to 400,000 animals most years other than during Hurricanes Katrina and Rita. The inability

Fig. 8 Nutria damage (acres) by marsh type for each year based on LDWF helicopter surveys. *Black line* indicates start of Coastwide Nutria Control Program

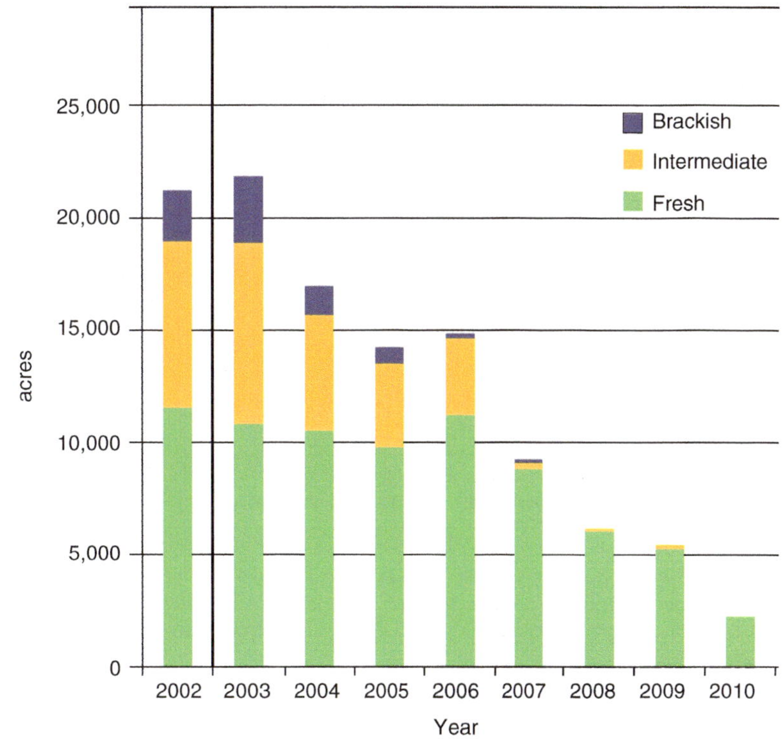

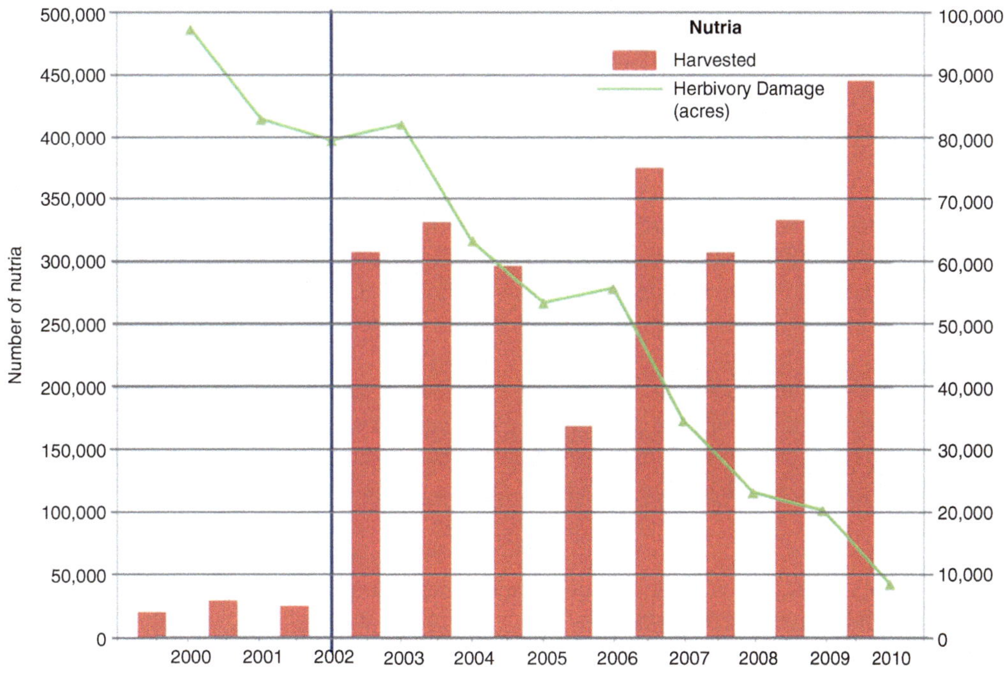

Fig. 9 Louisiana Dept. of Wildlife and Fisheries estimates of nutria damage to coastal marshes (*right axis*, *green line*) and the harvest of nutria from 1999–2010 for all the coastal parishes (*left axis*). A pilot Demonstration Program was initiated in 2000 and continued to 2002. The Coastwide Nutria Control Program has offered $5.00 for each nutria tail since the 2006–07 season. The *black line* indicates the start of the Coastwide Nutria Control Program

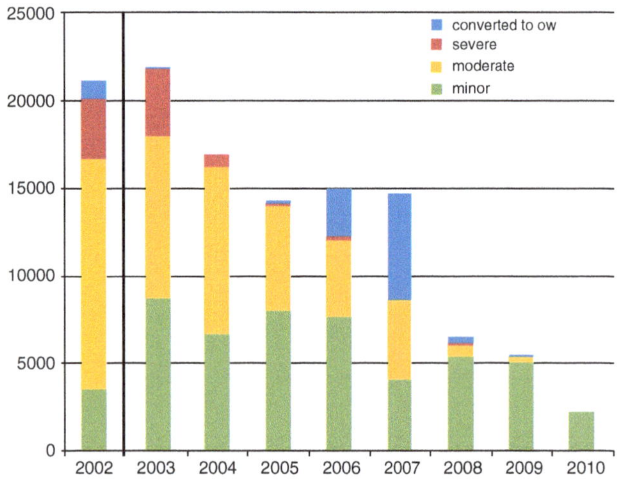

Nutria damage by damage type for each year. Black line separates pre-control program and control program years.

Fig. 10 Severity assessment of nutria damage by year based on LDWF helicopter surveys. The *black line* indicates start of the Coastwide Nutria Control Program (Data from LDWF Coastwide Nutria Control Program)

to separate damage due to hurricane effects and herbivory in 2006 accounts for the increase in damage that year. During the 2009–2010 season, a peak harvest was recorded with 445,963 nutria tails collected. The steady decline in nutria damaged wetlands while nutria harvest has exceeded 300,000 animals each year is a strong indication that the incentive program is a success.

Nutria damage intensity by year from 2002 to 2010 (from LDWF monitoring of CNCP) is indicated in Fig. 10. The information in this graph indicates a clear trend of severe and moderate damage in the early 2000s to mostly minor marsh damage in 2008–2010. This trend follows that discussed previously of decreasing acreage of damage as the CNCP continued to remove nutria from the marsh ecosystem from 2002 through 2010. This is again clear evidence of the remarkable success of this program, with the severity of marsh damage decreasing from 16,660 acres of moderate to severe damage in 2001–2002 to only 2260 acres of minor damage in 2010.

The Importance of Alligators as Predators on Nutria

The historical population of alligators in coastal Louisiana is unknown, but by the 1950s overharvesting was leading to an unsustainable population. Alligator hunting was suspended in the early 1960s. In 1970, LDWF began monitoring the population recovery and by 1981 alligator hunting resumed statewide. Alligator nests significantly increased from 1970 to 1993 (McNease et al. 1994). Over this 24-year period, average nest numbers increased by 13% each year coastwide. The highest nest densities (36.5 ha nest^{-1}, 1984–1993) occurred in southwest Louisiana; nest density in southeast Louisiana was 51.9 ha nest^{-1}.

An early study on Sabine NWR by Valentine et al. (1972) showed that nutria density contributed to diet of alligators. During periods when nutria populations were estimated at 74,000, nutria remains were identified in 56% of the stomachs of the alligators examined. In contrast, when nutria populations were estimated to be <10,000 animals on the refuge, <7.0% of alligator stomachs contained nutria. Valentine et al. (1972) suggested that the decline of nutria in the alligator diet was due to the nutria control program during that period of study, which accounted for a 30% reduction in the nutria population.

With the inception of the CNCP, there was concern that nutria harvesting would deplete an important alligator food item. A study comparing the diet of alligators from areas where nutria control was present or absent failed to reveal any trend in the likelihood that an alligator stomach would contain nutria (Gabrey et al. 2009). Among the several parishes studied and 550 stomachs analyzed, overall, about 1/3 of the alligator stomachs contained nutria remains.

Based on a population modeling exercise of alligator-nutria interactions, Keddy et al. (2009) suggest that controlling the harvest of large alligators may have a significant impact on nutria populations and serve as a means to help control nutria impacts to wetlands. Their conceptual model provides insight to the behavior of nutria population dynamics with different nutria-alligator ratios and predation rates. Given the lack of empirical data from controlled experiments, the authors could not prescribe the number of alligators required to control a nutria population.

Based on LDWF surveys of marsh damage from nutria and local alligator nest densities, there has not been a strong indication of top down control of nutria by alligators. McNease pointed out that since 1994 Terrebonne Parish "outranked all the other coastal parishes in terms of alligator nesting density for fresh and intermediate marsh types. It is important to note that at the same time, Terrebonne Parish was also number one in terms of nutria herbivory damage and CNCP harvest" (Marx et al. 2004). Nonetheless, there are anecdotal observations of low nutria abundance where alligators are known to be numerous. Two of these areas are the freshwater maidencane-dominated marshes around Lake Boeuf and the Jean Lafitte National Park in the upper Barataria Basin. Although no cause and effect information is available, the Lake Boeuf flotant marsh region is in an area known to have high alligator population (1978–1998 mean density of 81.8 alligators mi^{-1}; Visser et al. 1999) that could help explain the lack of visible nutria damage in these marshes.

Western, or Chenier Plain, marshes used to account for a high proportion of the nutria take in early years (Fig. 11). In

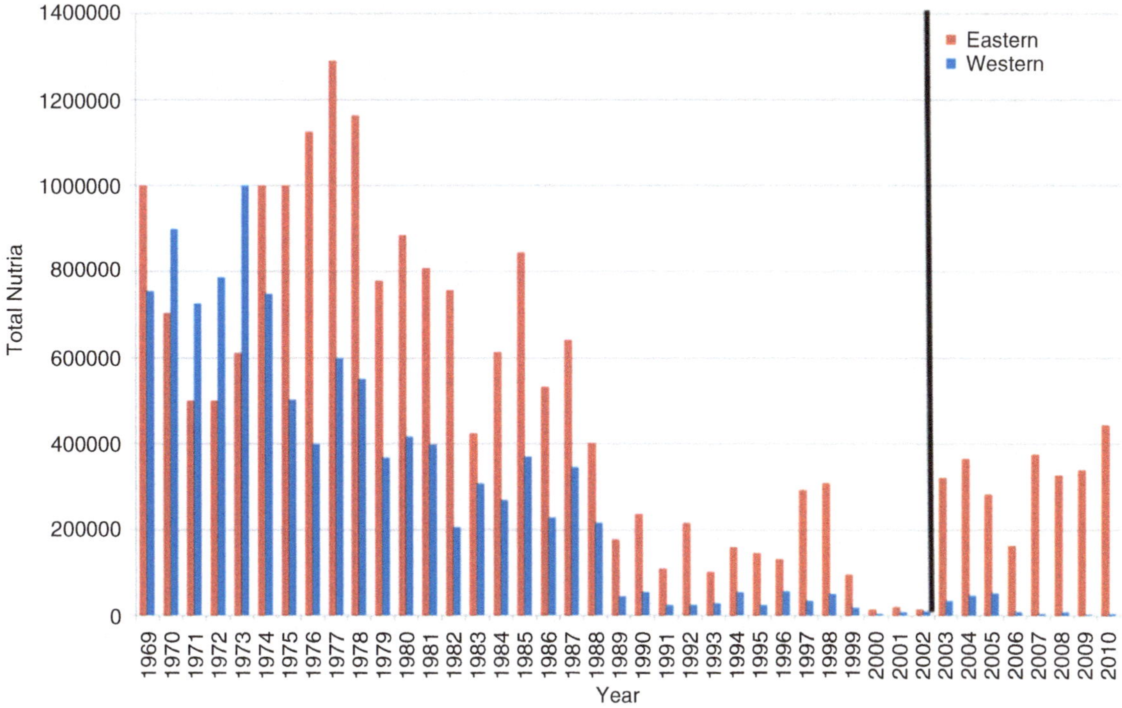

Fig. 11 Nutria harvests from 1969 through 2010 from west (Chenier Plain) and east (Delta Plain) Louisiana. The *black line* indicates the beginning of the Coastwide Nutria Control Program

recent years, even with a high incentive, there is not a significant harvest coming from the Chenier Plain marshes. For example, the total harvest of nutria in Terrebonne Parish (Delta Plain; 663,748 tails) far exceeds the number in Cameron Parish in the Chenier Plain (41,709 tails). A number of factors may be responsible, such as hydrologic changes, an increase in the proportion of saltwater to freshwater habitat, hurricanes, and a high density of alligators (McNease et al. 1994).

Other predators of nutria include coyote, eagles, domestic dogs, and humans. In addition, great horned owls, foxes, great blue herons, hawks, eagles, and raccoons prey on young nutria (Washington Dept. of Fish and Wildlife 2006). However, as observed by Edmond Mouton, LDWF (pers. comm. (04/05/2011), "While all of these might eat nutria, they are obviously not controlling the population."

Summary

Nutria is an exotic, aquatic rodent that was introduced to Louisiana wetlands during the early 1930s. From 1960 to 1990, greater than 36-million nutria were taken and the fur market was lucrative. By 2000, the worldwide fur market had entirely collapsed.

With favorable conditions, nutria can rapidly reproduce. Nutria reach sexual maturity at 4–8 months, may have 2–3 litters per year, and average 13.1 young per female per year.

Nutria have a small home range in marsh habitats, but long-distance dispersal (>32 km) to new areas can occur during tropical storms.

Nutria are pervasive across Louisiana coastal wetlands, but their ideal habitat is freshwater and intermediate marsh, where forage is most palatable. The organic nature of these soils makes them particularly susceptible to destruction with grazing.

Nutria are opportunistic feeders, with a broad diet comprising more than 60 plant species in Louisiana. Nutria are attracted to wetlands that contain a reliable source of nutrient-rich fresh water, such as river diversions and assimilation wetlands. It is imperative that restoration projects that increase input of nutrient-rich fresh water into wetlands have a nutria control program in place.

An average nutria can consume 26 kg of dry plant biomass per year. The entire amount of plant biomass that is ingested by an average nutria is equivalent to the biomass produced and available for consumption each year on approximately 24 m^2 of marsh. On one hectare of marsh, a density of 40 nutria could cause the permanent loss of 50% of all the biomass produced during a period of 20 years.

Decades of cumulative deleterious effects from nutria grazing on the coastal landscape have been difficult to quantify and separate from other factors producing wetland loss. Scientific studies investigating effects of nutria on marsh habitats consistently conclude that nutria grazing is damaging to marsh and young forest vegetation. It is generally accepted that nutria damage – in addition to larger scale subsidence, sea level rise, and salinity intrusion – can create an accelerated deterioration of wetlands.

Research studies conclude that nutria grazing damage limits regeneration in Louisiana swamp forests. Nutria grazing on baldcypress and water tupelo seedlings is extensive and remains a major factor in the inability of baldcypress-water tupelo forests to regenerate. Plantings of saplings have withstood grazing pressure better than seedlings and may be a good choice for restoration projects when feasible. Alternatively, seedlings can be protected with tree shelters. Projects designed to restore coastal swamp forests should include a nutria control component and suitable protection of transplants should be used to minimize mortality from grazing.

Eruptions of populations of nutria can cause severe wetland damage and subsequent loss. Some areas of the coast have persistent populations creating "Hot Spots" of severe damage.

The major "Hot Spots" of severe wetland damage occur in the freshwater-intermediate salinity areas of Terrebonne, Barataria, and Breton Sound basins. Nutria densities are relatively low in the Chenier Plain currently compared to historic observations and harvest records.

Damage from nutria was noted in 12 parishes in the Louisiana coastal zone.

The Coastwide Nutria Control Program (CNCP) was implemented in 2002–2003 by the LDWF, and since then there has been a reduction in 70,000 acres of marsh damage, from 80,000 acres in 2003 to less than 10,000 acres in 2010. Approximately 446,000 nutria were harvested in 2010 in the CNCP.

Nutria were harvested in 24 of the 63 Louisiana parishes from 2000 to 2010, but by far the greatest numbers were harvested in the Delta Plain.

When considering the costs of creating new wetlands (approximately $50,000–$70,000 per acre), the Coastwide Nutria Control Program can be viewed as a successful wetland conservation program that has produced measureable reduction in marsh damage. Since the program began with the 2002–2003 season, 2,571,480 nutria have been harvested under the CNCP. This program is a success and, from a resource management perspective, should be continued with improvement and expansion if possible.

Nutrient enrichment of coastal landscapes may cause nutria population growth and habitat damage. Coastal restoration projects with areas receiving nutrient enrichment should include nutria control to ensure plant productivity, establishment, and expansion.

The effect of predators on nutria population control is unclear. In particular, the predator/prey relationship between alligators and nutria is a worthy topic for further investigation. Anecdotal information from western Louisiana marshes, suggests that the recovery of the alligator population has significantly reduced the nutria population there. However, mortality from tropical storms and habitat modification also may be important.

Acknowledgements Several people were instrumental in helping us to pull together the information in this chapter. They were willing to discuss collected information as well as their understandings of the role of nutria in the Louisiana coastal area. We thank Edmond Mouton and Jillian Jordan (Louisiana Department of Wildlife and Fisheries) for providing information to help us understand the data collected through the Coastal Nutria Control Program. We also thank Greg Linscombe and Edmond Mouton, who have always been more than willing to share their wealth of knowledge about nutria and its habitat with us.

References

Adams WH Jr (1956) The nutria in coastal Louisiana. La Acad Sci 19:28–41

Addison JD (ed) (2000) Nutria: destroying marshes the old fashioned way. In: WaterMarks, New Orleans, LA. Louisiana Coastal Wetlands Conservation and Restoration Task Force, pages 5, 9

Baroch J, Hafner M, Brown TL, Mach JJ, Poche RM (2002) Nutria (Myocastor coypus) in Louisiana. Genesis Laboratories, Inc., Wellington, pp 1–155

Bernard SK (2002) M'sieu Ned's rat? Reconsidering the origin of nutria in Louisiana: the E. A. McIlhenny collection, Avery Island, Louisiana. La Hist: J La Hist Assn 43(3):281–293

Blair RM, Langlinais MJ (1960) Nutria and swamp rabbit damage Baldcypress plantings. J For 58:388–389

Bounds DL, Sherfy MH, Mollett TA (2003) Nutria. In: Feldhamer GA, Thompson BC, Chapman JA (eds) Wild mammals of North America: biology, management, and conservation. The Johns Hopkins University Press, Baltimore, pp 1119–1147

Brantley CG, Platt SG (1992) Experimental evaluation of nutria herbivory on baldcypress. La Acad Sci 55:21–25

Conner WH (1995) Cypress regeneration in coastal Louisiana and the impact of nutria upon seedling survival. Proceedings of the 13th Annual Conference, Society of Wetland Scientists. New Orleans, LA, pp. 130–136

Day JW Jr, Shaffer GP, Hunter R, Wood WB, Lane R, Lundberg CA, Day J, Hunter M (2011) An analysis of the Hammond Assimilation Wetland: system response, nutria herbivory, and vegetation recovery. White Paper, Louisiana Department of Environmental Quality

Evans J (1970) About nutria and their control. US Fish Wildl Ser Res Publ 86:1–65

Evers DE, Sasser CE, Gosselink JG, Fuller DA, Visser JM (1998) The impact of vertebrate herbivores on wetland vegetation in Atchafalaya Bay, Louisiana. Estuaries 21:1–13

Felsher JN (2000) Last days of the trapper. La Life 20:54–57

Ford MA, Grace JB (1998) Effects of vertebrate herbivores on soil processes, plant biomass, litter accumulation and soil elevation changes in a coastal marsh. J Ecol 86(6):974–982

Fuller DA, Sasser CE (eds) (1988) Vegetation and waterfowl use of islands in Atchafalaya Bay. Final report submitted to Louisiana

Board of Regents. 150 Riverside Mall, Suite 129, Baton Rouge, LA. Contract No. 86-LBR/018-B04

Fuller DA, Sasser CE, Johnson WB, Gosselink JG (1985) The effects of herbivory on vegetation on islands in Atchafalaya Bay, Louisiana. Wetlands 4:105–114

Gabrey SW, Kinler N, Elsey RM (2009) Impacts of nutria removal on food habits of American alligators in Louisiana. Southeast Nat 8:347–354

Geho EM, Campbell D, Keddy PA (2007) Quantifying ecological filters: the relative impact of herbivory, neighbours, and sediment on an oligohaline marsh. Oikos 116:1006–1016

Gosling LM (1974) The coypu in East Anglia. Trans Northfolk Norwich Nat Soc 23:49–59

Gosling LM, Baker SJ (1987) Planning and monitoring an attempt to eradicate coypus from Britian. Symp Zoo Soc Lond 58:99–113

Gough L, Grace JB (1998) Herbivore effects on plant species density at varying productivity levels. Ecology 79:1586–1594

Grace JB, Ford MA (1996) The potential impact of herbivory on the susceptibility of the marsh plant Sagittaria lancifolia to saltwater intrusion in coastal wetlands. Estuaries 19:13–20

Harris, V. T. and F. Webert. (1962) Nutria feeding activity and its effect. Special Scientific Report

Holm GO, Sasser CE, Evers E (2011) The Nutria in Louisiana: a current and historical perspective. Final report. Lake Pontchartrain Foundation, 68 pp

Howell SL (2003) Herbaceous aspect of a feasibility study concerning the Mississippi river diversion into the southern lake Maurepas Swamp. Master's thesis, Southeastern Louisiana University, 69 pp

Ialeggio JS, Nyman JA (2014) Nutria grazing preference as a function of fertilization. Wetlands 34:1–7. https://doi.org/10.1007/s13157-014-0557-7

Izdepski CW, Day JW Jr, Sasser CE, Fry BD (2009) Early floating marsh establishment and growth dynamics in a nutrient amended wetland in the lower Mississippi Delta. Wetlands 29 (3):1004–1013

Jojola S, Witmer G (2006) Unpubl Final Report, QA-1252. USDA/APHIS/WS-National Wildlife Research Center, Fort Collins, CO, 18 p

Kays CE (1956) An ecological study with emphasis on nutria (Myocastor coypus) in the vicinity of Price Lake, Rockefeller Refuge, Cameron Parish, Louisiana, Louisiana State Univ. M.S.: 145

Keddy PA, Campbell D, McFalls T, Shaffer GP, Moreau R, Dranguet C, Heleniak R (2007) The wetlands of lakes Ponchartrain and Maurepas: past, present and future. Environ Rev 15:43–77

Keddy PA, Gough L, Nyman JA, McFalls T, Carter J, Siegrist J (2009) Alligator hunters, pelt traders, and runaway consumption of gulf coast marshes. In: Silliman BR et al (eds) Human impacts on salt marshes, a global perspective. UC Berkeley Press, Berkeley, pp 115–133

Kinler N, Linscombe G (1998) A survey of nutria herbivory damage in Coastal Louisiana in 1998. Fur and Refuge Division Louisiana Department of Wildlife and Fisheries

Kinler NW, Linscombe G, Ramsey PR (1987) Nutria. In: Novok M, Baker JA, Obbard ME, Malloch B (eds) Wild furbearer managment and conservation in North America. The Ontario Trappers Association, Toronto, pp 327–343

LeBlanc DJ (1994) Nutria. In: Hygnstrom SE, Timm RM, Larson GE (eds) Prevention and control of wildlife damage, vol 2. University of Nebraska-Lincoln, Port Allen, pp B-71–B- 80

Linscombe G, Kinler N, Wright V (1981) Nutria population density and vegetative changes in brackish marsh in coastal Louisiana. In: Chapman JA, Purseley A (eds) Proceedings of the worldwide furbearer conference. Frostburg, MD, USA, pp 129–141

Linscombe G (1992) Nutria population, habitat use and damages. In: Proceedings of the Nutria and Muskrat management symposium. Louisiana State University Agricultural Center, Louisiana Cooperative Extension Service, pp. 27–23, 100 pp

Llewellyn DW, Shaffer GP (1993) Marsh restoration in the presence of intense herbivory: the role of *Justicia lanceolata* (Chapm.) small. Wetlands 13:176–184

Marx J, Mouton E, Linscombe G (2004) Nutria harvest distribution 2003–2004 and a survey of nutria herbivory damage in Coastal Louisiana in 2004, Fur and Refuge Division Louisiana Department of Wildlife and Fisheries: 1–45

McNease LL (2005) Evaluation of environmental variables influencing nutria populations, 26 p. plus appendix

McNease L, Kinley N, Joanen T, Richard D (1994) Distribution and relative abundance of alligator nests in Louisiana coastal marshes, pp 108–120. In: Crocodiles. Proceedings of the 12th working meeting of the Crocodile Specialist Group of the Species Survival Commission of IUCN- The World Conservation Union

Myers RS, Shaffer GP, Llewellyn DW (1995) Baldcypress (Taxodium distichum (L.) Rich.) restoration in southeast Louisiana: the relative effects of herbivory. Wetlands 15(2):141–148

Nagy KA, Girard IA, Brown TK (1999) Energetics of free-ranging mammals, reptiles, and birds. Annu Rev Nutr 19:247–277

Nolfo-Clements LE (2009) Nutria survivorship, movement patterns, and home ranges. Southeast Nat 8(3):399–410

O'Neil T (1949) The muskrat in the Louisiana coastal marshes. Louisiana Wildlife and Fisheries Commission, New Orleans. 159 p

O'Neil T (1963) Louisiana's fur industry today. Louisiana conservationist. New Orleans, LA, Louisiana's Wild Life and Fisheries Commission 15:3–6

O'Neil T (1968) The fur industry in retrospect. Louisiana conservationist. New Orleans, LA, Wild Life and Fisheries Commission 20:9–14

O'Neil T, Linscombe G (1977) The Fur Animals, the Alligator, and the Fur Industry in Louisiana. Baton Rouge, LA, Louisiana Wild Life and Fisheries Commission, Wildlife Education Bulletin 109:1–68

Palmisano AW (1972) Habitat preference of waterfowl and fur animals in the northern gulf coast marshes. Proc Coastal Marsh and Estuary Manage Symp 2:163–190

Ramsey PR, Edmunds RM, Linscombe G, Kinler NW (1981) Factors influencing blood chemistry in nutria. In: Chapman JA, Pursley D (eds) Proc. World Furbearer Conf., Frostburg, MD pp 325–342

Randall LA, Foote AL (2005) Effects of managed impoundments and herbivory on wetland plant production and stand structure. Wetlands 25(1):38–50

Robicheaux BL (1978) Ecological implications of variably spaced ditches on nutria in a brackish marsh, Rockefeller Refuge, Louisiana, Louisiana Tech. Univ. M.S.: 49

Sasser CE, Swenson EM, Evers DE, Visser JM, Holm GO, Gosselink JG (1994) Floating marshes in the Barataria and Terrebonne Basins, Louisiana, Louisiana State University, Coastal Ecology Institute, Baton Rouge, prepared for U.S. Environmental Protection Agency, Dallas, Texas, LSU-CEI-94-02

Sasser CE, Gosselink JG, Swenson EM, Evers DE (1995) Hydrologic, vegetation, and substrate characteristics of floating marshes in sediment-rich wetlands of the Mississippi river delta plain, Louisiana, USA. Wetl Ecol Manag 3(3):171–187

Sasser CE, Holm Jr.GO, Visser JM, Swenson EM (2004) Thin-mat marsh enhancement demonstration project TE-36. Final Report. School of the Coast and Environment, Louisiana State University, Baton Rouge

Sasser CE, Visser JM, Mouton E, Linscombe J, Hartley SB (2008) Vegetation types in coastal Louisiana in 2007. U.S. Geological Survey Open-File Report 2008–1224

Sasser CE, Gosselink JG, Holm GO Jr, Visser JM (2009) Freshwater tidal wetlands of the Mississippi River Delta. In: Barendregt A, Baldwin A, Meire P, Whigham D (eds) Tidal freshwater wetlands. Backhuys Publishers, Leiden, pp 167–178

Shaffer GP, Sasser CE, Gosselink JG, Rejmanek M (1992) Vegetation dynamics in the emerging Atchafalaya Delta, Louisiana, USA. J Ecol 80:677–687

Shaffer GP, Wood WB, Hoeppner SS, Perkins TE, Zoller JA, Kandalepas D (2009) Degradation of baldcypress – water tupelo swamp to marsh and open water in Southeastern Louisiana, USA: an irreversible trajectory? J Coast Res 54 (Special Issue):152–165

Shaffer GP, Day JW, Hunter RG, Lane RB, Jundberg CJ, Wood WB, Hamilton ER, Day JN, Strictland E, Kandalepas E (2015) System response, nutria herbivory, and vegetation recovery of a wetland receiving secondarily-treated effluent in coastal Louisiana. Ecol Eng 79:120–131

Slocum MG, Mendelssohn IA (2008) Effects of three stressors on vegetation in an oligohaline marsh. Freshw Biol 53:1783–1796

Touchette BW, Iannacone LR, Turner G, Frank A (2010) Ecophysiological responses of five emergent wetland plants to diminished water supply: an experimental mesocosm study. Aquat Ecol 44:101–112

Valentine JM Jr, Walther JR, McCartney KM, Ivy LM (1972) Alligators diets on the Sabine national wildlife refuge, Louisiana. J Wildl Manag 36:809–815

Visser JM, Sasser CE, Chabreck RA, Linscombe RG (1999) Long-term vegetation change in a Louisiana tidal marsh, 1968–1992. Wetlands 19(1):168–175

Warkentin MJ (1968) Observations on the behavior and ecology of the nutria in Louisiana. Tulane Stud Zool Bot 15:10–17

Wiebe J, Mouton E (2008) Nutria harvest and distribution 2007–2008 and a survey of nutria herbivory damage in coastal Louisiana in 2008. Fur and Refuge Division Louisiana Department of Wildlife and Fisheries, pp 1–39

Willner GR (1982) Nutria. In: Chapman JA, Feldhamer GA (eds) Wild mammals of North America: biology, management, and economics. The Johns Hopkins University Press, Baltimore/London, pp 1059–1076

Willner GR, Chapman JA, Pursley D (1979) Reproduction, physiological responses, food habits, and abundance of nutria on Maryland marshes. Wildl Monogr:6–43

Wilsey BJ, Chabreck RH, Linscombe RG (1991) Variation in nutria diets in selected freshwater forested wetlands of Louisiana. Wetlands 11:263–278

Optimum Use of Fresh Water to Restore Baldcypress – Water Tupelo Swamps and Freshwater Marshes and Protect Against Salt Water Intrusion: A Case Study of the Lake Pontchartrain Basin

Gary P. Shaffer, John W. Day, and Robert R. Lane

Abstract

Freshwater wetlands are important in the Mississippi Delta for habitat, water quality improvement, fisheries, carbon sequestration, and as a buffer against hurricane storm surge and waves. Forested wetlands are particularly important as hurricane buffers because of their 3-dimensional structure and their resistance to blow down during hurricanes. Fresh wetlands will be severely threatened by accelerated sea-level rise, more frequent stronger hurricanes, and intense drought that will lead to progressive inundation and saltwater intrusion. The Lake Pontchartrain Basin contains the largest area of tidal freshwater wetlands in the Delta. To ensure sustainability of fresh wetlands, a consistent source of fresh water is needed to counter increasing salinity levels. Sustainable restoration of baldcypress-water tupelo swamps in the Pontchartrain Basin can only be achieved through wise use of point and non-point sources of fresh water. In this paper, we identify potential sources of fresh water in the Pontchartrain Basin and determine the feasibility of engineering these to maximize sheet flow to enhance freshwater wetland health. Sources of fresh water include coastal plain rivers, Mississippi River diversions, non-point source runoff, direct rainfall on wetlands, storm water pumps, and treated municipal effluent. The latter is important because it is available even during drought periods.

Keywords

Freshwater resources • Forested wetlands • Water quality • Nutrients • Ecological engineering

Introduction and Statement of the Problem

Restoring and Maintaining Baldcypress-Water Tupelo Swamps and Freshwater Marshes

As documented in chapter "Energy and Climate – Global Trends and Their Implications for Delta Restoration", freshwater wetlands in the Mississippi Delta will be severely threatened by climate change. Accelerated sea-level rise and more frequent stronger hurricanes will lead to progressive inundation and saltwater intrusion. The impacts of saltwater intrusion will be exacerbated by more intense periodic droughts. Freshwater wetlands are important in the Delta for a number of reasons including habitat, water quality

G.P. Shaffer (✉) • R.R. Lane
Department of Biological Sciences, Southeastern Louisiana University, Hammond, LA, USA
e-mail: shafe@selu.edu; rlane@lsu.edu

J.W. Day
Department of Oceanography and Coastal Sciences, Louisiana State University, Baton Rouge, LA 70803, USA
e-mail: johnday@lsu.edu

J.W. Day, J.A. Erdman (eds.), *Mississippi Delta Restoration*,
Estuaries of the World, https://doi.org/10.1007/978-3-319-65663-2_5

improvement, and as a buffer against hurricane storm surge and waves (Shaffer et al. 2009b, 2016). Forested wetlands play a particularly important role as hurricane buffers because of their 3-dimensional structure that protects against both surge and waves. While relatively moderate hurricane surge can completely cover coastal marshes by several meters of water, forested wetlands are taller than the highest surges. Forested wetlands also are highly resistant to being blown down in hurricanes (Shaffer et al. 2009a, b, 2016). Dependable sources of fresh water are critical to protecting freshwater wetlands against saltwater intrusion. However, most freshwater input to the Mississippi Delta is channelized through the Delta directly to open water bodies. In this chapter, we present a case study of the western Pontchartrain Basin showing how freshwater resources can be more effectively used as a buffer against saltwater intrusion.

The Pontchartrain Basin is one of several interdistributary basins and is located between the Mississippi and Pearl Rivers and bordered by Pleistocene uplands north of Lakes Pontchartrain and Maurepas (see chapter "A Brief History of Delta Formation and Deterioration". Prior to levee construction there was regular flooding of river water into the Pontchartrain Basin especially during large floods. Numerous crevasse splays introduced large volumes of water (typically 2000 to 10,000 m^3/s (Davis 2000; Day et al. 2016a). There were crevasse splays (e.g., Bonnet Carré), minor (e.g., Bayou Manchac) and major (e.g., Gentilly and Metairie) distributaries and beach ridges in the Pontchartrain Basin (Saucier 1963; Davis 2000; Day et al. 2016a). Bayou Manchac was not always a shallow, dead end channel with low flow that it is today. During historical periods, Bayou Manchac carried steamboats between the Mississippi and Lake Pontchartrain. Kniffen (1935) described the upper Manchac as a raging torrent when fed by Mississippi floodwaters. Regular input of river water was critical to maintaining fresh and low salinity wetlands. Changes in coastal forests have been less documented (Chambers et al. 2005; Conner et al. 2014; Shaffer et al. 2009a, 2016) primarily because they can persist for long periods of time even as the forest degrades and slowly dies (Shaffer et al. 2009a, b; 2016; Day et al. 2012; Conner et al. 2014).

Until modified by human activity, many of the distributaries continued functioning, delivering fresh water, sediments, and nutrients to the Pontchartrain Basin. Fresh water forms a buffer against saltwater intrusion, and provides mineral sediments, nutrients, and other materials, such as iron, that sustain healthier more productive wetlands. The distributary network was very efficient in sediment retention and about 25–75% or more of sediment flux was retained in the Delta (Kesel et al. 1992; Blum and Roberts 2012; Tornqvist et al. 2008; Shen et al. 2015; Day et al. 2016b). The riverine input was important for coastal forested wetlands and marshes for several reasons, including as a buffer against saltwater intrusion, a source of nutrients and mineral sediments that stimulated primary productivity and strengthened soils, and iron that precipitated sulfides (Delaune and Pezeshki 2003; Delaune et al. 2003).

Baldcypress (*Taxodium distichum*)-water tupelo (*Nyssa aquatica*) swamps historically comprised 90% of the wetlands in the upper Lake Pontchartrain Basin of southeastern Louisiana (Saucier 1963). The area of swamp has been radically reduced and multiple stressors such as logging, development, hydrologic alteration, nutrient deprivation, subsidence, herbivory, and saltwater intrusion threaten most of the remaining area, which is in a state of non-sustainable deterioration (Fig. 1, Chambers et al. 2005; Effler et al. 2006; Shaffer et al. 2009a, 2016). Freshwater marshes in the region also have been impacted by these same stresses. Several projects have been proposed to help restore these forested wetlands, however, Hurricanes Katrina and Rita altered previous assumptions about wetland restoration in coastal Louisiana. In terms of flood- and wind-damage reduction, swamps are far superior to most other wetland habitat types (Shaffer et al. 2009a, b). The same is true for mangrove forests that were instrumental in protecting villages during the recent Asian tsunami (Danielsen et al. 2005). There is a need to re-think coastal restoration strategies, tailoring them to storm-protection alternatives as well as self-maintaining ecosystems (Boesch et al. 2006; Costanza et al. 2006; Day et al. 2007; Shaffer et al. 2009a, b).

Only live oak (*Quercus virginica*) and palms are more resistant to wind throw (being blown down) than baldcypress and water tupelo (Williams et al. 1999). Baldypress-water tupelo swamps faired far better than other forest types in Hurricanes Camille (Touliatos and Roth 1971), Andrew (Doyle et al. 1995), and Hugo (Gresham et al. 1991; Putz and Sharitz 1991). In addition, fresh, intermediate, and brackish emergent marshes suffered much greater losses from Hurricanes Katrina and Rita than did baldcypress-water tupelo swamps (Barras 2006). Katrina caused wind throws of up to 80% of the bottomland hardwood forests of the Pearl River Basin, with an estimated loss of 320 million trees (Chambers et al. 2007) and 900 million board feet of lumber in the Honey Island swamp alone, while contiguous baldcypress-water tupelo swamps remained largely intact.

Although baldcypress-water tupelo swamps are extremely resistant to wind throw and deep flooding, they are not very salt tolerant and are very susceptible to damage from saltwater intrusion events, and to stressors coupled with salinity stress, and thus require a reliable source of fresh water for system flushing following tropical storm events and during droughts. The same is true for freshwater wetlands. We include in this study a geographic information system (GIS) showing substantial point and non-point freshwater sources from Lake Maurepas to the Pearl River, which could be directed at degraded wetlands (see below). There

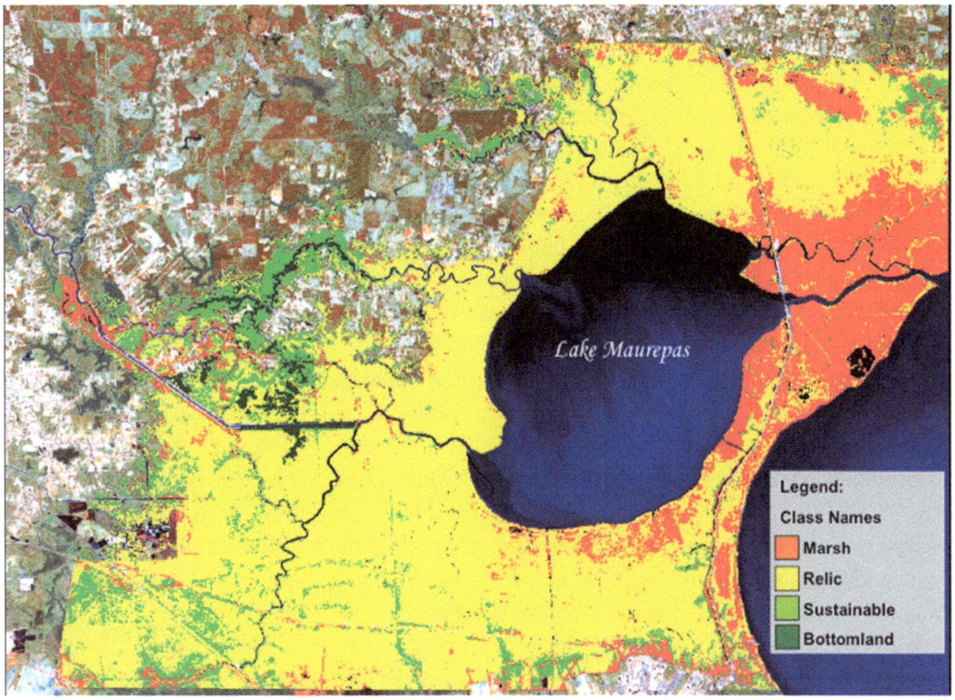

Fig. 1 Preliminary classification of wetland types in the Lake Maurepas swamp. *Red* areas indicate marsh, most of which was swamp in the mid 1950s. *Yellow* areas are classified as relic swamp in that the probability of regeneration following logging is very low. *Light green* areas indicate swamp that will likely regenerate if properly harvested. *Dark green* areas indicate bottomland hardwood forest or pine (Redrawn from Shaffer et al. 2009a)

are a number of freshwater resources that can be used to combat saltwater intrusion including local rivers, stormwater pumps, municipal wastewater treatment facilities, subdivisions treating their own sewage, non-point source runoff that often short circuits wetlands, industrial cooling water (Hyfield et al. 2007), and Mississippi River diversions. Only treated municipal effluent facilities offer consistent year round sources of fresh water. At present, most of these freshwater sources are configured to maximize drainage efficiency. Runoff is almost always routed into ditches and canals that carry it directly to lakes and bayous, bypassing surrounding wetlands and negating potential water quality improvement and wetland restoration. During dry periods, these same canals allow saltwater intrusion as was the case during the 1999–2000 drought (Day et al. 2012). This creates a 'loose-loose' situation as potential for eutrophication in water bodies is increased while the wetlands remain nutrient starved and more susceptible to saltwater impacts, and continue to degrade. In contrast, re-routing water into wetlands with maximized sheet flow will improve water quality, increase wetland primary productivity, enhance accretion to offset sea level rise, and provide a freshwater buffer to saltwater intrusion. In addition, implementation of proposed River diversions at Violet, Bonnet Carré, La Branche, Hope Canal (LDNR 1998), Sorrento (CPRA 2017), perhaps Bayou Manchac, and a mega-diversion as suggested in chapter "Large Infrequently Operated River Diversions for Mississippi Delta Restora tion" will greatly enhance restoration of historic salinity conditions.

In addition to the benefits mentioned above, increasing forested wetland acreage will enhance storm protection, increase carbon sequestration (Trettin and Jorgensen 2003; Day et al. 2004a, b; Rybczyk et al. 2002), improve water quality, enhance biodiversity, and augment several of the "multiple lines of defense" proposed by Lopez (2006). The closure of the Mississippi River Gulf Outlet (MRGO) in 2009 has helped decrease salinity in Lakes Pontchartrain and surrounding wetlands (Hillmann et al. 2015). However, salinity was reduced by an average of only about 3 psu while extreme droughts, such as occurred in 1999–2000, raised salinity levels in western Lake Pontchartrain and eastern Lake Maurepas by up to 12 psu (Day et al. 2012; Shaffer et al. 2016). One of the benefits of using treated municipal effluent is that it is always available, especially during droughts and periods of low river flow as occurred in 1999–2000.

One concern that managers and the general public have with restoring degraded forested wetlands is the amount of time required for cypress-tupelo swamp-like characteristics to emerge and manifest. Fortunately, given favorable hydrology and nutrient availability, baldcypress and water tupelo seedlings can reach greater than 10-meter in height within one decade (Shaffer et al. 2015). For example, a pilot planting of baldcypress seedlings at the Caernarvon diversion has yielded >10 m tall trees in a decade (Krauss et al. 2000), all of which resisted wind throw during the hurricanes of 2005. In addition, baldcypress growing along the outfall system of the Hammond Assimilation Wetland grew an average of 15–20 mm $year^{-1}$ in diameter, compared to 2–3 mm $year^{-1}$

elsewhere in the Manchac/Maurepas region (Shaffer et al. 2015). Similar growth rates of planted baldcypress seedlings have been documented at several other assimilation systems in coastal Louisiana (J.N. Day, pers. comm.) One of the main objectives of this study is to identify freshwater sources that potentially can be used for wetland restoration.

To understand the problems facing swamps and freshwater marshes in coastal Louisiana, it is helpful to put forested wetlands into the broader context of the development and deterioration of the Mississippi Delta as discussed in chapter "A Brief History of Delta Formation and Deterioration".

Most discussion of wetland loss has focused on marshes (i.e., Day et al. 2000, 2007) but recently more attention has been given to the issue of coastal forested wetland loss and restoration (Chambers et al. 2005; Shaffer et al. 2009a, 2016; Day et al. 2012). Forested wetland restoration is now a central part of the State Master Plan (see chapter "The Costs and Sustainability of Ongoing Efforts to Restore and Protect Louisiana's Coast" on the State Master Plan).

We believe that restoration of baldcypress-water tupelo swamps in the Pontchartrain Basin can only be achieved and sustained through wise use of point and non-point sources of fresh water. Over the past 20 years, several attempts have been made to restore baldcypress swamp in the Pontchartrain Basin and most have failed due to high salinity events. Ten thousand baldcypress seedlings had high survivorship on the Manchac land bridge for almost a decade, yet suffered nearly complete mortality caused by saltwater intrusion during the 1999–2000 drought (e.g., Shaffer et al. 2016). Similarly, greater than 70,000 baldcypress seedlings have been killed by saltwater intrusion events on Jones Island (just south of the Joyce wetlands) post 2000. *The key to best management and restoration of forested wetlands and freshwater marshes is reliable nutrient-rich, and if possible sediment-rich, fresh water.*

The main objective of this paper is to identify potential sources of fresh water currently being wasted in the Pontchartrain Basin and to determine the feasibility of engineering these to maximize sheet flow over degraded wetlands and to assess the potential of using each to convert degraded wetlands and open water back into healthy baldcypress-water tupelo swamps.

Sources of Fresh Water to the Pontchartrain Basin

Local River Input

There are a number of coastal plain rivers that discharge into Lakes Maurepas and Pontchartrain. These include the Amite, Blind, and Tickfaw that flow into Lake Maurepas and the Tangipahoa and Tchefunte and Bayous Lacombe and Liberty that discharge into Lake Pontchartrain. There are many smaller streams that discharge directly to Lake Pontchartrain. Historically many of these smaller streams discharged to wetlands before entering the lakes but have now been channelized to bypass wetlands (Lane et al. 2015a, see below for non-point source runoff). The Pearl River discharges to Lake Borgne/Mississippi Sound just outside the Rigoletes that connects Lake Pontchartrain to these estuarine waterbodies. Mean discharge of the larger rivers from west to east is as follows: Blind River mean flow 15.6 m^3/s, volume 0.49 km^3/year; Amite mean flow 45.0 m^3/s, volume 1.42 km^3/year; Tickfaw mean flow 10.8 m^3/s, volume 0.34 km^3/year; Tangipahoa mean flow 32.9 m^3/s, volume 1.04 km^3/year; Tchefuncta mean flow 4.5 m^3/s, volume 0.14 km^3/year (Xu and Wu 2006; Roy et al. 2016). The total mean flow is 108 m^3/s. The seasonality of these rivers follows the typical pattern of high flows in the winter and spring and low flows in the summer and early fall (see Fig. 2).

The discharge from these rivers is of the same order as that of the Caernarvon diversion which ranges from 0 to 253 m^3 s^{-1}, with a mean of 53.5 ± 0.9 m^3 s^{-1} from 2000 through 2012. The greatest annual volume (2.74 km^3) was discharged at Caernarvon during 2010 in response to the Deepwater Horizon oil spill, while the least annual volume (0.92 km^3) was discharged during 2005 due to a 4-month shutdown caused by Hurricane Katrina (Fig. 3).

Extreme discharges can be much higher. In March of 2016, about 50 cm (20 inches) of rain fell on the Hammond area resulting in high flows on the Tangipahoa. In August of 2016, 76 cm (30 inches) fell in the Baton Rouge and Denham Springs area, resulting in high discharge of the Amite River. Both rivers experienced record floods.

The rivers discharging into Lakes Maurepas and Pontchartrain should be managed so that as much flow as feasible flows into wetlands. This would introduce suspended sediments and nutrients to these wetlands, especially during floods as well as maintain fresh soil salinity. This would enhance the health of the wetlands and act as a buffer against periodic droughts.

Freshwater Diversions

Although freshwater diversions of the Mississippi River offer tremendous promise in restoring the wetlands of the Pontchartrain Basin, they are insufficient alone. One major limitation is that there exist few locations in the Basin where diversions are feasible due to human development, especially along the natural levee of the Mississippi River between Baton Rouge and New Orleans, which extends more than 1 km on either side of the River and is the highest elevation land in the region. Second, and most importantly,

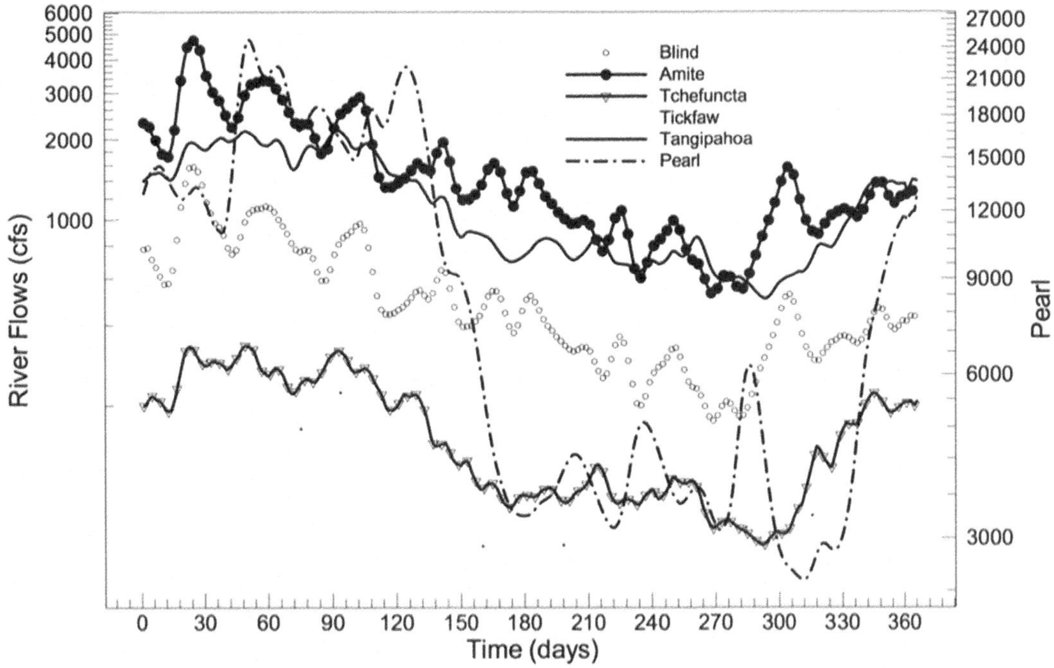

Fig. 2 Seasonal river discharge patters for rivers entering Lake Maurepas and Pontchartrain and the Pearl that discharges into Mississippi Sound (from LCA 2004, Appendix C)

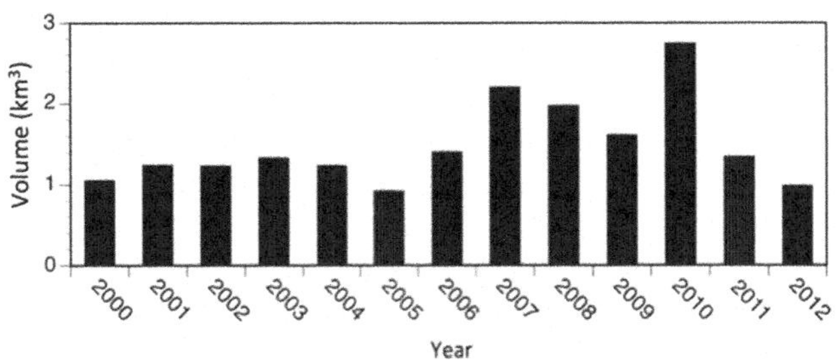

Fig. 3 Annual volume discharged from the Caernarvon river diversion

diversions are often inoperable, due to low river stage, during the season that they are most needed, namely late summer and fall, or during extreme droughts. It is during late summer and fall that the effects of drought are strongest and the frequency of tropical storms and hurricanes is highest, and coastal water levels are high. So, the probability of saltwater intrusion events is highest when the potential to deliver river water is lowest. The summer and early fall also is the time when there is little surplus runoff (see section below on water budgets). This is the paramount reason for finding other sources of nutrient-rich fresh water, such as treated municipal effluent, that are reliably available during the periods when they are most needed.

In 1998, EPA sponsored a Coastal Wetlands Planning, Protection, and Restoration Act (CWPPRA) project to divert 2000 cubic feet per second (cfs) or about 55 m^3/s of Mississippi River water into the southwestern Lake Maurepas swamp (LDNR 1998). For an 11-year period, Shaffer et al. (2016) conducted a feasibility study of this diversion to evaluate the current condition of these swamp forests and to assess the potential benefits the whole ecosystem would derive from the freshwater diversion. The wetlands are stressed and dying, and in need of restoration (Shaffer et al. 2009a, 2016). Fortunately, the first phase of this $150-million diversion is finally funded with BP settlement monies. While locally important, such a small diversion will do little to address the problems of most of the wetlands of Lake Maurepas and upper Lake Pontchartrain. Rutherford et al. (chapter "Large Infrequently Operated River Diversions For Mississippi Delta Restoration") modeled a very large infrequently operated diversion with a maximum discharge of 7000 m^3/s. By comparison, maximum discharge of the Bonnet Carré Spillway up to 2011 averaged 6425 m^3/s and ranged from 3115 to 9006 m^3/s for individual openings (Table 1, see Day et al. 2012).

As well as decreasing the detrimental effects of salinity throughout the Maurepas swamp, the proposed diversion also will increase the sediment load and nutrient supply to these wetlands. Hydrologic modeling has shown that due to the low water-holding capacity of the receiving waterway (Hope Canal and Dutch Bayou) most of the diverted water is likely to flow as sheet flow through the interior swamps (Kemp et al. 2001). The resulting, evenly distributed influx of sediments is expected to strengthen the highly organic soils of the swamp and to increase elevation in a small area

Table 1 Estimated fresh water resources for Lakes Pontchartrain and Maurepas

Diversions (proposed)	55 to 7000 $m^3 s^{-1}$
Total mean river input	108 $m^3 s^{-1}$
Ungauged runoff	20 $m^3 s^{-1}$
Direct rainfall on wetlands	60 $m^3 s^{-1}$
Stormwater pumps	5.1 $m^3 s^{-1}$
Treated municipal effluent	4 $m^3 s^{-1}$

Notes: See text for details. Population in 2012 of parishes surrounding Lakes Pontchartrain and Maurepas that discharge treated effluent that drains to the lakes: Ascension 112,126, Iberville 7861, Livingston 132,160, St. Charles 26,952, St. James 13,005, St. John the Baptist 41,229, St. Tammany 239,814, Tangipahoa 124,125 (Source: Lake Pontchartrain Basin Foundation, saveourlake.org)

sufficiently to make the natural regeneration of several wetland forest species possible.

The potential negative impacts of lake eutrophication due to the increase in nutrient loading to the swamp are unlikely to occur, as nutrient uptake analysis indicates high nutrient retention in the swamp with nutrient removal efficiencies of 94–99% as diverted water passes through the swamp and exits into Lake Maurepas (Day et al. 2001, 2004a, b; Lane et al. 2003). Experimental nutrient augmentation in the Maurepas swamp enhanced biomass production of the herbaceous vegetation by over 200% (Howell 2003; Shaffer et al. 2003; Shaffer et al. 2009a). Furthermore, several studies conducted over the past decade have demonstrated that nutrient augmentation to baldcypress seedlings at least doubles growth rates in the Manchac/Maurepas area (Greene 1994; Forder 1995; Myers et al. 1995; Boshart 1997; Effler et al. 2006; Shaffer et al. 2015). Results of studies at assimilation wetlands also show significant increases in vegetative productivity (Hesse et al. 1998; Day et al. 2004a, b, 2006; Brantley 2006; Lundberg et al. 2011; Shaffer et al. 2015). This enhanced productivity is essential for subsiding coastal wetlands to offset relative sea level rise (RSLR), as roots may contribute as much as 60% of the annual increment to soil organic matter (Megonigal and Day 1988).

An additional diversion is currently being considered for Blind River. We strongly disagree with this location as Blind River will easily accommodate the diverted water without significantly flooding the surrounding wetlands, essentially channeling the water directly to Lake Maurepas, bypassing wetlands and likely triggering algae blooms. A far better location would be further north near the crossing of I-10 and Hwy 61 near Sorrento. This relatively large (35,000 cfs) diversion is proposed in the 2017 State Master Plan. In addition, or alternatively, Bayou Manchac should be considered as a potential location for a Mississippi River diversion. Bayou Manchac was historically connected to the River and, because of its close proximity to the River, would be a relatively inexpensive project. Furthermore the diversion would be routed through the highly degraded Spanish Lake swamp that would greatly benefit from an infusion of River water. Bayou Manchac eventually ties into the Amite River, and then into the Amite River Diversion Canal (ARDC), which would yield higher benefits to the soon to be completed ARDC gapping project (see below). In chapter "Large Infrequently Operated River Diversions for Mississippi Delta Restoration" in this book, Rutherford (2017) model large, episodically operated diversions of 1000 and 7000 m^3/s. The larger diversion builds about an order of magnitude more land than the smaller diversion.

Opposition to diversions has developed based on a number of perceived threats. These include over-freshening of coastal estuaries, displacement of fisheries, perceived water quality problems, and assertions that nutrients in river water lead to wetland deterioration. In addition, growing climate impacts and increasing scarcity and cost of energy will make coastal restoration more challenging and limit restoration options. Day et al. (2016a, see chapter "Large Infrequently Operated River Diversions for Mississippi Delta Restora tion") proposed the use of very large but infrequently operated diversions to address these issues using an analysis of natural and artificial diversions, crevasse splays, and small sub-delta lobes. They suggest that episodic large diversions and crevasses (>5000 $m^3 s^{-1}$) can build land quickly while having transient impacts on the estuarine system. Small diversions (<200 $m^3 s^{-1}$) that are more or less continuously operated build land slowly and can lead to over-freshening and water level stress (Snedden et al. 2015). Day et al. (2016a) used land building rates for different sized diversions and impacts of large periodic inputs of river water to coastal systems in the Mississippi Delta to conclude that high discharge diversions operated episodically will lead to rapid coastal restoration and alleviate concerns about diversions. Single diversion events have deposited sediments more than 40 cm in depth over areas up to 130 to180 km^2.

Un-gauged Non-point Source Runoff

Almost all areas surrounding Lakes Maurepas and Pontchartrain and adjacent wetlands generate non-point source runoff. Runoff from leveed areas, such as the New Orleans metropolitan area, is pumped out to receiving water bodies, mainly Lake Pontchartrain. Runoff from non-leveed areas almost always flows through a series of ditches and canals that increase in size as they approach the wetlands. In most cases, these waterways cut through the wetlands and eventually discharge into the two lakes, almost completely

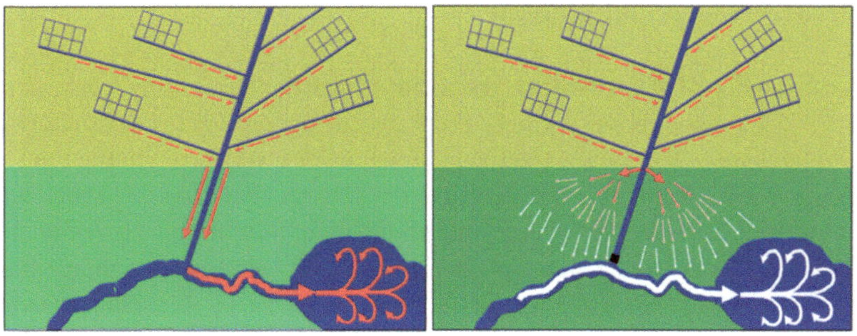

Fig. 4 Traditionally (*left*), non-point source runoff from neighborhoods collects in larger and larger ditches and short-circuits wetlands carrying eutrophied waters directly to rivers, streams, bayous, and lakes. Conversely, "Reverse Marsh Management" (*right*) places a plug at marsh elevation near drainage to water body, thereby neutralizing head. Non-point source runoff then sheetflows through wetlands that benefit from its nutrient and fresh water resources, simultaneously improving its water quality

short circuiting the wetlands. South Slough, located just south of Ponchatoula, is a example of this (Lane et al. 2015a). Not only does South Slough route runoff directly to Lake Maurepas via a canal that parallels Interstate Highway 55, its spoil bank prevents most flow into the Joyce wetlands. In addition, drainage canals through wetlands generally have spoil banks on one or both sides that further prevent sheet flow; these spoil banks need to be gapped at least every 0.5 km or degraded completely. Lane et al. (2015b) demonstrate that the distance between gaps and the width of the gaps have a large effect on the amount of sheet flow the wetland receives. In essence, we are proposing "reverse marsh management." This means removal of impediments that impede flow over wetlands. In the case of South Slough, the canal could be closed at the railroad or I-55 and large gaps could be put in the spoil bank to force all flow to the south through the wetlands. This could be done in an innovative way so that runoff and the effluent would be complimentary to one another.

To estimate non-gauged freshwater inputs to Lakes Pontchartrain and Maurepas, we calculated the area of these areas using Google maps and calculated the total amount of water falling on these areas using an average annual precipitation of 1.53 m. We used one third of this to account for evaporation and infiltration. These calculations resulted in a total un-gauged runoff from uplands equivalent to about 20 m^3/s.

Direct Rainfall on Wetlands Surrounding Lake Pontchartrain

We computed the area of wetlands not included in gauged river discharge and not under pump using the polygon feature of Google. This area totals 1009 km^2. Based on an average precipitation of 1.53 m, total precipitation is equivalent to about 60 m^3/s.

Storm Water Pumps

Large areas surrounding Lakes Pontchartrain and Maurepas do not drain by gravity and rainfall needs to be pumped out. The New Orleans metropolitan area is the largest forced drainage area. There also is forced drainage in adjacent St. Charles Parish. Because of rising water levels and the threat of hurricane flooding, there is active planning for more forced drainage area upriver from New Orleans. As sea levels rise, the pumped forced drainage area will have to increase significantly or low-lying developed areas will have to be abandoned.

Where possible, storm water pumps should discharge to wetlands rather than directly to waterbodies. In the case of the New Orleans, storm water could be pumped into the baldcypress swamps proposed in chapter "Raising New Orleans: The Marais Design Strategy" to help protect New Orleans from hurricane surge and waves. The runoff was calculated using the same approach used for un-gauged uplands. Currently, the total amount of water pumped is about 4.4 $m^3\ s^{-1}$ for the New Orleans metro area and 0.7 $m^3\ s^{-1}$ for St. Charles Parish.

In summary, at present due to drainage channels and spoil levees, surface runoff, both pumped and gravity drainage, from all parishes surrounding Lakes Pontchartrain and Maurepas is short-circuiting virtually all wetlands in the Lake Pontchartrain Basin. These channels are designed to maximize drainage efficiency, but in doing so convey pollutants to receiving water bodies, decreasing water quality (Fig. 4, left). Perhaps the cheapest management strategy to enhance wetland restoration in the Basin would involve plugging these ditches at marsh elevation near to the location of their connection to a river, stream, bayou or lake (Fig. 4, right). This would functionally eliminate any head and cause the water to sheet flow over wetlands. In doing so, the water would enhance wetland production and the wetland would enhance the quality of the water.

Treated Municipal Effluent

Conventional Wastewater Treatment Plants

As discussed above, when properly designed and implemented, wastewater treatment plants offer a reliable freshwater source that is disinfected and toxin-free that can be introduced into freshwater and oligohaline wetlands. This is addressed below in the section on wetland assimilation.

Decentralized Treatment systems

Decentralized treatment systems aim to have most wastewater treatment take place at or near where effluent is generated (i.e., rural subdivision developments), and to discharge the treated effluent near the source of effluent production. In these systems, effluent from a home or small business is discharged to a septic system on site. Solids are held on site and periodically collected by pump trucks and treated separately. The liquid portion of the waste is then pumped a short distance for further treatment. Such systems are generally much smaller than conventional treatment systems. For developments or small communities located near wetlands, these systems can relatively easily be modified to discharge to wetlands. In addition, since these discharges are small, constructed wetlands are also an option. The GIS analysis has identified over 1000 potential sources (see below). Though individually small, the manifold effect of many of these systems on wetland restoration and water quality improvement could be quite significant. What is needed is a careful study to determine if there are wetlands near these subdivisions and package plants to receive the discharge.

Wetland Assimilation

Several wetland assimilation systems discharging mainly to forested wetlands have been in operation for over 25 years and have enhanced productivity and continue to reduce nutrients to background including Breaux Bridge – 65–70 years (Hesse et al. 1998), Amelia - 43 years (Day et al. 2006), Gore-Riverbend – 34 years (Hunter et al. 2016), Mandeville – 27 years (Brantley et al. 2008), and Thibodaux – 25 years (Minor 2009; Hunter 2017). Most baldcypress in the 11,000 ha Central Wetland Unit were killed by saltwater intrusion from the MRGO in the early 1960s (Shaffer et al. 2009b; Hillmann et al. 2015; Hunter et al. 2016). The only place that baldcypress survived was near the discharge from the Gore and Riverbend plants. The Pontchatoula plant located north of the Hammond site has discharged to forested wetlands and freshwater marshes for over 50 years without apparent detrimental impact. For conventional treatment systems, water can be transported by pipe for relatively long distances. At Hammond, for example, water is pumped 11 km from the treatment plant to the Joyce wetlands.

There are three wetland assimilation systems that discharge to the Pontchartrian Basin. At Hammond, a manifold with a 900-discharge valve system is used to distribute effluent over a 1.2 km delivery width (Shaffer et al. 2015; Lane et al. 2015a). Because only 150 valves need to be opened at once, the outfall can be shifted from east to west, enabling substantial drawdown to occur where discharge is not concentrated. The Hammond treatment plant discharges about 14,500 m^3 day^{-1} of treated effluent into Four Mile Marsh of the northern Joyce wetlands, which began fall 2006. From February–March 2007, 5500 seedlings of baldcypress and water tupelo were planted with nutria protectors. We set up five 700 m permanent transects (four in the experimental area, one control) with paired permanent plots, ranging from 0 to 700 m from the outfall pipe at 50–100 m intervals, depending on distance from outfall (Shaffer et al. 2015). In addition, cover value and standing crop estimates of herbaceous vegetation were collected prior to initiation of discharge of the treated effluent (Lundberg 2008). During the year following initiation of discharge, herbaceous NPP nearly doubled compared to the control (Shaffer et al. 2015). In addition, seedling survival in the treatment wetlands is nearly complete and some of the seedlings have grown as tall as 10 m (Shaffer et al. 2015).

The Mandeville system presently discharges into Bayou Chinchuba, which Brantley et al. (2008) reported had increased wetland production and enhanced accretion. However, the loading rate has become too high to achieve desirable levels of nutrient assimilation so the city of Mandeville purchased approximately 1000 acres of wetland west of Mandeville and east of the Tchefuncte River for additional assimilation capacity. About two thirds of the current discharge is periodically routed to these wetlands, dispersing the effluent over a larger area, decreasing nutrient loading and increasing nutrient removal efficiency. Baldcypress and water tupelo have been planted in the immediate outfall area of the discharge to encourage the growth of desirable wetland species that are not killed by nutria and increase hurricane storm surge protection. These seedlings have high survival and growth rates.

Numerous studies have shown that wetlands can be effective tertiary processors of nutrients in secondarily treated effluent (Day et al. 2004a, b; Kadlec and Wallace 2009; Vyzamal 2007). Both natural and constructed wetlands have been shown to effectively reduce biochemical oxygen demand, total suspended solids, and nitrogen and phosphorus concentrations (Day et al. 2004a, b; Hunter et al. 2008a, b; Kadlec and Wallace 2009; Shaffer et al. 2015). The benefits of discharging municipal effluent into wetlands rather than the business-as-usual practice of discharging into surrounding rivers and streams include improved water quality (Blahnik and Day 2000; Boustany et al. 1997; Zhang et al. 2000; Day et al. 2006), financial and energy savings (Ko et al. 2004), increased primary production (Hesse et al. 1998; Day et al. 2004a, b; Brantley et al. 2008; Hunter et al. 2008a; Lundberg et al. 2011; Shaffer et al. 2015), and enhanced vertical accretion (Rybczyk et al. 2002; Brantley et al. 2008; Hunter et al. 2008b). Increases in aboveground biomass and root production enhance organic soil deposition and carbon sequestration that results in a healthier wetland (DeLaune et al. 2016; Mcleod et al. 2011; Morris et al. 2013). Geological subsidence of this organic soil results in significant permanent carbon burial. The introduction of treated municipal effluent into highly degraded wetlands of Louisiana is a major step towards their ecological restoration (Day et al. 2004a, b). The following wetland assimilation sites (and the average daily quantity of effluent being discharged) are currently functioning in Louisiana: Breaux Bridge (3800 m^3 day^{-1}; Hunter et al. 2008b), Amelia (3000 m^3 day^{-1}; Day et al. 2006), Mandeville (7200 m^3 day^{-1}; Brantley et al. 2008), Thibodaux (11,500 m^3 day^{-1}; Hunter et al. 2008b), Luling (6000 m^3 day^{-1}; Hunter et al. 2008b), Broussard (5700 m^3 day^{-1}) and Hammond (10,300 m^3 day^{-1} – dry weather flow, Hunter et al. 2008b; Shaffer et al. 2015; Lane et al. 2015a, b), as well as several smaller systems.

The high rate of burial due to subsidence and higher than average rates of denitrification due to warm temperatures are additional reasons for the use of wetland assimilation in Louisiana. Increasing vegetative productivity is especially critical for many parts of Louisiana where coastal subsidence of the Mississippi Delta results in a RSLR three times greater than eustatic sea level rise (Conner and Day 1988; Penland et al. 1988). An increase in productivity results in greater root production, which can lead to organic soil formation that enhances the accretion necessary to offset the subsidence that is contributing to wetland loss.

Continuous operation causes increased water levels in the wetlands (Lane et al. 2015a). To enable complete drawdown, discharge should be delivered to two independent wetlands as in Mandeville.

One issue that must be taken into consideration when introducing effluent into herbaceous wetlands is the potential for severe wetland grazing by nutria. Nutria have caused widespread marsh deterioration to marshes in the coastal zone (Sasser et al. chapter "The Nutria in Louisiana: A Current and Historical Perspective". At the Hammond Assimilation Wetland, high nutria populations caused deterioration to about 150 ha of marshes 1.5 years after discharge began. Subsequent studies have documented that the cause of the deterioration was largely due to nutria grazing and after they were controlled, the marsh began to recover (Shaffer et al. 2015; Lane et al. 2015a, b). Thus, strong nutria monitoring and control measures need to be undertaken when effluent or river water is discharged into herbaceous wetlands. Nutria also destroy unprotected baldcypress seedlings but they do not impact adult baldcypress (Shaffer et al. 2016).

We calculated the total volume of effluent generated in the parishes surrounding Lakes Pontchartrain and Maurepas. The most populous parishes are East Baton Rouge, Jefferson, and Orleans. However, almost all treated effluent in these three parishes is pumped to the Mississippi River. We used estimates of the population of the remaining parishes surrounding the lakes (Lake Pontchartrain Basin Foundation) and assumed a per capita daily water use of 5.7 m^3 (about 200 gallons per day) to estimate that total effluent flow is about 4 m^3/s.

Information on freshwater inputs to Lakes Pontchartrain are summarized in Table 1. The largest source of fresh water flowing into the lakes is river input that averages about 108 m^3/s. There is a high inter-annual variability and river discharge is low in the late summer and early fall when salinity is highest. River discharge is low during droughts when the highest salinities occur. Direct rainfall on wetlands is the second largest input at about 60 m^3/s and un-gauged runoff is third at about 20 m^3/s. Stormwater pumps contribute an average of about 5.1 m^3/s and treated municipal effluent flowing to the two lakes is about 4.0 m^3/s. Although municipal effluent is the lowest input, it is a consistent input that is flowing continuously. In a management system where all freshwater resources are used optimally, this small flow could be used to keep areas fresh that would be otherwise strongly impacted by drought (e.g., Lane et al. 2015a). Diversion inputs can potentially be the largest freshwater input and lead to much more sustainable freshwater wetlands ecosystems.

GIS Analysis of Freshwater Resources

We developed a geographic information system (GIS) containing substantial point and non-point freshwater sources along the North Shore of Lake Pontchartrain, from Lake Maurepas to the Pearl River, which potentially could be directed at degraded wetlands (Fig. 5). Thus far, we have

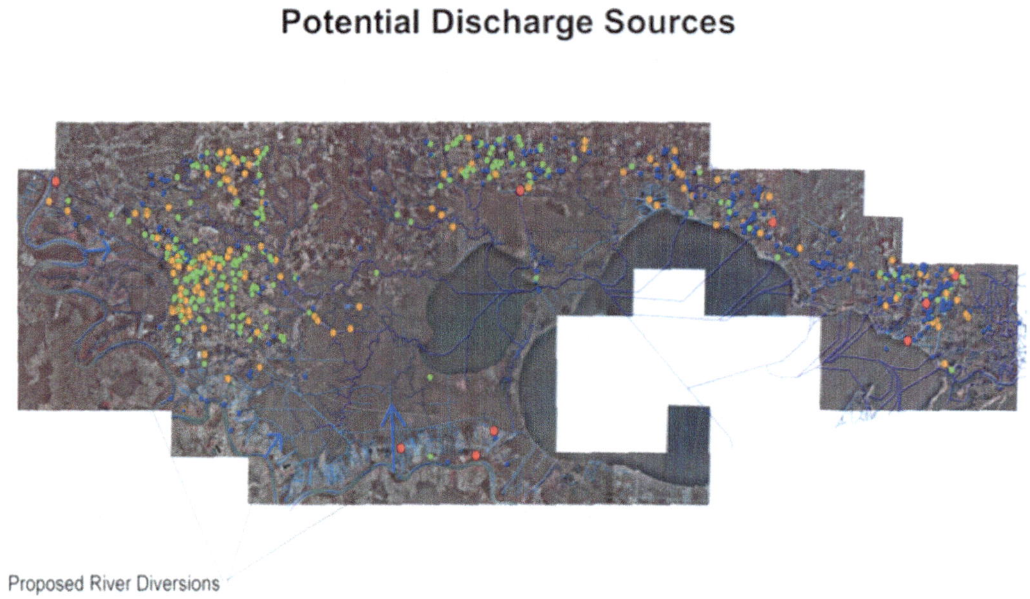

Fig. 5 GIS of wastewater sources in the Pontchartrain Basin (>100,000 GPD = Red, 50,000–100,000 = orange, 25,000–50,000 = green, < 25,000 = blue)

concentrated on municipal wastewater treatment facilities, subdivisions treating their own sewage, non-point source drainage features, and potential Mississippi River diversion sites. Eventually this GIS will include data on storm-water pumps and non-contact industrial cooling water. Most analysis, processing, and mapping were conducted in Arc GIS v9.2 by ESRI.

Raster Data All Digital Orthophoto Quarter Quadrangles (DOQQs) and LiDAR imagery were collected from the Louisiana State Wide GIS (ARC GIS 2006). All DOQQs in the area of interest (AOI) were stitched together for ease of viewing. All LiDAR imagery was combined to form one image of the AOI, namely the upper Lake Pontchartrain Basin.

Vector Data Road data was collected from the Louisiana State Wide GIS. Road information from Ascension, East Baton Rouge, Iberville, Livingston, St. James, St. John the Baptist, St. Tammany, and Tangipahoa parishes were merged and clipped to the AOI.

Rivers and lake water bodies were acquired from the 'United States DVD rivers & streams' file provided by ESRIs ArcGIS 9 Data and Maps media kit. Data was clipped to the AOI then set to display stream/rivers, canal/ditches, artificial paths, and connectors. Waste discharge facilities information was provided by the Louisiana Department of Environmental Quality, who provided the facility name, permit number, permit type, physical address, city, parish, latitude and longitude. The general permit types were as follows:

LAG53: discharges less than 5000 gallons per day (gpd)
LAG54: discharges between 5000 and 25,000 gpd
LAG56: discharges between 25,000 and 50,000 gpd
LAG57: discharges between 50,000 and 100,000 gpd
Major Sanitary Discharge: greater than 100,000 gpd

These were color and size coded and the level of detail in the GIS varies according to aerial coverage. That is, the overall image is in broad brush and increasing detail is automatically provided as the user increases magnification (Fig. 5).

Water Budgets

To prepare a water budget, daily precipitation and mean temperature values were obtained from the National Climate Data Center for the meteorological station at the New Orleans International Airport from 2004–2014. Using these data, potential evapotranspiration (PET) was calculated using the Thornthwaite equation (Thornthwaite 1948).

The water budget can be used to demonstrate climate variability during a year and in different years in the area

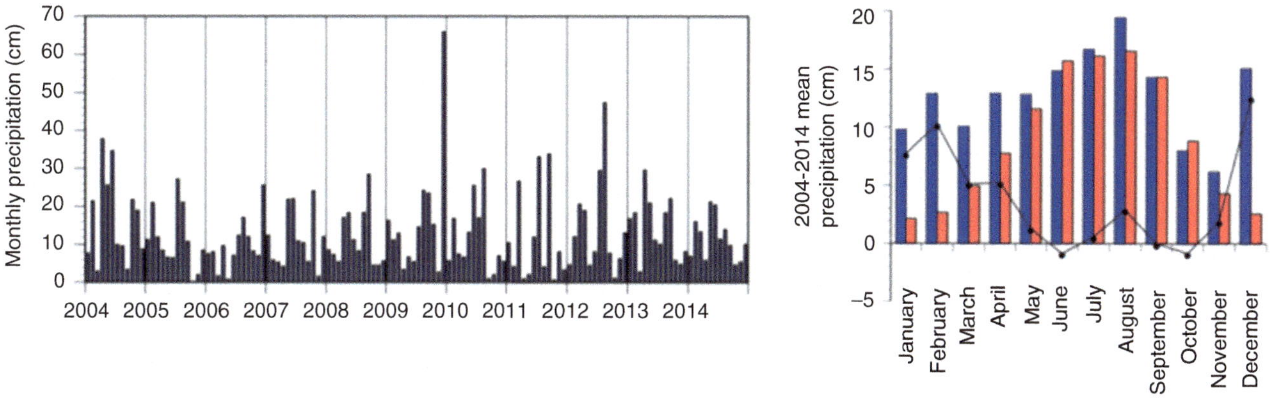

Fig. 6 Water budget for Mandeville (from Brantley et al. 2008). Monthly precipitation (*left panel*). Average rainfall (*right panel, blue*), potential evapotranspiration (*right panel, red*) and net surplus/deficit (*right panel, line*)

and to show the impacts of additional water loading. The components of the water budget are discussed below.

Figure 6 depicts a water budget for Mandeville. Monthly precipitation is highly variable over the 11-year time span from 2004–2014 (Fig. 6, left panel). The majority of rain occurs during the spring and late summer. On average, the greatest amount of precipitation was in August (19.4 cm) and the least in November (6.2 cm, Fig. 2, right panel, blue). Average annual precipitation was 153.1 cm. Potential evapotranspiration (PET) showed a typical trend of higher values during the warmer months and lower values during the winter months (Fig. 6, right panel, red). PET ranged from 2.1 to 16.6 cm mo^{-1}, with an average of 107.4 cm $year^{-1}$. Average precipitation exceeded average PET during all months except June, September and October, resulting in a surplus the majority of the year (Fig. 6, right panel, black line). Average net annual water surplus is 45.0 cm $year^{-1}$. Seasonal and annual variations of rainfall give rise to variability in water surplus/deficit (P-PE). Although rainfall is normally greatest during the warm weather months, high evapotranspiration rates during these months often lead to a net water deficit. Rainfall is generally somewhat lower during cold weather months, but net water surpluses are higher due to low evapotranspiration rates. Figures 7 and 8 show the effect of the 1999–2000 drought. The precipitation in 1999–2000 was significantly lower than that of 2001, a year with more normal precipitation. The water budget for this period shows that freshwater surplus was negative for most of 1999 and 2000 with surplus returning to a positive status in the second half of 2001 (Fig. 8) It is during the summer and early fall months when there is a water deficit coupled with southerly winds and tropical storm activity that saltwater intrusion can be a potential problem.

Hydrological Restoration

For freshwater resources to be used optimally, hydrological restoration is often needed. As mentioned above, riverine input to Lakes Pontchartrain and Maurepas should be managed to enhance river water flow into wetlands. An example of where hydrological restoration is needed is the Joyce Wildlife Management Area (JWMA). The JWMA contains the western-most reach in the USA of pondcypress (*Taxodium distichum* var. *imbricarum*). As for the majority of the Manchac/Maurepas swamp, the Joyce swamp is in a state of degradation. A large stand of pondcypress is located 1 km south of the Hammond effluent outfall pipe, but essentially all of the nutrients in the wastewater have been depleted at that distance from outfall (Shaffer et al. 2015). Currently plans are being developed to construct a second outfall system west of I-55. This will enable periods of drawdown in the current outfall area and also will enable swamp reforestation of a 3000 ha shrub-scrub area to the west.

Trunk-Line System An interesting idea has been proposed for collection and treatment of municipal effluent for a population that is spread out. This is a trunk-line collection system. In this approach, a trunk-line would be constructed to collect effluent from small cities, subdivisions, and individual houses. The trunk line would transport the waste to a centralized treatment system near wetlands. After secondary treatment, effluent would then be discharged through a distribution system, as in Hammond, to specified areas of wetland. The distribution system would be designed to optimally spread the effluent over target areas of wetland. As at Hammond, target treatment

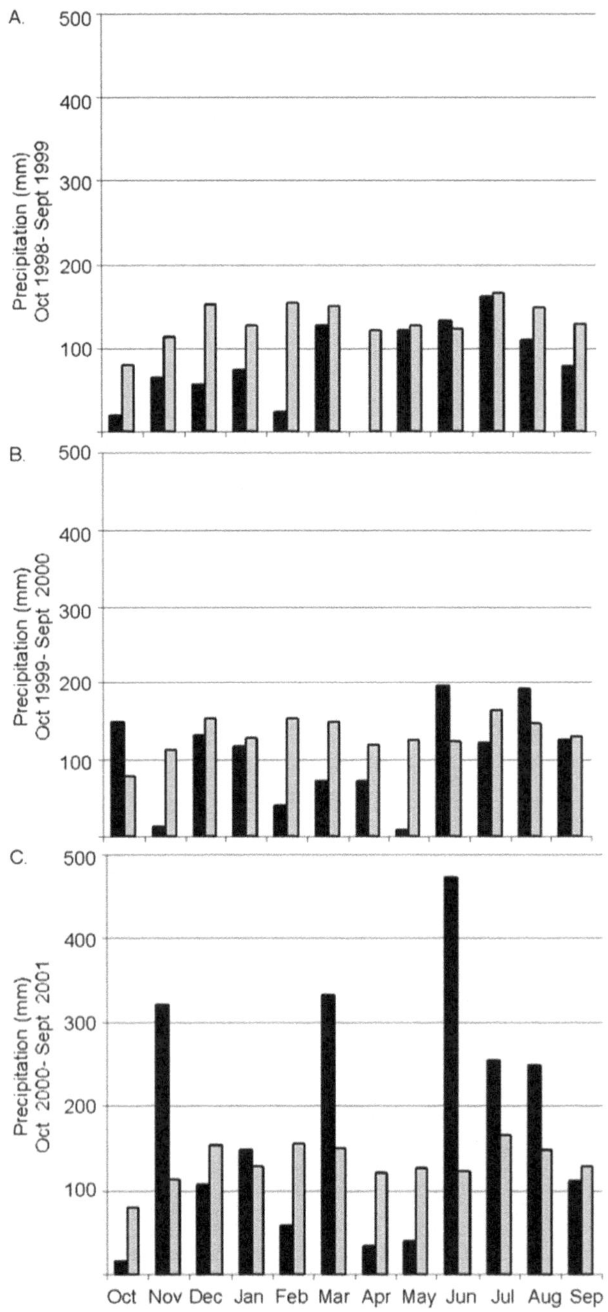

Fig. 7 Mandeville precipitation (*black bars*) vs. 30-year average precipitation at Covington, LA (*shaded bars*). (**a**) 1999 (**b**) 2000 (**c**) 2001 (from Brantley 2006)

levels would be maintained and the effluent would be disinfected. Unacceptable levels of toxic materials would not be discharged. Monitoring would be an integral part of such a project. The area west of the Tchefuncte River is an example of where such a system could be constructed to benefit an expanded Joyce WMA east of the Tangipahoa River. This technology also could be expanded west of the Tangipahoa River.

Gap Spoil Banks and Railroad Levees There are many spoil banks throughout the forested wetlands and freshwater marshes of the Pontchartrain Basin from oil and gas exploitation, railroad and highway impoundments, and from massive industrial logging that clear cut almost all old growth forests in Louisiana (Burns 1980; Mancil 1980; Keddy et al. 2007). These raised linear features often intersect leading to semi-impoundment of stagnant nutrient poor waters (Turner et al. 1982; Turner and Rao 1990). We recommend that these hydrological impediments be breeched (Turner et al. 1994; Turner 1997).

The primary benefits of the Maurepas diversion into Hope Canal will accrue only to the southern and southwestern portions of the wetlands surrounding the Lake. Consequently, the swamps north of Blind River will continue to degrade unless additional restoration efforts are devised and implemented, such as the Amite River Diversion Canal (ARDC) gapping project which is nearly complete. As mentioned above, River water needs to be reintroduced to one or more of the historic links to the northern Maurepas sub-basin such as Bayou Manchac or Sorento (Saucier 1963; Shaffer et al. 2005; Day et al. 2016a).

The hydrologic model for the Maurepas diversion (Day et al. 2004b) indicated that the spoil banks on Reserve Relief Canal would prevent eastward movement of diverted water, reducing wetland benefits, and that the footprint of the diversion would be substantially broadened by gapping these spoil banks. The spoil is generally placed only on one side of a canal, and that side then becomes impounded and more degraded than the wetlands on the other side. This is true of Hope Canal, Oil Field Canal, and Tobe Canal as well, and these would all benefit from gapping projects.

Other Issues

Carbon Sequestration The addition of river water or municipal effluent to wetlands increases wetland plant productivity and organic soil formation and accretion, which increases carbon sequestration (DeLaune et al. 2016; Mcleod et al. 2011; Morris et al. 2013; Lane et al. 2016). Geological subsidence and burial of this material results in a significant permanent carbon sink. For example, discharge of treated effluent at the Thibodaux assimilation wetland led to an increase in carbon burial by a factor of 2–3 fold (Rybczyk et al. 2002; Day et al. 2004a, b), and wetland assimilation systems are one of the best suited for carbon offset development and are believed to sequester some of the highest rates of carbon. Allowing entities to privately invest in wetland restoration projects to offset greenhouse gas emissions elsewhere holds promise as a new carbon offset sector (Mcleod et al. 2011;

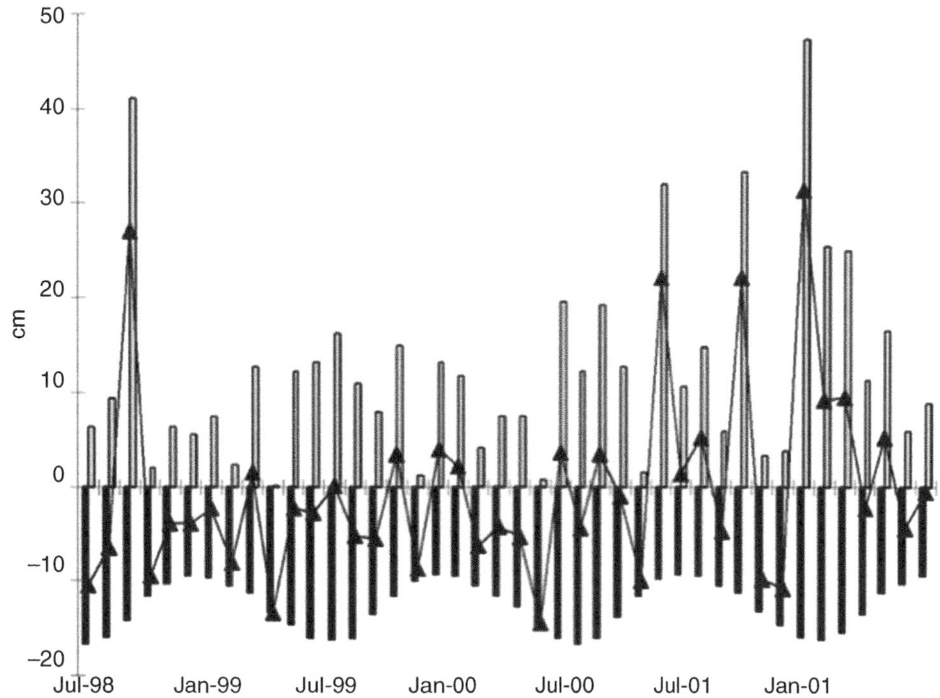

Fig. 8 Water budget for the Mandeville area during the drought of 1999–2000. *Black* bars are adjusted evapotranspiration, shaded bars are Mandeville precipitation and the *black* triangles indicate water surplus or deficit each month. Units are cm/month (from Brantley 2006)

Murray et al. 2011). Thus, there is the potential to sell carbon credits as a result of enhanced primary production due to wetland assimilation.

Acknowledgements Ecological baseline studies and monitoring at the different wetland assimilation sites were funded by the respective communities. A diversity of funding sources supported scientific studies on Maurepas and Pontchartrain wetlands, including EPA, Louisiana Dept. of Wildlife and Fisheries, Louisiana Dept. of Environmental Quality, and NOAA. JWD and RRL acknowledge that they carried out both ecological baseline studies and routine monitoring as employees of Comite Resources Inc., which received funding from the communities with assimilation projects.

References

Arc GIS 9 (2006): Using ArcGIS Desktop Redlands, CA, p 435

Barras JA (2006) Land area changes in coastal Louisiana after the 2005 Hurricanes: a series of three maps. U.S. Geological Survey Open-File Report -6–1274

Blahnik T, Day J (2000) The effects of varied hydraulic and nutrient loading rates on water quality and hydrologic distributions in a natural forested wetland. Wetlands 20:48–61

Blum MD, Roberts HH (2012) The Mississippi delta region: past, present, and future. Annu Rev Earth Planet Sci 40:655–683

Boesch DF, Josselyn MN, Mehta AJ, Morris JT, Nuttle WK, Simenstad CA, Swift DJ (1994) Scientific assessment of coastal wetland loss, restoration and management in Louisiana. J Coast Res, Special Issue No 2:1–103

Boshart WM (1997) The conservation and restoration of a baldcypress swamp: an investigation of macronurients, competition, and induced vegetation dynamics as related to nutria herbivory. Master's thesis. Southeastern Louisiana University, Hammond

Boustany RG, Crozier CR, Rybczyk JM, Twilley RR (1997) Denitrification in a South Louisiana wetland forest receiving treated sewage effluent. Wetl Ecol Manag 4:273–283

Brantley CG (2006) Nutrient interactions, plant productivity, soil accretion and policy implications of wetland enhancement in coastal Louisiana. PhD dissertation. Louisiana State University, Baton Rouge

Brantley CG, Day JW, Lane RR, Hyfield E, Day JN, Ko JY (2008) Primary production, nutrient dynamics, and accretion of a coastal freshwater forested wetland assimilation system in Louisiana. Ecol Eng 34:7–22

Burns AC (1980) Frank B. Williams, cypress lumber king. J For Hist 24:127–133

Chambers JL, Conner WH, Day JW, Faulkner SP, Gardiner ES, Hughes MS, Keim RF, King SL, McLeod KW, Miller CA, Nyman JA, Shaffer GP (2005) Conservation, protection and utilization of Louisiana's coastal wetland forests. Final report to the Governor of Louisiana from the Coastal Wetland Forest Conservation and Use Science Working Group. (special contributions from Aust WM, Goyer RA, Lenhard GJ, Souther-Effler RF, Rutherford DA, Kelso WE. p 121. Available from: Louisiana Governor's Office of Coastal Activities, 1051 N. Third St. Capitol Annex Bldg, Suite 138 Baton Rouge, LA 70802

Chambers JQ, Fisher JI, Eng HZ, Chapman EL, Baker DB, Hurtt GC (2007) Hurricane Katrina's carbon footprint on U.S. gulf coast forests. Science 16:1107

Coastal Protection and Restoration Authority of Louisiana (CPRA) (2017) Louisiana's comprehensive master plan for a sustainable coast. Coastal Protection Restoration Authority, Baton Rouge

Conner WH, Day JW (1988) Rising water levels in coastal Louisiana: implications for two coastal forested wetland areas in Louisiana. J Coast Res 4(4):589–559

Conner WH, Duberstein JA, Day JW, Hutchinson S (2014) Impacts of changing hydrology and hurricanes on forest structure and growth

along a flooding/elevation gradient in a south Louisiana forested wetland from 1986 to 2009. Wetlands 34(4):803–814
Costanza R, Mitsch WJ, Day JW (2006) A new vision for New Orleans and the Mississippi delta: applying ecological economics and ecological engineering. Front Ecol Environ 4(9):465–472
Danielsen F, Sorenson MK, Olwig MF, Selvam VF, Parish ND, Burgess T, Hiraishi VM, Karunagaran MS, Rasmussen LB, Hansen A, Quarto A, Suryadiputan N (2005) The Asian tsunami: a protective role for vegetation. Science 310:64
Davis DW (2000) Historical perspective on crevasses, levees, and the Mississippi River. In: Colten CE (ed) Transforming New Orleans and its environs. University of Pittsburgh Press, Pittsburgh
Day J, Shaffer G, Britsch L, Reed D, Hawes S, Cahoon D (2000) Pattern and process of land loss in the Mississippi delta: a spatial and temporal analysis of wetland habitat change. Estuaries 23:425–438
Day J, Shaffer G, Reed D, Cahoon D, Britsch L, Hawes S (2001) Patterns and processes of wetland loss in coastal Louisiana: a reply to Turner 2001. Estimating the indirect effects of hydrologic change on wetland loss: if the earth is curved, then how would we know it. Estuaries 24:647–651
Day JW, Kemp GP, Mashriqui HS, Dartez D, Lane RR, Cunningham R (2004a) Development plan for a diversion into the Maurepas swamp: water quality and hydrologic modeling components. Final Report. U.S. Environmental Protection Agency, Region 6, Dallas, p 197
Day J, Ko J-Y, Rybczyk J, Sabins D, Bean R, Berthelot G, Brantley C, Breaux A, Cardoch L, Conner W, Courville C, Day J, Englande A, Feagley S, Hyfield E, Lane R, Lindsey J, Mistich J, Mitsch W, Reyes E, Twilley R, Yáñez-Arancibia A, Zhang X (2004b) The use of wetlands in the Mississippi Delta for wastewater assimilation: a review. Ocean Coast Manag 47:671–691
Day JW, Westphal A, Pratt R, Hyfield E, Rybczyk J, Kemp GP, Day JN, Marx B (2006) Effects of long-term municipal effluent discharge on the nutrient dynamics, productivity, and benthic community structure of a tidal freshwater forested wetland in Louisiana. Ecol Eng 27:242–257
Day JW, Boesch DF, Clairain EJ, Kemp GP, Laska SB, Mitsch WJ, Orth K, Mashriqui H, Reed DJ, Shabman L, Simenstad CA, Streever BJ, Twilley RR, Watson CC, Wells JT, Whigham DF (2007) Restoration of the Mississippi Delta: lessons learned from hurricanes Katrina and Rita. Science 315:1679–1684
Day JW, Hunter R, Keim RF, DeLaune R, Shaffer G, Evers E, Reed D, Brantley C, Kemp P, Day J, Hunter M (2012) Ecological response of forested wetlands with and without large-scale Mississippi River input: implications for management. Ecol Eng 46:57–67
Day J, Cable J, Lane R, Kemp G (2016a) Sediment deposition at the Caernarvon crevasse during the great Mississippi flood of 1927: implications for coastal restoration. Water 8:1e12. https://doi.org/10.3390/w8020038
Day JW, Lane R, D'Elia C, Wiegman A, Rutherford J, Shaffer G, Brantley C, Kemp G (2016b) Large infrequently operated river diversions for Mississippi delta restoration. Estuar Coast Shelf Sci 135:106–115. https://doi.org/10.1016/j.ecss.2016.05.001
Delaune RD, Pezeshki SR (2003) The role of soil organic carbon in maintaining surface elevation in rapidly subsiding U.S. Gulf of Mexico coastal marshes. Water Air Soil Pollut 3:167–179
DeLaune RD, Jugsujinda A, Peterson GW, Patrick WH (2003) Impact of Mississippi River freshwater reintroduction on enhancing marsh accretionary processes in a Louisiana estuary. Estuar Coast Shelf Sci 58:653–662
DeLaune RD, Sasser CE, Evers-Hebert E, White JR, Roberts HH (2016) Influence of the Wax Lake Delta sediment diversion on aboveground plant productivity and carbon storage in deltaic island and mainland coastal marshes. Estuar Coast Shelf Sci 177:83–89
Doyle TW, Keeland BD, Gorham LE, Johnson D (1995) Structural impact of Hurricane Andrew on forested wetlands of the Atchafalaya Basin in south Louisiana. J Coast Res Spec Issue 21:354–364
Effler RS, Goyer RA, Lenhard GJ (2006) Baldcypress and water tupelo responses to insect defoliation and nutrient augmentation in Maurepas Swamp, Louisiana, USA. For Ecol Manag 236:295–304
Forder DR (1995) Timber harvesting, coppicing and artificial regeneration of a Cypress Muck Swamp. Master's Thesis, Southeastern Louisiana University, Hammond p 73
Greene MC (1994) The interacting effects of autogenic and Allogenic agents on selected swamp and Freshmarsh vegetation. Master's thesis, Southeastern Louisiana University, Hammond p 100
Gresham CA, Williams TM, Lipscomb DJ (1991) Hurricane Hugo wind damage to Southeastern U.S. coastal forest tree species. Biotropica 23:420–426
Hesse I, Doyle T, Day J (1998) Long-term growth enhancement of baldcypress (Taxodium distichum) from municipal wastewater application. Environ Manag 22(1):119–127
Hillmann E, Henkel T, Lopez J, Baker D (2015) Recommendations for restoration: Central Wetlands Unit, Louisiana; Lake Pontchartrain Basin Foundation, New Orleans, p 69
Howell SL (2003) Herbaceous aspect of a feasibility study concerning the Mississippi River diversion into the southern Lake Maurepas Swamp, Lake Pontchartrain Basin, Louisiana. Master's thesis, Southeastern Louisiana University, Hammond, p 70
Hunter RG, Day JW, Lane RR, Lindsey J, Day JN, Hunter MG (2008a) Impacts of secondarily treated municipal effluent on a freshwater forested wetland after 60 years of discharge. Wetlands 29:363–371
Hunter RG, Lane RR, Day JW, Lindsey J, Day JN, Hunter MG (2008b) Nutrient removal and loading rate analysis of Louisiana forested wetlands assimilating treated municipal effluent. Environ Manag 44:865–873
Hunter RG, Day JW, Shaffer GP, Lane RR, Englande AJ, Reimers R, Kandalepas D, Wood WB, Day JN, Hillman E (2016) Restoration and management of a degraded baldcypress swamp and freshwater marsh in Coastal Louisiana. Water 8:71–89
Hunter RG, Day JW, Lane RR, Shaffer GP, Day JN, Conner WH, Rybczyk JM (2017) Using natural wetlands for municipal effluent assimilation: a half-century of experience for the Mississippi River Delta and surrounding environs. In: Metcalf C (ed) Pollution abatement using natural and constructed wetlands. Springer Publishing Co. In press
Hyfield E, Day J, Mendelssohn I, Kemp P (2007) A feasibility analysis of discharging non-contact, once-through industrial cooling water to forested wetlands for coastal restoration in Louisiana. Ecol Eng 29:1–7
Kadlec RH, Wallace S (2009) Treatment wetlands. CRC Press, Boca Raton, p 1046
Keddy PA, Campbell D, McFalls T, Shaffer GP, Moreau R, Dranguet C, Heleniak R (2007) The wetlands of lakes Pontchartrain and Maurepas: origins, processes, and restoration. Environ Rev 15:43–77
Kemp GP, Mashriqui HS, Jones FW, Cunningham R (2001) Hydrologic modeling to evaluate the potential to divert Mississippi River water into the swamps south of Lake Maurepas. In: Diversion into the Maurepas Swamps: Complex Project Coastal Wetlands Planning, Protection, and Restoration Act. U. S. Environmental Protection Agency, Region Six, Dallas, TX, Report WA #5-02
Kesel RH, Yodis E, McCraw D (1992) An approximation of the sediment budget of the lower Mississippi River prior to major human modification. Earth Surf Process Landf 17:711–722
Kniffen FB (1935) Bayou Manchac: a physiographic interpretation. Geogr Rev 25(3):462–466

Ko J-Y, Day JW, Lane RR, Day JN (2004) A comparative evaluation of money-based and energy-based cost–benefit analyses of tertiary municipal wastewater treatment using forested wetlands vs. sand filtration in Louisiana. Ecol Econ 49:331–347

Krauss KW, Chambers JL, Allen JA, Soileau DM, DeBosier AS (2000) Growth and nutrition of baldcypress families planted under varying salinity regimes in Louisiana, USA. J Coast Res 16:153–163

Lake Pontchartrain Basin Foundation (n.d.) Saveourlake.org

Lane RR, Mashriqui HS, Kemp GP, Day JW, Day JN, Hamilton A (2003) Potential nitrate removal from a river diversion into a Mississippi delta forested wetland. Ecol Eng 20:237–249

Lane R, Day JW, Shaffer GP, Hunter R, Day JN, Wood W, Settoon P (2015a) Hydrology and water budget analysis of the East Joyce wetlands: past history and prospects for the future. Ecol Eng 87:34–44

Lane R, Huang H, Day JW, Justic D, DeLaune R (2015b) Water quality of a coastal Louisiana swamp and how dredging is undermining restoration efforts. Estuar Coast Shelf Sci 152:23–32

Lane RR, Day JW, Shaffer GP, Hunter RG, Day JN, Wood WB, Settoon P (2016) Hydrology and water budget analysis of the East Joyce wetlands: past history and prospects for the future. Ecol Eng 87:34–44

Lopez JA (2006) The multiple lines of defense strategy to sustain coastal Louisiana. White Paper, Lake Pontchartrain Basin Foundation, New Orleans, 21p

Louisiana Department of Environmental Quality (1988) Water quality management plan, vol 5 Part A

Louisiana Department of Natural Resources (LDNR) (1998) Coast 2050: toward a sustainable Louisiana. Louisiana coastal wetlands conservation and restoration task force. Louisiana Department of Natural Recources, Baton Rouge

LDNR (1998) Coast 2050: toward a sustainable coastal Louisiana. Louisiana Department of Natural Resources, Baton Rouge, p 161

Lundberg CJ (2008) Using secondarily treated sewage effluent to restore the baldcypress–water tupelo swamps of the Lake Pontchartrain Basin: a demonstration study. M.S. thesis, Southeastern Louisiana University. Hammond, p 71

Lundberg CJ, Shaffer GP, Wood WB, Day JW (2011) Growth rates of baldcypress (Taxodium distichum) seedlings in a treated effluent assimilation marsh. Ecol Eng 37:549–553

Mancil E (1980) Pullboat logging. J For Hist 24:135–141

Mcleod E, Chmura GL, Bouillon S, Salm R, Björk M, Duarte CM, Lovelock CE, Schlesinger WH, Silliman B (2011) A blueprint for blue carbon: toward an improved understanding of the role of vegetated coastal habitats in sequestering CO_2. Front Ecol Environ 9:552–560

Megonigal JP, Day FP (1988) Organic matter dynamics in four seasonally flooded forest communities of the Dismal Swamp. American J Bot 75(9):1334–1343

Minor A (2009) Forested freshwater wetland responses to secondarily treated municipal effluent discharge. M.S. thesis, Nicholls State University, Thibodaux , p 130

Morris JT, Shaffer GP, Nyman JA (2013) Brinson review: perspectives on the influence of nutrients on the sustainability of coastal wetlands. Wetlands 33:975–988

Murray BC, Pendleton L, Jenkins WA, Sifleet S (2011) Green payments for blue carbon: economic incentives for protecting threatened coastal habitats. Nicholas Institute for Environmental Policy Solutions, Report NI R 11-04

Myers RS, Shaffer GP, Llewellyn DW (1995) Baldcypress (Taxodium distichum (L.) Rich.) restoration in southeast Louisiana: relative effects of herbivory, flooding, competition, and macronutrients. Wetlands 15:141–148

Penland S, Ramsey KE, McBride RA, Mestayer JT, Westphal KA (1988) Relative sea level rise and delta-plain development in the Terrebonne parish region. Coastal geology technical report No. 4. Louisiana geological survey. Baton Rouge

Putz FE, Sharitz RR (1991) Hurricane damage to old-growth forest in Conaree Swamp National Monument, South Carolina, USA. Can J For Res 21:1765–1770

Roy ED, Smith EA, Bargu S, White JR (2016) Will Mississippi River diversions designed for coastal restoration cause harmful algal blooms? Ecol Eng 91:350–364

Rutherford, J (2017) Examining the benefits of a large, Intermittent river diversion into the Maurepas swamp. M.S. thesis, Louisiana State University, Baton Rouge

Rybczyk J, Day J, Conner W (2002) The impact of wastewater effluent on accretion and decomposition in a subsiding forested wetland. Wetlands 22(1):18–32

Saucier RT (1963) Recent geomorphic history of the Pontchartrain Basin. LSU Press, Baton Rouge, p 114

Shaffer, GP, Perkins TE, Hoeppner SS, Howell S, Bernard TH, and Parsons AC (2003) Ecosystem Health of the Maurepas Swamp: Feasibility and Projected Benefits of a Freshwater Diversion. Final Report. Environmental Protection Agency, Region 6, Dallas, Texas, 95p

Shaffer GP, Hoeppner SS, Gosselink JG (2005) The Mississippi River alluvial plain: characteriztion, degradation, and restoration. In: Fraser LH, Keddy PA (eds) The World's Largest Wetlands. Cambridge University Press, Cambridge, pp 272–315

Shaffer GP, Wood WB, Hoeppner SS, Perkins TE, Zoller JA, Kandalepas D (2009a) Degradation of Baldcypress – water tupelo swamp to marsh and open water in Southeastern Louisiana, USA: an irreversible trajectory? J Coast Res 54:152–165

Shaffer GP, Day JW, Mack S, Kemp GP, van Heerden I, Poirrier MA, Westpahl KA, FitzGerlad D, Milanes A, Morris C, Bea R, Penland PS (2009b) The MRGO navigation project: a massive human-induced environmental, economic, and storm disaster. J Coast Res 54:206–224

Shaffer GP, Day JW, Hunter RG, Lane RR, Lundberg CJ, Wood WB, Hillmann ER, Day JN, Strickland E, Kandalepas D (2015) System response, nutria herbivory, and vegetation recovery of an assimilation wetland in coastal Louisiana. Ecol Eng 79:120–131

Shaffer GP, Day JW, Kandalepas D, Wood WB, Hunter RG, Lane RR, Hillmann ER (2016) Decline of the Maurepas Swamp, Pontchartrain Basin, Louisiana, and approaches to restoration. Water 8:101–129

Shen Z, Tornqvist TE, Mauz B, Chamberlain EL, Nijhuis A, Sandoval L (2015) Episodic overbank deposition as a dominant mechanism of floodplain and delta-plain aggradation. Geology 43:875e878

Snedden GA, Cretini K, Patton B (2015) Inundation and salinity impacts to above- and belowground productivity in Spartina patens and Spartina alterniflora in the Mississippi River deltaic plain: implications for using river diversions as restoration tools. Ecol Eng 81:133e139

Törnqvist TE, Wallace DJ, Storms JEA, Wallinga J, van Dam RL, Blaauw M, Derksen MS, Klerks CJW, Meijneken C, Snijders EMA (2008) Mississippi Delta subsidence primarily caused by compaction of Holocene strata. Nat Geosci 1(3):173–176

Thornthwaite CW (1948) An approach toward a rational classification of climate. Geogr Rev 38:55–94

Touliatos P, Roth E (1971) Hurricanes and trees–ten lessons from Camille. J For 69(5):285–289

Trettin CC, Jorgensen MF (2003) Carbon cycling in wetland forest soils. In: Kimble JM et al (eds) The potential of U.S. forest soils to sequester carbon and mitigate the greenhouse effect. CRC Press, Boca Raton, pp 311–331

Turner RE (1997) Wetlands loss in the northern Gulf of Mexico: multiple working hypotheses. Estuaries 20:1–13

Turner RE, Rao YS (1990) Relationships between wetland fragmentation and recent hydrologic changes in a deltaic coast. Estuaries 13:272–281
Turner RE, Costanza R, Schaife W (1982) Canals and wetland erosion rates in coastal Louisiana. In: Boesch D (ed). In: Proceedings of the conference on coastal erosion and wetland modification in Louisiana: causes, consequences, and options. FWS/OBS-82/59, Office of Biological Services, US Fish and Wildlife Service, Slidell
Turner RE, Swenson EM, Lee JM (1994) A rationale for coastal wetland restoration through spoil bank management in Louisiana, USA. Environ Manag 18:271–282
Vyzamal J (2007) Removal of nutrients in various types of constructed wetlands. Sci Total Environ 380:48–65
Williams K, Pinzon ZS, Stumpf RP, Raabe EA (1999). Sea-level rise and coastal forests on the Gulf of Mexico. USGS-99-441, p 63
Xu YJ, Wu K (2006) Seasonality and interannual variability of freshwater inflow to a large oligohaline estuary in the Northern Gulf of Mexico. Estuar Coast Shelf Sci 68:619–626
Zhang X, Feagley S, Day J, Conner W, Hesse I, Rybczyk J, Hudnall W (2000) A water chemistry assessment of wastewater remediation in a natural swamp. J Environ Qual 29:1960–1968

Energy and Climate – Global Trends and Their Implications for Delta Restoration

Jeffrey S. Rutherford, Adrian R.H. Wiegman, John W. Day, and Robert R. Lane

Abstract

Climate change and energy scarcity pose challenges for the sustainability of the Mississippi River Delta and its future restoration. Projected trends for climate change suggest increasing risk in the coastal zone from sea-level rise, more frequent high-intensity hurricanes, and increased Mississippi River discharge. Simultaneously, analysis of energy return on investment suggests that energy is becoming more expensive and difficult to produce, which does not bode well for energy intensive activities like building river levees and pumping dredged sediment. For these reasons, while coastal restoration is becoming progressively urgent, its implementation will be both challenging and expensive. The objective of this chapter is to introduce the basic science behind climate change and energy, and discuss its relevance to the Mississippi River Delta. The evidence presented here makes it clear that the current management of the delta is unsustainable, and an aggressive new approach based on the natural functioning of the delta is required in the future.

Keywords

Climate change • Sea-level rise • Fossil fuels • Renewable energy • Energy return on investment • Ecological engineering

Introduction

In this chapter, we discuss two long-term trends that pose major threats to the sustainability of the Mississippi River Delta (MRD): climate change and energy scarcity. Climate change threatens the sustainability of the MRD for a number of reasons. Accelerating eustatic (global) sea-level rise (SLR) will exacerbate the growing difference between wetland vertical accretion and relative SLR (sum of eustatic SLR and subsidence). Energy scarcity threatens the economic sustainability of the MRD. Humans have been able to maintain habitability of the MRD with massive investments in built capital, but as inexpensive fossil energy becomes increasingly scarce, rising energy prices will make building and maintaining infrastructure increasingly more expensive. This chapter provides an introduction of the scientific information about climate change and energy scarcity. While the effects of climate change on deltaic management is relatively well understood, the importance of energy is less so (e.g. Day et al. 2016; Tessler et al. 2015). This chapter synthesizes the natural systems based worldview through which we believe that sustainable MRD management must be designed. The topics covered here underpin the subsequent sections on restoration and resilience of ecosystems and communities in the MRD.

J.S. Rutherford (✉) • A.R.H. Wiegman • J.W. Day • R.R. Lane
Department of Oceanography and Coastal Sciences, Louisiana State University, Baton Rouge, LA 70803, USA
e-mail: Jruthe@stanford.edu; adrian.wiegman@gmail.com; johnday@lsu.edu; rlane@lsu.edu

J.W. Day, J.A. Erdman (eds.), *Mississippi Delta Restoration*,
Estuaries of the World, https://doi.org/10.1007/978-3-319-65663-2_6

Climate Change

Few topics in recent science have sparked such heated debate and controversy as climate change. The debate has become increasingly polarized, and the perception of climate risk is increasingly connected with one's political stance or worldview (Hornsey et al. 2016). Despite the overwhelming scientific evidence (IPCC 2014), there is a large proportion of the public that doubts the validity for human-induced climate change. Unfortunately, with each passing year the stakes are raised and the costs of inaction grow. Mounting evidence indicates that human society is pushing the "climate system toward abrupt, unpredictable, and potentially irreversible changes with highly damaging impacts" (Molina et al. 2014). Here we summarize the compelling scientific evidence for climate change and its expected impacts for Louisiana, where coastal wetlands are already being influenced by high rates of subsidence and pervasive hydrologic change (Day et al. 2000, 2007, 2016; Kolker et al. 2011; Syvitski et al. 2009; Turner 1997). Relative SLR projections indicate that most of Louisiana's coastal zone will be underwater by 2100 if present trends continue and no restorative action is taken (Blum and Roberts 2012).

What Is Climate Change?

The current debate notwithstanding, Earth's climate has always changed without human influence. Climate refers to the average atmospheric conditions over a *long period of time* (Kaufmann and Cleveland 2008). Figure 1 demonstrates how the past 500,000 years have witnessed repeated ice ages separated by warmer interglacial periods (Petit et al. 1999).

Changes in the Earth's climate are caused by changes in its heat balance. The primary source of energy to Earth's surface is electromagnetic radiation from the Sun. A number of factors influence how much energy is captured by the Earth's surface and atmosphere. These include changes in the intensity of solar insolation, concentration of greenhouse gases (GHGs), and changes in the albedo of the Earth (snow and ice cover). Historically, the regularity of ice ages and interglacial periods were related to the Milankovitch cycles, which describe the periodic changes in orbital shape, wobble in the Earth's axis of rotation, and tilt of the Earth's axis which influence the intensity of solar insolation reaching Earth's surface. The landmark paper by Hays et al. (1976), describes these changes in Earth's orbital geometry as the "pacemaker of the ice ages".

Anthropogenic Effects – Global Warming and the Greenhouse Effect

The connection between CO_2 concentration and climate was made clear partly by measurements from Antarctic ice cores that illustrated a striking correspondence between CO_2 concentration and temperature during glacial cycles (Fig. 1) (Petit et al. 1999). Changes in solar irradiation due to orbital shifts created powerful feedbacks where changes in ocean chemistry, ocean circulation, and biological activity adjusted the CO_2 concentration in the atmosphere. The mechanism by which CO_2 concentration is connected to temperature change is the greenhouse effect. High energy, short wavelength radiation enters the atmosphere, some of this energy is absorbed by the Earth's surface (the rest is absorbed or reflected by the atmosphere), and the warmed Earth radiates low energy, longer wavelength radiation back towards

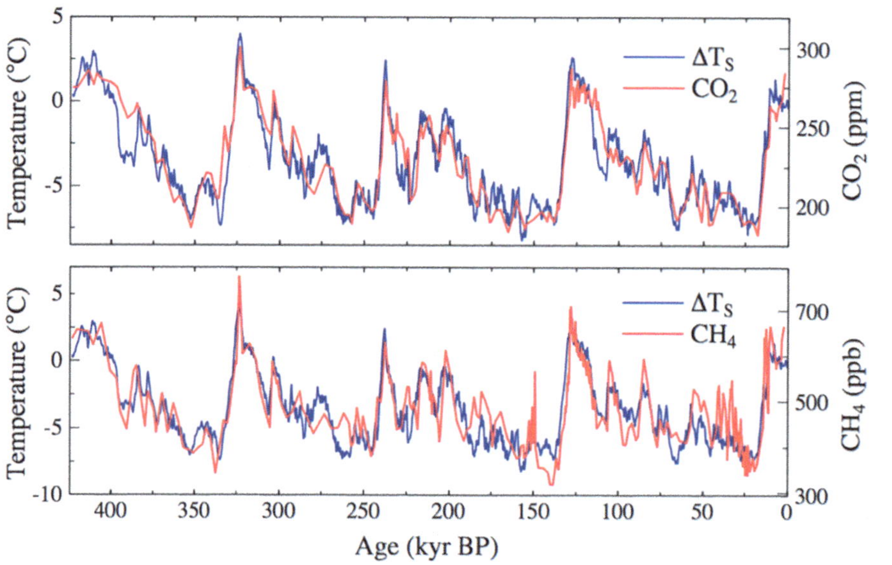

Fig. 1 Global temperature, CO_2, and CH_4 levels based on ice cores taken at Vostok, Antarctica. It is important to note that the current concentration of CO_2 is over 400 ppm even though the highest value shown on the graph is 280 ppm (From NASA, using data modified from Petit et al. 1999 and Vimeux et al. 2002 as described by Hansen and Soto 2004)

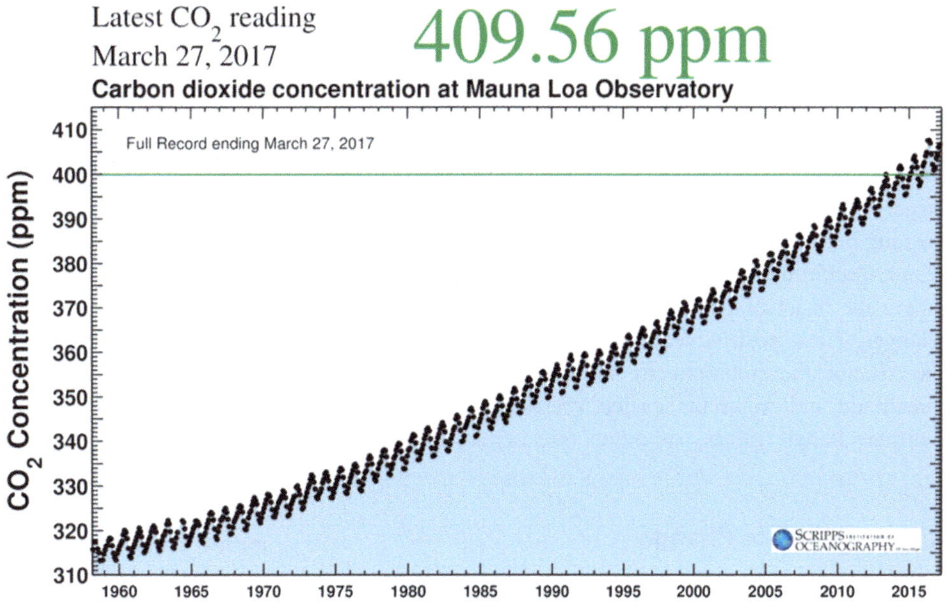

Fig. 2 Atmospheric CO_2 concentration measured at the Mauna Loa Observatory in Hawaii (Source: Scripps Institution of Oceanography, University of California San Diego, https://scripps.ucsd.edu/programs/keelingcurve/)

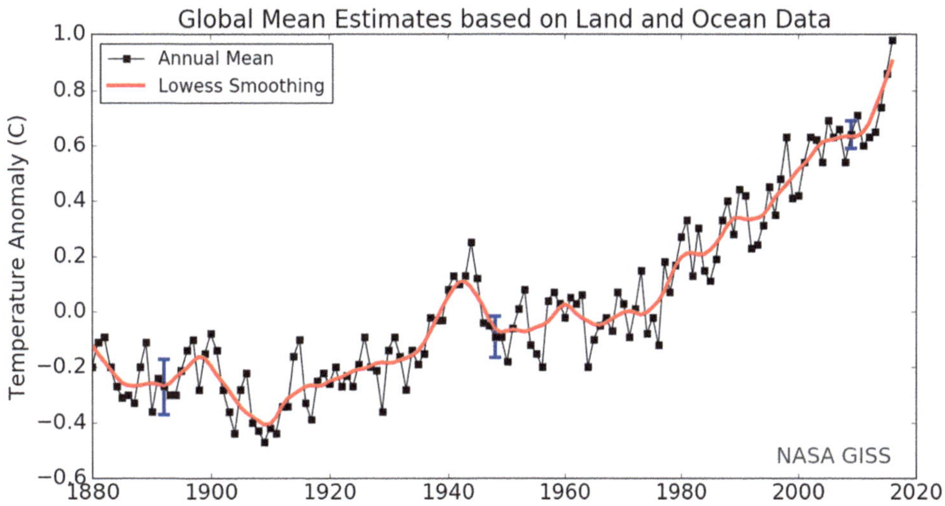

Fig. 3 Land-ocean temperature anomaly, 1880 to present, relative to the average from 1951–1980. The solid black line is the global annual mean and the solid red line is a smoothed data fit (lowess smoothing). The blue uncertainty bars (95% confidence limit) account only for incomplete spatial sampling. (From NASA: GISTEMP Team 2017; using data updated from Hansen et al. 2010)

space. But certain atmospheric gases (GHGs, such as CO_2) are capable of absorbing this long wavelength radiation emitted by the Earth and convert it to heat. The greenhouse effect describes the ability of the atmosphere to absorb energy with longer wavelengths and warm the Earth's surface. Historically, the Milankovitch orbital changes initiated the feedback loops and changes in GHG levels lagged a change in temperature by a few centuries. Today, the situation is different. Humans have accelerated the addition of gases to the atmosphere, and changes that once took centuries are now taking place in decades.

The scale of CO_2 emissions into the atmosphere from human activities has been high enough to initiate significant changes in the Earth's climate. Prior to the industrial revolution, atmospheric CO_2 concentrations were about 280 ppm (in 1750; IPCC 2014). Currently, the CO_2 concentration in the atmosphere exceeds 400 ppm and is rising (Fig. 2). The increased carbon in the atmosphere has come from burning of carbon stored in Earth's crust (coal, oil, natural gas, etc.), land use change which has reduced carbon stored in terrestrial biota, and methane emitted from industrial agriculture. This increase in CO_2 concentration has coincided with an increase of 0.85 °C in the global average temperature of the Earth (between 1880 and 2012, IPCC 2014) (Fig. 3).

There is a broad consensus in the scientific community that human activity is affecting global climate (IPCC 2014; Oreskes 2004). In the Fifth Assessment Report (AR5), the Intergovernmental Panel on Climate Change (2014) reported:

> Warming of the climate system is unequivocal, and since the 1950s, many of the observed changes are unprecedented over decades to millennia. The atmosphere and ocean have warmed, the amounts of snow and ice have diminished, sea-level has

risen, and the concentrations of greenhouse gases have increased... It is extremely likely that human influence has been the dominant cause of the observed warming since the mid-20th century.

The IPCC makes it clear that climate change is driven by an increasing concentration of GHGs in the atmosphere, largely related to economic and population growth and the ensuing burning of fossil fuels. Still uncertain is the point at which the Earth's system will (or has) reach a tipping point, where the onset of positive feedbacks initiates irreversible changes. These positive feedbacks could include collapse of the Atlantic thermohaline circulation, critical thresholds in Greenland and Antarctic sea ice, dieback of Amazon and Northern boreal forests, and others (see Lenton et al. 2008).

Risks of Climate Change

Temperature Rise

The IPCC (2014) publishes predictions of global temperature increase for various different scenarios of GHG emissions. These different scenarios are called Representative Concentration Pathways (RCP), and are based on different socioeconomic assumptions such as economic activity, energy sources, and population growth (Moss et al. 2008). Predictions range from as low as 0.3 °C of warming by 2100 for RCP2.6, and as high as 4.8 °C for RCP8.5. However, Friedlingstein et al. (2014) reported that CO_2 levels in the atmosphere are increasing at rates consistent with the highest IPCC scenarios, indicating that temperature change will likely also be on the high end.

A lack of surface temperature increase between 1998 and 2008 that has been called the global warming "hiatus" raised confusion and even prompted some to argue that the climate is cooling (Svensmark 2009). However, Kaufmann et al. (2011) reported that an increase in sulfur particles offset the rising GHG concentration. Church et al. (2011) calculated the Earth's total heat budget and showed that, while surface warming stagnated between 1998 and 2008, the total climate system continued to accumulate heat. Karl et al. (2015) conducted an analysis of surface temperatures and reported that global temperature trends are higher than reported by the IPCC. These analyses do not support the idea of a slowdown in the increase in global temperatures.

Sea-Level Rise

Rising global temperature contributes to eustatic SLR. Eustatic SLR is a global response in sea-level, primarily due to glacial mass loss and ocean thermal expansion (IPCC 2014). The historic rate of eustatic SLR can be constructed from a global network of tide gauges and more recent satellite altimetry data. Church and White (2011) assessed global average eustatic SLR between 1900 and 2009 and found a linear trend of 1.7 mm/year. By fitting a quadratic to the time series, Church and White (2011) also found a statistically significant acceleration in global sea-level of 0.009 mm/year^2. The most up to date measurement from the CU Sea Level Research Group in Colorado estimates the current rate of eustatic SLR at 3.4 mm/year (Nerem et al. 2010, http://sealevel.colorado.edu/).

Predicting the rate of future eustatic SLR has been done in two ways: using mechanistic, process based models, and using semi-empirical approaches (see Jones 2013 for a discussion of both). Process based modelling is typically more conservative since it attempts to characterize the physics of every factor. In their Fourth Assessment Report (AR4), for example, IPCC researchers added up all contributing physical processes and could only describe 60% of the change in sea-level between 1961 and 2003 (Jones 2013). Semi-empirical models, on the other hand, do not require an understanding of why a process occurs, but instead attempt to model what will occur in the future based on trends observed in the past. Vermeer and Rahmstorf (2009) used past relations between temperature and SLR to project sea-level response to future temperature change. The IPCC (2013) contended that "Despite the successful calibration and evaluation of semi-empirical models against the observed twentieth century sea level record, there is no consensus in the scientific community about their reliability, and consequently low confidence in projections based on them". Both process based and semi-empirical modelling, however, predicted that the rate of eustatic SLR will continue to accelerate due to global warming (IPCC 2014; Jevrejeva et al. 2012; Kopp et al. 2016; Parris et al. 2012; Pfeffer et al. 2008; Rahmstorf 2007; Vermeer and Rahmstorf 2009).

IPCC (2014) predicted that eustatic SLR will be about 26–55 cm in the best-case scenario (RCP2.6) and 45–82 cm in the worst-case scenario (RCP8.5) by the end of the twenty–first century. Despite the fact that projected eustatic SLR is significantly higher in the AR5 compared to the AR4, semi-empirical researchers still contested that the estimates are too conservative (Hausfather 2013). Most semi-empirical models project eustatic SLR higher than the IPCC AR5 projections (Jevrejeva et al. 2012; Rahmstorf 2007; Vermeer and Rahmstorf 2009). Vermeer and Rahmstorf (2009), for example, predicted 75–190 cm of eustatic SLR for the period 1990–2100. A semi-empirical model developed by Kopp et al. (2016) claims to "largely reconcile previous differences between semi-empirical twenty–first century GSL (global sea level) projections and the process model-based projections summarized in the [IPCC]'s [AR5]" with projections of 52–131 cm under RCP8.5. Given the substantial uncertainty that exists in predicting eustatic SLR, projects affected by SLR should consider both models. The NOAA National Climate

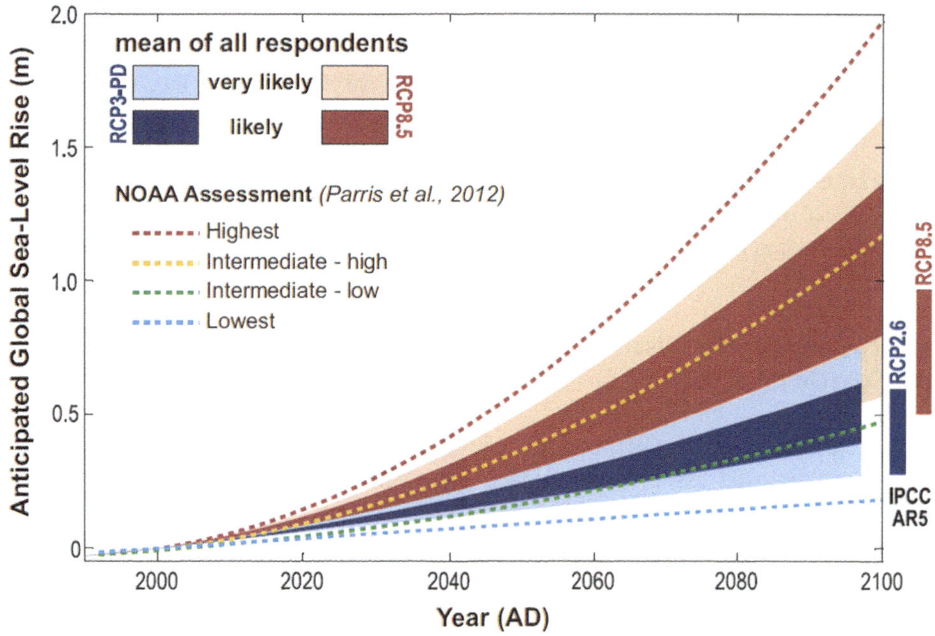

Fig. 4 Projections for eustatic SLR (1990–2100) comparing IPCC AR5 projections (vertical bars on right) with NOAA published estimates (dotted lines) (Parris et al. 2012) and survey response of ninety sea-level experts (thick shaded lines) (Horton et al. 2014). NOAA published estimates and expert opinion both predict greater SLR compared to the IPCC AR5 (Source: Horton et al. 2014)

Assessment (Parris et al. 2012) assigns plausible lower and upper limits of 0.2 and 2.0 m, where the upper limit considers the maximum possible glacier and ice sheet loss (Pfeffer et al. 2008). An expert elicitation by Horton et al. (2014) found that the general opinion of SLR experts falls between IPCC projections and the highest projections by semi-empirical models (Fig. 4).

Recently, there has been increasing concern over the vulnerability of the West Antarctica and Greenland ice sheets to global warming (Khan et al. 2014; Rignot et al. 2011, 2014; Tollefson 2017), which might suggest future rates of eustatic SLR closer to the upper end of projections. DeConto and Pollard (2016) report that if CO_2 emissions continue unabated, Antarctic contribution to sea-level rise may add an additional meter resulting in a total eustatic SLR of more than 2 m by 2100. It is worth noting that during the last interglacial period, about 115,000–130,000 years ago when global temperatures were only slightly warmer than today, the sea-level rose 3–5 m. This has been attributed to subsurface ocean warming and rapid ice loss in Antarctica (Hansen et al. 2015).

Tropical Storms and Hurricanes

While the IPCC (2014) has expressed low confidence in a long-term trend in tropical storm activity, it is noted by the IPCC and others that tropical cyclones have become more frequent in the North Atlantic and globally (Webster et al. 2005). Emanuel (2005) reported that a 1 °C increase in sea surface temperature in the tropics coincided with an 80% increase in total hurricane power. Hoyos et al. (2006) analyzed multiple factors contributing to hurricane intensity and concluded that the strong increasing trend in number of category 4 and 5 storms is directly linked to the increase in sea surface temperature between 1970 and 2004. Knutson et al. (2010) reported that the frequency of tropical cyclones will decrease or remain unchanged, but the frequency of the most intense hurricanes will increase. Mei et al. (2015) reported that typhoon intensity has increased in the Pacific region and predicted climate change will increase average typhoon intensity in the Pacific area by 14% by 2100.

Increase in Mississippi River Discharge

Tao et al. (2014) modeled the interactive effects of climate change, land use, and increased CO_2 levels in the atmosphere on discharge of the Mississippi River. Their results suggest that river discharge will increase by as much as 10–60% during this century. Such large increases may compromise the flood control system on the Mississippi River in times of increasing energy scarcity (Kemp et al. 2014). Large episodic diversions could reduce this threat especially if placed strategically to also offer hurricane protection.

Other Extreme Weather Events

Evidence indicates that the intensification of extreme weather events will continue in a warming climate (Coumou and Rahmstorf 2012; IPCC 2014). Most intuitively, a shift towards warmer conditions will increase evaporation and surface drying, therefore increasing drought events (Dai 2011). Warmer air also holds more moisture; heavy precipitation events, and consequently flooding, are expected to increase in intensity with climate change (Groisman et al. 2005; Min et al. 2011; Pall et al. 2011; Prein et al. 2016). Several recent examples are available for South Louisiana.

In 2000–2001, an extreme drought raised salinities in western Lake Pontchartrain and Lake Maurepas from an average of 2–3 psu to 10–12 psu and led to mortality of cypress over a wide area (Shaffer et al. 2016). Such droughts, in combination with increasing sea level and strong hurricanes, will lead to salinity stress on broad areas of freshwater wetlands. In August, 2016, nearly a meter of rain in three days led to extensive flooding east of the Mississippi River.

Implications for the Mississippi River Delta and Coastal Louisiana

Climate change affects the MRD both directly and indirectly through the drainage basin. Direct impacts are related to relative SLR and storm surge from hurricanes. In the MRD, increasing eustatic SLR occurs in addition to an extremely high rate of subsidence (an average of 9.4 mm/year for the lower Mississippi River Basin) (Zou et al. 2015), which produces relative SLR among the highest in the world. Blum and Roberts (2009) conclude that, even if sediment loads are restored to the basin, loss of land is inevitable given that relative SLR is about three times larger than that during delta-plain construction. Basin level impacts are related to precipitation which also affects river discharge. Increased precipitation and river discharge threaten flooding of low lying areas and the integrity of the Mississippi River control system (Tao et al. 2014).

Under threat is significant population and infrastructure. For example, in New Orleans 47% of the population lives below sea-level (Campanella 2007). SLR is already affecting communities, so much so that residents of a coastal Louisiana community are being called the first American "climate refugees" (Davenport and Robertson 2016). $48 million of federal funding has been allocated to aid the relocation of the Native American population of Isle de Jean Charles, who have lost 98% of their land to open water (www.coastalresettlement.org/). Significant infrastructure is located on the coast that provides important services for the entire country. Port Fourchon supports 18% of the nation's oil and gas production (Carter et al. 2014) and the Port of South Louisiana is the largest commodity port by tonnage in the United States (AAPA 2015). The population and infrastructure is protected from hurricane storm surge and river floods by a flood control system of levees and pumps and the deteriorating swamps and marshes of the delta. Hurricanes Katrina and Rita demonstrated in a devastating way how fragile the human engineered structures are to the forces of nature (Day et al. 2007). The risks from hurricane events will only worsen with climate change. Figure 5 presents flood risks for coastal Louisiana 50 years in the future from a 100-year storm event.

The heightened level of risk from climate change is already evident. As mentioned above, during the writing of this chapter a "1000-year flood" struck Baton Rouge, Louisiana. Extraordinary levels of precipitation broke records for rainfall and local river levels in a number of areas (Ferris 2016; Nobles 2016). Places that had never flooded before flooded, despite being on what owners believed to be high enough ground. Disasters without precedent will be the new normal with climate change. During the recovery process, there is a heavy reliance on federal funding and volunteer groups. As disasters occur simultaneously across the country (wildfires ravaged California as Baton Rouge recovered from flooding), the capacity of the emergency management system becomes strained (Montano 2016). Resilience does not mean a perpetual state of response and recovery. With accelerating climate change, there is an urgent need for change to the current system, where the focus is on mitigation and adaptability rather than response. These types of changes will be detailed in the rest of this book. The next section of this chapter presents another long-term trend adding pressure to the MRD management system.

Energy

Energy is the measure of the ability to do work or cause change. Throughout history, humans have achieved work through the controlled use of energy obtained from natural resources. Before the industrial revolution, the sun provided most energy for food, fodder, and wood fuel. Today, most energy is provided by fossil fuels. From propane for crawfish boils to electricity for football stadiums, energy is needed to power just about everything humans do in the economy.

It is through price that energy is connected to the rest of the economy, and serious economic and societal displacements occur when the energy system is disrupted. Examples of disruptions to the energy supply include the 1973 Arab oil embargo and the 2005 hurricane season in Louisiana, which both resulted in spikes in retail gasoline prices. When energy prices are high, high costs for transportation influence other critical goods like food and other fossil fuels. This affects individual discretionary income, business profitability, and government finances. In general, economic performance is significantly related to the availability and cost of energy, and each major U.S. recession in the late twentieth century is linked to oil price spikes (Aucott and Hall 2014; Hamilton 2009). Low oil prices also have significant negative impacts on the economy by stressing the profit margins of oil producers. The drop in oil prices from over $100 per barrel in 2014 to around $30 per barrel in 2016 demonstrated that low oil prices may be good for the welfare of consumers, but can wreak havoc on the economies of oil

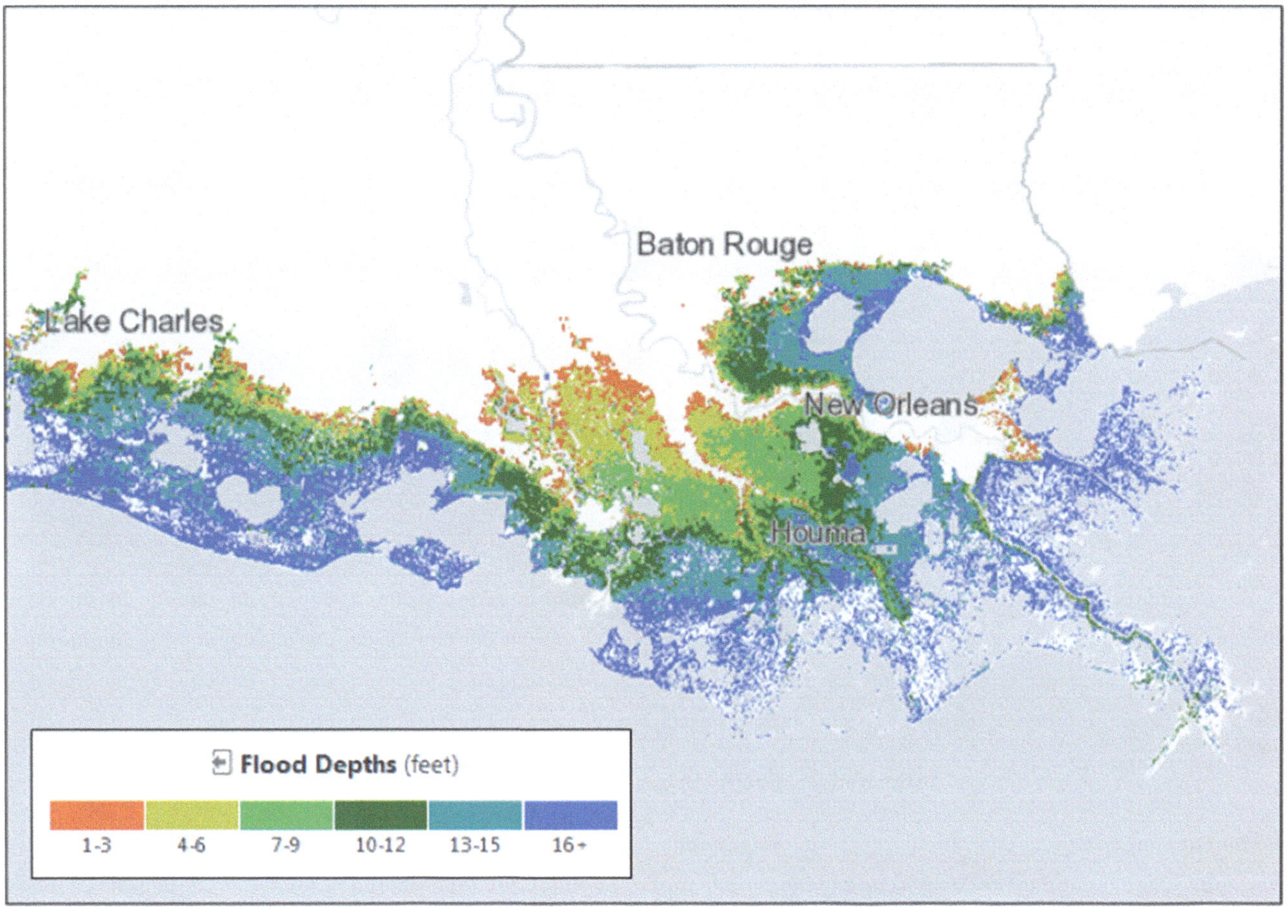

Fig. 5 Projected flood depths for a 100-year storm event 50 years in the future under the CPRA 2017 CMP high scenario. The high scenario considers the most severe possible eustatic SLR, subsidence, storm intensity, and storm frequency (CPRA 2017)

producing nations like Russia and Venezuela and states/provinces like Louisiana and Alberta, Canada.

It is apparent that abundant, useful energy is a prerequisite for the foundation, growth and maintenance of modern society, and therefore ought to be a concern for everyone. The objective of this section is to evaluate the prospects for a future energy supply, with special consideration given to the associated cost of obtaining energy. We will also explore the implications of energy supply for coastal Louisiana.

World Energy Production and Consumption

A good place to begin in predicting future energy supply is historical patterns of energy production. Global annual energy consumption was about 550 exajoules in 2015, a 3.5-fold increase from 1965 (Fig. 6). During this same period, a remarkable acceleration in other activities such as fertilizer consumption, urban migration, and water usage occurred (Steffen et al. 2015). Today, fossil fuels (oil, coal, natural gas) are overwhelmingly the most important energy source for the world. Renewable energy, which includes hydropower, biomass and waste, solar, and wind, makes up slightly more than 10% of the total energy supply. The "new" renewables, wind, solar, and some new forms of hydropower (wave, tidal, and hydrothermal), which are advocated by many as the solution to climate change and energy security, currently account for less than 2% (British Petroleum 2016).

As shown in Fig. 7, most primary energy sectors are still heavily reliant on fossil fuels. Oil is the primary fuel for transportation (over 90%), and oil derivatives are used as feedstocks for nonfuel products such as plastics, pharmaceuticals, fertilizers, and pesticides. Natural gas is used for heating and to generate electricity and is also used as a feedstock for fertilizers and plastics. Coal is used to produce electricity and for metal smelting.

Many in the energy industry project demand for liquid fuels to remain at or above current levels at least until the end of the century (Bentley 2016), in part due to the lack of suitable substitutes for liquid fuels for transportation. Oil is lightweight, energy dense, transportable, and relatively safe. Increasing demand for fossil fuels will also be driven by a rapidly growing population that is projected to reach 9–-13 billion by the end of the century (UN 2015), with the

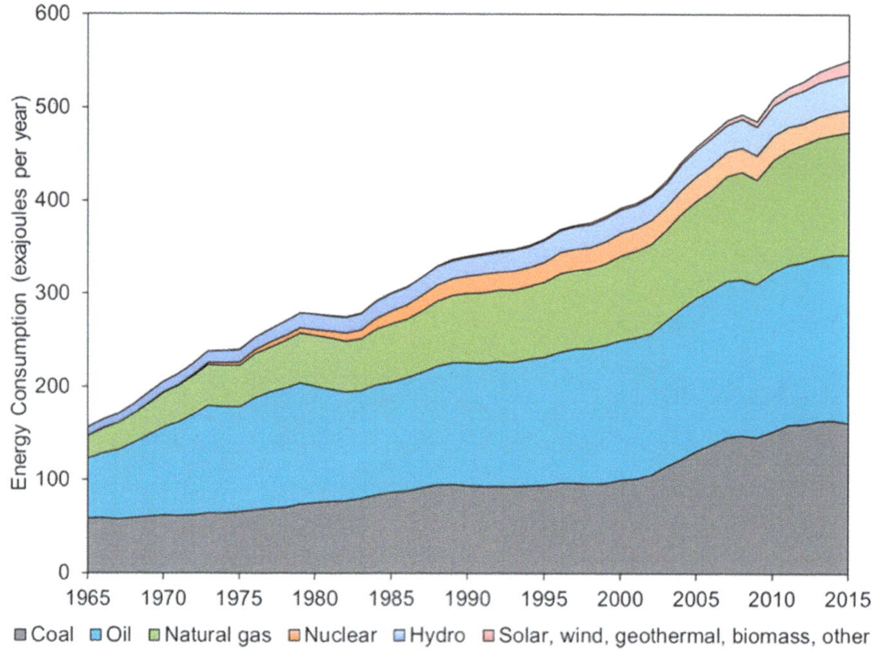

Fig. 6 Total world energy consumption by energy source (data from British Petroleum 2016)

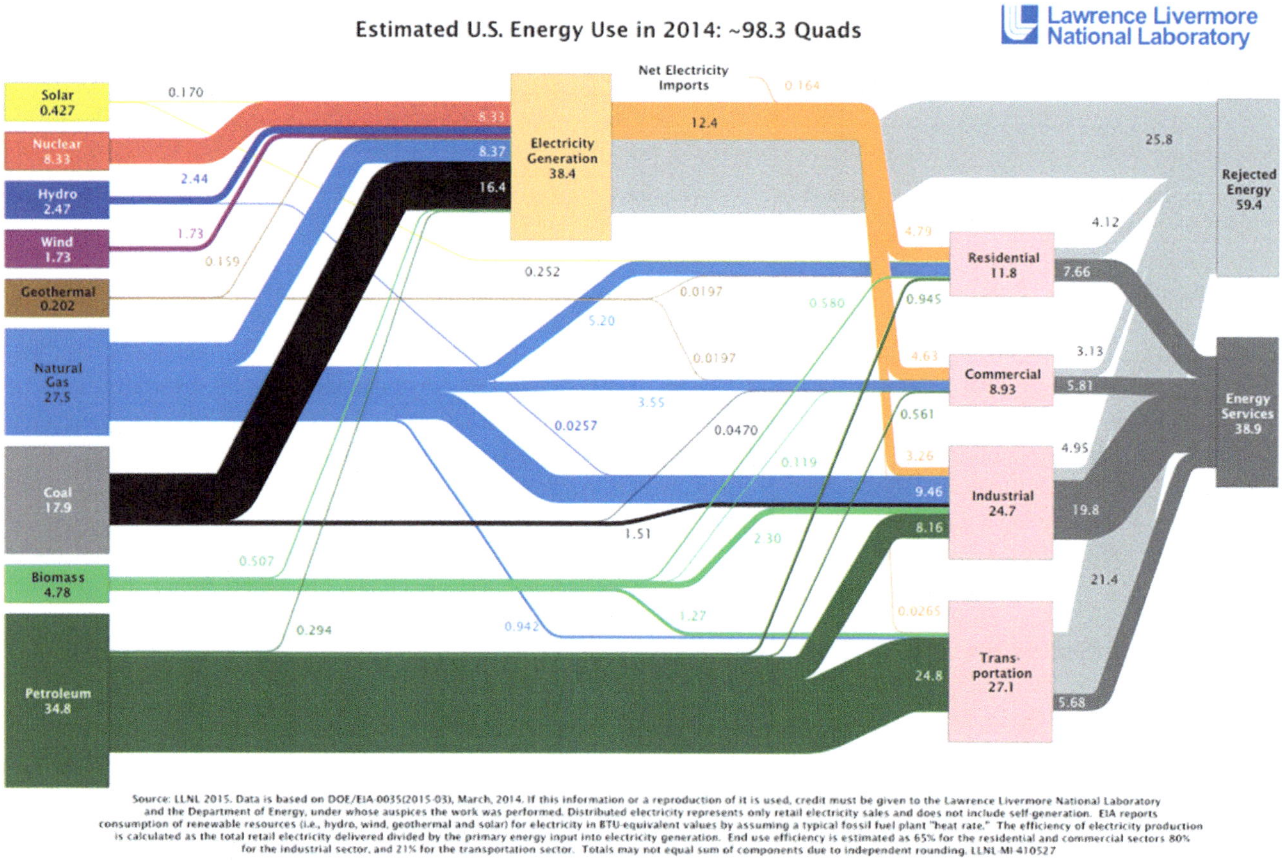

Fig. 7 A detailed "Sankey diagram" showing energy flow for the U.S. in 2014 (Lawrence Livermore National Laboratory 2016)

major portion of this growth in African and Asian countries that aspire to provide their citizens higher standards of living. For example, the number of cars in China increased by over 50 times between 1985 and 2016 (NBSC 2016). Such growth in developing countries will put pressure on future supplies of fossil fuels to meet demand, especially liquid fuels.

Future Production of Fossil Fuels

Governments, industry, and academics have developed models to forecast the future energy trajectory. The emphasis is on fossil fuels, as fossil fuels will remain ubiquitous across our economy for many years to come. The 'best-first' principle states that in an economy the easiest resources to find and extract will be developed first, meaning that over time finite resources like fossil fuels will become increasingly expensive and difficult to produce (Ricardo 1891). This raises the question of how long society can afford to maintain fossil fuel production. Geologists have developed forecasting approaches to shed light on this dilemma. There are two aspects to forecasting fossil fuel supply: the *availability*, or quantity of a resource, and the *accessibility*, or difficulty of extracting the resource in various forms and locations. Modelling *availability* involves predicting how much energy can be produced in each year, and if the rate of production can meet demand. Peak production refers to the point at which production reaches a maximum and then declines due to limitations of resource *availability* (Bentley 2016). Resource production models utilize a range of methods in determining the peak of global fossil fuel production, with some models having detail down to the drilling and production of individual wells (McGlade 2014). Because there is an associated cost with obtaining energy, and it is not just enough to have energy available, the *accessibility* of our future energy supply is a serious consideration as well. The *accessibility* of energy can be defined using energy return on investment (EROI), the ratio of energy outputs to energy inputs (Hall et al. 2014; Hall 2017). EROI is analogous to ROI (return on investment) and related to profit (sales minus expenditures) in finance. EROI helps to explain the quality of the resource and the effectiveness of the technology used to exploit the resource. EROI has yet to be incorporated in energy modelling on a wide scale.

Resource Availability

One method used in estimating fossil fuel availability is the aggregate approach, which models the behavior of global or regional production using a general function (Bentley 2016). The well-known logistic model developed by M. King Hubbert falls under this category (Hubbert 1956). With this approach, the early production history of the reservoir and an estimate of the ultimate recoverable resource (URR) are used to forecast the time of peak production and total oil production. For example, Maggio and Cacciola (2012) used a Hubbert approach to forecast total fossil fuel production based on global estimates of URR. Peak production is projected at about 2025 for oil, 2040 for natural gas, 2055 for coal, and 2035 for all fossil fuels (Fig. 8). Similarly, other authors used this method to forecast the production of many metals peaking prior to 2050 (Sverdrup et al. 2015).

A different type of model, using mechanistic rather than empirical techniques, is being used by energy companies and public institutions like the International Energy Agency (IEA). The IEA uses the TIMES Integrated Assessment Model to simulate the response of energy markets to GDP and population growth and publishes forecasts for the energy industry in the World Energy Outlook (WEO) (IEA 2015). The WEO presents scenarios that test the influence of energy and climate policy on production. The IEA projects that production will continue to increase into the future, but increasing energetic costs of extraction and increasing energy prices are not included as limiting factors of GDP growth (IEA 2015; McGlade 2014). EROI, which describes the energetic cost of a resource, will be discussed next.

Resource Accessibility

EROI is the ratio of energy outputs to energy inputs, and is a measure of energy *accessibility*, or how easily energy is obtained. The greater the EROI, the smaller the relative cost of production of the energy resource, and the greater the benefit to society. The profound implication of EROI is that it is the net energy that remains after investments, rather than the gross energy, that it is of value to society (Odum 1973). Because a certain amount of a society's energy is required for basic needs and maintenance of infrastructure, it is the surplus beyond this energetic cost that is reinvested into new capital production, or economic growth. There is a positive correlation between EROI and the production cost of oil (King and Hall 2011). As EROI decreases the price of energy goes up and the portion of energy available for reinvestment and GDP growth decreases. Historically, GDP growth in the US was positive only if the portion of GDP allocated to energy was less than 11%, which corresponds to an EROI of society of about 10:1 (Fizaine and Court 2016). If EROI of US energy falls below this threshold, it is questionable whether GDP growth could occur and a modern industrial society (with energy intensive coastal protection and restoration, for example) could be maintained.

EROI analysis indicates that there are diminishing returns on investments for energy resources (Cleveland 2005; Hall

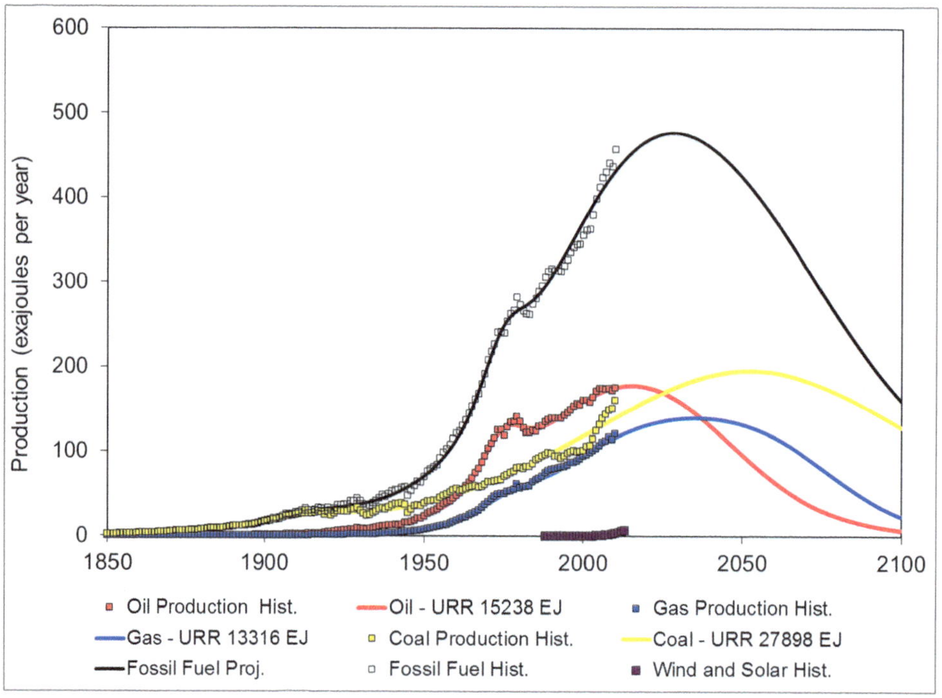

Fig. 8 Historical fossil fuel and renewables production and forecasts using a logistic calculation based on global resource estimates (Adapted from Maggio and Cacciola 2012)

et al. 2014; Garcia-Olivares and Ballabrera-Poy 2015). Oil, natural gas and coal had very high EROI up until the mid-twentieth century, but the EROI for domestic and imported oil and gas has declined considerably (from 30:1 to 10:1 or so) since 1950 (Hall et al. 2014). The EROI of coal is still relatively high and stable at mostly 25–50:1 depending on type. Compared to oil and gas at their highest, the alternative energy sources being proclaimed as substitutes for high quality fossil fuels, nonconventional oil and gas and renewables, have considerably lower EROI (Fig. 9). The trend towards a decline in EROI is well established, and does not bode well for future economic growth. The next section will take a closer look at the potential alternatives for conventional fossil fuels.

The Future of Energy

Failing unprecedented increases in efficiencies and reservoir recovery rates, conventional oil and gas, the cleaner burning of the fossil fuels, are projected to peak soon (Maggio and Cacciola 2012; Mohr et al. 2015). There are multiple differing viewpoints on what this means for the future. Some maintain that nonconventional oil substitutes will allow society to maintain the status quo far into the future (e.g. Lynch 2016). Others argue, given the rapid growth in renewable energy and the present climate dangers, that a 100% renewables energy system is needed and possible by 2050 supporting continued economic growth into the future (e.g. Jacobson and Delucchi 2011). Others are more pessimistic, and believe that biophysical constraints will not allow maintenance of the status quo, will limit any renewables revolution, and will lead to economic recession (e.g. Georgescu Roegen 1975; Lambert et al. 2014; Trainer 2007). These three viewpoints can be scrutinized through a closer look at the prospects for nonconventional oil substitutes and renewables.

Nonconventional Oil and Gas Substitutes

Exploration is currently focused on finding and developing nonconventional sources of oil and gas to meet increasing demand. The expectation of many is that conventional oil and gas will soon be replaced relatively easy by these substitutes, which include tight oil (sometimes called shale oil), tight and shale natural gas, oil sands, biofuels, extra heavy oil, and ultra-deep water (BP 2016; EIA 2015a; IEA 2015; McGlade 2014). Nonconventional oil and natural gas substitutes currently constitute 5% of global fossil fuel production, but this proportion is growing (see Fig. 10).

Both oil and natural gas can be found in low permeability shale, sandstone, and carbonate formations. These resources are grouped into the categories tight oil, shale natural gas, and tight natural gas, based on the type of formation. Through successful implementation of horizontal drilling and hydraulic fracking technologies the U.S is producing significant amounts of nonconventional oil and gas, and China and Argentina are beginning to exploit their large reserves (EIA 2013, 2015b). However, the long-term sustainability of tight oil and shale gas production is still uncertain. Brandt et al. (2015) estimated

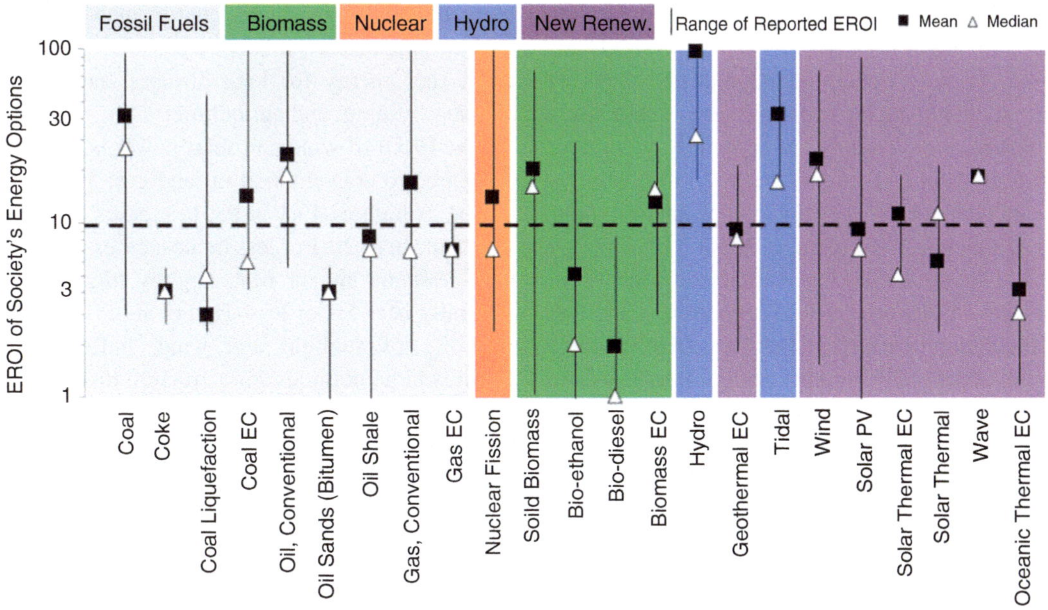

Fig. 9 Summary of published EROI estimates with mean (black squares), median (white triangles), and range (vertical lines) (Adapted from Carbajales-Dale et al. 2012). An EROI of 10:1 or greater is required to maintain a modern industrial society (Fizaine and Court 2016)

Fig. 10 Historical and projected energy production from conventional and nonconventional oil and natural gas. Here, nonconventional oil includes kerogen, bitumen, tight, and extra heavy. Nonconventional gas includes shale gas and gas hydrates (Adapted from Mohr et al. 2015)

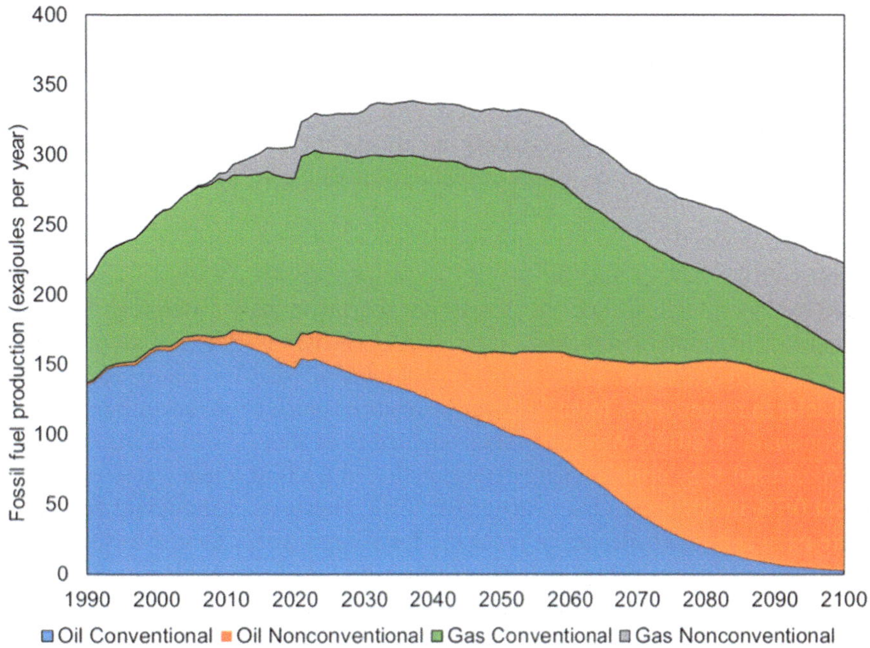

the median net energy return (an EROI equivalent) of the Bakken shale formation – for all wells, the majority of which are located near "sweet spots" – to be 29:1, and in decline. This decline in net energy return is related to a decrease in estimated ultimate recovery which could be the result of drillers moving out of early "sweet spot" locations, or due to well designs favoring initial well production over lifetime productivity. Waggoner (2013) calculated a much lower EROI of 10:1 using financially based EROI metrics.

Oil found in Northwest Alberta, Canada, referred to as oil sands, exists as bitumen, a tar-like substance mixed with sand and clay. The oil sands are recovered through either surface mining (followed by heating and hydrogen injection) or liquefied in place with water superheated with natural gas (or steam assisted gravity drainage – SAGD). Estimates of reserves in the oil sands are massive, as high as 1500 billion barrels (for perspective, the world has produced 1291 billion barrels of conventional oil to date), but there are significant costs which hamper production. Brandt et al. (2013)

estimated the EROI of oil sands operations to be 3:1, while cost of production ranges from $40–90 per barrel (McGlade and Ekins 2015). Next to Arctic production, the oil sands are one of most expensive sources of oil in the world. As a marginal producer, the oil sands are especially vulnerable to falling oil prices. As of February 2016, 16 new projects have been cancelled or put on hold since the oil price plummeted beginning in 2014 (Struzik 2016).

There are few forecasts available that include nonconventional oil and gas. Mohr et al. (2015) modeled fossil fuel trajectories including nonconventional resources using a Hubbert logistic approach by country and fuel source. Conventional oil was estimated to be at peak production currently, and is expected to be completely replaced by nonconventional sources by 2100 (Fig. 10). Post growth, Mohr et al. (2015) projected that oil sands, Venezuelan extra heavy oil, and U.S. kerogen (also known as oil shale, not to be confused with shale oil) will be unable to make up for the losses in conventional oil. In addition, nonconventional fuels are more expensive and have associated environmental and social issues. These higher costs are directly related to decreasing net energy.

Renewable Energy

Environmental groups and many from government and industry are encouraging a transition from fossil fuels to renewable energy in order to meet future energy needs while controlling carbon emissions. Renewables include hydro, solar, wind, and biofuels. But, given the low EROI of biofuels (often barely greater than 1:1) (Hopkinson and Day 1980; Murphy et al. 2011) and competition for food production (Pimentel et al. 2009) and the lack of remaining hydropower sites in the developed world (Höök et al. 2012), solar and wind hold the most promise for meeting society's future energy needs.

Many suggest, based on the rapid growth rates of wind and solar in recent years, fossil fuels can be replaced quickly (Jacobson et al. 2015). Wind and solar increased from about 0.1% of total U.S. primary energy consumption in 2000 to about 0.7% in 2010 and 1.8% in 2015 (BP 2016). However, there exist serious challenges in replacing fossil fuels with wind and solar power. Since 2010, the growth rates for wind and solar PV have dropped from about 74% to 28% for solar PV, and about 25% to 17% for wind power (REN21 2015). As the installed capacity becomes larger, it will become increasingly hard to keep up a high growth rate partly because integrating intermittent power into the grid becomes increasingly difficult. Sunlight and wind are often not abundant in areas with the greatest demand (Palmer 2014). The proposal that renewables could provide all of the energy for current electricity use (about 16% of world energy use) is plausible but challenging as this would be a "like-for-like" swap of electricity. But beyond electricity it becomes much more difficult for renewable energy, in the form of electricity or low grade heat (concentrated solar power), to replace fossil fuel energy for long distance transportation, heavy industry, mining, and agriculture.

The EROI of wind and solar power is, on average, lower compared to conventional oil and gas. The EROI for wind can be as high as 40:1 in the best cases for larger turbines, with a mean of 20:1 or less in most cases. Solar photovoltaic (PV) systems are, at best, slightly higher than 10:1, but generally near 5:1 or less (Hall et al. 2014). To manage the volatility of sunlight and wind, buffering and storage requires large complex infrastructure that can reduce EROI by over 50% (Weißbach et al. 2013). In high latitudes with lower insolation it is uncertain if solar energy provides any net energy at all. An extended analysis of PV production in Germany and Switzerland reported EROI values of less than 1:1 (Ferroni and Hopkirk 2016).

Replacing existing fossil fuel infrastructure with a wind and solar system would require enormous amounts of resources. The rapid transition to and sustaining of a renewable energy system will have high demands for energy and bulk materials such as copper and steel (Davidsson et al. 2014). Supply would be hard-pressed to meet demand; the rapid increase in demand for input commodities could inflate prices, potentially resulting in a halting of economic growth (Carbajales-Dale et al. 2012).

Renewables have the potential to produce a large amount of energy for society, and have had success in many countries. As an example, in several instances over the past two years Germany, Denmark, and Portugal have nearly met or exceeded electricity demand with renewable energy (Neslen 2015, 2016). However, given the differences with fossil fuels that we have described, they do not seem poised to provide for the same lifestyle that western societies have enjoyed for the past century. Transforming the energy system to lower EROI renewables will mean society must allocate more energy to investing in renewable energy systems, electric grids, energy storage, and liquid fuel substitutes, meaning that less net energy is available for society (i.e. a lower societal EROI). The remaining energy for society will be used to cover infrastructure maintenance and consumer staples, and discretionary spending and investments could likely be squeezed out (Hall et al. 2008). Discretionary investments that get squeezed out could include environmental projects like coastal restoration. Therefore, despite innovations in renewable energy technology, it will be difficult to realize a future powered by renewables that looks anything like the industrial society of today. Claims that a transition to renewables will be easy ignore the many difficulties involved in a shift to less energy dense and less flexible energy sources.

Implications for Coastal Louisiana

We have presented evidence for fossil fuels peaking before mid-century, with the peak in oil coming first. Nonconventional oil and gas substitutes and renewables show promise, but it will be difficult for them to meet the demands brought on by diminishing conventional oil and gas supplies. This does not bode well for nations and states whose economies are fundamentally dependent on fossil fuels. Louisiana is one such state, depending on navigation, tourism, and oil and gas and petrochemical industries for most of its revenue (Barnes et al. 2015). Deep draft cargo ships on the Mississippi River run on diesel fuel and tourists arrive to New Orleans by planes running on jet fuel or by cars running on gasoline. The busy Mississippi River corridor between Baton Rouge and New Orleans is lined with oil refineries, petrochemical facilities, and ports shipping rice and grains grown with fossil fuel derived fertilizers and diesel powered tractors. It is difficult to imagine a productive Louisiana coast without high net yield fossil fuels.

Louisiana's coastal management is vulnerable to energy availability and cost (Tessler et al. 2015; Day et al. 2016). Running an industrial economy in the delta requires dredging of navigation channels, building and maintaining levees on the Mississippi River, pumping dredged sediments in pipelines, building and maintaining large water control structures, and maintaining flood protection systems for areas near or below sea level. To restore the delta and protect against flooding given relative SLR will require a massive upscaling of such endeavors. The two main strategies for wetland restoration are marsh creation via dredged sediments and sediment diversions from the Mississippi River. Both marsh creation and river diversions are energy intensive, though marsh creation is likely more energy intensive in the long-term since re-nourishing is required. Coastal restoration will be discussed in greater detail in chapter "The Costs and Sustainability of Ongoing Efforts to Restore and Protect Louisiana's Coast".

Based on the importance of fossil fuel energy to Louisiana's economy and the tremendous energy requirements of the current coastal management plan, an understanding of future energy trajectories is critical to any discussion of the sustainability of coastal Louisiana.

Conclusions

In this chapter we summarized projected trends for climate change and energy scarcity and discussed how they threaten the sustainability of the MRD. Climate change threatens the geomorphological sustainability of the MRD. Projections indicate that if aggressive action is not taken very soon most coastal wetlands will disappear by 2100, largely due to subsidence and accelerating eustatic SLR (Blum and Roberts 2009). Energy scarcity threatens the economic sustainability of the MRD. Evidence suggests that total fossil fuel production will peak before mid-century, with oil peaking first, followed by natural gas and coal. Renewables have the potential to produce large amounts of energy for society, but at a much higher cost compared to conventional fossil fuels. Declining EROI of the energy supply over time means that energetic and monetary costs will increase. This will eventually make the highly energy intensive flood control and navigation systems of the MRD extremely expensive.

Based on climate and energy trends, it will be necessary to implement management plans that allow the coast to become more resilient to climate change impacts, but also at the lowest operating energy cost. Day et al. (1997) suggest that "only management that is based on the functioning of deltas is sustainable" in the long run. The dynamic pulsing nature of the historic MRD was described in chapter "Levees and the Making of a Dysfunctional Floodplain". The current management of the flood control system, which involves building and maintaining increasingly larger levees, is unsustainable because it has suppressed the natural pulsing of the delta. Tessler et al. (2015) and Day et al. (2016) report how the severity of climate change and energy scarcity will make coastal restoration and management much more challenging and expensive during this century, especially for first world deltas where energy intensive management activities are used. Climate change and declining availability and accessibility of energy necessitate the need for an aggressive new approach to management of the MRD. Chapter "The Costs and Sustainability of Ongoing Efforts to Restore and Protect Louisiana's Coast" will discuss different approaches to coastal restoration, and how climate change and energy scarcity must inform future management approaches.

References

American Association of Port Authorities (AAPA). World Port Rankings 2015. Accessed on 5 Jan 2017 at: www.aapa-ports.org/unifying/content.aspx?ItemNumber=21048

Aucott M, Hall C (2014) Does a change in price of fuel affect GDP growth? An examination of the US data from 1950–2013. Energies 7(10):6558–6570

Barnes S, Bond C, Burger N, Anania K, Strong A, Weilant S, Virgets S (2015) Economic evaluation of coastal land loss in Louisiana. Available online: coastal.la.Gov/economic-evaluation-of-land-loss-in-louisiana/

Bentley RW (2016) Introduction to peak oil. Springer, New York

Blum MD, Roberts HH (2009) Drowning of the Mississippi Delta due to insufficient sediment supply and global sea-level rise. Nat Geosci 2(7):488–491

Blum MD, Roberts HH (2012) The Mississippi delta region: past, present, and future. Annu Rev Earth Planet Sci 40:655–683

Brandt AR, Englander J, Bharadwaj S (2013) The energy efficiency of oil sands extraction: energy return ratios from 1970 to 2010. Energy 55:693–702

Brandt AR, Yeskoo T, Vafi K (2015) Net energy analysis of Bakken crude oil production using a well-level engineering-based model. Energy 93:2191–2198

British Petroleum (BP) (2016) BP statistical review of world energy 2016. British Petroleum, London

Carbajales-Dale M, Krumdieck S, Bodger P (2012) Global energy modelling—a biophysical approach (GEMBA) Part 2: Methodology. Ecol Econ 73:158–167

Campanella R (2007) Above sea-level New Orleans: the residential capacity of Orleans Parish's higher ground. Center for Bioenvironmental Research, New Orleans

Carter LM, Jones JW, Berry L, Burkett V, Murley JF, Obeysekera J, Schramm PJ, Wear D (2014) Ch. 17: Southeast and the Caribbean. In: Melillo JM, Richmond TC, Yohe GW (eds) Climate change impacts in the United States: the third national climate assessment. U.S. global change research program, Washington, D.C, pp 396–417. https://doi.org/10.7930/J0NP22CB

Church JA, White NJ (2011) Sea-level rise from the late 19th to the early 21st century. Surv Geophys 32(4–5):585–602

Church JA, White NJ, Konikow LF, Domingues CM, Cogley JG, Rignot E, Gregory JM, van den Broeke MR, Monaghan AJ, Velicogna I (2011) Revisiting the Earth's sea-level and energy budgets from 1961 to 2008. Geophys Res Lett, 38(18):L18601. https://doi.org/10.1029/2011gl048794

Cleveland CJ (2005) Net energy from the extraction of oil and gas in the United States. Energy 30(5):769–782

Coastal Protection and Restoration Authority of Louisiana (CPRA) (2017). 2017 master plan data viewer Accessed 15 Feb 2017 at: cims.coastal.louisiana.gov/masterplan/

Coumou D, Rahmstorf S (2012) A decade of weather extremes. Nat Clim Chang 2(7):491–496

Dai A (2011) Drought under global warming: a review. Wiley Interdiscip Rev Clim Chang 2(1):45–65

Davenport C, Robertson C (2016) Resettling the first American climate refugees, New York Times, 3 May, Available online: www.nytimes.com/2016/05/03/us/resettling-the-first-american-climate-refugees.html?_r=1

Davidsson S, Grandell L, Wachtmeister H, Höök M (2014) Growth curves and sustained commissioning modelling of renewable energy: investigating resource constraints for wind energy. Energ Policy 73:767–776

Day JW, Martin JF, Cardoch L, Templet PH (1997) System functioning as a basis for sustainable management of deltaic ecosystems. Coast Manag 25(2):115–153

Day JW, Britsch LD, Hawes SR, Shaffer GP, Reed DJ, Cahoon D (2000) Pattern and process of land loss in the Mississippi Delta: a spatial and temporal analysis of wetland habitat change. Estuar Coasts 23(4):425–438

Day JW, Boesch DF, Clairain EJ, Kemp GP, Laska SB, Mitsch WJ, Orth K, Mashriqui H, Reed DJ, Shabman L, Simenstad CA (2007) Restoration of the Mississippi Delta: lessons from hurricanes Katrina and Rita. Science 315(5819):1679–1684

Day JW, Agboola J, Chen Z, D'Elia C, Forbes DL, Giosan L, Kemp P, Kuenzer C, Lane RR, Ramachandran R, Syvitski J (2016) Approaches to defining deltaic sustainability in the 21st century. Estuar Coast Shelf Sci 183:275–291

DeConto RM, Pollard D (2016) Contribution of Antarctica to past and future sea-level rise. Nature 531(7596):591–597

EIA (2015a) Annual Energy Outlook 2015. U.S. Energy Information Administration, Washington

EIA (2015b) Today in Energy. Energy Information Agency, 26 June, Available online: www.eia.gov/todayinenergy/detail.php?id=21832

Emanuel K (2005) Increasing destructiveness of tropical cyclones over the past 30 years. Nature 436(7051):686–688

Energy Information Agency (EIA) (2013) Today in Energy. U.S. Energy Information Agency, 10 June, Available online: www.eia.gov/todayinenergy/detail.php?id=11611

Ferris R (2016) Baton Rouge floods broke a record many thought would never be broken , CNBC, 17 August, Available online: www.cnbc.com/2016/08/17/baton-rouge-floods-broke-a-record-many-thought-would-never-be-broken.html

Ferroni F, Hopkirk RJ (2016) Energy return on energy invested (ERoEI) for photovoltaic solar systems in regions of moderate insolation. Energ Policy 94:336–344

Fizaine F, Court V (2016) Energy expenditure, economic growth, and the minimum EROI of society. Energ Policy 95:172–186

Friedlingstein P, Andrew RM, Rogelj J, Peters GP, Canadell JG, Knutti R, Luderer G, Raupach MR, Schaeffer M, Van Vuuren DP, Le Quéré C (2014) Persistent growth of CO2 emissions and implications for reaching climate targets. Nat Geosci 7(10):709–715

García-Olivares A, Ballabrera-Poy J (2015) Energy and mineral peaks, and a future steady state economy. Technol Forecast Soc Chang 90:587–598

Georgescu-Roegen N (1975) Energy and economic myths. Southern Economic Journal

GISTEMP Team (2017) GISS surface temperature analysis (GISTEMP). NASA Goddard Institute for Space Studies. Accessed 16 April 2017 at: data.giss.nasa.gov/gistemp/

Groisman PY, Knight RW, Easterling DR, Karl TR, Hegerl GC, Razuvaev VN (2005) Trends in intense precipitation in the climate record. J Clim 18(9):1326–1350

Hamilton JD (2009) Causes and Consequences of the Oil Shock of 2007-08(No. w15002). Natl Bur Econ Res

Hall CA (2017) Energy return on investment. Springer International Publishing, Cham

Hall CAS, Powers R, Schoenberg W (2008) Peak oil, EROI, investments and the economy in an uncertain future. In: Pimentel D (ed) Renewable energy systems: environmental and energetic issues. Elsevier, London

Hall CA, Lambert JG, Balogh SB (2014) EROI of different fuels and the implications for society. Energ Policy 64:141–152

Hansen J, Sato M (2004) Greenhouse gas growth rates. Proc Natl Acad Sci 101:16109–16114

Hansen J, Ruedy R, Sato M, Lo K (2010) Global surface temperature change. Rev Geophys 48(4). doi:10.1029/2010RG000345

Hansen J, Sato M, Hearty P, Ruedy R, Kelley M, Masson-Delmotte V, Russell G, Tselioudis G, Cao J, Rignot E, Velicogna I (2015) Ice melt, sea level rise and superstorms: evidence from paleoclimate data, climate modeling, and modern observations that 2 C global warming is highly dangerous. Atmos Chem Phys 16(6):3761–3812

Hausfather Z (2013) IPCC's new estimates for increased sea-level rise. Yale Climate Connections, 23 October, Available online: www.yaleclimateconnections.org/2013/10/ipccs-new-estimates-for-increased-sea-level-rise/

Hays JD, Imbrie J, Shackleton NJ (1976) Variations in the Earth's orbit: pacemaker of the ice ages. American Association for the Advancement of Science, Washington, DC

Höök M, Li J, Johansson K, Snowden S (2012) Growth rates of global energy systems and future outlooks. Nat Resour Res 21(1):23–41

Hopkinson CS, Day JW (1980) Net energy analysis of alcohol production from sugarcane. Science 207(4428):302–304

Hornsey MJ, Harris EA, Bain PG, Fielding KS (2016) Meta-analyses of the determinants and outcomes of belief in climate change. Nat Clim Chang 6(6):622–626

Horton BP, Rahmstorf S, Engelhart SE, Kemp AC (2014) Expert assessment of sea-level rise by AD 2100 and AD 2300. Quat Sci Rev 84:1–6
Hoyos CD, Agudelo PA, Webster PJ, Curry JA (2006) Deconvolution of the factors contributing to the increase in global hurricane intensity. Science 312(5770):94–97
Hubbert MK (1956) Nuclear energy and the fossil fuels drilling and production practice (95). American petroleum institute, Shell Development Co
International Energy Agency (IEA) (2015) Key world energy statistics 2015. International Energy Agency, Paris
IPCC Stocker TF, Qin D, Plattner GK, Tignor M, Allen SK, Boschung J, Nauels A, Xia Y, Bex V, Midgley PM (2013) Climate change 2013: the physical science basis: working group I contribution to the fifth assessment report of the intergovernmental panel on climate change
IPCC, Pachauri RK, Allen MR, Barros VR, Broome J, Cramer W, Christ R, Church JA, Clarke L, Dahe Q, Dasgupta P, Dubash NK (2014) Climate change 2014: synthesis report. Contribution of working groups I, II and III to the fifth assessment report of the intergovernmental panel on climate change
Isle de Jean Charles Resettlement. Available online: www.coastalresettlement.org/
Jacobson MZ, Delucchi MA (2011) Providing all global energy with wind, water, and solar power, part I: technologies, energy resources, quantities and areas of infrastructure, and materials. Energ Policy 39(3):1154–1169
Jacobson MZ, Delucchi MA, Bazouin G, Bauer ZA, Heavey CC, Fisher E, Morris SB, Piekutowski DJ, Vencill TA, Yeskoo TW (2015) 100% clean and renewable wind, water, and sunlight (WWS) all-sector energy roadmaps for the 50 United States. Energy Environ Sci 8(7):2093–2117
Jevrejeva S, Moore JC, Grinsted A (2012) Sea level projections to AD2500 with a new generation of climate change scenarios. Glob Planet Chang 80:14–20
Jones N (2013) Rising tide. Nature 501(7467):300
Karl TR, Arguez A, Huang B, Lawrimore JH, McMahon JR, Menne MJ, Peterson TC, Vose RS, Zhang HM (2015) Possible artifacts of data biases in the recent global surface warming hiatus. Science 348 (6242):1469–1472
Kaufmann RK, Cleveland CJ (2008) Environmental science. McGraw-Hill College, Boston
Kaufmann RK, Kauppi H, Mann ML, Stock JH (2011) Reconciling anthropogenic climate change with observed temperature 1998–2008. Proc Natl Acad Sci 108(29):11790–11179
Kemp GP, Day JW, Freeman AM (2014) Restoring the sustainability of the Mississippi River Delta. Ecol Eng 65:131–146
Khan SA, Kjær KH, Bevis M, Bamber JL, Wahr J, Kjeldsen KK, Bjørk AA, Korsgaard NJ, Stearns LA, Van Den Broeke MR, Liu L (2014) Sustained mass loss of the northeast Greenland ice sheet triggered by regional warming. Nat Clim Chang 4(4):292–299
King CW, Hall CA (2011) Relating financial and energy return on investment. Sustainability 3(10):1810–1832
Knutson TR, McBride JL, Chan J, Emanuel K, Holland G, Landsea C, Held I, Kossin JP, Srivastava AK, Sugi M (2010) Tropical cyclones and climate change. Nat Geosci 3(3):157–163
Kolker AS, Allison MA, Hameed S (2011) An evaluation of subsidence rates and sea-level variability in the northern Gulf of Mexico. Geophys Res Lett 38(21):L21404
Kopp RE, Kemp AC, Bittermann K, Horton BP, Donnelly JP, Gehrels WR, Hay CC, Mitrovica JX, Morrow ED, Rahmstorf S (2016) Temperature-driven global sea-level variability in the common era. Proc Natl Acad Sci 113(11):E1434–E1441
Lam NSN, Arenas H, Brito PL, Liu KB (2014) Assessment of vulnerability and adaptive capacity to coastal hazards in the Caribbean region. J coast Res 70(sp1):473–478
Lambert JG, Hall CA, Balogh S, Gupta A, Arnold M (2014) Energy, EROI and quality of life. Energ Policy 64:153–167
Lawrence Livermore National Laboratory (LLNL) 2016 Energy Flow Charts. http://flowcharts.llnl.gov. (Accessed 25 Jan 2016)
Lenton TM, Held H, Kriegler E, Hall JW, Lucht W, Rahmstorf S, Schellnhuber HJ (2008) Tipping elements in the Earth's climate system. Proc Natl Acad Sci 105(6):1786–1793
Lynch MC (2016) The "Peak Oil" scare and the coming oil flood. Praeger, Santa Barbara
Maggio G, Cacciola G (2012) When will oil, natural gas, and coal peak? Fuel 98:111–123
McGlade CE (2014) Uncertainties in the outlook for oil and gas. Doctoral dissertation, UCL (University College London)
McGlade C, Ekins P (2015) The geographical distribution of fossil fuels unused when limiting global warming to 2°C. Nature 517 (7533):187–190
Mei W, Xie SP, Primeau F, McWilliams JC, Pasquero C (2015) Northwestern Pacific typhoon intensity controlled by changes in ocean temperatures. Sci Adv 1(4):e1500014
Min SK, Zhang X, Zwiers FW, Hegerl GC (2011) Human contribution to more-intense precipitation extremes. Nature 470(7334):378–381
Mohr SH, Wang J, Ellem G, Ward J, Giurco D (2015) Projection of world fossil fuels by country. Fuel 141:120–135
Molina M, McCarthy J, Wall D, Alley R, Cobb K, Cole J, Das S, Diffenbaugh N, Emanuel K, Frumkin H, Hayhoe L (2014) What we know: the reality, risks and response to climate change. American Association for the Advancement of Science, Washington, DC
Montano S (2016) The Louisiana floods are devastating, and climate change will bring more like them. We're not ready. Vox, 23 August, available online: http://www.vox.com/2016/8/18/12522036/louisiana-flood-climate-change-emergency-management
Moss R, Babiker M, Brinkman S, Calvo E, Carter T, Edmonds J, Elgizoul I, Emori S, Erda L, Hibbard K, Jones R, Kainuma M, Kelleher J, Lamarque JF, Manning M, Matthews B, Meehl J, Meyer L, Mitchel J, Nakicenovic N, O'Neill B, Pichs R, Riahi K, Rose S, Runci P, Stouffer R, van Vuuren D, Weyant J, Wilbanks T, van Ypersele JP, Zurek M (2008) Towards new scenarios for analysis of emissions, climate change, impacts, and response strategies technical summary. Intergovernmental Panel on Climate Change, Geneva
Murphy DJ, Hall CAS, Powers B (2011) New perspectives on the energy return on (energy) investment (EROI) of corn ethanol. Environ Dev Sustain 13(1):179–202
National Bureau of Statistics of China (NBSC). 2016. China statistical yearbook. Available online: http://www.stats.gov.cn/tjsj/ndsj/2016/indexeh.htm
Nerem RS, Chambers DP, Choe C, Mitchum GT (2010) Estimating mean sea level change from the TOPEX and Jason altimeter missions. Mar Geod 33(S1):435–446
Neslen A (2015) Wind power generates 140% of Denmark's electricity demand. The Guardian, 10 July, available online: https://www.theguardian.com/environment/2015/jul/10/denmark-wind-windfarm-power-exceed-electricity-demand
Neslen A (2016) Portugal runs for four days straight on renewable energy alone. The Guardian, 18 may, available online: https://www.theguardian.com/environment/2016/may/18/portugal-runs-for-four-days-straight-on-renewable-energy-alone
Nobles WP (2016) Louisiana flooding records were broken on these four rivers, The Times-Picayune, 14 August, Available online: http://www.nola.com/weather/index.ssf/2016/08/4_rivers_break_crest_records_i.html
Odum HT (1973) Energy, ecology, and economics. Ambio 2:220–227
Oreskes N (2004) The scientific consensus on climate change. Science 306(5702):1686–1686
Pall P, Aina T, Stone DA, Stott PA, Nozawa T, Hilberts AG, Lohmann D, Allen MR (2011) Anthropogenic greenhouse gas

contribution to flood risk in England and Wales in autumn 2000. Nature 470(7334):382–385
Palmer G (2014) Energy in Australia: peak oil, solar power, and Asia's economic growth. Springer, New York
Parris A, Bromirski P, Burkett V, Cayan D, Culver M, Hall J, Horton R, Knuuti K, Moss R, Obeysekera J, Sallenger A, Weiss J (2012) Global sea level rise scenarios for the US National Climate Assessment. NOAA Tech Memo OAR CPO-1
Petit J, Jouzel J, Raynaud D, Barkov NI, Barnola JM, Basile I, Bender M, Chappellaz J, Davis M, Delaygue G, Delmotte M (1999) Climate and atmospheric history of the past 420,000 years from the Vostok ice core, Antarctica. Nature 399(6735):429–436
Pfeffer WT, Harper J, O'Neel S (2008) Kinematic constraints on glacier contributions to 21st-century sea-level rise. Science 321 (5894):1340–1343
Pimentel D, Marklein A, Toth MA, Karpoff MN, Paul GS, McCormack R, Kyriazis J, Krueger T (2009) Food versus biofuels: environmental and economic costs. Hum Ecol 37(1):1–12
Prein AF, Rasmussen RM, Ikeda K, Liu C, Clark MP, Holland GJ (2016) The future intensification of hourly precipitation extremes. Nat Clim Chang 7(1):48–52
Rahmstorf S (2007) A semi-empirical approach to projecting future sea-level rise. Science 315(5810):368–370
REN21 (2015) Renewables 2016 global status report, REN21 secretariat, Paris
Ricardo D (1891) Principles of political economy and taxation. G. Bell and sons, London
Rignot E, Velicogna I, van den Broeke MR, Monaghan A, Lenaerts JT (2011) Acceleration of the contribution of the Greenland and Antarctic ice sheets to sea level rise. Geophys Res Lett 38(5):L18601
Rignot E, Mouginot J, Morlighem M, Seroussi H, Scheuchl B (2014) Widespread, rapid grounding line retreat of Pine Island, Thwaites, smith, and Kohler glaciers, West Antarctica, from 1992 to 2011. Geophys Res Lett 41(10):3502–3509
Scripps Institution of Oceanography. The keeling curve. Available online: https://scripps.ucsd.edu/programs/keelingcurve/
Shaffer GP, Day JW, Kandalepas D, Wood WB, Hunter RG, Lane RR, Hillmann ER (2016) Decline of the Maurepas swamp, Pontchartrain Basin, Louisiana, and approaches to restoration. Water 8(3):101
Steffen W, Broadgate W, Deutsch L, Gaffney O, Ludwig C (2015) The trajectory of the Anthropocene: the great acceleration. Anthropocene Rev 2(1):81–98
Struzik E (2016) Once unstoppable, Tar Sands now battered from all sides. Yale environment 360, 1 February, available online: http://e360.yale.edu/feature/once-unstoppable_tar_sands_now_battered_from_all_sides/2954/
Svensmark H (2009) Mens Solen sover, Jyllands- Posten, 9 September, Available online: http://jyllands-posten.dk/protected/debat/kronik/article4170671.ece
Sverdrup HU, Ragnarsdottir KV, Koca D (2015) An assessment of metal supply sustainability as an input to policy: security of supply extraction rates, stocks-in-use, recycling, and risk of scarcity. J Clean Prod 140:359–372
Syvitski JP, Kettner AJ, Overeem I, Hutton EW, Hannon MT, Brakenridge GR, Day J, Vörösmarty C, Saito Y, Giosan L, Nicholls RJ (2009) Sinking deltas due to human activities. Nat Geosci 2 (10):681–686
Tao B, Tian H, Ren W, Yang J, Yang Q, He R, Cai W, Lohrenz S (2014) Increasing Mississippi river discharge throughout the 21st century influenced by changes in climate, land use, and atmospheric CO2. Geophys Res Lett 41(14):4978–4986
Tessler ZD, Vörösmarty CJ, Grossberg M, Gladkova I, Aizenman H, Syvitski JPM, Foufoula-Georgiou E (2015) Profiling risk and sustainability in coastal deltas of the world. Science 349 (6248):638–643
Tollefson J (2017) Larsen C's big divide. Nature 543(7643):402–403
Trainer T (2007) Renewable energy cannot sustain a consumer society. Springer Science & Business Media, New York
Turner RE (1997) Wetland loss in the northern Gulf of Mexico: multiple working hypotheses. Estuar Coasts 20(1):1–13
United Nations, Department of economic and social affairs, Population Division (2015). World population prospects: The 2015 revision, key findings and advance tables. Working paper No. ESA/P/WP.241
Vermeer M, Rahmstorf S (2009) Global sea level linked to global temperature. Proc Natl Acad Sci 106(51):21527–21532
Vimeux F, Cuffey KM, Jouzel J (2002) New insights into southern hemisphere temperature changes from Vostok ice cores using deuterium excess correction. Earth Planet Sci Lett 203:829–843
Waggoner EG (2013). Sweet spots, EROI, and the limits to Bakken production. Thesis (M.S.), State University of New York College of environmental science and forestry
Webster PJ, Holland GJ, Curry JA, Chang HR (2005) Changes in tropical cyclone number, duration, and intensity in a warming environment. Science 309(5742):1844–1846
Weißbach D, Ruprecht G, Huke A, Czerski K, Gottlieb S, Hussein A (2013) Energy intensities, EROIs (energy returned on invested), and energy payback times of electricity generating power plants. Energy 52:210–221
Zou L, Kent J, Lam NSN, Cai H, Qiang Y, Li K (2015) Evaluating land subsidence rates and their implications for land loss in the lower Mississippi River basin. Water 8(1):10

The Costs and Sustainability of Ongoing Efforts to Restore and Protect Louisiana's Coast

Adrian R.H. Wiegman, Jeffrey S. Rutherford, and John W. Day

Abstract

The coast is the most vital economic region in Louisiana and is being affected greatly by land loss. The Coastal Protection and Restoration Authority (CPRA) is an overarching agency responsible for planning, coordinating, and implementing restoration and protection projects in the Louisiana Coastal Zone. CPRA publishes the Louisiana Coastal Master Plan (LACMP), a living document that is revised every five years according to new information from science and models. We overview the major processes involved with coastal restoration planning in Louisiana and the various types of restoration approaches and tradeoffs between them. The LACMP has co-evolved with numerical modeling tools that allow managers to identify tradeoffs between objectives that are sometimes at odds, such as: restoring natural processes to sustain ecosystems in the delta, and promoting economic development in the working coast. Marsh creation and river sediment diversions are two contrasting approaches that have the greatest potential for land building within the coastal zone. Both marsh creation and river diversions require a vast amount of energy and capital during implementation. Marsh creation requires large amounts of fuel for each unit of sediment delivered to a marsh and the costs are subject to changes in energy prices. River diversions, however, use gravitational energy to move sediment and are more sustainable in the long term, especially in an energy-constrained future. There are significant financial limitations on the LACMP that could be exacerbated by fluctuations in energy markets. An appropriate next step for the CPRA would be to incorporate energy forecasts into the planning process. Doing so might illuminate sustainability risks not yet considered by coastal planners.

Keywords

Coastal planning • Energy costs • Marsh creation • River diversions • Hydraulic dredging • Decision support tools

Coastal Restoration and Protection in Louisiana

The coast is the most vital economic region in Louisiana and is threatened by the negative impacts of land loss and flooding (Barnes et al. 2015), mostly due to relative sea level rise, hydrologic modification in wetlands, and isolation from the Mississippi River (Day et al. 2000). At risk areas include: the major commodity ports on the Mississippi River

A.R.H. Wiegman (✉) • J.S. Rutherford • J.W. Day
Department of Oceanography and Coastal Sciences, Louisiana State University, Baton Rouge, LA 70803, USA
e-mail: adrian.wiegman@gmail.com; Jruthe@stanford.edu; johnday@lsu.edu

J.W. Day, J.A. Erdman (eds.), *Mississippi Delta Restoration*, Estuaries of the World, https://doi.org/10.1007/978-3-319-65663-2_7

and the Gulf Intracoastal Waterway and elsewhere, highway infrastructure, cities in the coastal zone (e.g. New Orleans, Houma, Lake Charles, Morgan City, the north shore of Lake Pontchartrain), small fishing communities, the Mississippi River and hurricane flood protection levee system, and deltaic ecosystems that provide $12–47 billion dollars in annual services (Batker et al. 2010). Currently, flooding from storm surge in Louisiana inflicts an average of $2.7 billion per year in annual damages. If no further restorative action is taken, in 50 years annual damages from storm surge are projected to be $12.1 billion per year (CPRA 2017a). An additional annual loss of $5.8–7.4 billion per year is projected from reduced economic activity associated with storm related inactivity and abandonment of infrastructure (e.g. businesses leaving) (Barnes et al. 2015). In response to these socioeconomic threats, the state of Louisiana has initiated a 50-year plan to restore and protect coastal Louisiana ecosystems, communities, and infrastructure (CRPA 2007, 2012a, 2017a).

Objectives and Overview

The objective of this chapter is to familiarize the reader with ongoing coastal restoration and protection efforts aimed at reducing the economic risks of deltaic inundation and deterioration in Louisiana. The focus will be on restoration activities that are proposed by the Coastal Protection and Restoration Authority (CPRA), the agency that oversees the development and execution of the State's Coastal Master Plan (LACMP). We overview the major processes involved with restoration planning and the various types of restoration approaches and tradeoffs between them. Special focus is placed on marsh creation and river sediment diversions, two contrasting approaches that have the greatest potential for land building within the coastal zone. In the footnotes we provide references to online public resources, and we encourage readers to seek out these materials for more information. Below we review the evolution of policy and the LACMP. Later in the chapter we discuss restoration approaches and highlight some key omissions of the LACMP, most notability the consideration of changing energy costs. This chapter concludes with insights on how energy and climate megatrends that were discussed in chapter "Energy and Climate – Global Trends and Their Implications for Delta Restoration" will limit options for restoration in Louisiana and will necessitate a reimagining of life and management in the future coastal zone. In a way, this chapter serves as a primer for the remaining chapters of the book, which focus more closely on alternative restoration and protection approaches, some of which are not proposed in the LACMP.

Review of Recent Coastal Restoration and Protection Policies

As far back as the 1970s and early 1980s, state and federal agencies were aware of the socioeconomic and environmental threat of coastal submergence and flooding threats in the Mississippi Delta. In response to a growing scientific understanding and documentation of coastal wetland deterioration and subsidence (Hatton et al. 1983; Gosselink et al. 1984; Delaune et al. 1987, 1989, 1994; Chmura 1992; Webb and Mendelssohn 1996; USGS 1997; Day et al. 2000), a series of initiatives were undertaken: The Breaux Act or Coastal Wetlands Planning, Protection and Restoration Act (CWPPRA),[1] Coast 2050 (LCWCRTF 1998), and Louisiana Coastal Area Ecosystem Restoration Plan (USACE 2004).[2] The various plans above were designed to address high rates relative sea level rise, wetland loss, and associated flood risk hazards in the coastal zone. These policies helped implement a range of projects (including freshwater diversions at Caernarvon and Davis Pond), but fell considerably short of reaching the level of funding needed to fully mitigate wetland loss and the collateral risks statewide, which is estimated at $1 billion per year (Davis and Vorhoff 2014).

In 2005, shortly after a devastating hurricane season which saw hurricanes Rita and Katrina devastate the coast, the state decided to take a more aggressive approach to protect its citizens and coastal resources. Act 8 was passed by the Louisiana legislature in an extraordinary session,[3] and established the CPRA as an overarching agency responsible for planning, coordinating, and implementing restoration and protection projects in the coastal zone. The mandate of the CPRA is "to develop, implement, and enforce a comprehensive coastal protection and restoration master plan" (CPRA 2007).[4] As a part of this objective the CPRA is required to publish the LACMP and update it every 5 years based on the latest science so that the plan remains an adaptable, "living document". The first master plan,

[1] 16 U.S. Code Chapter 59A – WETLANDS, https://www.law.cornell.edu/uscode/text/16/chapter-59A, https://lacoast.gov/new/Pubs/Report_data/Caring.aspx

[2] https://www.lca.gov/

[3] http://coastal.la.gov/wp-content/uploads/2016/03/Introduction-1.pdf

[4] Louisiana's 2007 Comprehensive Master Plan for a Sustainable Coast http://sonris-www.dnr.state.la.us/dnrservices/redirectUrl.jsp?dID=4063376

"Louisiana's Comprehensive Master Plan for a Sustainable Coast", was published in 2007 (CPRA 2007) and later updated in 2012 (CPRA 2012a, b, c, d, e, f, g, h, i, j, k).[5] With a $50 billion dollar proposal, the 2012 LACMP provided the comprehensiveness that previous policies have lacked. In 2017, a new update was published that incorporates the scientific research conducted and learning outcomes of projects constructed since 2012.[6]

In many ways the LACMP's adaptive planning process is a model for the integration of policy and science. In developing plans and selecting projects, CPRA depends on coastal scientists, engineers and consulting firms. For example, the Water Institute of the Gulf (established in 2011)[7] is a major partner in the LACMP scientific effort to model restoration and the RAND corporation has developed a planning tool that optimizes selection of restoration projects to best match a set of ecological and socioeconomic criteria (Groves and Sharon 2013; CPRA 2012h). As a result, the projects proposed in the LACMP summarize the work of many scientific reports that incorporate the most up to date information on restoration, engineering, ecology, and socioeconomics (CPRA 2012e).

On the implementation side, barrier island, headland beach and dune restoration, and marsh creation projects are moving along well, with about $1.5 billion of projects implemented between 2007 and 2016 (CPRA 2015a). The engineering of large-scale river diversions, however, has gone much slower than anticipated. Aside from reopening sections of Lafourche, river freshwater and sediment diversions, which are essential for creating sustainable deltaic wetlands (see chapter "Levees and the Making of a Dysfunctional Floodplain" and chapter "Large Infrequently Operated River Diversions for Mississippi Delta Restora tion"), have yet to be constructed under the LACMP (CPRA 2015a). This is in part because proposed in the plan were the dual objectives of building sustainable land and promoting the economies of the working coast (CPRA 2007). Designing a river diversion that satisfies these criteria in the eyes of politicians and their constituents has proved to be a difficult hurdle.[8] In the search for diversion designs and operation schedules that satisfy all stakeholders (see chapter "Large Infrequently Operated River Diversions for Mississippi Delta Restoration"), the complex interactions between salinity, fish, water quality, and plant production, and sediment flow for land building have necessitated ever more complicated numerical models and planning tools (CPRA 2012h, 2014; Das et al. 2012; Meselhe et al. 2013; Peyronnin et al. 2013).

Coevolution of the Coastal Master Plan and Numerical Models

With the massive scale of the LACMP, decision makers must attempt to minimize the risk of using public funding on projects that do not have long-term sustainability. Thus, it is crucial to consider and model an array of potential projects with a variety of future scenarios (Twilley et al. 2008). Accordingly, the 2007 LACMP specifically called on the use of models to analyze the impacts and uncertainties of restoration strategies. Therefore, a major component of the LACMP's development involves the use of numerical models to simulate the ecological, hydrogeological and economic tradeoffs of different combinations of restoration strategies (CPRA 2012e, f, g, h; Groves et al. 2013; Peyronnin et al. 2013; Reed 2015). Numerical models can be used to examine the tradeoffs associated with different restoration projects (i.e. maximizing land building (restoration) verses levee construction (protection), or ecosystem services as related to land building) (CPRA 2012h). The LACMP's models are used to answer questions such as: how many diversions should be built, what size do they need to be and where should they be located (CPRA 2012f, g)? The outputs of model simulations give policy makers an idea of the opportunity costs of tradeoffs identified by these different research questions (CPRA 2012h).

In the 2007 LACMP, river diversions were identified as a key component for land building (CPRA 2007). However, at the time of the plan there was still a great deal of uncertainty and controversy surrounding the logistics and limitations of diverting flow specifically for land building.

The 2007 LACMP put forth two general concepts for diversion strategies that identified an important tradeoff between land building and maintaining economic benefits from ecosystem services. As it turns out, the objectives of land building, supporting ecosystems and maintaining a working coast are sometimes at odds. The 2007 LACMP proposed two contrasting scenarios that illustrated this tradeoff. One scenario had diversions placed further inland near the upper reaches of the Barataria Basin and Breton sound (CPRA 2007). These diversions were aimed to maximize land building, but the freshwater introduced from these diversions would cause significant reductions in salinity on state water bottoms where fishermen hold oyster leases. Oysters are a key source of income to the local economy in

[5] Louisiana's 2012 Comprehensive Master Plan for a Sustainable Coasthttp://sonris-www.dnr.state.la.us/dnrservices/redirectUrl.jsp?dID=4379731

[6] DRAFT Louisiana's 20,017 Comprehensive Master Plan for a Sustainable Coasthttp://coastal.la.gov/wp-content/uploads/2016/08/2017-MP-Book_2-page-spread_Combined_01.05.2017.pdf

[7] http://thewaterinstitute.org/

[8] http://www.nola.com/environment/index.ssf/2013/06/louisiana_fishers_and_coastal.html

southern Louisiana (Barnes et al. 2015) and can't survive at salinities below 10 ppt for an extended period of time (Das et al. 2012). Operating diversions continuously would likely kill the oysters in certain areas and reduce the value of oyster leases (Das et al. 2012), which was a cause for oyster lease holders to sue via the takings clause of the U.S. constitution (Ko et al. 2017). In an alternative 2007 scenario, diversions were placed in the lower reaches of the delta. This plan aimed to have minimal impacts on salinity and ecosystem services (oyster leases and estuarine fisheries), but had significantly reduced potential for land building because a higher proportion of the diverted sediment was lost to the open Gulf and deposited in areas of high subsidence.

After the 2007 LACMP was reviewed by technical panels, it was clear that there was not enough information for any of the proposed river diversions to be authorized. More complex modeling and scenario analyses were needed, and in response the CPRA ramped up its scientific analysis and modeling efforts. In preparation for the 2012 plan, the CPRA expanded a mass balance (eco-hydrology) compartment model to simulate sediment transport, nutrient dynamics, salinity, and vegetation change under various restoration scenarios (CPRA 2012f). Prior to the publication of the 2012 LACMP a special issue in the Journal of Coastal Research outlined the different models and empirical analysis undertaken as part of the LACMP (Cobell et al. 2013; Couvillion et al. 2013; Johnson et al. 2013; Nyman et al. 2013; Rivera-Monroy et al. 2013; Visser et al. 2013).[9] The CPRA also documented these models in the appendices of the 2012 LACMP (see bibliography).

The 2012 LACMP used a newly developed "Integrated Compartment Model" to test the tradeoffs between upstream land building and salinity impacts to fisheries that were identified in the 2007 LACMP. In addition, expert opinion and scientific literature regarding size and location of diversions (e.g. Allison and Meselhe 2010) were taken into account. The 2012 LACMP included four scenarios: future without action, a few large diversions, many small diversions, and a scenario with no diversion activity (CPRA 2012a). The simulation outputs in these scenarios found very little change in ecosystem services compared to future without action. However, the model runs in the report indicated that without diversions, it would not be possible to sustain restored coastal land in the long-term (CPRA 2012a).

Ten new river diversions were proposed in the 2012 plan. All had discharges between 1416 and just over 7000 m^3/s and the largest diversions were proposed in the uppermost inland reaches of Breton Sound and Barataria Basin. Still, the controversy over diversion placement and operation continued after publication of the 2012 LACMP (see chapter "Large Infrequently Operated River Diversions for Mississippi Delta Restoration").[10] The Integrated Compartment Model was used to analyze the proposed diversions efficacy of land building with respect to size and location of sediment deposition (Meselhe et al. 2013). The results indicated that deposition was likely to occur nearest to the diversion, making it difficult for basins to keep up with RSLR let alone increase land area; though it was acknowledged the model was likely underestimating elevation gains due to accretion of organic matter from plant productivity (Wang et al. 2014). The findings of the compartment model were subject to criticism because the low resolution of the model reduces its ability to predict fine scale interactions pertaining to salinity, accretion and land elevation change (Das et al. 2012). Recently, research has focused on the negative impacts of nutrients and inundation period on brackish wetland species, demonstrating that net land loss could occur from wetland plant mortality due to diversion operation (Snedden et al. 2015, see chapter "Large Infrequently Operated River Diversions for Mississippi Delta Restoration").

In preparation for the 2017 LACMP, CPRA initiated a model improvement plan (MIP), based on review from expert panels. The developments for 2017 are highlighted in the MIP, which includes improving resolution of the Integrated Compartment Model, and further refining of some parameters (i.e. organic accretion rates). In the MIP specific details are given regarding updates to each model component (CPRA 2014). One of the most important developments was the addition of high-resolution hydrodynamic models to assess the sediment and salinity dynamics of specific diversion projects. The vegetation models were updated to better simulate productivity, mortality, and succession of plant species caused by changes in inundation and salinity (CPRA 2014). Another addition was the consideration of fine grain sediment re-suspension and transport during cold fronts and storms and the effects on coastal geomorphology (e.g. Roberts et al. 2015) (CPRA 2014). All of these developments are crucial to improving the understanding of the dynamic coastal environment and the processes controlling land loss. Further, they clarify the limits on the ability to build and sustain land, while outlining the tradeoffs between land building and supporting ecosystem services and coastal industry.

With the release of the 2017 LACMP, the state acknowledged that net land gain was not achievable on a 50-year time scale. Even under the moderate environmental scenario the LACMP models predict the Mississippi Delta will lose

[9] http://www.jcronline.org/toc/coas//67?seq=67

[10] http://www.theadvocate.com/baton_rouge/news/politics/legislature/article_0dea5912-b540-5b4c-8cb6-f3e6a91af2f9.html http://www.theadvocate.com/new_orleans/news/environment/article_807a4176-9232-11e6-9a41-5f9bc5ae605c.html

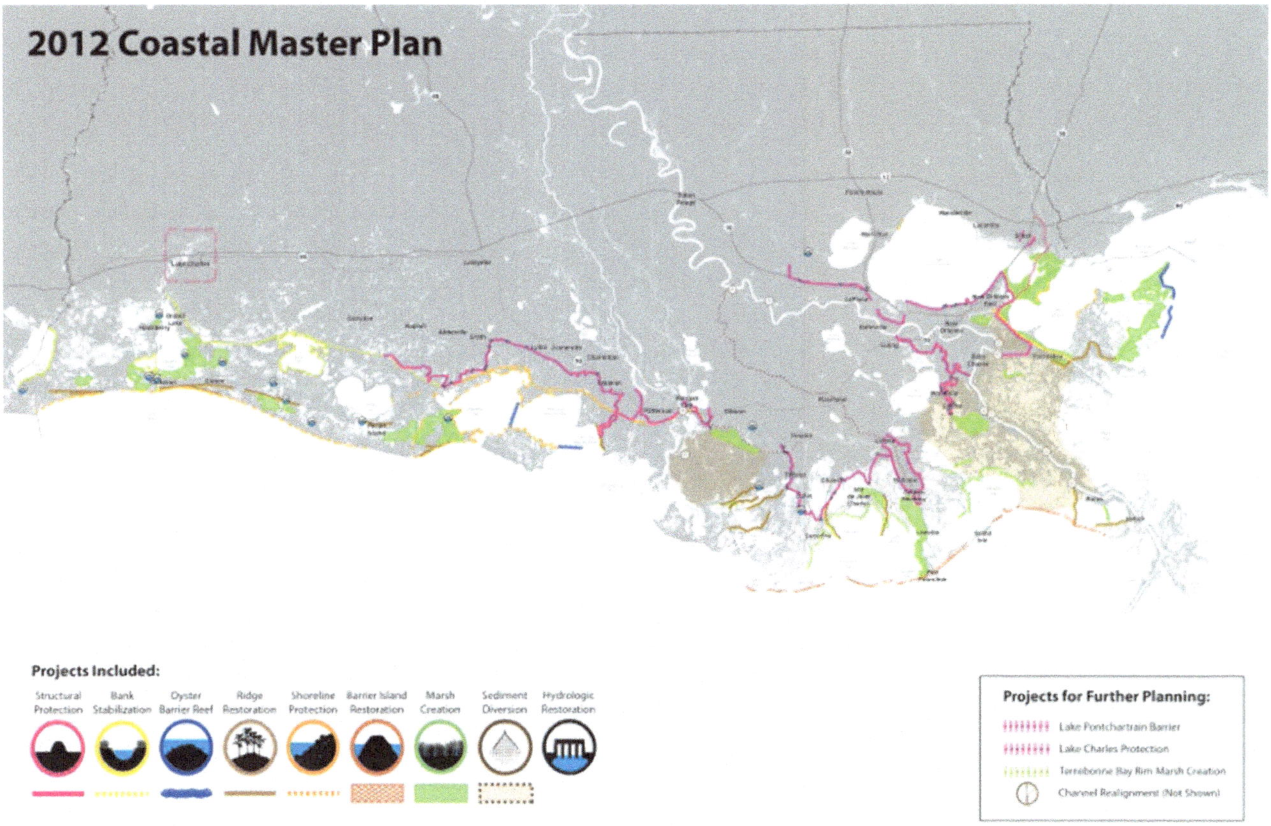

Fig. 1 A map of south Louisiana and the projects proposed in the 2012 LACMP (CPRA 2012a)

land. The 2017 LACMP aims to reduce land loss. Diversions and structural protection are concentrated around urban centers of greater New Orleans and the cities of the Gulf Coast Intracoastal Water Way (e.g. Houma, Morgan City). The major goal of the 2017 LACMP report is to select a few key diversion projects and bring them into the engineering and design phase and model their specific land building potential (CPRA 2017a, b, c, d, e). The project proposals for the 2017 LACMP required site specific and design related information so that the project specifications can be integrated as precisely as possible into the most recent predictive models. After vetting proposals, the two major diversions being focused on are the Mid Barataria and Mit Breton sound diversions. These diversions will undergo the environmental impact statement process and engineering and design between 2016 and 2022 (CPRA 2017a, b), and construction is anticipated to begin in the early 2020s.

Restoration Projects

In this section we discuss the types of projects proposed in the LACMP, the amount of funding dedicated to different types of projects, and the general procedures necessary to get a project approved. There are many different types of projects that are proposed in the plan including: bank stabilization, barrier island/headland restoration, diversion and channel realignment, hydrologic restoration, marsh creation, oyster barrier reef creation, ridge creation/restoration, levee construction, and shoreline protection (Fig. 1). For more on project types see LACMP 2017 Appendix A (CPRA 2017b). The CPRA classifies project types as restoration or risk reduction. There is also a third type, resiliency, which includes raising houses and roadways, building pumping stations, and leaky levees (which allow some water flow).[11] The 2017 LACMP will spend about 50% of funds on restoration, and 50% on risk reduction. Restoration spending will include \$17.8 billion on marsh creation (down from \$20 billion in the 2012 LACMP) and \$5.1 billion on sediment diversions (up from \$3.8 billion in 2012) (Table 1). The sediment diversions proposed in the 2017 LACMP are generally more expensive per unit of discharge capacity than the diversions proposed in 2012. The risk reduction category includes mainly the construction of more flood control levees in the southern parishes. A map of the proposed projects in the 2017

[11] http://coastal.la.gov/our-work/projects/project-types/

Table 1 2017 Louisiana coastal master plan funding allocation by project type

Class	Project type	Funding ($Billions)	Percent of funds (%)	Prime mover
Restoration	(Total)	25.8	50.8	N/A
	Barrier Island	1.5	3	Hydraulic Dredge, Bulldozer
	Hydrologic	0.4	0.8	Pump or gravity[a]
	Marsh creation	17.8	35	Hydraulic Dredge, Bulldozer
	Ridges	0.1	0.2	Excavator, Dragline or Bucket Dredge
	Sediment diversion	5.1	10	Gravity[a]
	Shoreline protection	0.9	1.8	Barge, Crane or N/A[b]
Risk reduction	(Total)	25	49.2	N/A
	Structural (levees)	19.0	37.4	Excavator, Dragline or Bucket Dredge
	Nonstructural	6.0	11.8	Various
Total		50.8	100	N/A

Adapted From CPRA (CPRA 2017a)
[a]Various machinery is required to build the control structures, after which the displacement of water or sediment is controlled by gravity (and pumps in some cases for hydrological restoration); [b]Oyster reefs have various methods of creation; Rock armor shorelines and jetties require barges and cranes

LACMP is given at the end of chapter "Large Infrequently Operated River Diversions for Mississippi Delta Restoration". Resiliency is a very important concept that we will return to in chapter "Raising Buildings: The Resilience of Elevated Structures", chapter "Raising New Orleans: The Marais Design Strategy" and chapter "Eroding Communities and Diverting Populations: Historical Population Dynamics in Coastal Louisiana".

Coastal restoration and protection projects can be classified into three basic physical activities: building protective (natural or hardened) structures (bank stabilization, oyster barrier reef creation, ridge creation/restoration, shoreline protection, levees), dredging and/or pumping to displace sediment and water (barrier island/headland restoration, marsh creation), and taking advantage of head differential or gravity power to engineer the displacement of water and sediment (hydrologic restoration, diversion and channel realignment). All are energy intensive to implement, typically requiring heavy machinery to displace quantities of material during construction and are built using energy intensive cement and steel. Later, we will return to this concept when we evaluate the tradeoffs of land building.

Trends in Restoration Funding and Allocation

Several billion dollars have already been spent on new and reinforced flood protection structures, such as the $1.1 billion surge barrier near the closed MRGO channel, which was responsible for the levee break and subsequent flooding of New Orleans during Katrina (Fig. 2).[12] The 2012 LACMP allocated 50% of its spending to restoration (land building) and 50% to (flood) protection. $20 billion was allocated for dredging projects, which were expected to restore and maintain 200 square miles of wetlands over the 50-year plan. $3.8 billion was allocated for river diversion projects that are expected to build and maintain over 300 square miles of wetlands (CRPA 2012a). Barrier island and headland dune restoration were prioritized in 2007 and as of the end of 2016, $1 billion has been spent (Fig. 2) and nearly all of the barrier islands and dunes in eastern Louisiana were restored (CPRA 2015a, b).

Funding for coastal restoration comes from a variety of state and federal funding pools. Presently, the state budget is affected by revenues from oil produced on the outer continental shelf region, but this amount is expected to increase significantly as part of Phase II of the Gulf of Mexico Energy and Security Act of 2006 (GOMESA). An increasing share of energy revenues will be available for coastal restoration and funding of the LACMP. In 2018, GOMESA revenue for coastal restoration and protection will increase from roughly $80,000 to $140 million. As of the end of the 2015 fiscal year, $18 billion had been secured from state and federal funding (CPRA 2015a). In addition, Louisiana will receive an estimated $8.7 billion from state claims due to the BP Deepwater Horizon Spill (CPRA 2017a), and CWPPRA provides about $75 million per year in federal funding. The operating revenue and expenditures for 2017 are expected to be about $735 million. After 2017, project revenue will drop significantly to $495 million in 2018 and $526 million in 2019 (CPRA 2016).

From Planning to Project Completion

There are many steps that must be completed before a project can be constructed. First the project must be

[12] http://www.npr.org/2015/08/28/432059261/billions-spent-on-flood-barriers-but-new-orleans-still-a-fishbowl

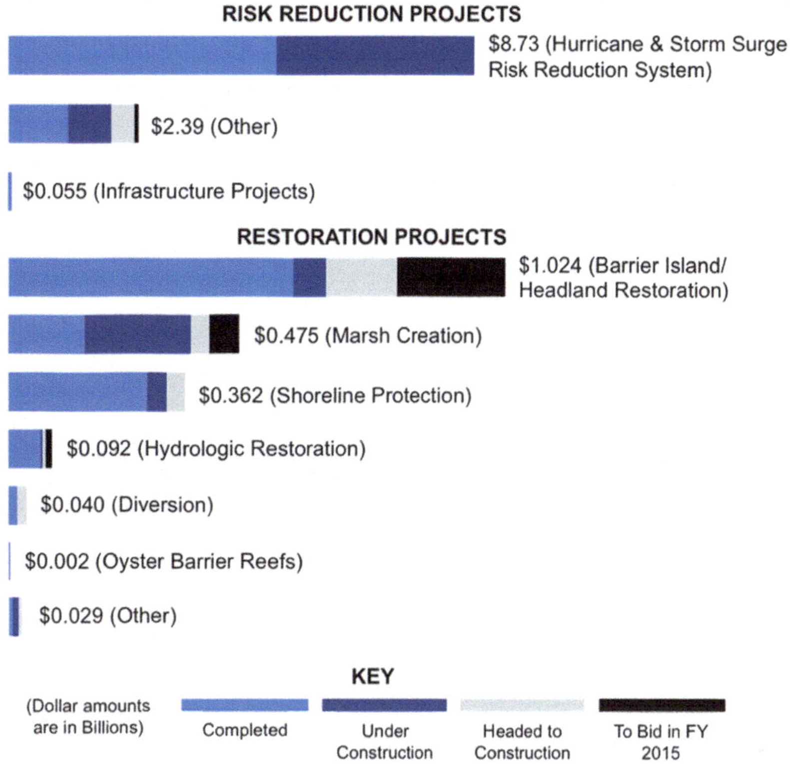

Fig. 2 Allocation of funds to coastal restoration and protection as of 2016 (Source: modified from CPRA 2015a)

proposed, which includes cost engineering, construction drawings, and submission to CPRA. Once a project is submitted, it then goes through a round of scrutiny from technical advisory panels and is evaluated using a priority checklist. If it meets the criteria of the CPRA then the project makes it into a project priority list. Before any project gets approved for funding, projects are ranked using a set of criteria. After this process only a few of the most suitable projects are selected (CPRA 2012c). Once a project is approved, the project bidding commences and a variety of construction permits must be obtained. Getting such permits approved typically includes a series of intensive geotechnical surveys to evaluate structural integrity and a survey of surrounding biota and hydrology for development of an environmental impact statement.

Virtually all restoration activities require approval from a suite of governing agencies. For example, the national environmental protection act requires an environmental impact statement for all projects appropriated with state or federal funds.[13] Because proposed LACMP projects fall within political boundaries of the Louisiana Coastal Zone and the navigable waters of the Unites States, most projects are subject to permitting and comment, under the State and Local Coastal Resources Management Act of 1978,[14] Section 404 of the Clean Water Act, the Rivers & Harbors act, the Endangered Species Act, etc. This means that in addition to state and local permitting, the Army Corps of Engineers, Environmental Protection Agency, Fish and Wildlife Service, National Marine Fisheries Service and perhaps others, must sign off on every project. For each agency that must sign off on the project there is potential for a delay or costly redesign.

The Endangered Species Act is often problematic and expensive to deal with when endangered species are found onsite. For example, barrier island and headland beach and dune restoration projects in Louisiana have suffered frequent delays due to nesting of the least tern (*Sternula antillarum*),[15] a migratory shorebird. The least tern prefers to nest on an exposed open beach with a good view of predators; after mating the birds will lay eggs and guard their nest for up to 2 months. If a least tern settles on the beach of an ongoing project, construction cannot commence within a 300-foot radius of the nest. Project managers have resorted to extraordinary proactive measures to prevent nesting. For example, existing beaches are plowed like a sugarcane field prior (Fig. 3); the ridges make for poor visibility and deter the terns from nesting on the beach.

In addition to permitting and comment, the state is required to seek imminent domain purchase agreements and to give just compensation for private landowners

[13] https://en.wikipedia.org/wiki/National_Environmental_Policy_Act
[14] http://data.dnr.la.gov/lcp/lcphandbook/lcphandbook.pdf

[15] https://en.wikipedia.org/wiki/Least_tern

Fig. 3 Caminada Headland Beach and Dune Restoration Project (BA-45). The beach has been plowed to deter least tern (*Sternula antillarum*) nesting while the contractors await the arrival of dredged sand from a far-off shoal. Offshore oil rigs can be seen on the horizon (Photo Credit: Adrian Wiegman)

located in areas directly impacted by diversion pathways. This compensation includes physical takings from potential changes in state water bottoms and economic takings via losses of cash flow on oyster leases and other fishing activities.

Once a project is underway, the contractor logs daily construction and monitoring reports. Upon completion, the contractor files a project completion report, which summarizes the location and type of project, the final as-built features, the key project cost elements (construction, engineering, land rights, monitoring, operations, and maintenance), the items of work (a list of costs for construction activities), the major equipment used, and the sequence of activities (including problems and solutions). In some cases as-built drawings are attached or submitted separately. Once the completion report is given the final acceptance the project is considered complete. After completion, each restoration project is monitored and surveyed over time (Folse et al. 2008). The monitoring process helps gauge the success of the project and can inform future project selection based on patterns that emerge from successful and unsuccessful designs. These reports are usually published online by the agency that oversaw the project (e.g. CWPPRA or CPRA).[16]

[16] CWPPRA projects: https://www.lacoast.gov/new/Projects/Default.aspxCPRA projects: http://cims.coastal.louisiana.gov/outreach/OPL_Full_page.html

Tradeoffs of Land Building Approaches

The two most significant restoration strategies for land building are marsh creation via pumped sediments and river diversions. Both marsh creation and river diversions are examples of projects that require a vast amount of energy and capital during implementation (Day et al. 2005; Day and Moershbaecher 2014; Tessler et al. 2015). Dredging sediment for marsh creation builds land quickly, but requires periodic re-nourishment to be sustainable. After construction, diversions have minimal recurring costs and may be operable for over a century (e.g. Bonnet Carré Spillway), but build land gradually (see chapter "Large Infrequently Operated River Diversions for Mississippi Delta Restora tion"). The LACMP states that selection of restoration projects must achieve a balance between building land over the short term, building future land, providing ecosystem services benefits and supporting a working coast (CPRA 2012a). Deciding what mix of marsh creation and river diversion projects should be funded, while balancing these objectives is a challenging task that has necessitated the use of multicriteria decision support tools (Groves et al. 2013).

The state has an obligation to protect people now, but must also try to secure a sustainable future. These objectives can be contradictory. For example, coastal communities rely on oyster leases and fishing grounds that occur in specific salinity ranges and a mixture of wetlands and open water. While river diversions may be a cost effective long-term

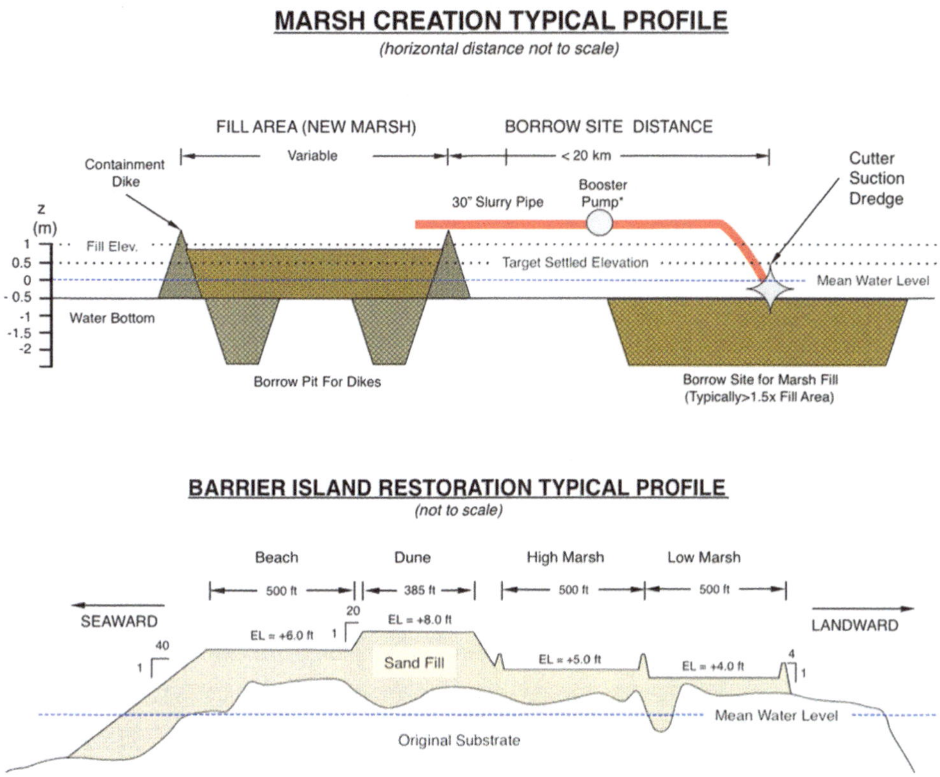

Fig. 4 Typical profiles of an idealized marsh creation project (*top*) and a barrier island beach and dune restoration project (*bottom*). In the mud rich Mississippi Delta, sediment for marsh creation projects can be sourced from nearby bays, while beach and dune restoration requires sand to be pumped or shipped from 10s of km away (Adapted from CPRA 2017c)

approach, by impacting water levels, currents and salinity, diversions have potential to disrupt the current livelihoods of coastal Louisianans that rely on fishing for income. On this merit, using marsh creation to build wetlands, which does not generate a perceivable, long-term, local environmental disturbance (other than carbon emissions), conforms more generally to the goal of supporting the working coast. When marsh creation and river diversions are analyzed with traditional economics (net present value with a discount rate), the speed at which marsh creation projects build land makes them more valuable than river diversions (Caffey et al. 2014). The heavy weight put on present day gains from this type of financial valuation runs in direct opposition to sustainable management of the delta (Tessler et al. 2015). There are two main reasons for this: (1), the river built the delta and the river's sediment will be needed to sustain it; (2), the state must pay for the energy, labor and machines for each acre built with marsh creation, while gravity flow of sediment from the river is nearly free once the diversion is built. Clearly, the CPRA must rely on both river diversions and marsh creation, but how and when should the two be implemented? To answer this question, the state looks at the following factors: support for navigation, support for traditional fishing communities, support for oil and gas activities and communities, support for agricultural communities, use of natural processes, flood protection of historic properties, flood protection of strategic assets, and social vulnerability index (Groves et al. 2013). Below, we delve deeper into the economic and environmental tradeoffs of marsh creation and river diversions and look at potential synergies between them.

Hydraulic Dredging (Marsh Creation and Dune Restoration)

Marsh creation is the process of pumping sediment in order to establish wetlands in open water such as a bay (conventional marsh creation) or existing fragmented marsh (typically called marsh nourishment). Marsh creation uses hydraulic dredges (typically powered by diesel generators) to pump a sediment-rich slurry into a fill area from a borrow source generally located from 1 to 20 miles away (CPRA 2012c). The fill area is raised to a target elevation, typically 1 m above the mean water level. The sediments settle after several months resulting in a final elevation of 0.3 to 0.5 m (Fig. 4). After settlement, the area is planted and reseeded – sometimes with local volunteers – in order to reduce erosion and accelerate plant colonization and succession. Beach and dune restoration on barrier islands and headlands uses a variant of this process. The difference being that dunes are restored to several feet above the tidal frame and the sediment source must be sand of a particular grain size distribution.

Fuel, labor and capital must be allocated for every area of land restored using hydraulic dredging (Clark et al. 2015). While hydraulic dredging for marsh creation and

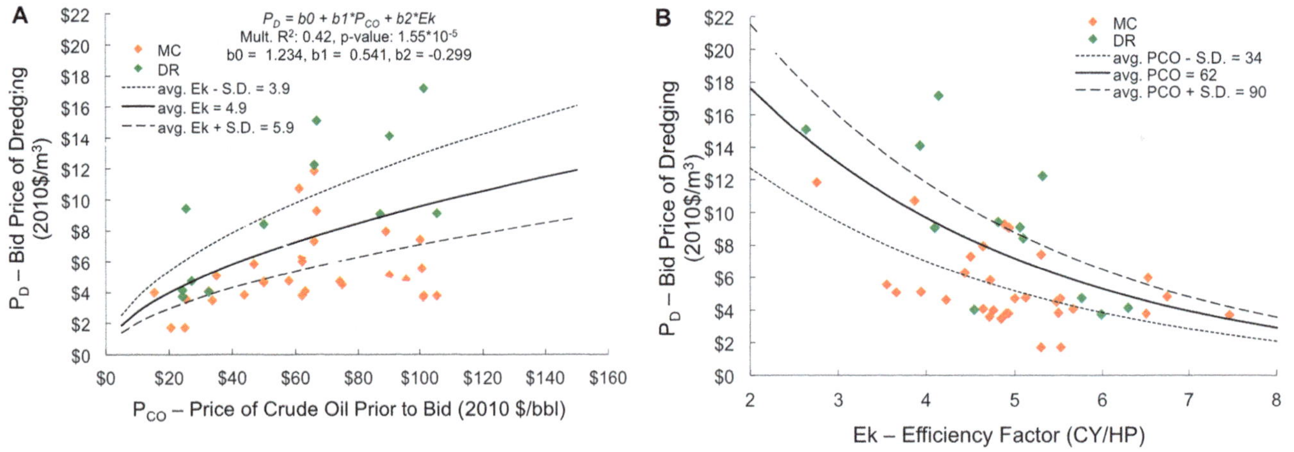

Fig. 5 Scatter plot and multiple regression results for the bid price of cutter suction dredging versus mean price of crude oil 12 mo. prior to bid (**a**) and an efficiency factor, equal to the log of borrow volume (CY) divided by horsepower (**b**). Data is for cutter suction dredging projects completed in the Louisiana coastal zone between 1994 and 2012 (Source: data is reported in Wiegman 2017, and was compiled from https://www.lacoast.gov/new/Projects/List.aspx, http://cims.coastal.louisiana.gov/outreach/OPL_Full_page.html, http://louisianadigitallibrary.org/)

barrier island restoration provides immediate benefits, projects restore only several hundred acres at a time and are limited by sources of nearby sediment. Sand is hard to find in the Mississippi Delta and must be taken from far away sources, often 10s of km away. For this reason, dune restoration is much more expensive (Figs. 5 and 6). Typically a dredging project is designed for a minimum project lifetime of 20 years (CPRA 2017b). For marsh creation, target fill heights and project lifetime estimates are based on geotechnical surveys that estimate compaction and subsidence of local sediment. For dune restoration, these estimates are based on long-shore sediment transport models (CPRA 2017c). In twenty years time, if needed, the dune will be raised to restore the area again. This is, however, expensive and thus is dependent on funding. Beach re-nourishment makes sense for high value beach front property where an acre can be worth very much to a developer, but most dune restoration in Louisiana is in areas that will never be developed. Re-nourishing marsh will also be very expensive.

The cost of marsh creation is influenced primarily by the cost of fuel and onsite labor hours. Typically 60–70% of marsh creation project cost is spent on hydraulic dredging (CPRA 2012b). Of the total cost of dredging, fuel and lubricants likely represents between 20% and 35%, though this depends on a variety of factors including price of fuel and operation efficiency (Belesimo 2000). Because fuel and lubricant make up a significant portion of the dredging cost, the unit costs of dredging are sensitive to changes in the price of crude oil (Fig. 5). Hollinberger (2010) modeled the response of dredging costs to changes in operations and fuel costs, and found that there was a 2 to 1 ratio between

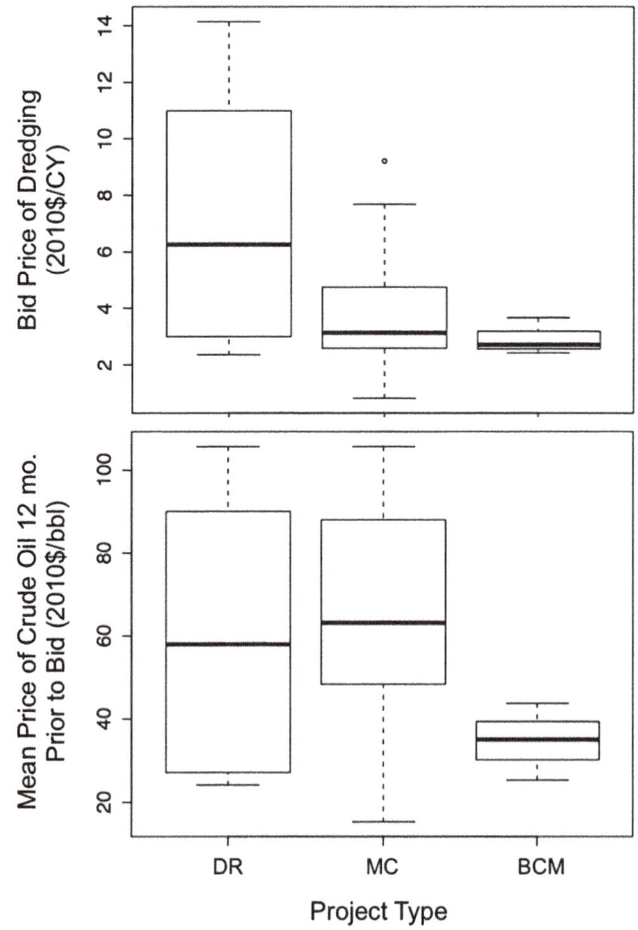

Fig. 6 Box and whisker plot of price of crude oil and bid price of dredging versus project type. *DR* dune restoration, *MC* marsh creation, *BCM* beneficial channel maintenance (Data source: see details in Fig. 5 and Wiegman 2017)

increases in fuel costs and increases in total project costs over the short term. If fuel costs were to double then project costs would increase by 50% (Hollinberger 2010). Mobilization-demobilization costs are usually 20–30% of a total marsh creation project cost. Stoppages and delays either from storms and faulty equipment are unpredictable and a 15% markup is usually added to account for such occurrences (CPRA 2012b).

The efficiency of dredging can be broken down into two elements: fuel intensity and time efficiency (Bray et al. 1997). Several factors influence the fuel intensity and resulting unit costs of pumping sediment, including the type of dredge, the type of borrow substrate (e.g., presence of oyster reefs, inshore verses off shore), and the distance from the borrow source to the fill area (Clark et al. 2015; Bray et al. 1997; Hollinberger 2010). The variables needed to estimate fuel requirements are fuel usage rates for a given engine size and rate of work (gal/hour), dredge horsepower, average engine rate (rpm or percent of max), boat speed, dredging distance and rate of sediment pumped under these parameters (CY or lb./hour) (Belesimo 2000). Most of these parameters are more or less fixed for an individual project; however, borrow source distance ranges from less than a mile to over 20 miles (CPRA 2017c; Clark et al. 2015; Murphy 2012). Borrow distance is much shorter for marsh creation than barrier island restoration, which require large quantities of sand of specific grain size. Fuel use increases with borrow site distance and so does cost (Fig. 7); a general rule of thumb in the dredging industry is that an additional booster pump is required every five miles of pipeline (CPRA 2012c).

The time efficiency of marsh creation projects controls the labor hours on site. Each project employs 20 or more workers at a time, with wages ranging from $11.50 to over $30 per hour. These costs add up to over 50% of dredging costs (Belesimo 2000, CPRA 2012j). Therefore, reducing the time to complete the project reduces costs. The prospects for increasing the time efficiency of marsh creation are not well documented. Operation time is a function of pumping rate (CY/day) and down time (days) (Bray et al. 1997); both are influenced by site location and weather and are quite variable according to project completion reports for marsh creation. Marsh creation projects proposed in the 2017 LACMP are very large (in most cases pumping greater than 1 million CY), and likely have already achieved economy of scale (CPRA 2017b). The cutter suction dredges used for marsh creation are the among largest in the world, with capacities in excess of 10,000 HP and 2500 CY/hr. through 30″ pipe (Coastal Engineering Consultants 2010). To put this in perspective, a typical 10,000 HP dredge pumping at 1500 CY/hr. could fill a typical 3000 sq. ft. home with mud in about 20 h. The point is that it is more likely that efficiency improvements will come from better project design/phasing and more skilled/coordinated dredge operators and construction crews, than from improvements in dredging technology (Clark et al. 2015).

River Diversions

Diversion projects have, perhaps, the greatest potential to sustain wetlands in the long term (>50 years) because they will be able to provide sediment to lands adjacent to the river using natural forces (Day et al. 2005) and once constructed, diversion structures are likely to last for well

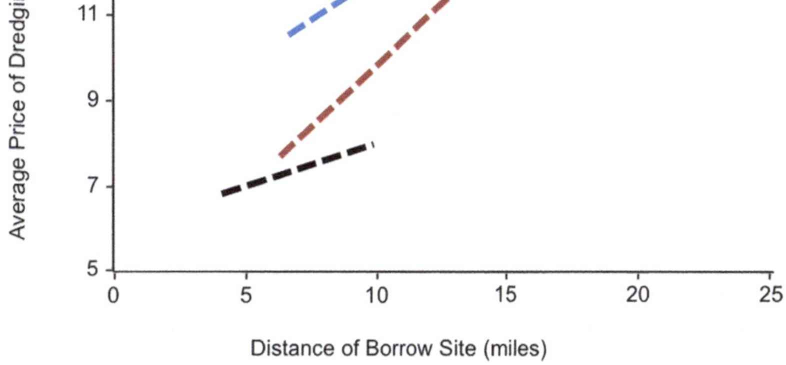

Fig. 7 Unit costs of sand as a function of borrow site distance (Source: Redrawn from CPRA 2017b)

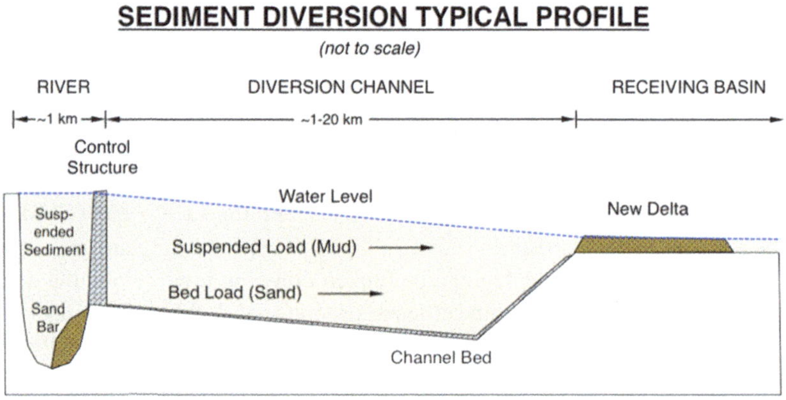

Fig. 8 Typical profile of a sediment diversion structure and channel. Suspended sediments and bed load are diverted from the main river channel into the diversion channel and deposited in the receiving basin. The channel dimensions and slope of water on the diversion channel determines the amount of sediment that can be carried into the receiving basin

over a century. Figure 8 shows a cross section of typical diversion structure.

River sediment diversions simulate the natural cycle of crevasses and flooding by diverting water and sediment from the main channel into an adjacent area (Allison and Meselhe 2010, Day and Moerschbaecher 2014). This method builds land incrementally as seasonal flood cycles progress and provide a long-term source of attendant freshwater, sediments, and nutrients to the wetlands within a diversion's area of impact. Diversion science is discussed in more detail in chapter "Large Infrequently Operated River Diversions for Mississippi Delta Restoration". The costs of river diversions are front-loaded (Caffey et al. 2014), and the land gains are not necessarily immediate or linear. The 2017 LACMP assumes that over 50 years the operation and maintenance cost of a river diversion will be roughly 15.6% of the construction cost of the diversion (CPRA 2017c). Using this factor, the average annual operation and maintenance costs over the lifetime of a $1 billion river diversion would be roughly $3,120,000 per year.

For a diversion of any particular discharge capacity there are several factors that influence both the cost and the sediment load. The dimensions, location, and operational strategy are contributing factors to the cost and land building potential of a river diversion. The potential sediment load is determined by the positioning of the diversion on the river, angle of alignment, conveyance channel dimensions, and mean hydraulic head between the river and outfall basin (Allison and Meselhe 2010). The cost is related to the dimensions of the conveyance channel, the built infrastructure between the river and the outfall area (which impacts complexity and cost of land rights), and the cost of energy and materials at the time of bidding and construction. Kenney et al. (2013) fit a production function to box-culvert river diversions (e.g. Canaervon, Davis Pond), which have low variance in channel length and found that the cost of the diversion correlated strongly with the channel's width and depth. What this implies is that the cost of construction is related to the volume of material used/displaced to build the diversion control structure and channel.

It is imperative to study the design and placement of diversions to maximize sediment capture early in the restoration program (Allison and Meselhe 2010, Meselhe et al. 2016). Simpler designs in areas with less existing infrastructure and greater hydraulic head will tend to have lower costs per unit of discharge capacity; the elevation of the river head can be greater than 8 m near Sorrento and less than 3 m for Breton and Lower Barataria. The ability to manage flow adds to the cost of a diversion. Using explosives to blow a hole in a levee to form a crevasse (as was done at Caernarvon during the 1927 flood) is probably the simplest and most cost effective type of sediment diversion. Sediment retention is highest in areas with existing wetlands and further upriver where there is more development (Roberts et al. 2015; Xu et al. 2016). However, uncontrolled crevasse flow is not desirable in areas with human development. In addition, there is an economic incentive to divert less water per unit sediment to minimize impacts on salinity and fisheries. As a result of these factors, more complex and costly designs are being investigated by CPRA.

From the 2012 to 2017 LACMP, the average discharge capacity of proposed diversions decreased while the estimated average cost and net 50 year land benefit increased. In the 2012 LACMP, the average discharge capacity was 2356 m^3/s, the average cost was $360,000,000, and the average 50 year land benefit was 8092 ha (CPRA 2012c, nominal 2012 dollars). In the 2017 LACMP, the average discharge capacity was 836 m^3/s, the average cost was $501,170,000, and the average 50 year land benefit was 17,175 ha (CPRA 2017c, nominal 2017 dollars).

The relationships between discharge capacities, cost and land building changed greatly between 2012 and 2017. Figure 9 compares metrics for the diversions proposed in the 2012 and 2017 plans. In the 2012 LACMP, cost and land building potential of modeled sediment diversions were positively correlated to discharge capacity (Fig. 9a, c). The cost per unit area of land built for diversions modeled in the LACMP decreased non-linearly with discharge (Fig. 9d). In

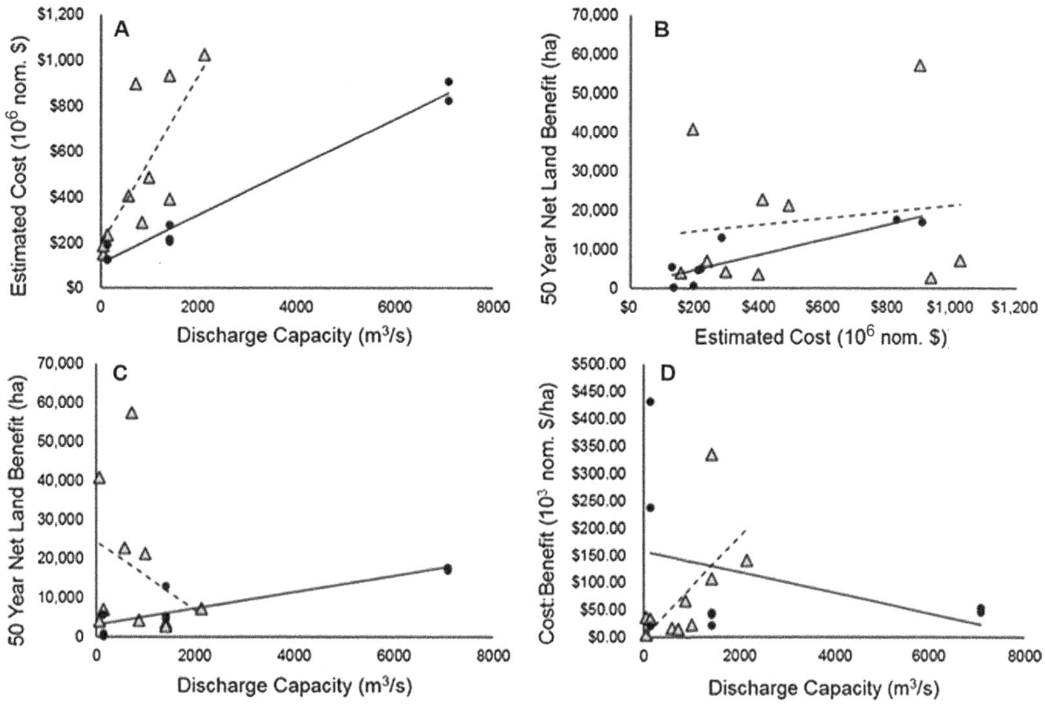

Fig. 9 Cost versus discharge capacity and land building for diversions proposed in the 2012 (*grey triangles and dashed trend lines*) and 2017 (*black circles and continuous trend lines*) LACMP. (a) cost (million nominal dollars) versus discharge (cubic meters), (b) 50 year net land benefit (ha) versus cost (million nominal dollars), (c) 50 year net land benefit (ha) versus discharge, (d) cost per unit area of 50 year net land benefit (thousand nominal dollars per ha) versus discharge (Data from CPRA 2012c, 2017c)

the 2017 LACMP, the cost increased more steeply with discharge than in 2012 (Fig. 9a), land benefit did not show a strong relation to cost (Fig. 9b), land benefit decreased with discharge (Fig. 9c), and cost benefit ratio increased with discharge (Fig. 9c).

The costs and benefits of diversions are greatly impacted by diversion design and location (Allison et al. 2012). The changes in anticipated outcomes between the 2012 and 2017 LACMP (Fig. 9) are likely due to improving resolution of models and the CPRA's focus on maximizing sediment delivery per unit of discharge. An improved understanding of the dynamics of both the Mississippi River and outfall basin (e.g. Allison et al. 2014, Gaweesh and Meselhe 2016, Meselhe et al. 2016) has allowed CPRA to develop diversion designs with higher sediment load per unit discharge. Although the new diversion designs are more costly, they more efficiently use water to deliver sediment (Gaweesh and Meselhe 2016), which reduces impacts on salinity gradients and water levels in outfall basins (See chapter "Large Infrequently Operated River Diversions for Mississippi Delta Restoration"). Having less disruptive sediment diversions will help the CPRA achieve its dual objective of land building and support of coastal industry.

Comparing the Options

Compared to the slow rate of land building provided by river diversions, hydraulic dredging for building land is a high-powered approach to restoration that builds land quickly, having immediate tangible benefits. However, over the long-term river diversions are likely to be more cost effective for building land than marsh creation, especially when considering increases in the cost of energy. Marsh creation projects listed on the CWPRA database[17] had an average cost of $182,040 per hectare (net area benefited after 20 years). Proposed water diversion (including siphons, hydrologic restoration, freshwater and sediment diversion) projects listed on Lacoast.gov have an average cost of $16,584 per hectare (net area benefit after 20 years). In the 2012 LACMP, marsh creation projects had an average cost of about $360,000 per hectare over a 50-year time span, while sediment diversions had an average cost (including engineering, operation and maintenance) of about $45,000 per hectare. The 2017 LACMP did not report land benefits estimates for river diversion and marsh creation separately as many marsh creation projects are located in the outfall of river diversions. Instead the plan estimates that the

[17] https://lacoast.gov/new/Projects/List.aspx

combined effect of proposed marsh creation and diversion projects will result in a net land benefit of 300,000 ha over 50 years, which equates to roughly $86,000 ha.

The timing of benefits received from restoration projects is a key factor that should be taken into account. The CPRA does not use a discount rate in its decision tool (CPRA 2012g), which prevents biasing of cost and benefits based on timing. Traditional economic thinking suggests that the state should employ a positive discount rate. Positive discount rates undervalue future costs and benefits, biasing decision making towards projects with rapid returns on investment, such as march creation, and away from projects with gradual returns, such as river diversions (Gowdy and Ericson 2005; Beckerman and Hepburn 2007). Net present value (NPV) and other economic analyses have been used to compare the cost effectiveness of diversion projects and marsh creation with respect to timing of received benefits independently of the LACMP. When a 5% discount rate is included, the life-time project benefits of marsh creation are calculated to be more valuable than diversion benefits (Caffey et al. 2014). This view (biased towards present benefits) is counter to the view of the great conservationist Aldo Leopold, who made the argument that economic measures were not able to properly quantify or protect the integrity of an ecosystem. According to Leopold's "land ethic", an action is right if it works to preserve the integrity of ecosystem, and wrong if it does otherwise (Leopold 1970). Daly (1994) stresses that ecosystems underwrite all economic activity. By not placing value on activities that protect the integrity of the delta ecosystem, using conventional cost benefit analysis to inform restoration could result in a decision pathway that may ultimately undermine the economy.

Society has degraded marshlands in the Mississippi Delta by isolating the river from wetlands and causing pervasive hydrologic change. River diversions are the closest approach to restoring the delta's natural processes, and model simulations suggest that sustained coastal restoration cannot be achieved without them (Wang et al. 2014; CPRA 2012a, 2017a). The State's use of the planning tool (Groves et al. 2013), and the bias that CPRA puts on future verses present cost and benefits will play an important role in which projects get selected. Conventional cost benefit analysis, which utilizes positive discount rates, is simply not appropriate for sustainable delta management. Certainly the state should build structures to protect dense areas of people and capital as soon as possible, and the LACMP does include proposals for this, but if these structures also compromise the natural processes that built and sustained the delta in the first place then the approach must be reconsidered. CPRA is beginning to look at synergy between river diversions and hydraulic dredging. It is likely that more complex modeling of ecosystem interactions will be needed to estimate the true benefits of and tradeoffs between river diversions and marsh creation.

Sustainable Steps Forward for the Mississippi Delta

The accelerating rates of sea-level rise, more strong hurricanes, and more extreme weather projected in the latter half of the twenty-first century (see "Energy and Climate – Global Trends and Their Implications for Delta Restoration") will require enormous effort to overcome if the delta is to be sustained. But climate change is not the only forcing that will limit restoration. Given the LACMP's 50-year planning horizon and the cost and energy intensity of restoration, it is reasonable to assume that changes in the global economy and energy availability will influence the long term affordability of coastal restoration. In other words, as fossil fuels deplete and energy becomes more expensive (see "Energy and Climate – Global Trends and Their Implications for Delta Restoration") the cost per unit of restoration effort will likely go up (Figs. 5 and 10). The combination of climate forcings and energy cost increases may lead to significant components of the LACMP being unaffordable in a few decades (see chapter "Large Infrequently Operated River Diversions for Mississippi Delta Restoration"). Studying global economic trends will shed light on the most prudent pathways for investment of public resources in coastal Louisiana. For example, the effect of fossil fuel scarcity on the future cost of restoration was not considered in the 2007, 2012, and 2017 LACMPs, but must be considered in the future.

Considering the long-term trends that will affect the delta, the approaches to coastal restoration should be reimagined. For example, hydraulic dredging must be used very carefully and effectively. Marsh creation projects must focus on maximizing the benefits from dredged sediment. This could be achieved by placing projects near diversion sites (see "Large Infrequently Operated River Diversions for Mississippi Delta Restoration") or by using alternative dredging designs, such as thin deposition on failing marsh (Ford et al. 1999) or a latticed approach (Clark et al. 2015). One possible way to increase the sustainability of barrier islands is to armor beaches using timber. Timber that is carried down stream during floods has been identified as a major contributor to shoreline morphology in undisturbed river basins (Kramer and Wohl 2015). The beaches of the Misssissippi Delta were reportedly covered in timber when explorers first arrived at its shores (Condrey et al. 2014). Timber can also be used as a structural protection tool in coastal communities of industries and major ports. Cypress forests are also a robust structural protection tool in freshwater basins (see chapter "Optimum Use of Fresh Water to Restore Baldcypress – Water Tupelo Swamps and Freshwater Marshes and Protect Against Salt Water Intrusion: A Case Study of the Lake Pontchartrain Basin"). Future studies should examine these sorts of approaches in further detail.

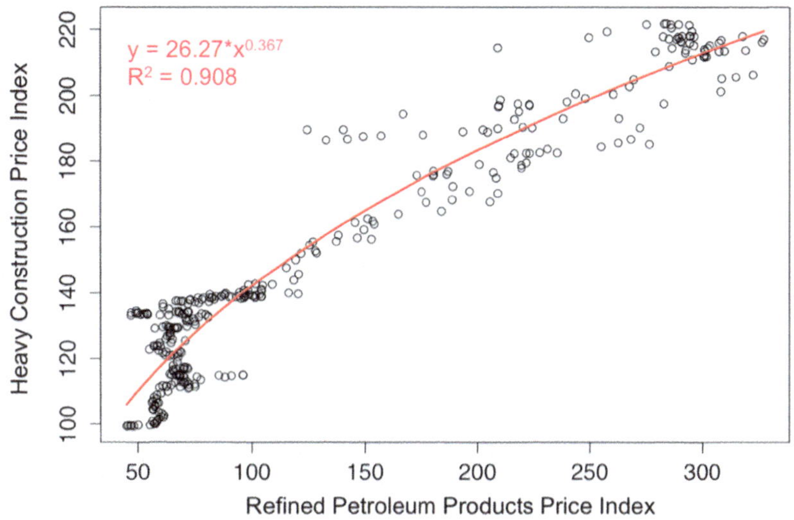

Fig. 10 Monthly nominal price index data (June 1986–Oct 2014), for heavy construction (code: BCON) versus refined petroleum products (code: WPS0571). Nonlinear regression line and equation are shown in *red* (Data from Bureau of Labor and Statistics https://www.bls.gov/data/#prices)

Financial Constraints

The importance of energy on the affordability of coastal restoration in the twenty-first century is stressed in chapter "Energy and Climate – Global Trends and Their Implications for Delta Restoration", earlier this chapter, and in Day et al. (2005), Day and Moerschbaecher (2014), and Tessler et al. (2015). But even without consideration of energy, there are serious financial constraints on Mississippi Delta restoration. A 2015 study by the Tulane Institute on Water Resources Law and Policy analyzed the funding availability and costs for the 2012 LACMP. After including critical omissions from the LACMP, such as maintenance of flood control structures, the study estimated that the cost to restore Louisiana's coastline and flood protection infrastructure would be about $90 billion, greatly exceeding the $50 billion estimate of the 2012 LACMP (Davis et al. 2015). The report stated:

> *Furthermore, [the LACMP] openly and explicitly excludes from its scope several vital responsibilities related to the sustainability of the coast and its communities. ...[Including] navigation channel bank maintenance, Mississippi River Gulf Outlet (MRGO) ecosystem restoration, the operation, maintenance, and rehabilitation of certain flood risk reduction infrastructure, and rainwater (and related subsidence) management within polders. ...[CPRA] does not... have sufficient funds lined up to fully implement [the] plan. The sum of the [identified funding] sources is approximately $20.617 billion over the 50 year planning period... the currently identified funding sources fall well short of the $91.693 billion inflation adjusted total cost [which includes costs excluded from the scope of the plan]. This $71.076 billion gap means that 77.5% of the total cost still needs to be secured.*

Quite bluntly, the state is in need of an extraordinary windfall if the LACMP goals are to be met at even current energy prices (Davis and Vorhoff 2014). There is a great amount of embodied energy in the Mississippi flood control system. The levee system must be maintained in order to protect coastal populations and commerce, but doing so will be extraordinarily energy intensive. We have already shown that oil prices and dredging costs have a positive correlation (Fig. 5). The energy intensity of coastal restoration is likely similar to the energy intensity of heavy construction (Day and Moerschbaecher 2014) and the price of heavy construction is highly correlated to the price of petroleum (Fig. 10). From this, we can presume that future increases in energy costs will cause the amount of funding needed to restore the coast to rise. This is a sobering reality for many of the world's developed deltas. Industrialized deltas that rely on energy-intensive structural protection, such as the Mississippi and the Rhine delta will be at the highest risk of failure in a future with high energy costs (Tessler et al. 2015). This is because the Rhine and Mississippi deltas contain two of the world's largest ports and have complex urbanized flood control systems. A high percentage of the population in these deltas live in high-risk areas that are below sea level. This type of development is increasingly becoming the case in many of the world's deltas, such as the Po, Nile, Mekong and Yangtze (Syvitski et al. 2009). Industrialized deltas experienced higher rates of subsidence after drainage and high levels of development (Syvitski et al. 2009). In addition, the flood control systems of these deltas are integrated within complex urban, agricultural, and industrial areas along the coast that make interventions expensive and politically and technically difficult.

To date, no large river diversion has been authorized, while nearly $541 million has been spent or approved for marsh creation (CPRA 2015a). In summer 2016 a milestone for diversion implementation was achieved. The state legislature settled on a program to move oyster leases from grounds where diversions would affect salinity.[18] So slowly diversions are gaining ground in the public arena. River

[18] http://www.theadvocate.com/baton_rouge/news/politics/legislature/article_0dea5912-b540-5b4c-8cb6-f3e6a91af2f9.html

diversions were not going to be implemented without a proactive policy to protect and fairly distribute new oyster leases to those affected by diversions. Still, as of the end of 2016 no river diversion had passed beyond the engineering and design phase. Construction of any new diversion is not likely to start until 2020 at the earliest.

Future tends in world energy markets will affect Louisiana's ability to execute the LACMP (see chapter "Energy and Cli mate – Global Trends and Their Implications for Delta Restora tion"). If the diversion planning process is delayed beyond 2020, there is a chance that any diversion built will have greater costs and reduced benefits due to increasing energy costs and rates of sea level rise. NPV analyses apply a discount on future returns from investments made in the present, primarily because the future is not guaranteed. Just as land built in the future is not guaranteed to benefit people who spent money in the present, the state isn't guaranteed to have money to spend in the future, nor is the same amount of money spent in the future guaranteed to have as much benefit as money spent today (e.g. higher energy prices resulting in higher unit costs of restoration; higher rates of sea level rise diminishing the impact of restoration projects). Therefore, it can be argued that if the future outlook is negative, a negative discount rate that would inflate the cost and benefit of future investments, should be used for NPV (Beckerman and Hepburn 2007).

Volatility and long-term increases in oil price may impact the ability of the state to pay for hydraulic dredging on the massive $17.8 billion scale proposed in the 2017 LACMP. From 2001–2016, oil prices have been more volatile, compared to the period between 1986–2000, when prices averaged about $25/bbl (EIA 2017). Oil influences the prices of most commodities due to its near ubiquitous use as an input in supply chains (Sadorsky 2014; Shafiee and Topal 2010; Ji and Fan 2012; Gardebroek and Hernandez 2013). Volatility induced price changes integrate through the economic system and result in changes in spending patterns on the macro scale and the ability to pay back debt (Kerschner et al. 2013; Tverberg 2012; Ebrahim et al. 2014; Chen and Hsu 2012). When oil prices increase or decrease rapidly, economic growth is stunted (Hamilton 2009; Aucott and Hall 2014). Therefore, as resources of cheap oil are depleted and oil becomes more expensive, high price volatility and poor economic performance can be expected (Dittmar 2016). Restoration funding is dependent upon oil and gas revenues, and as these decline so will the restoration budget (Davis et al. 2015). When oil prices are high, state and GOMESA funds may increase but restoration will also be more expensive; conversely, when oil prices are low restoration costs may decrease (Fig. 5) but so will restoration funding (Davis et al. 2015; CPRA 2015a).

There are significant financial limitations to the LACMP that will be magnified if energy prices go up as they are projected to do (EIA 2017). The tradeoffs between the timing of restoration projects and the relation of these projects to energy markets needs to be understood and incorporated by coastal planners. For example, how do funding and costs of restoration change when oil prices are high or low; can financing be structured to take advantage of price fluctuations rather than being a victim of them? The timing of restoration investments in relation to global trends in economics and energy must be considered. Costs of marsh creation and river diversions under different scenarios for oil prices must be studied further both from direct cost and macroeconomic perspectives (chapter "Large Infrequently Operated River Diversions for Mississippi Delta Restoration" gives an example of this type of analysis).

Summary and Conclusions

The Louisiana Coastal Master Plan was designed to be a living document and has been revised according to information from models and evolving science. 2012 LACMP models indicated that long-term restoration and land building were not possible without large river diversions located in middle and upper estuaries. The 2017 LACMP includes more sophisticated diversion designs placed further upstream and have more efficient delivery of sediment per unit of discharge capacity; however, the 2017 plan concedes that net land gain is not possible even with maximized use of river sediment. While it is clear that river diversions will be a necessary and vital aspect of coastal restoration, uncertainty still persists about the effectiveness of river diversions and their economic benefit in the short term. It is possible that they could be a detriment to the economy in the short term, as fishermen struggle to adapt to changes in the Breton Sound and Barataria estuaries, and land building benefits from diverted sediment are not yet realized. These uncertainties regarding benefits from diversions support the necessity of marsh creation projects in the near term, and continued study of diversions. Adaptive management can be used to reap the maximum benefits from diversions and this may include pulsed operation and putting marsh creation projects near the outfall of river diversions (see chapter "Large Infrequently Operated River Diversions for Mississippi Delta Restoration"). State of the art numerical models will continue to be a critical component of the planning process. An appropriate next step for the LACMP would be to incorporate global economic and energy considerations (e.g. EIA 2017) into the planning process.

References

Allison MA, Meselhe EA (2010) The use of large water and sediment diversions in the lower Mississippi River (Louisiana) for coastal restoration. J Hydrol 387(3):346–360

Allison MA, Ramirez MT, Meselhe EA (2014) Diversion of Mississippi River Water Downstream of New Orleans, Louisiana,

USA to Maximize Sediment Capture and Ameliorate Coastal Land Loss. Water Resour Manag 28(12):4113–4126
Aucott M, Hall C (2014) Does a Change in Price of Fuel Affect GDP Growth? An Examination of the US Data from 1950–2013. Energies 7(10):6558–6570
Barnes S, Bond C, Burger N, Anania K, Strong A, Weilant S, Virgets S (2015) Economic evaluation of coastal land loss in Louisiana. Lousiana State University and the Rand Corporation. Published online. http://coastal.la.gov/economic-evaluation-of-land-loss-in-louisiana/
Batker D, Torre ID, Costanza R, Swedeen P, Day JW, Boumans R, Bagstad K (2010) Gaining ground: wetlands, hurricanes, and the economy: the value of restoring the Mississippi River Delta. Environ Law Rep News Anal 40(11):11106–11110
Beckerman W, Hepburn C (2007) Ethics of the discount rate in the Stern Review on the economics of climate change. World Econ-Henley Thames 8(1):187
Belesimo FJ (2000) Cost estimating projects for large cutter and hopper dredges. Doctoral dissertation, Texas A&M University
Bray RN, Bates AD, Land JM (1997) Dredging: a handbook for engineers
Caffey RH, Wang H, Petrolia DR (2014) Trajectory economics: Assessing the flow of ecosystem services from coastal restoration. Ecol Econ 100:74–84
Chen SS, Hsu KW (2012) Reverse globalization: Does high oil price volatility discourage international trade? Energy Econ 34 (5):1634–1643
Chmura G (1992) Modeling coastal marsh stability in response to sea level rise: a case study in coastal Louisiana, USA. Ecol Model 64:47–64
Clark FR, Bienn HC, Willson CS (2015) Assessing the cost of coastal land creation using dredged material. The water institute of the Gulf. Funded by the Coastal Protection and Restoration Authority under the Science & Engineering Plan–Project implementation support task. Baton Rouge
Coastal Engineering Consultants (2010) Pass Chaland to Grand Bayou Pass Barrier Shoreline Restoration Project (BA-35) Project Completion Report. Prepared for: State of Louisiana Office of Coastal Protection and Restoration New Orleans Field Office and NOAA Fisheries US Department of Commerce
Cobell Z, Zhoa H, Roberts HJ, Clark R, Zou S (2013) Storm surge and wave modeling for the Louisiana 2012 Coastal Master Plan. In: Peyronnin NS, Reed D (eds) Louisiana's 2012 Coastal Master Plan Technical Analysis. J Coast Res, Special Issue 67
Condrey RE, Hoffman PE, Evers DE (2014) The last naturally active delta complexes of the Mississippi River (LNDM): discovery and implications. In: Day JW et al (eds) Perspectives on the restoration of the Mississippi delta, Estuaries of the World. Springer, Dordrecht
Couvillion BR, Steyer GD, Wang H, Beck HJ, Rybczyk JM (2013) Forecasting the effects of coastal protection and restoration projects on wetland morphology in coastal Louisiana under multiple environmental uncertainty scenarios. In: Peyronnin NS, Reed DJ (eds) Louisiana's 2012 Coastal Master Plan Technical Analysis. Journal of Coastal Research, Special Issue 67
CPRA (2007) Louisiana's comprehensive master plan for a sustainable coast. Coastal Protection and Restoration Authority of Louisiana (CPRA), Baton Rouge
CPRA (2012a) Louisiana's Comprehensive master plan for a sustainable coast. 2012 Coastal master plan. CPRA, Baton Rouge
CPRA (2012b) Appendix A: project definitions. 2012 Coastal master plan. CPRA, Baton Rouge
CPRA (2012c) Appendix A: 2 Project fact sheets. 2012 Coastal master plan. CPRA, Baton Rouge
CPRA (2012d) Appendix C: Environmental scenarios. 2012 Coastal Master Plan. CPRA, Baton Rouge
CPRA (2012e) Appendix D: decision support tools–predictive models. CPRA, Baton Rouge
CPRA (2012f) Appendix D-1 Eco-hydrology model technical report. CPRA, Baton Rouge
CPRA (2012g) Appendix D: 2 Wetland morphology model technical report. CPRA, Baton Rouge
CPRA (2012h) Appendix E: decision support tools–planning tool. 2012 Coastal master plan. CPRA, Baton Rouge
CPRA (2012i) Appendix F: 1 Adaptive management. CPRA, Baton Rouge
CPRA (2012j) Appendix K: Davis-Bacon prevailing wage determinations. CPRA, Baton Rouge
CPRA (2012k) Bid tool – "PPL25 nominee west vermilion MC SP final cost 4–01-15". CPRA, Baton Rouge
CPRA (2014) 2017 coastal master plan: model improvement plan. Version II (March 2014), prepared by The Water Institute of the Gulf. Coastal Protection and Restoration Authority, Baton Rouge, p 52
CPRA (2015a) Fiscal year 2016 annual plan: integrated ecosystem restoration and hurricane protection in coastal Louisiana. CPRA, Baton Rouge
CPRA (2015b) Barrier Island status report draft fiscal year 2016 annual plan February 2015. CPRA, Baton Rouge
CPRA (2016) Fiscal year 2017 annual plan: integrated ecosystem restoration and hurricane protection in coastal Louisiana. CPRA, Baton Rouge
CPRA (2017a) Louisiana's comprehensive master plan for a sustainable coast. 2017 Coastal master plan. Louisiana Coastal Protection and Restoration Authority, Baton Rouge
CPRA (2017b) Appendix A: project definitions. 2017 Coastal Master Plan. Louisiana Coastal Protection and Restoration Authority, Baton Rouge
CPRA (2017c) Appendix A: 2 project fact sheets. 2017c Coastal Master Plan. Louisiana Coastal Protection and Restoration Authority, Baton Rouge
CPRA (2017d) 2017 draft coastal master plan. http://cimscoastalloui-sianagov/masterplan/. Accessed 10 Feb 2015
CPRA (2017e) Appendix C: environmental scenarios. 2017 Coastal master plan. Louisiana Coastal Protection and Restoration Authority, Baton Rouge
Daly HE (1994) Fostering environmentally sustainable development: four parting suggestions for the World Bank. Ecol Econ 10 (3):183–187
Das A, Justic D, Inoue M, Hoda A, Huang H, Park D (2012) Impacts of Mississippi River diversions on salinity gradients in a deltaic Louisiana estuary: ecological and management implications. Estuar Coast Shelf Sci 111:17–26
Davis M, Vorhoff H (2014) Financing the future. Turning coastal restoration and protection plans into realities: the cost of comprehensive coastal restoration and protection. First in an occasional series. An issue paper of the Tulane Institute on Water Resources Law and Policy
Davis M, Vorhoff H, Boyer D (2015) Financing the future. Turning coastal restoration and protection plans into realities: the cost of comprehensive coastal restoration and protection. second in an occasional series. An issue paper of the Tulane Institute on water resources law and policy
Day JW, Moerschbaecher M (2014) The impact of global climate change and energy scarcity on Mississippi delta restoration. In: Perspectives on the Restoration of the Mississippi Delta. Springer Netherlands, Dordrecht, pp 175–184
Day JW, Britsch LD, Hawes SR, Shaffer GP, Reed DJ, Cahoon D (2000) Pattern and process of land loss in the Mississippi Delta: a spatial and temporal analysis of wetland habitat change. Estuaries 23(4):425–438
Day JW Jr, Barras J, Clairain E, Johnston J, Justic D, Kemp GP et al (2005) Implications of global climatic change and energy cost and availability for the restoration of the Mississippi delta. Ecol Eng 24 (4):253–265

DeLaune RD, Pezeshki SR, Patrick WH Jr (1987) Response of coastal plants to increase in submergence and salinity. J Coast Res 3 (4):535–546

DeLaune RD, Whitcomb JH, Patrick WH Jr, Pardue JH, Pezeshki SR (1989) Accretion and canal impacts in a rapidly subiding wetland. I. 137Cs and 210Pb techniques. Estuaries 12(4):247–259

DeLaune RD, Nyman JA, Patrick WH Jr (1994) Peat collapse, ponding and wetland loss in a rapidly submerging coastal marsh. J Coast Res 10(4):1021–1030

Dittmar M (2016) Regional oil extraction and consumption: a simple production model for the next 35 years Part I. arXiv preprint arXiv:1601.07716

Ebrahim Z, Inderwildi OR, King DA (2014) Macroeconomic impacts of oil price volatility: mitigation and resilience. Front Energy 8 (1):9–24

EIA (2017) Annual energy outlook 2017 with projections to 2050. United States Energy Information Administration (EIA). January 5, 2017 www.eia.gov/aeo

Folse TM, West JL, Hymel MK, Troutman JP, Sharp L, Weifenbach D, McGinnis T, Rodrigue LB (2008) A Standard Operating Procedures Manual for the Coastwide Reference Monitoring System; Wetlands Methods for Site Establishment, Data Collection, and Quality Assurance/Quality Control. Louisiana Coastal Protection and Restoration Authority, Office of Coastal Protection and Restoration, Baton Rouge, p 191

Ford M, Cahoon DR, Lynch JC (1999) Restoring marsh elevation in a rapidly subsiding salt marsh by thin layer deposition of dredged material. Ecol Eng 12:189–205

Gardebroek C, Hernandez MA (2013) Do energy prices stimulate food price volatility? Examining volatility transmission between US oil, ethanol and corn markets. Energy Econ 40:119–129

Gaweesh A, Meselhe E (2016) Evaluation of sediment diversion design attributes and their impact on the capture efficiency. J Hydraul Eng 142(5):04016002

Gosselink JG, Hatton R, Hopkinson CS (1984) Relationship of organic carbon and mineral content to bulk density in Louisiana Marsh soils. Soil Sci 137(3):177–180

Gowdy J, Erickson JD (2005) The approach of ecological economics. Camb J Econ 29(2):207–222

Groves DG, Sharon C (2013) Planning tool to support planning the future of coastal Louisiana. In: Peyronnin NS, Reed D (eds) Louisiana's 2012 Coastal master plan technical analysis. J Coast Res, Special Issue No. 67

Groves DG, Sharon C, Knopman D 2013. Planning tool to support louisiana's decision making on coastal protection and restoration technical description. Technical report. The RAND Corporation

Hatton RS, DeLaune RD, Patrick WH Jr (1983) Sedimentation, accretion, and subsidence in marshes of Barataria Basin, Louisiana. Limnol Oceanogr 28(3):494–502

Hollinberger TE (2010) Cost estimation and production evaluation for Hopper Dredges. Doctoral dissertation, Texas A & M University

Ji Q, Fan Y (2012) How does oil price volatility affect non-energy commodity markets? Appl Energy 89(1):273–280

Johnson DR, Fischbach JR, Ortiz DS (2013) Estimating surge-based flood risk with the coastal Louisiana risk assessment model. In: Peyronnin NS, Reed D (eds) Louisiana's 2012 Coastal master plan technical analysis. J Coast Res, Special Issue 67

Kerschner C, Prell C, Feng K, Hubacek K (2013) Economic vulnerability to peak oil. Glob Environ Chang 23(6):1424–1433

Kenney MA, Hobbs FJ, Mohrig D, Huang H, Nittrouer JA, Kim W, Parker G (2013) Cost analysis of water and sediment diversions to optimize land building in the Mississippi River delta. Water Resour Res 49(6):3388–3405

Ko JY, Day JW, Wilkins JG, Haywood J, Lane RR (2017) Challenges in collaborative governance for coastal restoration: lessons from the caernarvon river diversion in Louisiana. Coast Manag 45 (2):125–142

Kramer N, Wohl E (2015) Driftcretions: The legacy impacts of driftwood on shoreline morphology. Geophys Res Lett 42 (14):5855–5864

LCWCRTF (Louisiana Coastal Wetlands Conservation and Restoration Task Force and the Wetlands Conservation and Restoration Authority). (1998) Coast 2050: Toward a Sustainable Coastal Louisiana. Louisiana Department of Natural Resources. Baton Rouge, p 161. Available online: http://www.lawater.lsu.edu/docs/2943_coast2050.pdf. Accessed 3 Mar 2017

Leopold A (1970) A sand county almanac. Ballantine, New York

Meselhe E, McCorquodale JA, Shelden J, Dortch M, Brown TS, Elkan P, ..., Wang Z (2013) Ecohydrology component of Louisiana's 2012 coastal master plan: mass-balance compartment model. J Coast Res 67(sp1):16–28

Meselhe EA, Sadid KM, Allison MA (2016) Riverside morphological response to pulsed sediment diversions. Geomorphology 270 (2016):184–202

Murphy JT (2012) Fuel provisions for dredging projects. Proceedings of the Western Dredging Association (WEDA XXXII) Technical Conference and Texas A&M University (TAMU 43) Dredging Seminar, San Antonio, Texas, June 10–13, 2012

Nyman JA, Baltz DM, Green M, Kaller MD, Leberg PL, Richards CP, Romaire RP, Soniat TM (2013) Likely Changes in habitat quality for fish and wildlife in coastal Louisiana during the next 50 years. In: Peyronnin NS, Reed DJ (eds) Louisiana's 2012 coastal master plan technical analysis. J Coast Res, Special Issue 67

Peyronnin N, Green M, Richards CP, Owens A, Reed D, Chamberlain J, … Belhadjali K (2013) Louisiana's 2012 coastal master plan: overview of a science-based and publicly informed decision-making process. J Coast Res 67(sp1):1–15

Roberts HH, DeLaune RD, White JR, Li C, Sasser CE, Braud D, Weeks E, Khalil S (2015) Floods and cold front passages: Impacts on coastal marshes in a river diversion setting (Wax Lake Delta Area, Louisiana). J Coast Res 315:1057–1068

Rivera-Monroy VH, Branoff B, Meselhe E, McCorquodale A, Dortch M, Steyer GD, Visser J, Wang H (2013) Landscape-level estimation of nitrogen loss in coastal Louisiana wetlands: potential sinks under different restoration scenarios. In: Peyronnin NS, Reed D (eds) Louisiana's 2012 Coastal master plan technical analysis. J Coast Res, Special Issue 67

Sadorsky P (2014) Modeling volatility and correlations between emerging market stock prices and the prices of copper, oil and wheat. Energy Econ 43:72–81

Shafiee S, Topal E (2010) An overview of global gold market and gold price forecasting. Resour Policy 35(3):178–189

Snedden GA, Cretini K, Patton B (2015) Inundation and salinity impacts to above- and belowground productivity in Spartina patens and Spartina alterniflora in the Mississippi River deltaic plain: implications for using river diversions as restoration tools. Ecol Eng 81:133–139

Syvitski JP, Kettner AJ, Overeem I, Hutton EW, Hannon MT, Brakenridge GR, Day J, Vörösmarty C, Saito Y, Giosan L, Nicholls RJ (2009) Sinking deltas due to human activities. Nat Geosci 2 (10):681–686

Tessler ZD, Vörösmarty CJ, Grossberg M, Gladkova I, Aizenman H, Syvitski JPM, Foufoula-Georgiou E (2015) Profiling risk and sustainability in coastal deltas of the world. Science 349 (6248):638–643

Tverberg GE (2012) Oil supply limits and the continuing financial crisis. Energy 37(1):27–34

Twilley RR, Couvillion BR, Hossain I, Kaiser C, Owens AB, Steyer GD, Visser JM (2008) Coastal Louisiana ecosystem assessment and restoration program: the role of ecosystem forecasting in evaluating restoration planning in the Mississippi River deltaic plain. Am Fish Soc Symp 64:29–46

USACE (US Army Corps of Engineers) (2004) Louisiana coastal area comprehensive coast wide ecosystem restoration study. US Army Corps of Engineers, New Orleans, LA

USGS (1997) Global warming, sea level rise, and coastal marsh survival, FS: 091–97. United States Geological Survey, Reston, p 2

Visser JM, Duke-Sylvester SM, Carter J, Broussard WP III (2013) A computer model to forecast wetland vegetation changes resulting from restoration and protection in coastal Louisiana. In: Peyronnin NS, Reed DJ (eds) Louisiana's 2012 Coastal master Plan technical analysis. J Coast Res, Special Issue 67

Wang H, Steyer GD, Couvillion BR, Rybczyk JM, Beck HJ, Sleavin WJ et al (2014) Forecasting landscape effects of Mississippi River diversions on elevation and accretion in Louisiana deltaic wetlands under future environmental uncertainty scenarios. Estuar Coast Shelf Sci 138:57–68

Webb EC, Mendelssohn IA (1996) Factors affecting vegetation dieback of an oligohaline marsh in coastal Louisiana: field manipulation of salinity and submergence. Am J Bot 83:1429–1434

Wiegman ARH (2017) Modeling the influence of energy and climate megatrends on future costs and benefits of marsh creation in the Mississippi delta. Master's Thesis, Louisiana State University, Baton Rouge

Xu K, Bentley SJ, Robichaux P, Sha X, Yang H (2016) Implications of texture and erodibility for sediment retention in receiving basins of coastal Louisiana diversions. Water 8(1):26

Large Infrequently Operated River Diversions for Mississippi Delta Restoration

John W. Day, Robert R. Lane, Christopher F. D'Elia, Adrian R.H. Wiegman, Jeffrey S. Rutherford, Gary P. Shaffer, Christopher G. Brantley, and G. Paul Kemp

Abstract

Currently the Mississippi delta stands as a highly degraded and threatened coastal ecosystem having lost about 25% of coastal wetlands during the twentieth century. To address this problem, a $50 billion, 50 year restoration program is underway. A central component of this program is reintroduction of river water back into the deltaic plain to mimic natural functioning of the delta. However, opposition to diversions has developed based on a number of perceived threats. These include over-freshening of coastal estuaries, displacement of fisheries, perceived water quality problems, and assertions that nutrients in river water leads to wetland deterioration. In addition, growing climate impacts and increasing scarcity and cost of energy will make coastal restoration more challenging and limit restoration options. We address these issues in the context of an analysis of natural and artificial diversions, crevasse splays, and small sub-delta lobes. We suggest that episodic large diversions and crevasses (>5000 m^3 s^{-1}) can build land quickly while having transient impacts on the estuarine system. Small diversions (<200 m^3 s^{-1}) that are more or less continuously operated build land slowly and can lead to over-freshening and water level stress. We use land building rates for different sized diversions and impacts of large periodic inputs of river water to coastal systems in the Mississippi delta to conclude that high discharge diversions operated episodically will lead to rapid coastal restoration and alleviate concerns about diversions. Single diversion events have deposited sediments up to 40 cm in depth over areas up to 130–180 km^2. This approach should have broad applicability to deltas globally.

Keywords

Wetlands • Mississippi delta • River diversions • Climate change • Energy scarcity

J.W. Day (✉) • R.R. Lane • A.R.H. Wiegman • J.S. Rutherford
G. Paul Kemp
Department of Oceanography and Coastal Sciences, Louisiana State University, Baton Rouge, LA 70803, USA
e-mail: johnday@lsu.edu; rlane@lsu.edu; adrian.wiegman@gmail.com; Jruthe@stanford.edu; gpkemp@lsu.edu

C.F. D'Elia
College of the Coast and Environment, Louisiana State University, Baton Rouge, LA, USA
e-mail: cdelia@lsu.edu

G.P. Shaffer
Department of Biological Sciences, Southeastern Louisiana University, Hammond, LA, USA
e-mail: shafe@selu.edu

C.G. Brantley
U.S. Army Corps of Engineers, New Orleans District, Bonnet Carré Spillway, Norco, LA, USA
e-mail: Christopher.G.Brantley@usace.army.mil

J.W. Day, J.A. Erdman (eds.), *Mississippi Delta Restoration*,
Estuaries of the World, https://doi.org/10.1007/978-3-319-65663-2_8

Introduction: Historical Perspective

Currently the Mississippi delta stands as a highly degraded and threatened coastal ecosystem. Before human activities impacted the delta, primarily in the twentieth century, it was a healthy functioning ecosystem (Kolb and Van Lopik 1958; Condrey et al. 2014; Muth 2014). In this paper we discuss how the delta was formed and sustained, the causes of deterioration, and the potential use of very large but episodic diversions for wetland creation while minimizing negative impacts that arise when diversions are operated continuously.

Since the stabilization of sea level approximately 5000 years ago and prior to massive human impact, mainly in the twentieth century, the Mississippi River formed a vast deltaic wetland complex encompassing about 25,000 km^2 in the north central Gulf of Mexico (Roberts 1997; Day et al. 2007, 2014). One need only look at the northern Gulf coast to recognize that riverine inputs formed the delta, which protrudes out into the Gulf more than a degree of latitude compared to coastlines east and west of the delta. The coastlines on either side of the delta tend to be coastal bays fronted by linear barrier islands. It is a truism that it was the river that built the delta, but this question bears looking into in more detail.

Condrey et al. (2014) used maps and journals of early European explorers to describe what they called the last natural delta of the Mississippi that existed just prior to European settlement. The delta was a vast seaward-advancing arc that occupied, through four distributaries, all of the five most recent delta complexes of the Mississippi River (Teche, St. Bernard, Lafourche, Modern, and Atchafalaya) and extended across the deltaic plain. It was characterized by plumes of fresh water that extended for more than 10 km into the GOM during the spring flood of the river and by a vast offshore oyster reef that functioned as both an impediment to navigation and an offshore harbor. Insofar as possible, restoration should attempt to replicate conditions that existed before massive human impact. Condrey et al. (2014) suggest that much of Louisiana's coast was advancing into the sea at the onset of European colonization, that colonial and post-colonial modification of the Mississippi resulted in the loss of much of this potential.

The natural delta described by Condrey et al. (2014) was formed and sustained by a hierarchical series of energetic forcings or events that occurred over a wide range of temporal and spatial scales. The delta formed as a series of overlapping delta lobes (Blum and Roberts 2009). The energetic forcings included the shifting deltaic lobes, but also crevasse formation, great river floods, hurricanes, annual river floods, frontal passages, and tides (Roberts 1997; Roberts et al. 2015; Boesch et al. 1994; Day et al. 1997, 2000, 2007; Vorosmarty et al. 2009). As the delta developed, it formed a skeletal framework of interconnected natural levee ridges and barrier islands that enhanced sediment trapping and served to protect the delta from storm surge and salinity intrusion (e.g., Day et al. 2007; Xu et al. 2016).

During the twentieth century, there was massive deterioration of the delta, with about 25%, or 4800 km^2, of coastal wetlands lost (Barras et al. 1994; Britsch and Dunbar, 1993; Couvillion et al. 2011). A variety of factors led to this wetland loss, including reduction of sediment input from the basin, pervasive alteration of the hydrology of the deltaic plain, enhanced subsidence due mainly to petroleum extraction, saltwater intrusion, creation of impoundments, and barrier island deterioration (Day et al. 2000; Nyman 2014; Olea and Coleman 2014; Kemp et al. 2014). However, the most important factor leading to deterioration of the delta have been the near complete elimination of river water and sediment input to the deltaic plain due to flood control levee construction and closure of distributaries that connected the river to the wetlands (Day et al. 2000, 2007), as well as the reduction in sediment flux from the Mississippi River basin (Meade and Moody 2010), and increased rate of sea-level rise experienced during the last several decades (Blum and Roberts 2012).

There is broad agreement that the river must be reconnected to the delta if restoration is to succeed (Allison and Meselhe 2010; Day et al. 2007, 2012; DeLaune et al. 2013; Lopez et al. 2014; Morris et al. 2013a; Nittrouer et al. 2012; Paola et al. 2010; Shen et al. 2015; Twilley and Rivera-Monroy 2009). The river provides fresh water to reduce salinity stress, iron to complex with sulfide and reduce sulfide toxicity, mineral sediments to promote accretion, and nutrients to stimulate wetland productivity, which leads to organic soil formation (Mendelssohn and Morris 2000; Nyman 2014; Morris et al. 2013a; Day et al. 2014). Combating coastal erosion and restoring coastal wetlands is now a main component of State and Federal policy (CPRA 2012a), and the construction of river diversions to reintroduce Mississippi River water and sediments into coastal basins is planned for coming decades (Wang et al. 2014; CPRA 2012a; Day et al. 2014). Understanding how historical floods and crevasses built land informs future restoration work on diversions as scientific research and engineering converge on the best approaches for coastal land building. The objective of this paper is to discuss the potential of using large but infrequently opened diversions for wetland creation and delta restoration while addressing issues related to fisheries and water quality as a result of over-freshening in the context of twenty-first century mega-trends.

Existing Natural and Artificial Diversions as Models

The State of Louisiana's Master Plan for restoration and flood protection of the Mississippi delta involves a suite of restoration activities including wetland creation using pumped dredged sediments, hydrologic restoration, barrier island restoration, and structural and non-structural flood protection (CPRA 2012a). A central element of the plan is the reintroduction of river water back into the delta plain to create and sustain wetlands. River diversions involve the construction of water control structures along the river and hydraulic management is often needed to direct water flow after it enters the coastal wetland complex. Successful river diversions should be informed by the history of natural sub-deltas and crevasse-splay deposits. The design and operation of diversions should be based on an understanding of the history of natural and constructed projects where river water flows into coastal systems (i.e., Allison and Meselhe 2010). Here we review a number natural and artificial crevasses, diversions, and sub-deltas.

Davis (1993) reported that between 1850 and 1927 there were more than a thousand crevasses along the lower Mississippi levee. These crevasses overlapped to form a continuous band of crevasse deposits essential to the formation and maintenance of both natural levees and coastal wetlands (Fig. 1; Saucier 1963; Davis 2000; Allison and Meselhe 2010; Shen et al. 2015). Day et al. (2016) reported that an artificial crevasse opened at Caernarvon during the great flood of 1927 had a peak discharge of nearly 10,000 $m^3\ s^{-1}$ and deposited a crevasse splay of about 130 km^2 in about 3 months (Fig. 2) with sediment deposition was as high as 42 cm.

The river diversion at Caernarvon, the first of the modern era, began discharging during 1991 into Big Mar - a water body that resulted from a failed agricultural impoundment – before entering the upper Breton Sound estuary (Lane et al. 1999, 2006). A small new sub-delta of almost 250 ha has formed in Big Mar, with about 235 ha since Hurricane Katrina in 2005 (Fig. 3; Lopez et al. 2014). However, under historical conditions of prolonged high sediment yield discharge during a single flood event, such as occurred in 1927 in the same location as the modern diversion, the impact on land-building is much more dramatic, as discussed below.

The artificial 1927 crevasse at Caernarvon was similar in size and duration to naturally occurring historical crevasses. For example, the Davis Crevasse, which formed in 1884 and is located on the west bank of the river 32 km upriver of New Orleans, had an area of about 150 km^2 and is still clearly visible on photos of the area (Fig. 4). Davis (1993) reported that the crevasse was directly attributed to an old rice flume that was abandoned and removed a few months prior to the high water, although a muskrat hole may have contributed to the break. Davis reported that "the crevasse was by far the single most destructive crevasse in the history of overflows in Louisiana". It led to extensive flooding with water levels as high as 0.6–2.4 m over much of the affected area.

Just east of the historical Davis Crevasse, the Davis Pond river diversion was initiated in 2002. The insert in Fig. 4 shows the new sub-delta that is forming, which is similar in size to the Caernarvon sub-delta. The Caernarvon, Davis, and Bonnet

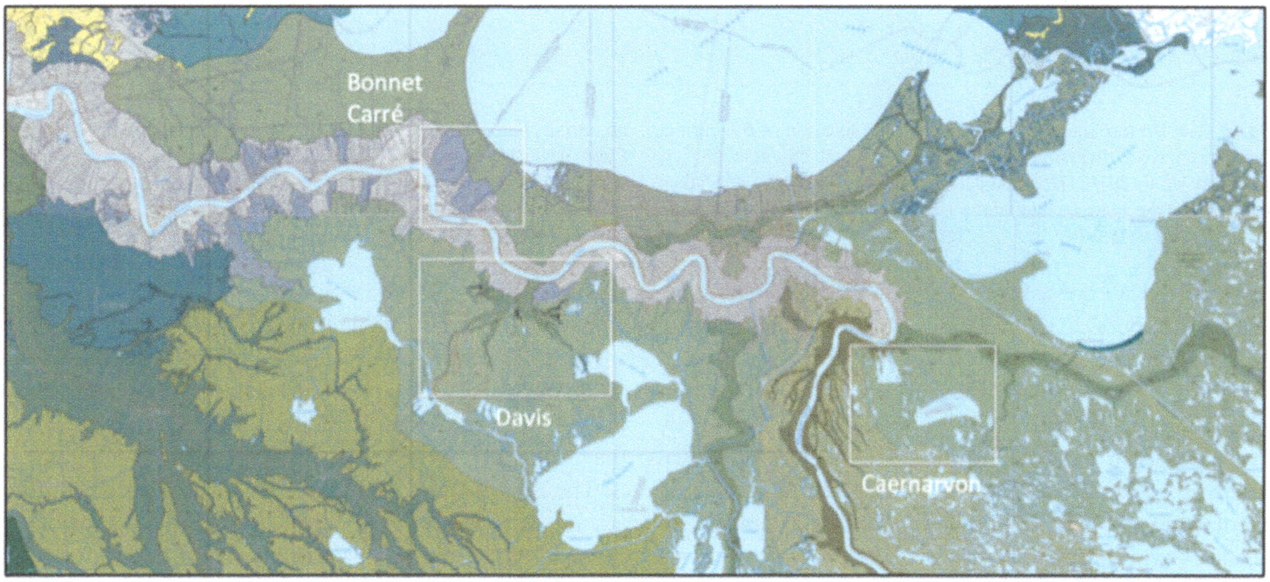

Fig. 1 Soils map of the Mississippi River natural levee between Baton Rouge and New Orleans showing former crevasse splays with surface expression (*purple*; Louisiana Geological Survey). The areas in *white boxes* show the location of former crevasses and active diversions at Bonnet Carré, Davis, and Caernarvon discussed in this chapter

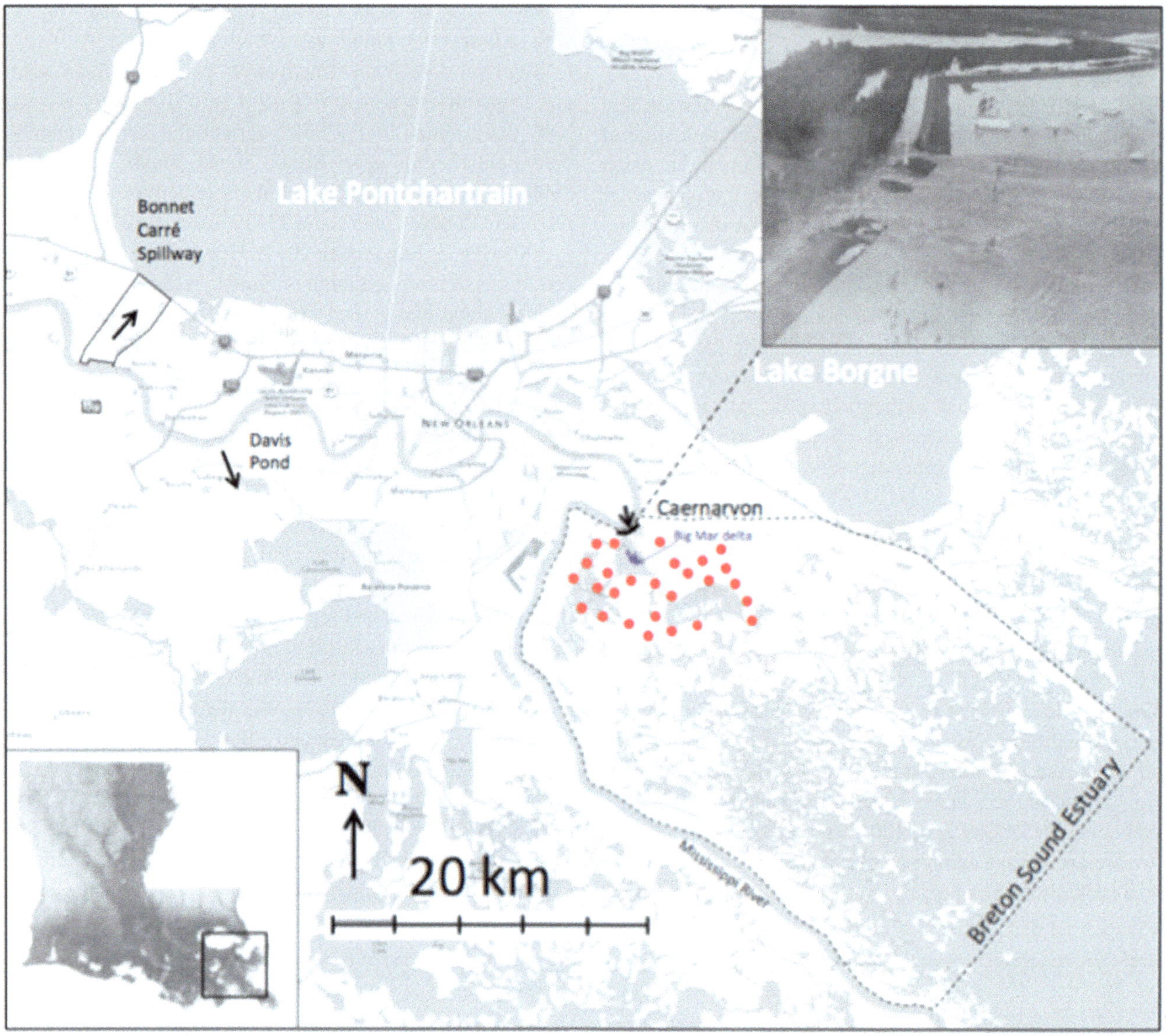

Fig. 2 The Breton Sound estuary showing the location of the current Caernarvon river diversion and the 1927 artificial crevasse (*arrow*, from Day et al. 2016). *Dots* indicate where core samples were taken to determine the spatial extent of the 1927 crevasse and the *inset* shows an aerial photo of Mississippi River water flowing through the 1927 Caernarvon levee breach. The *purple area* shows the size of the Big Mar subdelta (shown in Fig. 3, Lopez et al. 2014) that took more than two decades to form

Carré historical crevasse splays and the newly forming small delta lobes at Caernarvon and Davis Pond illustrate the land building capacities of vary large diversions (e.g., >5000 m^3 s^{-1}) and small diversions (mean discharge about 50 m^3 s^{-1}) (See Table 1). The crevasse splays were formed by one (Caernarvon and Davis) to several (Bonnet Carré) annual discharge events. In contrast the small river diversions at Caernarvon and Davis Pond have been in operation for 25 and 14 years, respectively. Thus, large crevasses can create up to two orders of magnitude more land in a few events than is created over several decades by small diversions.

The Bonnet Carré crevasse was active in the 2nd half of the nineteenth century (Saucier 1963; Davis 1993). The crevasse functioned in 1849, 1857, 1867, 1871, and 1874 with flow discharge ranging from about 2000 to 6500 m^3 s^{-1}. The 1874 crevasse remained open for 10 years. These discharges created a large crevasse splay of about 70 km^2 in existing wetlands and filled in a large area in western Lake Pontchartrain with up to 2 m of deposition.

The Bonnet Carré Spillway is probably the closest modern analog to the 1927 crevasse (Fig. 1) and other historical crevasses. The floodway has been opened 12 times since the 1930s (about once a decade and representing about 1% of the time since it was completed in 1933; the last opening was in early 2016) with flows ranging from 3000 to 9000 m^3 sec^{-1}, and accretion rates in the spillway average about 25 mm year

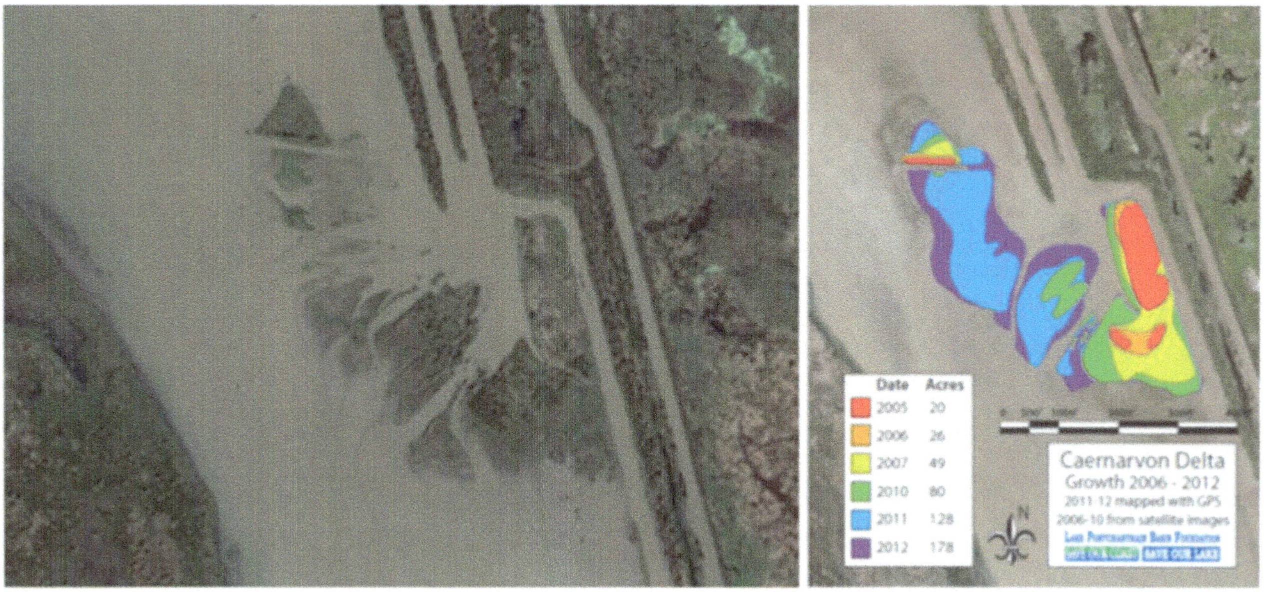

Fig. 3 Land gain at the Caernarvon Delta in Big Mar Pond from 2005 through 2012. See Fig. 2 for comparison with the large crevasse splay formed during the 1927 flood (Modified from Lopez et al. 2014)

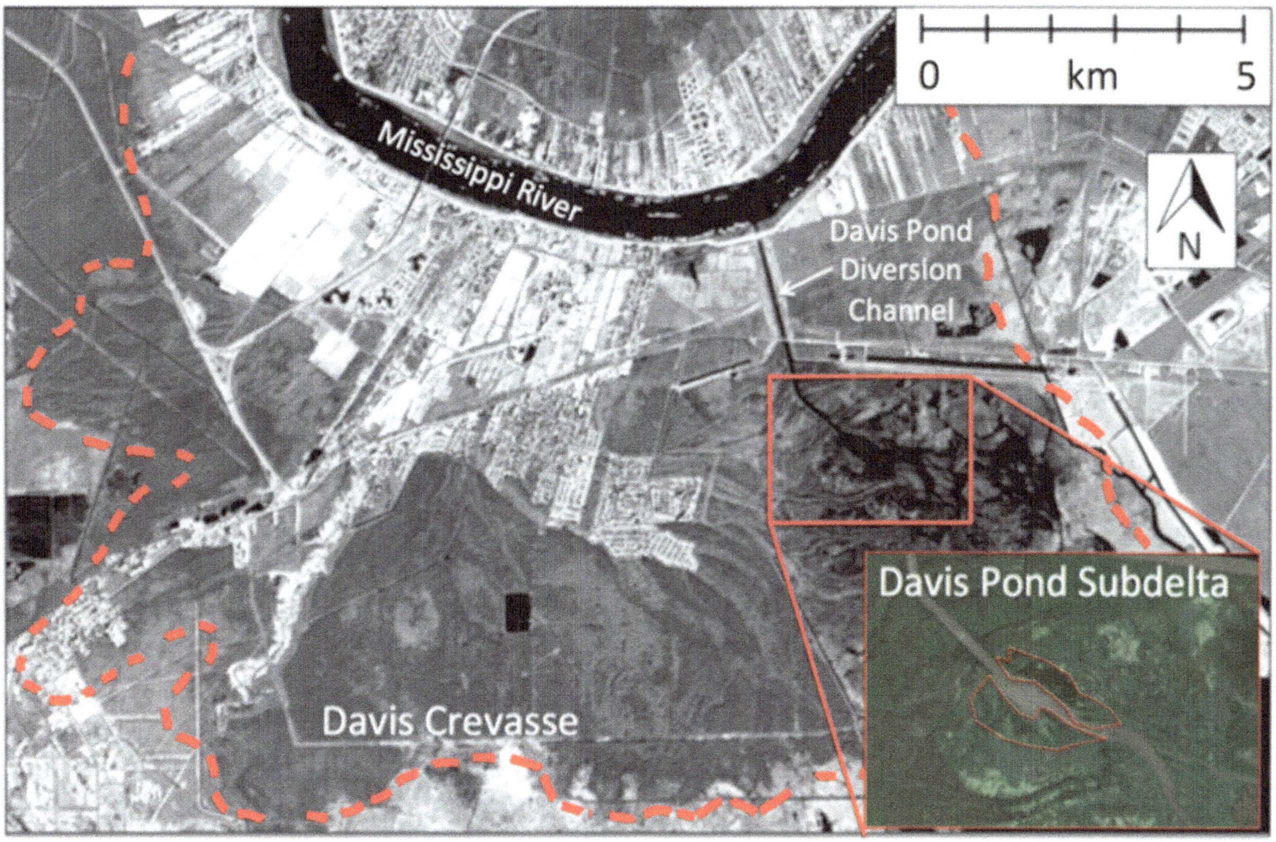

Fig. 4 Aerial image of the Davis Crevasse splay. The Davis Pond river diversion that began operations in 2002 is indicated. River water (*blue* in this image) is spreading out from the diversion channel. The *red dotted line* shows the outline of the crevasse splay. The *insert image* on the *bottom right shows* the newly forming subdelta from the Davis Pond Diversion (image from Hampton Peele, LSU)

Table 1 Characteristics of historical diversions and crevasses in the Mississippi Delta

Location	Diversion/ crevasse	Discharge environment	Maximum discharge (m^3/s)	Average discharge (m^3/s)	Max annual land gain (ha/year)	Land Built (ha)	Years Active (years)	Rate of accretion (mm/year)	Citations
Caernarvon	Diversion	Discharge is into Big Mar, an artificial lake, followed by fresh and saline marsh	227	71	34	243	1991–present	11.7	Andrus (2007), Lopez et al. (2014) and Lane et al. (2006)
Bonnet Carré Spillway	Diversion	50% of the spillway area is forested wetlands	3100–9000	N/A	N/A	N/A	1932–present	26–27	Andrus (2007), Day et al. (2012) and Nittrouer et al. (2012)
Davis Pond[a]	Diversion	Fresh and brackish marsh	302	68	N/A	36[b]	2003–present	5.9–10.3	DeLaune et al. (2013); waterdata.usgs.gov
West Bay	Diversion	Open bay	765[c]	628[d]	N/A	400–2498[e]	2003–present	7–15	Kolker et al. (2012), Yuill et al. (2016) and National Audubon Society (2012)
1927 at Caernarvon	Crevasse	Discharge is directly into Big Mar, an artificial lake, followed by fresh and saline marsh	10,000	N/A	N/A	13,000	Three months during the flood of 1927	0–450	Day et al. (2016)
Davis	Crevasse	Fresh and brackish marsh	N/A	N/A	N/A	18,000[b]	1884	N/A	Davis (1993)
Bonnet Carré	Crevasse	Mostly cypress swamp	2000–6500	N/A	N/A	7000[f,g]	1849–1874	N/A	Saucier (1963) and Davis (1993)
Napoleonville	Crevasse	Low salinity open water to cypress swamps	N/A	N/A	N/A	6000	600 and 1500[h]	10–40	Shen et al. (2015)
Wax Lake	Delta Lobe	Open bay	N/A	8495	100–300	10,000 (by 2005)[i]	Became subaerial in 1973	N/A	Roberts et al. (2015), Kim et al. (2009) and Wells and Demas (1977)
Atchafalaya	Delta Lobe	Open bay	N/A	33,980	650	16,000[b]	Became subaerial in 1973	N/A	Roberts et al. (2015)
Baptiste Collette, Cubit's Gap, West Bay, Garden Island Bay	Subdelta Lobe	Open bay (birdsfoot delta)	N/A	3.9%	110	5700	(1874–1946)	N/A	Wells and Coleman (1987)
				14.5%	630	19,300	(1862–1946)		
				4.8%	700	29,700	(1845–1932)		
				4.1%	690	10,400	(1891–1922)		

[a]Vertical accretion in these marshes is governed by organic accumulation; [b]Estimate based on a measurement of Google Earth imagery; [c]Max measured flow capacity as of 2008; [d]On average the West Bay diversion captured 7.07% of the Lower Mississippi River, which had a mean discharge of 8877m^3/s, between 2004 and 2014; [e]400 ha is based on a preliminary estimate of deltaic islands formed by 2011, and 2498 ha is based on the area experiencing net aggradation between 2009 and 2011; [f]Up to 2 m of deposition in western Lake Pontchartrain; [g]Estimated based on figure in Saucier (1963); [h]Active episodically during these time periods; [i]Includes subaqueous delta toe

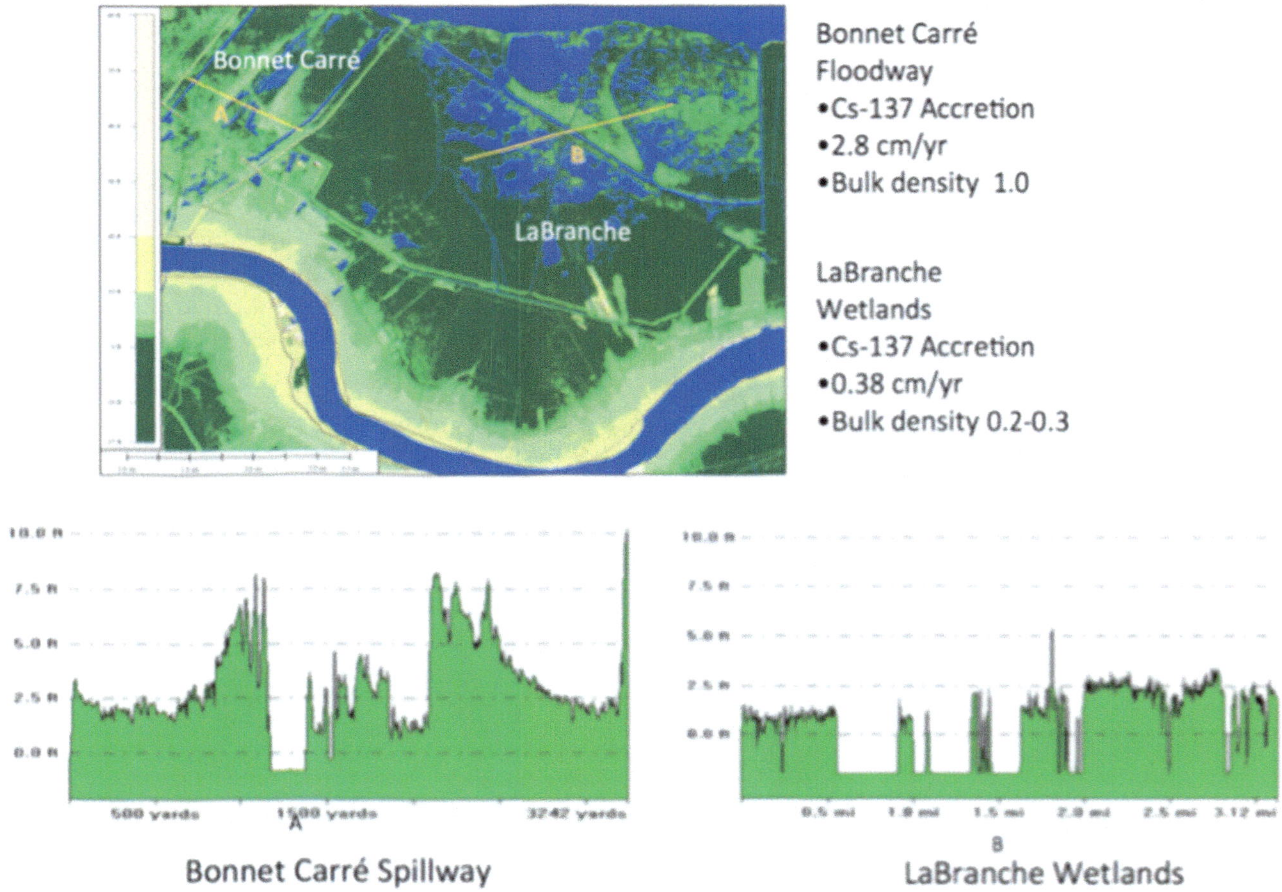

Fig. 5 Digital LIDAR elevation maps showing elevation (NGVD1984) in the Bonnet Carré Spillway (A–A′) and the LaBranche wetlands (B–B′; modified from Day et al. 2012)

$^{-1}$ compared to about 4 mm year^{-1} in adjacent wetlands without river input (Fig. 5; Day et al. 2012). Total fine-grained sediment deposition in wetlands within spillway wetlands near Lake Pontchartrain is as high as 2 m or an average of about 20 cm for each flood event. Wetland forests in the spillway are 2–3 times more productive than in adjacent wetlands where forests are not sustainable (Day et al. 2012).

Shen et al. (2015) reported on a crevasse splay that developed off the Bayou Lafourche distributary just downstream of its juncture with the Mississippi River (Fig. 6). The splay formed in fresh to low salinity wetlands. They concluded that "discrete, episodically deposited sediment bodies dominate overbank stratigraphy" and that periodic aggradation rates of 1–4 cm/year could persist for centuries. The total area of this splay was at least 60 km^2.

The West Bay diversion into a large shallow open bay near the modern Birds Foot delta has led to from 400 to nearly 2500 ha since 2003. Some of this land area created during this period was due to dredged spoil deposit, which formed land directly but also accelerated land-building in quiescent zones sheltered from wave action.

Major Delta Lobe Development vs. Crevasse Formation

It is important to distinguish diversions and crevasse splay formation that we discuss in this paper from the Atchafalaya and Wax Lake delta complex, which are part of the beginning stages of a major new deltaic lobe development fed by the Atchafalaya River distributary that carries about a third of the combined lower Mississippi River discharge (Roberts 1997; Roberts et al. 2015; Delaune et al. 2016). Subaqueous deposition occurred in Atchafalaya Bay for much of the twentieth century, first via the lower Atchafalaya River and then beginning in 1941 via the dredged Wax Lake Outlet. Both delta lobes first became subaerial during the large flood in 1973. The Atchafalaya delta lobe has a dredged navigation channel running through it but the Wax Lake lobe has developed naturally, though at the end of a 26 km long artificial

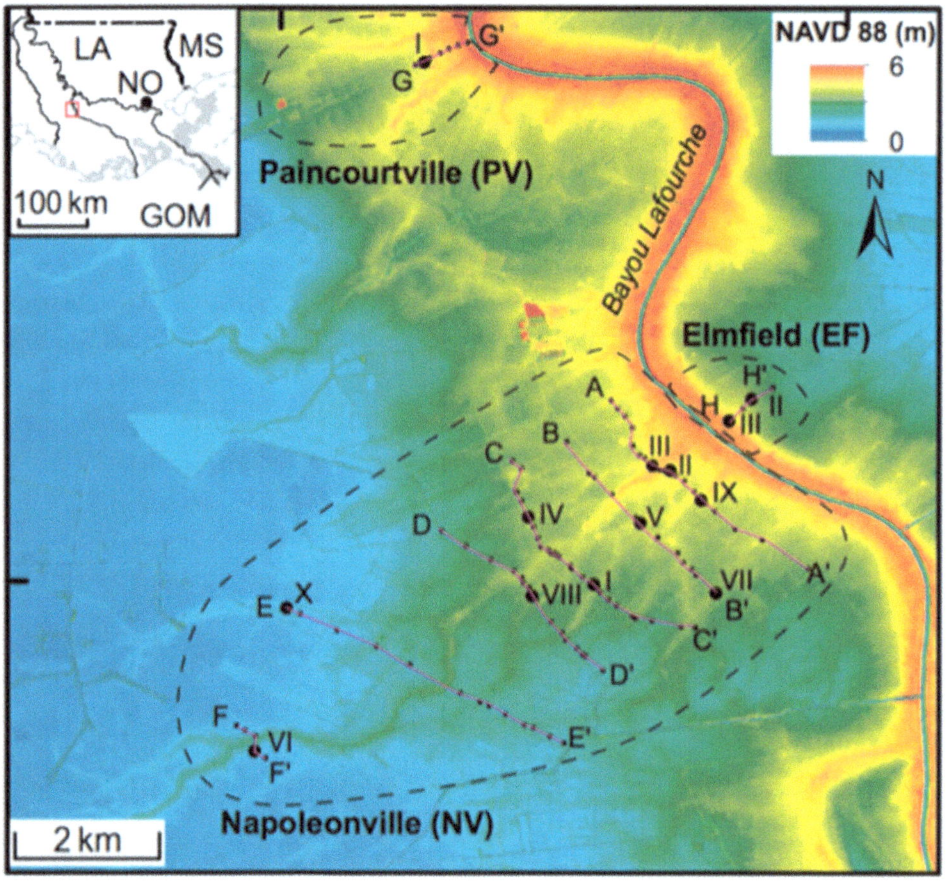

Fig. 6 Digital elevation model (DEM) derived from LIDAR data of a crevasse splay deposit along Bayou Lafourche. Elevations are with reference to the North American Vertical Datum of 1988 (NAVD 88; from Shen et al. 2015)

conveyance channel. The delta lobes have grown at about 3 km^2/year to achieve total areas of 160 and 100 km^2, respectively, by 2015 (Roberts et al. 2015). Atchafalaya River discharge has also led to mineral sediment deposition in wetlands in a broad arc from Fourleague Bay to the east to Vermillion Bay to the west (e.g., Day et al. 2011) and these wetlands have the lowest rates of land loss in the Mississippi delta. It is important to note that the Atchafalaya delta lobe complex developed in open waters of the Atchafalaya-Cote Blanc-Vermillion Bay complex. By contrast, almost all diversions in the State Master Plan are planned for the Mississippi River below Baton Rouge and discharge to coastal basins with significant amounts of wetlands. Thus, these receiving basins have a much lower accommodation space and much higher friction and are expected to retain a higher percent of introduced sediments (Blum and Roberts 2009).

The data from the survey of diversions and crevasses is summarized in Table 1. Peak discharge ranged from nearly 10,000 $m^2\ s^{-1}$ for the Bonnet Carré Spillway and the 1927 Caernarvon crevasse, to 6500 for the Bonnet Carré crevasse, to 300 or less for the David Pond and Caernarvon diversions. Total land gain ranged from 36 ha for the Davis Pond diversion to an estimated 18,000 ha for the Davis Crevasse. Accretion ranged from <1 to 4 cm $year^{-1}$. Deposition in a single year averaged about 20 cm for the Bonnet Carré Spillway to as high as 43 cm for the 1927 Caernarvon Crevasse. These data show that sustained accretion rates of 1–4 cm $year^{-1}$ are possible and that single discharge events can deposit from 20–40 cm or more.

Benefits of Large Infrequent Diversions

The use of very large but infrequent diversions would have a number of benefits and address many of the issues related to diversions. They would build land quickly. A diversion the size of the 1927 crevasse at Caernarvon into existing but degrading wetlands would lead to significant elevation gain and wetlands being created after the first operation. In more inland parts of the delta, such as the Maurepas Swamp (Fig. 7), considerably smaller discharges would likely lead to similar sized crevasse splays because of shallower water and the high trapping efficiency of forested wetlands (e.g., Xu et al. 2016; Blum and Roberts 2009). As reported by Saucier (1963), the Bonnet Carré crevasse flowed about

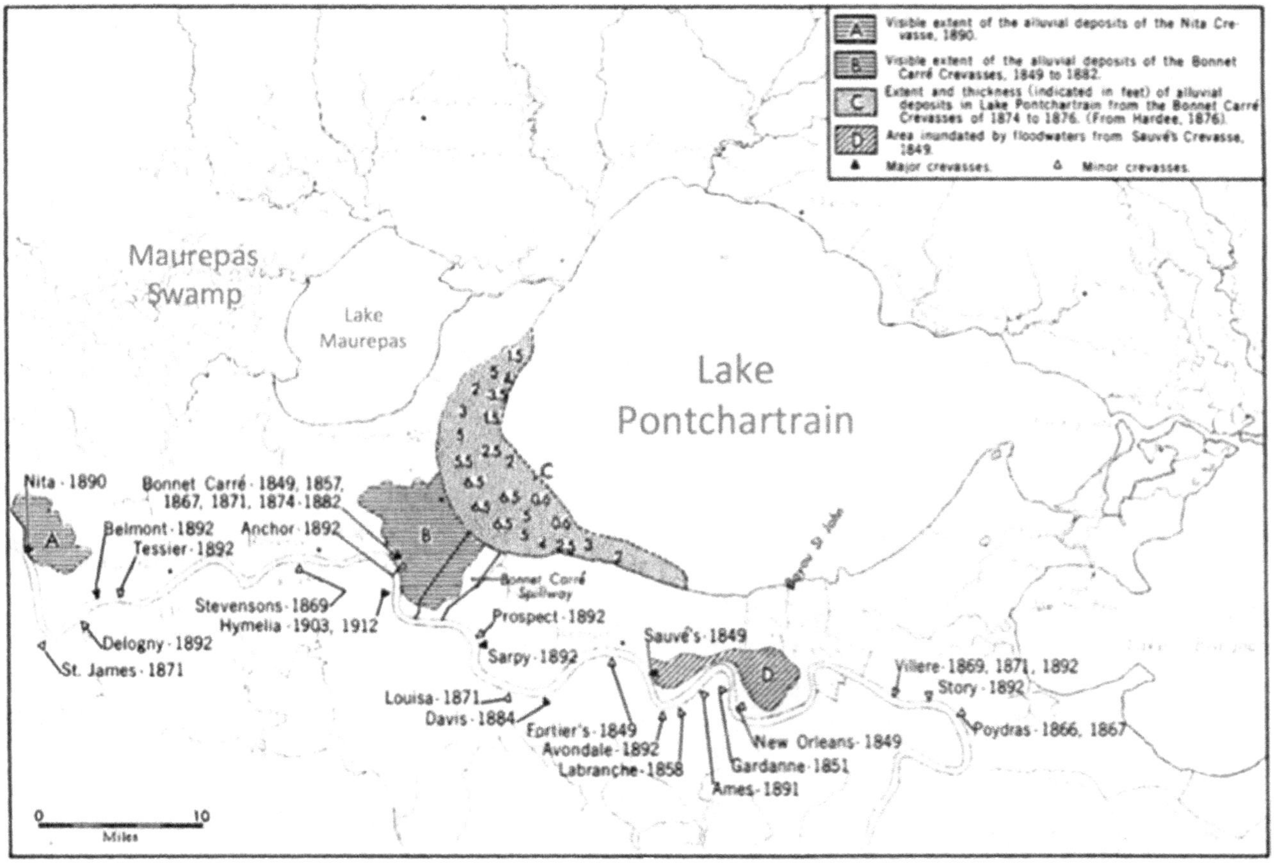

Fig. 7 Location of crevasses along the Mississippi River from 1849 to 1927 including areas affected by three major crevasses (From Saucier 1963)

6–8 times in from 1849 to 1882 and built a large crevasse splay of about 7000 ha and added up to 2 m of sediments in western Lake Pontchartrain (Fig. 7).

Large diversions would self-design and form a newly emerging sub-delta without large management expense, as has been the case for the Wax Lake delta (Roberts et al. 2015; DeLaune et al. 2016). Large episodic diversions would continue to build land rapidly and consistently over time, thus potentially offsetting sea-level rise. For example, Shen et al. (2015) reported that periodic aggradation rates of 1–4 cm/year could be maintained for more than a century. Because they would be operated infrequently, they would not lead to persistent over-freshening and displacement of fishery species. Water quality problems due to algal blooms would be short lived. The costs of such diversions would occur up front, so that increasing cost for energy-intensive activities due to energy scarcity would be minimized. The operation of gravity-driven diversions is financially and energetically inexpensive so that the diversion could be operated far into the future without incurring large costs. The Bonnet Carré Spillway is an example of this. The 2011 budget for operation and maintenance of the spillway was about $4.9 million (U.S. Army Corps of Engineers 2011). Much of this cost was for environmental monitoring and routine maintenance. Less that 10% of this budget was for actual opening and closing of the structure during this record flood year.

Large diversions could be designed to have multiple benefits (Nittrouer et al. 2012). For example, a large diversion into the Maurepas Swamp would compliment the Bonnet Carré Spillway as a flood relief outlet. One critical problem for the Bonnet Carré is that during each operation, a large amount of material is deposited in the spillway. This material must be removed to extend the life of the flood outlet. A large Maurepas diversion would help extend the life of the Bonnet Carré Spillway. If properly designed, the outlet channel from the river could be designed to be self-scouring thus eliminating the need for dredging. The Wax Lake Outlet channel serves as an example of such a self-maintaining channel. Properly designed, such diversions could last a century or more and require minimal operation and maintenance costs, especially when compared with other restoration approaches such as wetland creation using dredged and pumped sediments.

The Relative Costs of Restoration

Over the long-term, river diversions are more cost effective for building land than marsh creation (MC). In the 50 year 2012 State Master Plan (SMP), MC projects had an average cost of about \$360,000 per hectare of net land benefit, while sediment diversions had an average cost (including engineering, operation and maintenance) of about \$47,000 per hectare per hectare of net land benefit (CPRA 2012a). The SMP estimates that over 50 years the operation and maintenance cost of a river diversion will be roughly 18% of the total cost of the diversion (CPRA 2012b). Thus, the average annual operation and maintenance costs over the lifetime of the \$1 billion, 7000 m^3 s^{-1} Mid-Barataria Diversion modeled in the 2012 SMP would be \$3.8 million per year (CPRA 2012b). Areas restored via hydraulic dredging will need to be renourished periodically due to relative sea-level rise at an increasingly higher cost due to increasing cost of energy.

Issues Related to Diversions: Inundation, Salinity, Fisheries, and Water Quality

It is widely accepted that a supply of sediment and freshwater is critical to maintain the health of wetlands, but large scale diversion plans have raised questions about the risks of increased flooding, rapid decreases in salinity, impacts on fisheries and the consequences of added nutrients. Below are discussions of the most pertinent issues related to river diversions.

Inundation

Concern has been expressed that diversions lead to increased flooding duration that makes marshes more susceptible to hurricane disturbance as occurred at Caernarvon during Hurricane Katrina (Howes et al. 2010; Kearney et al. 2011). Snedden et al. (2015) measured biomass production at Caernarvon of *Spartina alterniflora*, *S. patens,* and *Sagittaria lancifolia* under different flooding regimes using marsh organs (Morris et al. 2013b). Above- and below-ground biomass was highest for both *Spartina* species at higher elevations where inundation was minimal, and decreased exponentially with decreased elevation and increased flood duration. The percent of time marshes at Caernarvon were flooded due to the diversion increased from less than 10% in 1997 to between 60% and 90% after 2000 (Snedden et al. 2015). Peak discharges in the 1990s were generally less than 113 m^3 s^{-1} and generally occurred during winter when vegetation was dormant. After 2005, peak discharges were generally between 170 and 227 m^3 s^{-1} and occurred mainly during the spring growing season. Piazza and LaPeyre (2007) also reported water inundations times twice as long at the Caernarvon outfall area compared to reference areas. The results of Snedden et al. (2015) as well as the other studies cited above and in the section below demonstrate that increased flooding was most likely the cause of lower biomass production, and not excessive nutrients. Increased flooding decreased above- and below-ground biomass production in interior marshes at Caernarvon for *S. alterniflora* and *S. patens* marshes, and likely increased susceptibility of marshes to disturbance during Hurricane Katrina. However, about 4 km^2 of new wetlands have formed in the Big Mar pond since 2005 due to freshwater and fine-grained sediment input from the river via the diversion (Lopez et al. 2014; Hillman et al. 2015). Large episodic diversions would address excessive flooding because extensive inundation would occur infrequently and allow for consolidation and strengthening of soil (Day et al. 2011).

Flooding Threat to Human Infrastructure

The introduction of large volumes of river water may lead to flood threats to low-lying infrastructure. This can and is being alleviated by levee construction. Flood control on the lower Mississippi is accomplished by continuous levees adjacent to the Mississippi from Baton Rouge south to nearly the mouth of the river to prevent flooding from the river. There are also "back" levees that protect developed areas from hurricane surge. These extend from near the mouth of the river past the Bonnet Carré Spillway on the east bank and through the New Orleans metropolitan area on the west side of the river. There are current plans to extend the east bank back levee nearly 20 km further upstream. Ultimately, the back levees will be extended as far upstream as needed to protect against hurricane surge, which has the potential to push water far higher than any diversion.

Sediment Retention and Consolidation

Consolidation of sediments is a key aspect of building sustainable wetlands (Day et al. 2011). Xu et al. (2016) reported that during the dry season, when bottom sediment has had a chance to consolidate after deposition in the spring, the minimum shear stress required to erode sediment in Big Mar and West Bay was about 0.2 pa which is about 4 times less erodible than freshly deposited sediment. Similarly, Day et al. (2011) compared soil properties of Old Oyster Bayou (OOB) marshes, which were inundated for only 15% of the year, to Bayou Chitique (BC) marshes, which were inundated for 85% percent of the year. The bulk density of OOB soils was double that of BC and fall cone strength of OOB was more than ten times higher than BC. BC is not located near a river sediment source and the surrounding marshes rapidly deteriorated during the 1990s (Wang et al. 1994; Day et al. 2011). Old Oyster Bayou is located near Fourleague Bay, which receives fresh sediment from the Atchafalaya. Aerial photography records of OOB indicates marsh ponds and channel morphology that has been stable for over 60 years (Day et al. 2011).

Xu et al. (2016) concluded that in order to maximize retention of mud, which is the largest fraction of sediment carried by the Mississippi, operations strategies should allow for consolidation and reduced sediment loss/bypass. This can be achieved by building diversions in fresher portions of estuarine basins that are less fragmented, shallower and sheltered from oceanic and wind driven erosional forcing, and by diverting sediment intermittently in high energy pulses (Xu et al. 2016). Large intermittent diversions allow for consolidation of a large deposited sediment load that is otherwise not possible during constant inundation that allows shear stress from wind waves, tidal and riverine fluxes. Thus, large infrequent diversions located in upper basins will minimize the long-term impacts of flooding on deterioration of marsh soils while promoting development of more consolidated sediments.

Salinity

The impact of river diversions on salinity in receiving basins has been shown to be widespread but episodic (Lane et al. 1999, 2004, 2007). For example, Lane et al. (2004) reported a rapid decrease in salinity over a two-week period for the entire Breton Sound estuary due to high experimental discharges from the Caernarvon diversion, but salinities returned to pre-diversion estuarine conditions just two-weeks after the discharge ceased. Over-freshening of Lake Pontchartrain due to openings of the Bonnet Carré Spillway is short-lived and the Lake returns to normal quite rapidly after the structure is closed. Diversions could be used to reduce salinity intrusion during droughts as occurred in 1999–2000 and led to death of baldcypress-water tupelo swamps in the western portion of Lakes Pontchartrain and Maurepas (Shaffer et al. 2009, 2016; Day et al. 2012).

Fisheries

A number of questions have been raised about the impact of diversions on fisheries (e.g., Bowman et al. 1995). These concerns are related primarily to over-freshening and displacement of fishery species. Members of both recreational and commercial fishing communities are concerned that historical fishing grounds will be freshened by diversions such that fishery species such as oysters, shrimp, and saltwater fish will be displaced. Commercial fishers especially are worried that costs will increase substantially due to travel times to displaced fishing grounds. Many in the fishing community suggest that fishers should be paid by the state for additional costs due to longer travel times.

However, Piazza and LaPeyre (2007) reported increased nekton density and biomass in marshes receiving high-volume freshwater flow from the Caernarvon river diversion compared to marshes not receiving pulsed flow. These differences in nekton communities can be attributed to differences in submerged aquatic vegetation density in areas receiving river water compared to reference areas (Rozas et al. 2005). The isotopic signature of the river has been found in common local consumers such as grass shrimp (*Palaemonetes* sp.), barnacles (*Balanus* sp.), and small plankton-feeding fish (bay anchovies, *Anchoa mitchilli*), with the influence strongest closest to the diversion structure and along known flow pathways (Wissel and Fry 2005).

Large diversions with infrequent openings would address over-freshening by allowing estuarine conditions to become reestablished within a few months of closure of the diversion. This would also lead to a rapid return to estuarine conditions and enhanced fishery production following cessation of freshwater input. The value of diversions to fisheries resources has been historically demonstrated. Openings of the Bonnet Carré Spillway have resulted in subsequent increases of oysters, finfish, and shrimp (Viosca 1938; Gunter 1953; Dugas and Perret 1975; Browder and Moore 1981; Bowman et al. 1995).

Water Quality

It has been reported that following openings of the Bonnet Carré Spillway, large algal blooms developed in Lake Pontchartrain (Fig. 8). But these blooms are short lived and are not dominated by harmful algal species (White et al. 2009; Roy et al. 2013, 2016).

Nutrients Cause Marsh Deterioration

Over the past several years, there has been considerable discussion about the impacts of nutrients on the below-ground productivity, soil organic matter decomposition, and soil strength of coastal marshes (Darby and Turner 2008a, b, c; Fox et al. 2012; Anisfeld and Hill 2012; Day et al. 2013; Deegan et al. 2012; Graham and Mendelssohn 2014; Morris et al. 2013a; Nyman 2014; Swarzenski et al. 2008; Turner 2010; VanZomeren et al. 2011). Part of these concerns have resulted from observations that Breton Sound

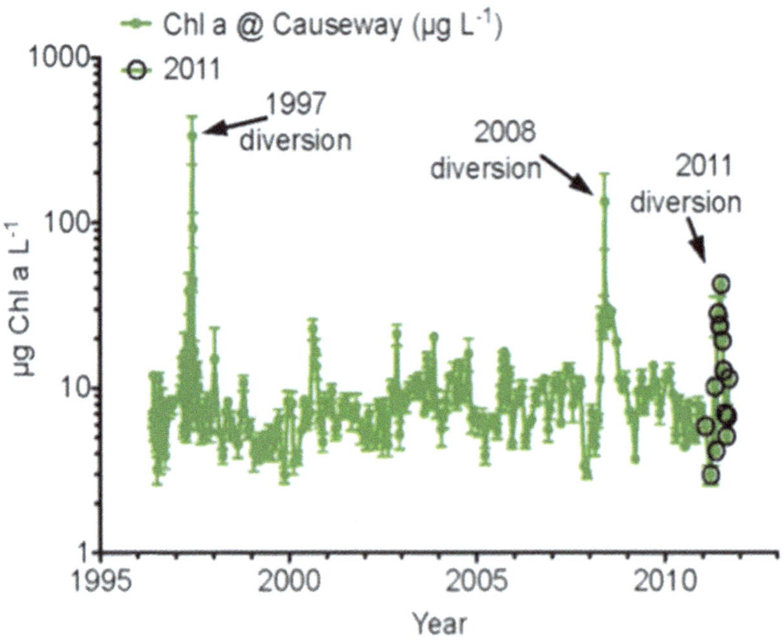

Fig. 8 Chlorophyll *a* concentrations in Lake Pontchartrain (Source: Lake Pontchartrain Basin Foundation)

marshes affected by the Caernarvon river diversion suffered extensive damage from Hurricane Katrina surge and waves (Howes et al. 2010; Kearney et al. 2011).

Various studies at Caernarvon do not support the conclusion that nutrients were an important factor in marsh deterioration after Hurricane Katrina. Studies documented very low dissolved inorganic nitrogen (DIN) concentrations of surface waters in interior marshes at Caernarvon, and rapid decreases in DIN as river water flowed through the estuary (Day et al. 2009; Lane et al. 1999, 2004; Snedden et al. 2015). Nitrate concentrations of diverted river water generally ranged between 1.5 and 2.0 mg/L while ammonium levels were an order of magnitude lower at about 0.1 mg/L (Lane et al. 1999). Field measurements show that nitrate concentrations decreased rapidly with distance from the diversion (Lane et al. 1999, 2004). Just south of Big Mar, a square water body resulting from a failed agricultural impoundment, NO_3 concentrations were 106 uM, while four km away levels decreased to less than 0.2 uM as river water flowed through the estuary (Day et al. 2009; Snedden et al. 2015). Yet high wetland loss was observed over a much larger area including mostly interior marsh regions.

VanZomeren et al. (2011) discussed the fate of NO_3 in a brackish marsh affected by the Caernarvon diversion. They conducted a 3-month mass balance study where ^{15}N–labeled nitrate was added to plant sediment cores collected from a *S. patens* marsh at Caernarvon. Approximately 65% of all added labeled NO_3 was unaccounted for after 12 weeks which suggests gaseous loss. The remaining ^{15}N was incorporated in plant and soil compartments, with most sequestered in the aboveground plant organic matter. There were no significant differences in belowground biomass production between the NO_3 loaded and the control cores after three months. These results indicate that although denitrification was the main removal mechanism for NO_3, soil organic matter reserves were not significantly affected. Stoichiometric calculations demonstrated that maximum rates of denitrification would require just 0.11, 0.22, and 0.33% of soil carbon in the top 20-cm for 30, 60, and 90 days, respectively, of continuous nitrate loading. The contention that tight coupling of denitrification and oxidation of soil carbon stores decreases soil strength and, hence, marsh stability, was not supported by this 12-week study.

Implications of Climate Change and Energy Scarcity for Coastal Restoration

Climate Change

Sustaining deltas will become more difficult in the face of increasingly severe climate impacts (Tessler et al. 2015). CO_2 levels are now tracking at the highest IPCC scenarios (Friedlingstein et al. 2014), with CO_2 concentrations in the atmosphere now regularly exceeding 400 ppm, a dramatic increase over levels in the late nineteenth century of 280 ppm (IPCC 2007). These are the highest CO_2 levels of the past three million years (Friedlingstein et al. 2014).

The effects of climate change on the Mississippi delta include high rates of eustatic sea-level rise (FitzGerald et al.

2008; Meehl et al. 2007; Pfeffer et al. 2008; Vermeer and Rahmstorf 2009; IPCC 2013), more frequent severe hurricanes (Emanuel 2005; Goldenberg et al. 2001; Hoyos et al. 2006; Kauffman et al. 2011; Webster et al. 2005; Mei et al. 2015), floods (Tao et al. 2014), and droughts (IPCC 2007), and more erratic weather in general (Min et al. 2011; Pall et al. 2011). These climate impacts are affecting the delta directly and at the level of the entire drainage basin. For example, Tao et al. (2014) modeled the interactive effects of climate change, land use, and increased CO_2 levels in the atmosphere on discharge of the Mississippi River. Their results suggest that river discharge will substantially increase by as much as 10 to 60% during this century. Such large increases may compromise the flood control system on the Mississippi in times of increasing energy scarcity (Kemp et al. 2014). Large episodic diversions could reduce this threat especially if placed strategically to also offer hurricane protection.

Between 1901 and 2010, eustatic sea-level rise averaged 1.7 mm/year compared to 3.2 mm/year between 1993 and 2010 (IPCC 2013). Thus, the rate of sea-level rise over the past two decades is nearly twice the average of the twentieth century as a whole. The IPCC (2013) projects that sea-level rise will reach between 0.3 and 1.0 m by 2100. The lower end of this range implies that no further acceleration will occur. Semi-empirical model projections indicate sea level will rise from about 0.3 to nearly 2 m by 2100 (Horton et al. 2014; Kopp et al. 2016; DeConto and Pollard 2016). Sea level will continue rising over the next couple of centuries after 2100 by 2–3 m or higher depending on different climate scenarios (IPCC 2013). Thus, continued use of large diversions into the future will be necessary.

Energy Scarcity

Less energy availability and higher energy prices will also limit options for restoration of deltas and complicate the response to climate change. The implication of emerging energy scarcity is that the cost of energy will rise on average and in addition to becoming volatile in coming decades (Bentley 2002; Campbell and Laherrère 1998; Deffeyes 2001; Hall and Day 2009; Murphy and Hall 2011; Day et al. 2014, 2016; EIA 2015). This means that the cost of energy intensive activities will increase significantly. Much of the current management of the Mississippi delta is based on large-scale energy intensive activities (Day et al. 2014). These include dredging, maintenance of navigation channels, building and maintaining levees and dikes, transporting dredged sediments in pipelines, and building large water control structures. As energy prices escalate in coming decades, the cost of maintaining such systems will likely become very expensive, perhaps prohibitively so. Tessler et al. (2015) reported that the combination of land use changes, climate change, and increasing energy costs will increase risks significantly of non-sustainable outcomes, especially in first world deltas such as the Mississippi and Rhine, where energy intensive restoration and management activities are common. Because of the increasing severity of climate change and energy scarcity, coastal restoration and management will become much more challenging and expensive and restoration options more limited in this century (Tessler et al. 2015). For these reasons, delta restoration plans put in place in the short term will likely have to be largely self-maintaining within a few decades.

Construction of diversions is expensive. But once constructed, the structures have a long life and operation and maintenance of these systems is much less costly than construction. The Bonnet Carré Spillway is an example of this. The structure was completed in 1933 and still is functional and will likely remain so for years to come. Operation and maintenance costs of the structure are rather low. As noted above, operation of the spillway during the 2011 flood was about $4.9 million and this included environmental monitoring (http://www.usace.army.mil/Portals/2/docs/civilworks/budget/workplan/fy11wp_mrandt.pdf). By comparison, construction of new diversions costs hundreds of millions of dollars (CPRA 2012a). Thus, it is imperative that diversions be constructed soon to avoid the impact of rising costs. But once constructed, diversions are inexpensive to operate.

Summary

The use of very large but infrequently opened diversions would have a number of benefits. They would build land quickly. A diversion the size of the 1927 crevasse at Caernarvon would lead to wetlands being created after the first operation. These diversions would self-design and form a newly emerging subdelta without large management expense. Because they would be operated infrequently, they would not lead to persistent over-freshening and displacement of fishery species. Water quality problems due to algal blooms would be short lived. The cost of such diversion would occur up front, so that increasing cost due to energy scarcity would be avoided. Large diversions also would build land quickly and consistently over time, thus offsetting sea-level rise. Because operation of large diversions on opposite sides of the river could be alternated, diversion conveyance efficiency, which diminishes with the number of structures simultaneously open, can be increased. Although this paper addresses coastal restoration in the Mississippi delta, we believe that it can serve as a model for many deltaic areas around the world where rising sea level is leading to wetland loss.

Addendum

Investigation 1: A Large, Intermittent River Diversion into the Maurepas Swamp

We present here modeling results for a large and intermittent river diversion into the Maurepas Swamp, a 57,000 ha baldcypress tupelogum (*Taxodium distichum – Nyssa aquatica*) shallow, forested freshwater wetland system in southeast Louisiana (Fig. A1, see Rutherford 2017). The Maurepas Swamp is currently suffering from subsidence, permanent flooding, lack of sediments and nutrients from the Mississippi River, nutria herbivory, and saltwater intrusion from increasing sea-level rise and drought (Shaffer et al. 2009, 2016). The Maurepas Swamp is located further upstream from large diversions in the Coastal Protection and Restoration Authority's Coastal Master Plan, and will more efficiently build land due to shallower water and flat topography. To estimate land building, we used a 2-dimensional delta progradation model originally developed by Dr. Gary Parker (Parker et al. 1998; Parker and Sequeiros 2006; Kim et al. 2009). The delta progradation model assumes radially symmetric delta formation, a flat basin floor, and negligible hydrodynamic effects. In addition to sediment accretion from the river diversion, relative wetland elevation is affected by accelerating sea-level rise and wetland subsidence.

The amount of mud retained in the delta and the fan spreading angle must be specified by the modeler. To approximate the mud retention in the swamp, we used data from the Day et al. (2016) study of the 1927 Caernarvon crevasse. Using sediment cores, Day et al. (2016) estimated that 55–75% of the total sediment was retained during the crevasse at Caernarvon. If 10% of the river discharge is diverted, and assuming that 30% of sand discharge and 100% of mud discharge is delivered by the crevasse into the basin, a total sediment retention of 55–75% corresponds to a mud retention of 44–69% (for details see Rutherford 2017). We assume that the fan spreading angle within the Maurepas Swamp will be 120°, matching the fan spreading angle of the Wax Lake delta. Sediment is transported to the swamp via an 18 km long, 177 m wide, and 20 m deep diversion channel, which we assume is 30% efficient at capturing sediment from the river.

Rates of eustatic (global) sea level rise used in this study cover the full breadth of scientific projections. The "no-change" scenario (SLR-1) assumes a constant rate of sea-level rise equivalent to the current rate, which is about 3.5 mm/year (Nerem et al. 2010, http://sealevel.colorado.edu/). This is near the low end reported by IPCC process-based models; Church et al. (2013) report a minimum value of 0.31 m of sea-level rise by 2100, relative to 1992. The SLR-5 scenario is consistent with the uppermost sea-level rise reported by semi-empirical models and new findings

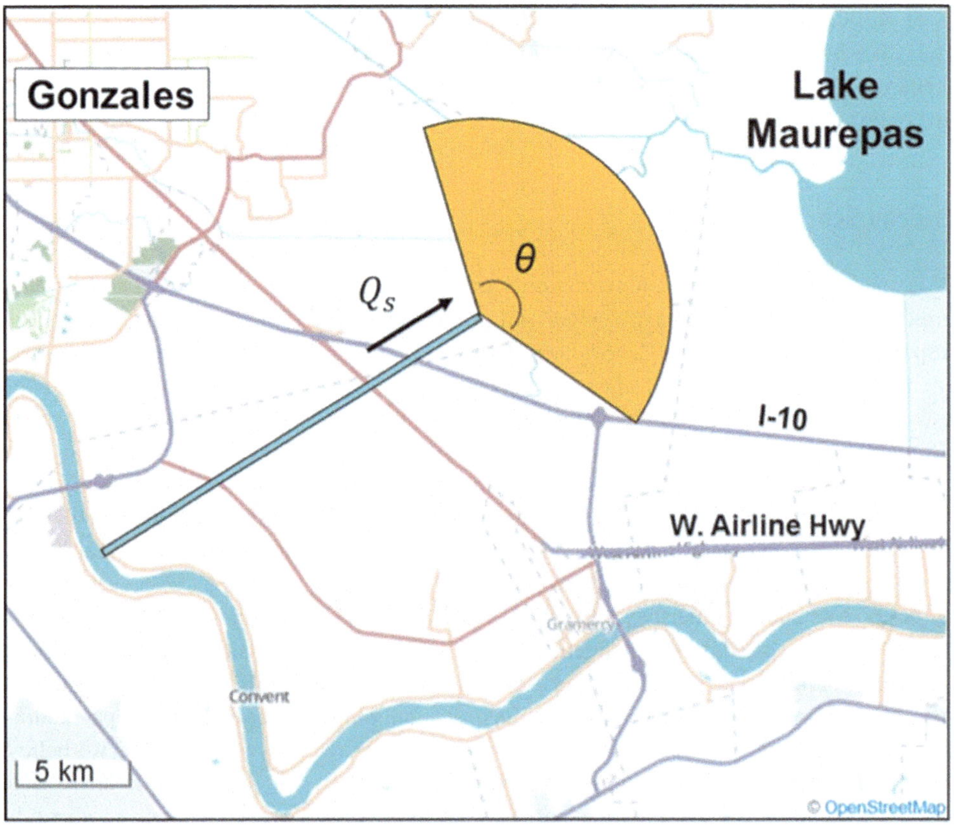

Fig. A1 Generalized location of diversion channel discharging sediment at a rate Q_s into the Maurepas Swamp and building a delta with spreading angle θ

which indicate greater contributions from polar ice sheets. These studies suggest up to 2 m of sea-level rise (relative to 1992) by 2100 (Pfeffer et al. 2008; Vermeer and Rahmstorf 2009; Parris et al. 2012; Deconto and Pollard 2016). For this study (in addition to SLR-1), we developed four scenarios (SLR-2, SLR-3, SLR-4, and SLR-5); each begins with 3.5 mm/year of sea-level rise in 2016, and accelerates according to a second order exponential function towards a specified sea-level in 2100 relative to 2016 (0.57 m, 1.03 m, 1.45 m, and 1.83 m, respectively).

Figure A2 presents potential land gain trajectories for a 7079 m^3/s river diversion into the Maurepas Swamp, operated for 16 weeks every 2 years. Simulations for the large, intermittent diversion begin in 2030 based upon documentation from the 2017 Coastal Master Plan which estimates river diversion projects require up to 6 years of engineering and design and up to 7 years of construction (McMann et al. 2016). Land gain was simulated for 50 years, also to be consistent with the Coastal Master Plan. Land gain is rapid, as the Maurepas Swamp is shallow (the water depth is on average 26 cm in the depositional zone, and we assume an additional 10 cm of water from diversion discharge) and the topography is flat. Based on these results, impressive land building is possible with a large and intermittent river diversion. However, there is significant variation in land building with respect to sea-level rise trajectory. Compared to the total land gain over 50 years in the SLR-3 scenario (143 km^2), the total land gain in the SLR-1 scenario is about 50% greater (214 km^2) and the total land gain in the SLR-5 scenario is about 28% less (103 km^2). The most striking result is that continued land gain past 50 years is possible in all sea-level rise scenarios but SLR-5. In the SLR-5 scenario, the sediment supplied is no longer able to fill the increasing accommodation space at 44 years. In each scenario, annual land gain decreases due to the increasing amount of sediment required to maintain land gain with greater delta size, and due to the increasing accommodation space with increasing sea level.

Day et al. (2016 and reprinted here in this chapter) demonstrated that a large, intermittent river diversion has many advantages compared to a continuously operated river diversion. This example shows that although land building will still be affected by climate trajectories and future sea--level rise, continued land gain is still possible in all but the worst-case scenario (Table A1).

Investigation 2: The Impacts of Oil Price and Suspended Sediments on Marsh Creation Costs

Here we show the potential synergy between river diversions and marsh creation and how the cost of sustaining wetlands with marsh creation is affected by both sea level rise and energy price. As we discussed in chapter "The Costs and Sustainability of Ongoing Efforts to Restore and Protect Louisiana's Coast", dredging is highly fuel intensive and it's costs are sensitive to energy price. About 60-70% of a marsh creation project is typically spent on hydraulic dredging alone (Caffey et al. 2014, CPRA 2012a), and about 20–30% of dredging costs are spent directly on fuel depending on fuel price and project design (Belesimo 2000). The price of energy, crude oil in particular, also affects the price of dredging indirectly because energy is

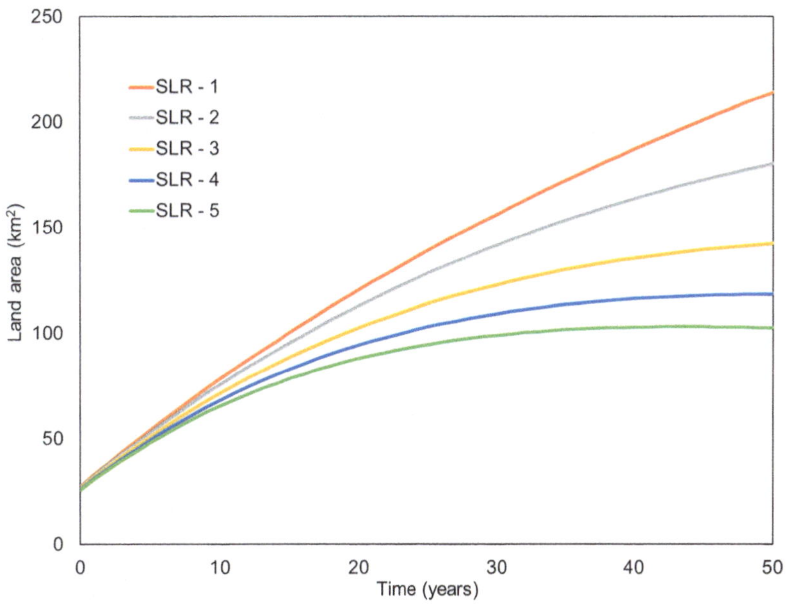

Fig. A2 Potential land gain for a 7079 m^3/s river diversion into the Maurepas Swamp, operated 16 weeks at a time every 2 years, with various SLR scenarios

Table A1 Key model parameters

Parameter description	Symbol	Value	Units
Duration of diversion operation	X	16	Weeks
Periodicity of diversion operation	N	2	Years
Diversion water discharge	Q_{bfw}	7079	m^3/s
Diversion sand discharge	Q_{bfs}	0.043	m^3/s
Mud retention	Λ	57	%
Grain size of bed material	D	0.235	mm
Submerged specific gravity of sediment	R	1.65	
Bed porosity	λ_p	0.6	
Chezy Resistance Coefficient	C_z	25	
Water depth	h_{ru}	−0.36	m
Subaqueous basement slope	S_b	0.00001	
Fan angle	θ_F	120	°
Rate of subsidence	σ	1.5	mm/year
Total subsidence of surface, high porosity soils	h	0.09	m
Channel length	L_c	18,000	m
Channel width	B_c	177	m

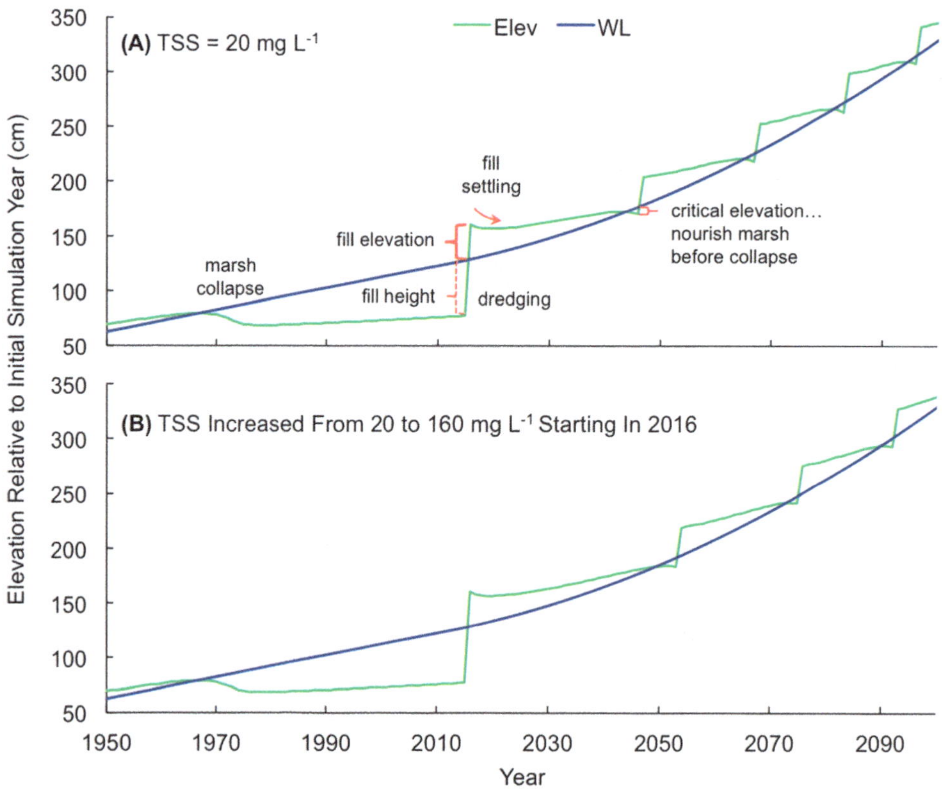

Fig. A3 Simulation of marsh elevation over time with input of dredged materials to sustain the marsh. Results are shown for total suspended sediment concentrations (TSS) of (**a**) 20 mg L^{-1} and (**b**) 160 mg L^{-1}. Fill elevation and the elevation at which restoration is triggered are adjustable parameters, set in this Fig. to 50 cm and −10 cm respectively (from Wiegman 2017)

needed at every transformation in the economy and is strongly correlated with the price of other major commodities (Murphy and Hall 2011; World Bank 2015).

We developed a model that simulated the costs of sustaining marsh with hydraulic dredging into the future. We used existing eco-geomorphic marsh elevation models calibrated to the physical setting of the Mississippi delta to simulate the cost of marsh creation given different scenarios of sea-level rise and energy costs (see Wiegman 2017). Sea-level rise was simulated using the same functions as in Investigation 1. The modeling objective was to simulate marsh creation in an open bay (initial depth of −50 cm) starting in 2016 and to sustain a productive marsh out to 2100 by periodically nourishing the marsh with dredged sediment (fill elevation of 100 cm) when the marsh fell more than 10 cm below mean sea level (Fig. A3). We built a subroutine in the model to estimate the cost of dredging as a function of crude oil prices. Crude oil price projections

Table A2 Future scenarios for oil prices

Scenario name	Generalized assumptions of the scenario	Projected price $bbl^{-1} (±0.5 SD.)	
		2025	2035
No change	Techno optimist scenario: sustained rapid renewable energy growth, high efficiency gains and changes in end use drastically decrease demand for oil in the residential, commercial, and transportation sectors, drilling technology improves recovery of unconventional fuels, a small amount of oil is used for chemical feedstock and heavy industry	55	55
Low	Stringent energy and climate policies: low oil demand and low short term GDP growth in developing countries, break up of OPEC low cost oil floods the market, high renewable energy investment, adoption of carbon tax curbs demand	$73 (±7)	$83 (±6)
Central	Moderate energy and climate policy: moderate oil demand and GDP growth, OPEC operates as a swing producer to control price, moderate renewable investment, stated emissions targets are upheld, no further climate policies initiated	$99 (±6)	$124 (±7)
High	Business as usual economics and failed climate policy: high oil demand and high short term GDP growth in developing countries, conservative investment practices from oil producers causes demand to exceed supply, low renewable investment, significant climate policies not adopted	$134 (±20)	$251 (±44)

See Wiegman (2017)
The projected price (real 2010 $/bl) in 2025 and 2035 is given with the half standard deviation value in parentheses

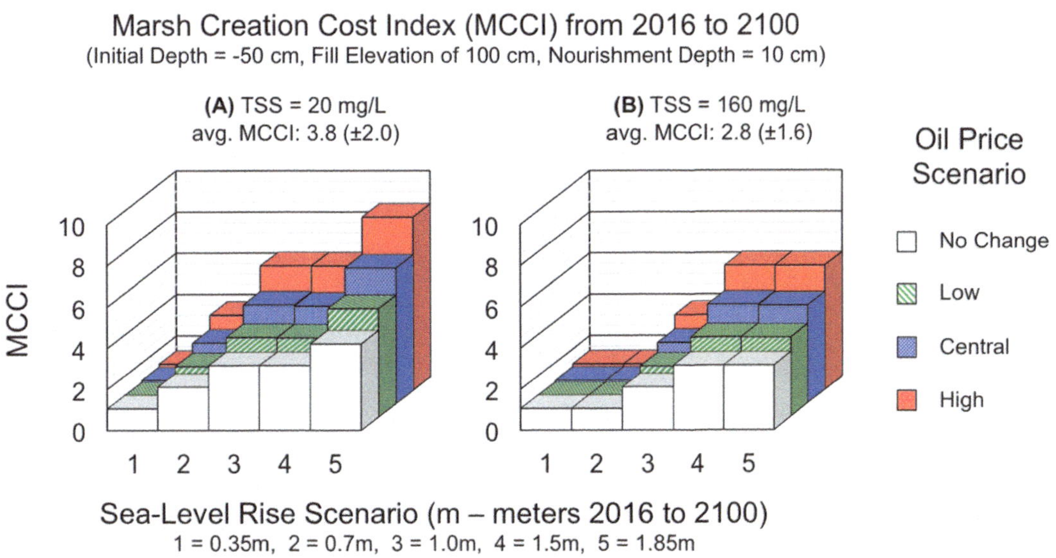

Fig. A4 The impact of oil price, sea-level rise and suspended sediment concentration (TSS), on the cost of sustaining coastal marsh. MCCI is the increase in cost above a no change scenario in sea level rise and energy costs (2016–2100)

were assembled from a suite of energy modeling agencies and published literature (EIA 2015; IEA 2015; McGlade 2014; Heun and De Wit 2012). A list of energy scenarios and assumptions are provided in Table A2.

We ran simulations for the energy and sea-level rise scenarios for a marsh with low total suspended sediment (TSS) input and high TSS input (Fig. A3). The low TSS marsh was set to a concentration of 20 mg/L, which on the low end of values reported for marshes in Barataria and Terrebonne bays that are isolated from sediment input from the Mississippi River by flood control levees. The high TSS marsh was set to a concentration of 160 mg/L, which is the rough mean concentration in the Mississippi and Atchafalaya rivers (Allison and Meselhe 2010). We calculated a marsh creation cost index (MCCI). The MCCI is the cost increase above a no change scenario in energy price and sea level rise, at TSS of 20 mg L^{-1}. An MCCI of 1 corresponds to a cost of ~121,600 2010$/ha, an MCCI of 2 would be double this amount, and an MCCI of 0.5 would be half. With TSS at 20 mg L^{-1} the average MCCI across all sea-level rise and energy scenarios is 3.8. With TSS at 160 mg L^{-1} the average MCCI is 2.8, a ~25% reduction in costs (Fig. A4). In the high sea-level rise and Energy scenarios the MCCI for both TSS levels exceeds 6, over 750,000 $2010/ha. Pumping distance and substrate type affect costs (see chapter "The Costs and Sustainability of Ongoing Efforts to Restore and Protect Louisiana's Coast"), while local subsidence rates affect marsh sustainability. We assumed a nearby borrow source (<5 miles) and a subsidence rate of 0.9 cm/year. MCCI

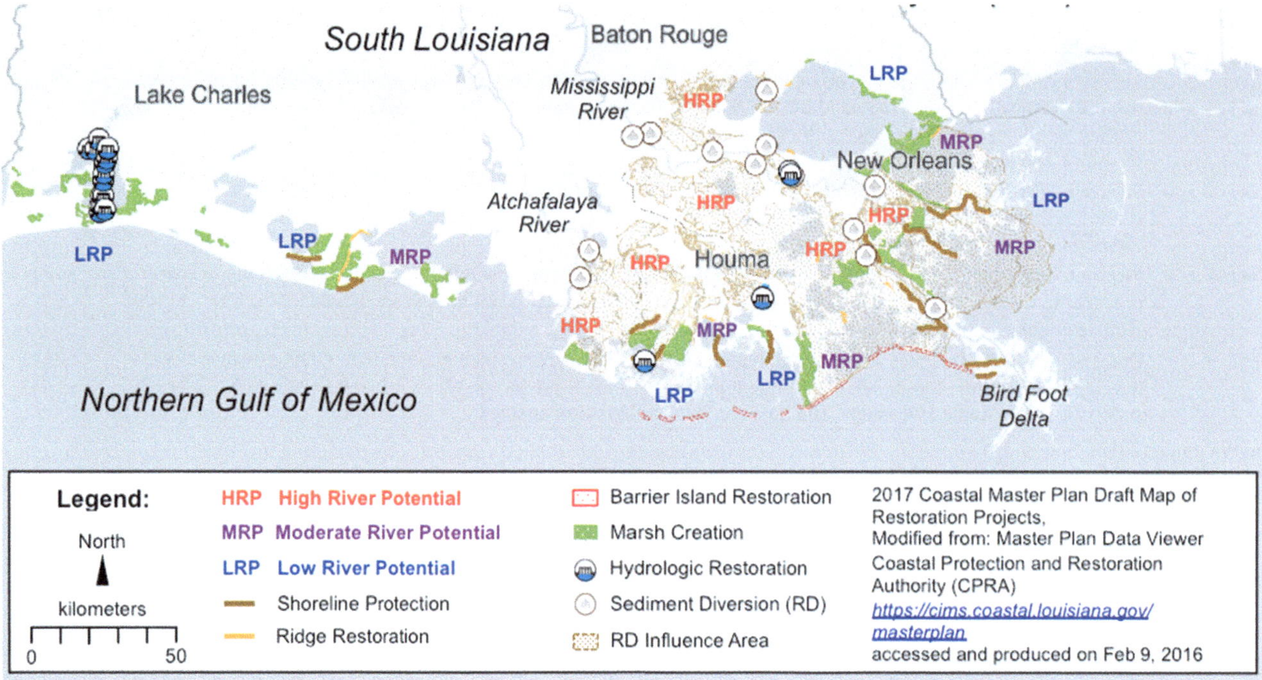

Fig. A5 Map of coastal Louisiana with planned wetland restoration projects and zones of river potential (see *legend*). River potential zones are classified by the approximate slope (S) between the wetland and the nearest major river; $S = (R-W)/d$, where R is river elevation, W wetland elevation, d is distance between the river and the wetland. Steeper S results in greater river potential (Source: Adapted from Coastal Master Plan)

would rise substantially if subsidence or pumping distance were to increase. Figure A5 shows areas where there is potential for river re-introduction and high TSS levels in the Mississippi delta; these areas will be more sustainable than regions with low river potential.

Acknowledgments Partial support for this project was provided by a grant from the Gulf Research Program of the National Academies of Sciences, Engineering, and Medicine. Additional support came from the Coastal Sustainability Studio and the Department of Oceanography and Coastal Sciences at LouisianaLouisiana State University (LSU). We thank Hampton Peele of LSU for satellite imagery of the Davis Crevasse and Don Davis for information on historical crevasses.

This chapter is a reprint of Day et al. 2016 (cited below), originally published in Estuarine Coastal and Shelf Science. Permission to reprint this was granted courtesy of elsevier. We provide an addendum at the end of this chapter. To reference the main text, cite the Estuarine Coastal and Shelf Science version. Cite this chapter to reference the materials from the addendum.

Day, J.W., Lane, R.R., D'Elia, C.F., Wiegman, A.R., Rutherford, J. S., Shaffer, G.P., Brantley, C.G. and Kemp, G.P., (2016) Large infrequently operated river diversions for Mississippi delta restoration. Estuarine, Coastal and Shelf Science, 183, pp. 292–303.

Literature Cited

Allison MA, Meselhe EA (2010) The use of large water and sediment diversions in the lower Mississippi River (Louisiana) for coastal restoration. J Hydrol 387(3):346–360

Andrus TM (2007) Sediment flux and fate in the Mississippi River diversion at West Bay: Observation study. Department of Oceanography & Coastal Sciences, Louisiana State University

Anisfeld SC, Hill TD (2012) Fertilization effects on elevation change and belowground carbon balance in a Long Island Sound tidal marsh. Estuar Coasts 35:201–211

Barras JA, Bourgeois PE, Handley LR (1994) Land loss in coastal Louisiana, 1956–1990. U.S. Geological Survey, National Wetlands Research Center Open File Report 94–01, 4p. 10 color prints

Bentley RW (2002) Global oil and gas depletion: an overview. Energ Policy 30:189–205

Blum MD, Roberts HH (2009) Drowning of the Mississippi Delta due to insufficient sediment supply and global sea-level rise. Nat Geosci 2(7):488

Blum MD, Roberts HH (2012) The Mississippi delta region: past, present, and future. Annu Rev Earth Planet Sci 40:655–683

Boesch DF, Josselyn MN, Mehta AJ, Morris JT, Nuttle WK, Simenstad CA, Swift DJP (1994) Scientific assessment of coastal wetland loss, restoration and management. J Coast Res, Special Issue No 20

Bowman PE, Perret WS, Roussel JE (1995) Freshwater introduction and implications for fisheries production in Louisiana. Department of Wildlife and Fisheries, Baton Rouge. 8 p

Britsch LD, Dunbar JB (1993) Land loss rates: Louisiana Coastal Plain. J Coast Res 9:324–338

Browder JA, Moore D (1981) A new approach to determining the quantitative relationship between fishery production and the flow of fresh water to estuaries. In: Cross R, Williams D (eds) Proceedings, national symposium on freshwater inflow to estuaries, vol 1, pp 403–430. FWS/OBS-81/04. US Fish Wildlife Service, Washington, DC

Campbell CJ, Laherrère JH (1998) The end of cheap oil. Sci Am:60–65

Church JA, Clark PU, Cazenave A, Gregory JM, Jevrejeva S, Levermann A, Merrifield MA, Milne GA, Nerem RS, Nunn PD,

Payne AJ (2013) Sea-level rise by 2100. Science 342 (6165):1445–1445
Condrey RE, Hoffman PE, Evers DE (2014) The last naturally active delta complexes of the Mississippi River (LNDM): discovery and implications. In: Day JW, Kemp GP, Freemen AM, Muth DP (eds) Perspectives on the restoration of the Mississippi Delta. Springer, Dordrecht, pp 33–50
Couvillion BR, Barras JA, Steyer GD, Sleavin W, Fischer M, Beck H, Trahan N, Griffin B, Heckman D (2011) Land area change in coastal Louisiana from 1932 to 2010: U.S. Geological Survey Scientific Investigations Map 3164, scale 1:265,000, 12 p. pamphlet
CPRA (Coastal Protection and Restoration Authority of Louisiana) (2012a) Louisiana's comprehensive master plan for a sustainable coast. CPRA, Baton Rouge. 190 p
CPRA (Coastal Protection and Restoration Authority of Louisiana) (2012b) Appendix A2: project fact sheets. 2012 coastal master plan. CPRA, Baton Rouge. 446 p
Darby FA, Turner RE (2008a) Below- and aboveground Spartina alterniflora production in a Louisiana salt marsh. Estuar Coasts 31:223–231
Darby FA, Turner RE (2008b) Below- and aboveground biomass of Spartina alterniflora: response to nutrient addition in a Louisiana salt marsh. Estuar Coasts 31:326–334
Darby FA, Turner RE (2008c) Effects of eutrophication on salt marsh root and rhizome accumulation. Mar Ecol Prog Ser 363:63–70
Davis DW (1993) Crevasses on the lower course of the Mississippi River. Coast Zone 1:360–378
Davis DW (2000) Historical perspective on crevasses, levees, and the Mississippi River. In: Colten CE (ed) Transforming New Orleans and its environs: centuries of change. University of Pittsburgh Press, Pittsburgh, pp 84–106
Day JW, Martin J, Cardoch L, Templete P (1997) System functioning as a basis for sustainable management of deltaic ecosystems. Coast Manag 25:115–153
Day JW, Britsch LD, Hawes S, Shaffer G, Reed DJ, Cahoon D (2000) Pattern and process of land loss in the Mississippi Delta: a spatial and temporal analysis of wetland habitat change. Estuaries 23:425–438
Day JW, Yáñez-Arancibia A, Horton BP (2007) Emergence of complex societies after sea level stabilized. Eos 88:169–170
Day JW, Cable JE, Cowan JH, DeLaune R, Fry B, Mashriqui H, Justic D, Kemp P, Lane RR, Rick J, Rick S, Rozas LP, Snedden G, Swenson E, Twilley RR, Wissel B (2009) The impacts of pulsed reintroduction of river water on a Mississippi Delta coastal basin. J Coast Res 54:225–243
Day JW, Kemp GP, Reed DJ, Cahoon DR, Boumans RM, Suhayda JM, Gambrell R (2011) Vegetation death and rapid loss of surface elevation in two contrasting Mississippi delta salt marshes: The role of sedimentation, autocompaction and sea-level rise. Ecol Eng 37(2):229–240
Day J, Hunter R, Keim RF, DeLaune R, Shaffer G, Evers E, Reed D, Brantley C, Kemp P, Day J, Hunter M (2012) Ecological response of forested wetlands with and without large-scale Mississippi River input: implications for management. Ecol Eng 46:57–67
Day JW, Lane R, Moerschbaecher M, DeLaune R, Mendelssohn I, Baustian J, Twilley R (2013) Vegetation and soil dynamics of a Louisiana estuary receiving pulsed Mississippi River water following hurricane Katrina. Estuar Coasts 36:1–18
Day JW, Moerschbaecher M, Pimentel D, Hall C, Yáñez-Arancibia A (2014) Sustainability and place: How emerging mega-trends of the 21st century will affect humans and nature at the landscape level. Ecol Eng 65:33–48
Day JW, Cable JE, Lane RR, Kemp GP (2016) Sediment deposition at the Caernarvon Crevasse during the Great Mississippi Flood of 1927: implications for coastal restoration. Water 3(38):1–12
Deegan LA, Johnson DS, Warren RS, Peterson BJ, Fleeger JW, Fagherazzi S, Wollheim WM (2012) Coastal eutrophication as a driver of salt marsh loss. Nature 490:388–392
Deffeyes KS (2001) Hubbert's Peak – the impending world oil shortage. Princeton University Press, Princeton. 208 p
DeLaune RD, Kongchum M, White JR, Jugsujinda A (2013) Freshwater diversions as an ecosystem management tool for maintaining soil organic matter accretion in coastal marshes. Catena 107:139–144
DeLaune RD, Sasser CE, Evers-Hebert E, White JR, Roberts HH (2016) Influence of the Wax Lake Delta sediment diversion on aboveground plant productivity and carbon storage in deltaic island and mainland coastal marshes. Estuar Coast Shelf Sci 177:83–89
Dugas RJ, Perret WS (1975) The effects of 1973 spring floodwaters on oyster populations in Louisiana. In: Proceedings of the 29th annual conference of the South Eastern Association of Game and Fish Commissioners October 12–15, 1975, pp 208–214
EIA (Energy Information Administration) (2015) Annual energy outlook with projections to 2040. EIA Independent Statistics and Analysis, United States Department of Energy, Washington, DC. 154 p
Emanuel K (2005) Increasing destructiveness of tropical cyclones over the past 30 years. Nature 436:686–688
FitzGerald DM, Fenster MS, Argow BA, Buynevich IV (2008) Coastal impacts due to sea-level rise. Annu Rev Earth Planet Sci 36:601–647
Fox L, Valiela I, Kinney EL (2012) Vegetation cover and elevation in long-term experimental nutrient-enrichment plots in Great Sippewissett Salt Marsh, Cape Cod, Massachusetts: Implications for eutrophication and sea level rise. Estuar Coasts 35:445–485
Friedlingstein P, Meinshausen M, Arora VK, Jones CD, Anav A, Liddicoat SK, Knutti R (2014) Uncertainties in CMIP5 climate projections due to carbon cycle feedbacks. J Clim 27(2):511–526
Goldenberg SB, Landsea CW, Mestas-Nuñez AM, Gray WM (2001) The recent increase in Atlantic hurricane activity: causes and implications. Science 293:474–479
Graham SA, Mendelssohn IA (2014) Coastal wetland stability maintained through counterbalancing accretionary responses to chronic nutrient enrichment. Ecology 95:3271–3283
Gunter G (1953) The relationship of the Bonnet Carré Spillway to Oyster Beds in Mississippi Sound and the "Louisiana Marsh", with a Report on the 1950 Opening 70: 22–71
Hall C, Day J (2009) Revisiting the limits to growth after peak oil. Am Sci 92:230–237
Hillman E, Henkel T, Lopez J, Baker D (2015) Recommendations for restoration: Central Wetlands Unit, Louisiana, Lake Pontchartrain Basin Foundation, New Orleans, LA, USA. 69 p
Horton BP, Rahmstorf S, Engelhart SE, Kemp AC (2014) Expert assessment of sea-level rise by AD 2100 and AD 2300. Quat Sci Rev 84:1–6
Howes NC, FitzGerald DM, Hughes ZJ, Georgiou IY, Kulp MA, Miner MD, Smith JM, Barras JA (2010) Hurricane-induced failure of low salinity wetlands. Proc Natl Acad Sci 107:14014–14019
Hoyos CD, Agudelo PA, Webster PJ, Curry JA (2006) Deconvolution of the factors contributing to the increase in global hurricane intensity. Science 312:94–97
IPCC (Intergovernmental Panel on Climate Change) (2007) Climate change 2007: the scientific basis, Contribution of Working Group I to the third assessment report. Cambridge University Press, Cambridge
IPCC (Intergovernmental Panel on Climate Change) (2013) Climate change 2013: the physical science basis. Contribution of Working Group 1 to the Fifth Assessment Report of the Intergovernmental Panel on Climate Change (Stocker TF, Qin D, Plattner GK, Tignor M, Allen SK, Boschung J, Nauels A, Xia Y, Bex V, Midgley PM (eds)). Cambridge/New York, 1535 p

Kauffman JB, Heider C, Cole TG, Dwire KA, Donato DC (2011) Ecosystem carbon stocks of Micronesian mangrove forests. Wetlands 31(2):343–352

Kearney MS, Riter JCA, Turner RE (2011) Freshwater river diversions for marsh restoration in Louisiana: twenty-six years of changing vegetative cover and marsh area. Geophys Res Lett 38:L16405

Kemp GP, Day JW, Freeman AM (2014) Restoring the sustainability of the Mississippi River Delta. Ecol Eng 65:131–146

Kim W, Mohrig D, Twilley R, Paola C, Parker G (2009) Is it feasible to build new land in the Mississippi River delta? Eos Trans AGU 90 (42):373–374

Kolb CR, van Lopik JR (1958) Geology of the Mississippi deltaic plain – Southeastern Louisiana. U.S Army Corps of Engineers Waterways Experiment Station. Technical Report no. 3-483, 120 p

Kolker AS, Miner MD, Weathers HD (2012) Depositional dynamics in a river diversion receiving basin: The case of the West Bay Mississippi River Diversion. Estuar Coast Shelf Sci 106:1–12

Kopp RE, Kemp AC, Bittermann K, Horton BP, Donnelly JP, Gehrels WR, Hay CC, Mitrovica JX, Morrow ED, Rahmstorf S (2016) Temperature-driven global sea-level variability in the Common Era. Proc Natl Acad Sci 113(11):E1434–E1441

Lane RR, Day JW, Thibodeaux B (1999) Water quality analysis of a freshwater diversion at Caernarvon, Louisiana. Estuaries 2A:327–336

Lane RR, Day JW, Justic D, Reyes E, Day JN, Hyfield E (2004) Changes in stoichiometric Si, N and P ratios of Mississippi River water diverted through coastal wetlands to the Gulf of Mexico. Estuar Coast Shelf Sci 60:1–10

Lane RR, Day JW Jr, Day JN (2006) Wetland surface elevation, vertical accretion, and subsidence at three Louisiana estuaries receiving diverted Mississippi River water. Wetlands 26:1130–1142

Lane RR, Day JW Jr, Marx B, Hyfield E, Day JN, Reyes E (2007) The effects of riverine discharge on temperature, salinity, suspended sediment and chlorophyll a in a Mississippi delta estuary measured using a flow-through system. Estuar Coast Shelf Sci 74:145–154

Lopez JA, Henkel TK, Moshogianis AM, Baker AD, Boyd EC, Hillmann ER, Connor PF, Baker DB (2014) Examination of deltaic processes of Mississippi River outlets—Caernarvon delta and Bohemia Spillway in southeastern Louisiana. Gulf Coast Assoc Geol Soc J 3(2014):79–93

McMann B, Schulze M, Sprague H, Smyth K (2016) 2017 Coastal master plan: Appendix A: Project definitions. Version I. Coastal Protection and Restoration Authority, Baton Rouge

Meade RH, Moody JA (2010) Causes for the decline of suspended-sediment discharge in the Mississippi River system, 1940–2007. Hydrol Process 24(1):35–49

Meehl GA, Stocker TF, Collins WD, Friedlingstein P, Gaye AT, Gregory JM, Kitoh A, Knutti R, Murphy JM, Noda A, Raper SCB, Watterson IG, Weaver AJ, Zhao ZC (2007) Global climate projections. In: Solomon S, Qin D, Manning M, Chen Z, Marquis M, Avery KB, Tignor M, Miller HL (eds) Climate change 2007: the physical science basis. Contribution of Working Group I to the fourth assessment report of the Intergovernmental Panel on Climate Change. Cambridge University Press, Cambridge/New York

Mei W, Xie S-P, Primeau F, McWilliams JC, Pasquero C (2015) Northwestern Pacific typhoon intensity controlled by changes in ocean temperatures. Sci Adv 1:e1500014

Mendelssohn IA, Morris JT (2000) Eco-physiological controls on the productivity of Spartina alterniflora loisel. In: Weinstein MP, Kreeger DA (eds) Concepts and controversies in tidal marsh ecology. Kuwer Acedemic Publishers, Boston, pp 59–80

Min S-K, Zhang X, Zwiers FW, Hegerl GC (2011) Human contribution to more-intense precipitation extremes. Nature 470:378–381

Morris JT, Shaffer GP, Nyman JA (2013a) Brinson review: perspectives on the influence of nutrients on the sustainability of coastal wetlands. Wetlands 33:975–988

Morris JT, Sundberg K, Hopkinson CS (2013b) Salt marsh primary production and its responses to relative sea level and nutrients in estuaries at Plum Island, Massachusetts, and North Inlet, South Carolina, USA. Oceanography 26:78–84

Murphy DJ, Hall CAS (2011) Adjusting the economy to the new energy realities of the second half of the age of oil. Ecol Model 223:67–71

Muth D (2014) The once and future delta. In: Day JW, Kemp GP, Freemen AM, Muth DP (eds) Perspectives on the Restoration of the Mississippi Delta. Springer, Dordrecht, pp 9–28

National Audubon Society (2012) West Bay Delta development since 2009. National Audubon Society, Baton Rouge

Nittrouer JA, Best JL, Brantley C, Cash RW, Czapiga M, Kumar P, Parker G (2012) Mitigating land loss in coastal Louisiana by controlled diversion of Mississippi River sand. Nat Geosci 5:534–537

Nyman JA (2014) Integrating successional ecology and the delta lobe cycle in wetland research and restoration. Estuar Coasts 37:1490–1505

Olea RA, Coleman JL (2014) A synoptic examination of causes of land loss in southern Louisiana as they relate to the exploitation of subsurface geologic resources. J Coast Res 30:1025–1044

Pall P, Aina T, Stone DA, Stott PA, Nozawa T, Hilberts AGJ, Lohmann D, Allen MR (2011) Anthropogenic greenhouse gas contribution to flood risk in England and Wales in autumn 2000. Nature 470:382–385

Paola C, Twilley RR, Edmonds DA, Kim W, Mohrig D, Parker G, Viparelli E, Voller VR (2010) Natural processes in delta restoration: application to the Mississippi delta. Annu Rev Mar Sci 3:67–91

Pfeffer WT, Harper JT, O'Neel S (2008) Kinematic constraints on glacier contributions to 21st-century sea-level rise. Science 321 (5894):1340–1343

Piazza BP, LaPeyre MK (2007) Restoration of the annual flood pulse in Breton Sound, Louisiana, USA: habitat change and nekton community response. Aquat Biol 1:109–119

Roberts HH (1997) Dynamic changes of the holocene Mississippi River delta plain: the delta cycle. J Coast Res 13:605–627

Roberts HH, DeLaune RD, White JR, Li C, Sasser CE, Braud D, Weeks E, Khalil S (2015) Floods and cold front passages: impacts on coastal Marshes in a River Diversion Setting (Wax Lake Delta Area, Louisiana). J Coast Res 31(5):1057–1068

Roy ED, White JR, Smith EA, Bargu S, Li C (2013) Estuarine ecosystem response to three large-scale Mississippi River flood diversion events. Sci Total Environ 458:374–387

Roy ED, Smith EA, Bargu S, White JR (2016) Will Mississippi River diversions designed for coastal restoration cause harmful algal blooms? Ecol Eng 91:350–364

Rozas LP, Minello TJ, Munuera-Fernandez I, Fry B, Wissel B (2005) Macrofaunal distributions and habitat change following winter-spring releases of freshwater into the Breton Sound estuary, Louisiana. Estuar Coast Shelf Sci 65:319–336

Saucier RT (1963) Recent geomorphic history of the Pontchartrain basin. Louisiana State University Press, Baton Rouge

Shaffer GP, Wood WB, Hoeppner SS, Perkins TE, Zoller J, Kandalepas D (2009) Degradation of Baldcypress-Water Tupelo Swamp to marsh and open water in southeastern Louisiana, USA: an irreversible trajectory? J Coast Res 54:152–165

Shaffer GP, Day JW, Kandalepas D, Wood WB, Hunter RG, Lane RR, Hillmann ER (2016) Decline of the Maurepas Swamp, Pontchartrain Basin, Louisiana, and approaches to restoration. Water 8(3):101

Shen Z, Törnqvist TE, Mauz B, Chamberlain EL, Nijhuis AG, Sandoval L (2015) Episodic overbank deposition as a dominant mechanism of floodplain and delta-plain aggradation. Geology 43:875–878

Snedden GA, Cretini K, Patton B (2015) Inundation and salinity impacts to above-and belowground productivity in Spartina patens and Spartina alterniflora in the Mississippi River deltaic plain:

implications for using river diversions as restoration tools. Ecol Eng 81:133–139
Swarzenski CM, Doyle TW, Fry B, Hargis TG (2008) Biogeochemical response of organic-rich freshwater marshes in the Louisiana delta plain to chronic river water influx. Biogeochemistry 90:49–63
Tao B, Tian H, Ren W, Yang J, Yang Q, He R, Cai W, Lohrenz S (2014) Increasing Mississippi river discharge throughout the 21st century influenced by changes in climate, land use, and atmospheric CO2. Geophys Res Lett 41:4978–4986
Tessler ZD, Vörösmarty CJ, Grossberg M, Gladkova I, Aizenman H, Syvitski JPM, Foufoula-Georgiou E (2015) Profiling risk and sustainability in coastal deltas of the world. Science 349:638–643
Turner R (2010) Beneath the salt marsh canopy: Loss of soil strength with increasing nutrient loads. Estuar Coasts 34:1084–1093
Twilley RR, Rivera-Monroy V (2009) Sediment and nutrient tradeoffs in restoring Mississippi river delta: restoration vs. eutrophication. Contemp Water Res Educ 141:39–44
U.S. Army Corps of Engineers (2011) Available online. http://www.usace.army.mil/Portals/2/docs/civilworks/budget/workplan/fy11wp_mrandt.pdf
U.S. Army Corps of Engineers. National Water Information System. Available online: http://waterdata.usgs.gov/la/nwis/
VanZomeren CM, White JR, DeLaune RD (2011) Fate of nitrate in vegetated brackish coastal marsh. Soil Sci Am J 76:1919–1927
Vermeer M, Rahmstorf S (2009) Global sea level linked to global temperature. Proc Natl Acad Sci 106(51):21527–21532
Viosca P Jr (1938) Effect of the Bonnet Carré Spillway on Fisheries. LA Conserv Rev 6:51–53
Vorosmarty C, Syvitski J, Day J, Sherbinin A, Giosan L, Paola C (2009) Battling to save the world's river deltas. Bull At Sci 65:31–43
Wang FC, Sikora WB, Wang M (1994) Hydrologic regimes of tidal channel-salt marshes flow systems, Fourleague Bay, Louisiana, USA. J Coast Res:809–824
Wang H, Steyer GD, Couvillion BR, Rybczyk JM, Beck HJ, Sleavin WJ, Meselhe EA, Allison MA, Boustany RG, Fischenich CJ, Rivera-Monroy VH (2014) Forecasting landscape effects of Mississippi River diversions on elevation and accretion in Louisiana deltaic wetlands under future environmental uncertainty scenarios. Estuar Coast Shelf Sci 138:57–68
Webster PJ, Holland GJ, Curry JA, Chang H-R (2005) Changes in tropical cyclone number, duration, and intensity in a warming environment. Science 309:1844–1846
Wells JT, Coleman JM (1987) Wetland loss and the subdelta life cycle. Estuar Coast Shelf Sci 25(1):111–125
Wells FC, Demas CR (1977) Hydrology and water quality of the Atchafalaya River basin. In: Cooperation with Louisiana Department of Transportation and Development, Baton Rouge, Office of Public Works Water Resources Technical Report, 14
White JR, Fulweiler RW, Li CY, Bargu S, Walker ND, Twilley RR, Green SE (2009) Mississippi river flood of 2008: observations of a large freshwater diversion on physical, chemical, and biological characteristics of a shallow estuarine lake. Environ Sci Technol 43 (15):5599–5604
Wissel B, Fry B (2005) Tracing Mississippi River influences in estuarine food webs of coastal Louisiana. Oecologia 144:659–672
Xu K, Bentley SJ, Robichaux P, Sha X, Yang H (2016) Implications of texture and erodibility for sediment retention in receiving basins of coastal Louisiana diversions. Water 8:26
Yuill BT, Khadka AK, Pereira J, Allison MA, Meselhe EA (2016) Morphodynamics of the erosional phase of crevasse-splay evolution and implications for river sediment diversion function. Geomorphology 259:12–29

References for Addendum

Belesimo FJ (2000) Cost estimating projects for large cutter and hopper dredges. Doctoral dissertation, Texas A&M University
Caffey RH, Wang H, Petrolia DR (2014) Trajectory economics: assessing the flow of ecosystem services from coastal restoration. Ecol Econ 100:74–84
Day JW, Lane RR, D'Elia CF, Wiegman AR, Rutherford JS, Shaffer GP, Brantley CG, Kemp GP (2016) Large infrequently operated river diversions for Mississippi delta restoration. Estuar Coast Shelf Sci 183:292–303
DeConto RM, Pollard D (2016) Contribution of Antarctica to past and future sea-level rise. Nature 531(7596):591–597
Heun MK, de Wit M (2012) Energy return on (energy) invested (EROI), oil prices, and energy transitions. Energ Policy 40:147–158
International Energy Agency (IEA) (2015) 2015 world energy outlook. International Energy Agency, Organization of Economic Coordination and Development, Paris, France. ISBN:978-92-64-24366-8
McGlade CE (2014) Uncertainties in the outlook for oil and gas. Doctoral dissertation, UCL (University College London)
Nerem RS, Chambers DP, Choe C, Mitchum GT (2010) Estimating mean sea level change from the TOPEX and Jason altimeter missions. Mar Geod 33(S1):435–446
Parker G, Sequeiros O (2006) Large scale river morphodynamics: application to the Mississippi Delta. In River Flow 2006: proceedings of the international conference on Fluvial Hydraulics, pp 3–11
Parker G, Paola C, Whipple KX, Mohrig D (1998) Alluvial fans formed by channelized fluvial and sheet flow. I: theory. J Hydraul Eng 124 (10):985–995
Parris A, Bromirski P, Burkett V, Cayan D, Culver M, Hall J, Horton R, Knuuti K, Moss R, Obeysekera J, Sallenger A, Weiss J (2012) Global sea level rise scenarios for the US National Climate Assessment. NOAA Tech Memo OAR CPO-1., 37 p
Rutherford JR (2017) Examining the benefits of a large, intermittent river diversion into the Maurepas Swamp. Master's thesis, Department of Oceanography and Coastal Sciences, Louisiana State University, Baton Rouge, LA, USA
Wiegman ARH (2017) Modeling the influence of energy and climate megatrends on future costs and benefits of marsh creation in the Mississippi Delta. Master's thesis, Department of Oceanography and Coastal Sciences, Louisiana State University, Baton Rouge, LA, USA
World Bank (2015) Commodities price forecast. Released October 20, 2015

Raising Urban Land: Historical Perspectives on Adaptation

Craig E. Colten

Abstract

The biophysical environment of coastal Louisiana has never been static since the last Ice Age, and since their arrival humans have continually adapted to changing conditions. This chapter reviews the historical adaptations to the floodprone site where New Orleans is situated, including a plan to raise the land level of portions of the city. In particular, it compares two successful examples of land raising in Chicago and Galveston to the project proposed in this volume to raise segments of New Orleans' land surface. It contrasts the underlying rationale for land raising efforts and the social, legal, and economic circumstances of these analogs as a basis for considering the potential viability of the ambitious plan for New Orleans.

Keywords

New Orleans • Floods • Adaptation • Raising land • Sea level rise

Introduction

Sometimes the best ideas are old ones, reworked and rejuvenated in a current context. In the mid-nineteenth century, a New Orleans engineer, Lewis De Russy, recommended an audacious plan to reduce flood risk in the low-lying city. He called it a "system of colmates" for "filling and elevating by river deposits, the large district of country in the rear of the City of New Orleans" (DeRussy 1859, 3). Rather than perpetuating the traditional methods of relying on levees to fend off high river water and a drainage system to remove excess water from the city, he sought a bold new solution – building land to a height above flood risk and above the zone of saturation. Not unlike current plans to build land with diversions, he sought to deploy the natural process of floodplain sedimentation in a controlled manner. By building a series of levee-enclosures between the existing city and Lake Pontchartrain, his plan would divert floodwaters from the river into a set polders where sediment would settle out, thereby creating a higher land surface and enabling safe expansion of the city on newly elevated terrain. This would have been a more expensive, and slower process, than levees and drains, but it had a certain appeal for its lasting benefits.

With our current understanding of coastal land loss and rising sea levels, perhaps it is time to reconsider this long-forgotten proposal for offsetting the loss of sediment due to levees for the urbanized territory of the lower river. While the extent of development in New Orleans is vastly different than it was in the mid-nineteenth century, there is precedent for massive land raising projects even in the midst of existing dense development. There have been bold and successful projects that might offer a historical analog for a partial solution to New Orleans's current dilemma.

This chapter explores the historical adaptations made to cope with excess water in coastal Louisiana since European settlement began, previous efforts around the world to reclaim wetlands, and projects to raise the elevation of the urban landscape. This analysis of historical examples sheds

C.E. Colten (✉)
Department of Geography and Anthropology, Louisiana State University, Baton Rouge, LA, USA
e-mail: ccolten@lsu.edu

J.W. Day, J.A. Erdman (eds.), *Mississippi Delta Restoration*,
Estuaries of the World, https://doi.org/10.1007/978-3-319-65663-2_9

light on the long-term awareness of the challenges faced by inhabiting places subject to high water tables and inundation from rivers and coastal waters. In particular, it will review the processes put in place in Chicago and Galveston as models for a comparable undertaking in New Orleans.

Human Adaptations to Wetlands in Louisiana

Native Americans coped with floodplain life in numerous ways. Upstream, they resided in settlements atop the terraces near the river – above the flood risk, but close to the resources of the alluvial plain. Mounds at places like Baton Rouge, Marksville, and Poverty Point reflect these choices. In the marshes near the coast, indigenous groups discarded clam shells to such an extent that their middens became the foundations for settlements. According to archaeologists, native inhabitants created what amount to islands in a monotonous marsh environment and these locations, in time, became sites used by Europeans who re-occupied a territory vacated after massive depopulation of indigenous people due to disease and conflict. Thus one of the first adaptations to a wetland territory was to elevate the land surface – both by unplanned and deliberate actions (Kidder 2000, 12–15). Archaeological sites in New Orleans like Little Woods, Little Oak Island, and Big Oak Island testify to this process.

In addition to Europeans exploiting the relict "islands" in the marshes, they also found building practices imported from the Caribbean well suited to a floodplain environment. The earliest houses in New Orleans used a poteaux-en-terre building technology practiced in most French colonial settlements from Canada to the Gulf Coast. Builders inserted large vertical posts directly into the ground and filled the spaces with a mixture of mud and moss known as bousillage. Such construction was ill-suited for a landscape that endured nearly annual floods. French planters in the Caribbean adapted to the tropical climate and built their homes on stilts several feet above the ground. This practice provided the benefits of cooling from beneath and also reduced some of the near-ground mosquito exposure. When introduced to Louisiana, raised houses largely replaced the poteaux-en-terre construction practices for the same advantages they provided in the Caribbean along with the added benefit of elevating the living quarters above floodwaters (Edwards 1988; Taylor 2015).

Initial settlement by Europeans in the lower river concentrated on the natural levees, elevated shoulders of better-drained soil adjacent to the river. These 10–15 ft high ridges graded toward the marshes which were near sea level. As the premium real estate atop the natural levees came into short supply, leaders of the city sought to expand the urban territory into the adjacent wetlands. This urban growth required drainage. The city built a series of canals that cut from the back side of the higher ground, through the marsh, to Lake Pontchartrain. With minimal gradient, these excavations served as open septic systems rather than as effective drains. By the mid-nineteenth century, steam driven pumps assisted gravity to expel excess water and "reclaim" additional acreage behind the city (Fenner 1849).

Sensing a limit to drains, DeRussy offered his proposal to build land in 1858. He argued that building land, rather than the common practice of merely draining it, would enable the city to expand the territory available for habitation, while also improving the site's salubrity, and also offer a means to reclaim "useless" wetlands across the state (DeRussy 1859, 3–4). He called for a series of parallel levees between the river and the lake, and cross levees at the lakefront to create enclosures. A system of canals, pumps, and drains would provide a means to deliver sediment-laden river water and then remove water after the sand and silt settled out. He estimated a cost of nearly $2 million dollars – a considerable cost at the time (De Russy 1859). Although offering no specific time table, he observed that the "filling and raising of the swamp land would be much sooner obtained than has been estimated" (DeRussy 1859, 4). Government officials were not persuaded and did not adopt this plan or even less expensive drainage plans proposed a few years earlier (Pile 1857). On the eve of the Civil War the city remained reliant on raised houses built atop the natural levee in tandem with modest levees and ineffective drainage.

As an indication of their ineffectiveness for expanding the city's footprint, there were few and only modest land-building efforts. By the 1850s, public health reports reveal that citizens seeking to raise the grade of individual lots dumped refuse on their property. Despite laws prohibiting such practices that dated back to 1819, lax enforcement allowed dumping to continue. In the 1870s, a report claimed that residents continued to fill lots with debris taken from dumping grounds (Holt 1879). How much land was actually raised is uncertain, but private actions sought to raise property above the high water table and periodic floods.

There were numerous projects, usually failures, to enclose wetlands within levees, pump them dry, and thereby promote agricultural development. Railroads and land developers crafted plans in the early twentieth century to recruit residents to suburban tracts where they could be small holders and raise fruit (Okey 1914). The LaBranche wetlands just west of Kenner is an example of such a failed venture. On a larger scale, Delta Farms in Lafourche Parish is indicative of the urge to cultivate drained wetlands. Such endeavors exposed one of the geophysical processes known to those who drained wetlands in Europe. When water is pumped out of marsh soils they subside. Delta Farms continued as an agricultural operation until the levee failed and the interior land flooded in the early 1970s (Merry 1910; Harrison and Kollmorgen 1947).

The first full-fledged efforts to build land in New Orleans by heaping sediment on wetlands occurred following the hurricane of 1915. This storm drove surge from Lake Pontchartrain into the rear of the city and caused extensive damage to lakefront properties. Over the next 19 years, the Orleans Levee Board worked to build a concrete seawall in the lake that stood 9.5 ft high some distance from the existing shoreline. This barrier was to protect against future hurricanes and also provide a retaining wall for a massive land-building project. Crews pumped lake-bed sediments into the area south of the seawall, filled the lakebed between the seawall and the previous shore, and created new real estate with the intent that its sale would pay for levee board activities. By the time of the next major hurricane in 1947, a sizable tract of urban land stood well above inland areas. It served as a buffer between the lake and areas that were now being drained by the city without the threat of inundation from the lake (Burke and Lamantia Architects 1962). The University of New Orleans and sizable suburban neighborhoods now occupy this area and the higher portions of this development did not suffer flooding after Katrina in 2005.

Draining and Building Land Beyond Louisiana

Without question, the Dutch pioneered the methods for draining wetlands that spread throughout Europe. The Dutch also came to understand the related processes of drainage of wetlands and subsequent subsidence. By the 1300s, land owners in the Netherlands had drained extensive fenlands. Following drainage, some of this land was lost to the sea due to subsidence. Construction of dikes and windmill driven pumps enabled the reclamation of many of these areas by the end of the sixteenth century. In addition to draining their own landscape, the Dutch exported their expertise across the low lands of France, Denmark and Germany, and also to England and eventually Italy. Efficient use of reclaimed land for grazing milk cows to supply dairy products to burgeoning urban markets underwrote this massive investment (Danner et al. 2005; van Dam and Petra 2002; Darby 1956).

In the Netherlands, studies of dike failures provide evidence of instances of land building within polders. Following deliberate flooding for strategic military purposes in the 1500s, some polders remained under water for extended periods of time. Silting occurred during the long interval between the flooding and subsequent reclamation. "A thick new layer of clay covered all remnants of buildings, roads, and ditches" (de Kraker 2015, 2677). Although there have been many technical publications on land drainage based on the experiences of the Dutch and other European engineers, no comparable literature on land building of the type noted in Holland exist.

The technology of land drainage arrived in the new world with European immigrants. One of the more successful examples took hold around the Bay of Fundy and has direct ties to Louisiana, although land drainage was practiced from New England to Florida (Hatvany 2003, 42 and Nesbit 1885). Acadians, in what is now Nova Scotia, undertook extensive reclamation of coastal marshes. Relying on family labor, the Acadians build *aboiteaux* to reclaim the salt marshes and established a sound agricultural society between 1607 and 1755 (Hatvany 2003; Butzer 2002). The aboiteaux held back the high tides of the Bay of Fundy and the Acadians farmed the rich marsh soils. Theirs was a reclamation process, not a land building one.

In numerous eastern seaboard port cities, however, there arose interest in creating new land to accommodate larger ships and industrial activity. Deliberate land building completely reshaped the waterfronts in cities like Boston, New York, and Baltimore (Seasholes 2003; Steinberg 2014; Olson 1980). The most extensive account of this process is the work by archaeologist Nancy Seasholes. She makes the case that "land building" is the most appropriate term for the process of making land by filling areas covered by water – as distinct from land reclamation, or draining wetlands, and land filling or dumping garbage on low land. To accomplish this, Bostonians built retaining walls in the bay and then dumped fill on the landward side until the level rose above the high tide level. Initially, the need for wharfs and warehouse space drove the impulse to make land. Later, the need for space for railroad rights-of-way propelled a second phase of land making in the first half of the nineteenth century and eventually the desire to expand parks contributed to land making by the end of the nineteenth century (Seasholes 2003, 1–11).

Far from the eastern seaboard, a similar process of making land unfolded in Chicago, with a distinct variation. Chicago coped with the dual problems of a low-lying waterfront without adequate harborage and the need for garbage dumps by allowing land building along its lakefront using municipal refuse as the fill. As in Boston, retaining walls were built in the lake and refuse heaped behind them to raise the level of the land. Much of the city's prime lakefront park lands were constructed this way by the end of the nineteenth century (Chrzastowski 1991; Colten 1994). The urge to enlarge an industrial district near the mouth of Lake Calumet followed a similar method, but relied on industrial wastes, mainly slag from the steel mills of South Chicago. Sizable tracks of land were built into Lake Michigan in Illinois, and also in northern Indiana during the early twentieth century. Similar land building took place in other cities such as San Francisco (Colten 1985; Wlasiuk 2014; Rubin 2011).

Chicago also provides an example of raising land in its densely developed central business district. Concern with frequent cholera outbreaks and the need for improved sanitation prompted the city to look into land raising. The city drew its water from Lake Michigan which also served as its sewage receptacle. During summer months, when weather and wind conditions allowed for sewage to reach the water intakes set some distance off shore in the lake, epidemics ensued with some frequency. Chicago set out to eliminate this public health problem and hired the prominent sanitary engineer Elis Chesbrough in 1855 to develop a remedy. He proposed the construction of massive new sewers to deliver sewage to the Chicago River. Given the city's minimal elevation above the lake level, it would have been impossible to provide adequate gradient from most of the city to the river if the sewers were installed below the streets. To ensure satisfactory flow, the city laid the sewers atop existing the streets and then mounded dirt around the drains and thereby raised the street levels. The main source of fill was sediment dredged from the Chicago River – another part of the plan to flush the city's sewage. In addition, raising the city's grade prompted the need to lift hundreds of structures so that their doorways opened up at the new street level. Even the largest structures in the central business district were raised (Cain 1972; Writers' Program 1941). The process took years to accomplish. Throughout the CBD, there is a common grade today, but there is also a subterranean street system (Lower Wacker Drive) beneath a portion of the city center that reflects the original land level. Beyond the CBD, private home owners, who continued to occupy their dwellings, had to finance raising their houses without city assistance. A flurry of law suits challenged the imposition of costs, but ultimately the courts sided with the city and declared it had a right to build streets where it chose – even if they were higher than originally designed. Some homeowners raised their houses and inserted basements at the old street grade. Others, with two-story houses, converted up-stairs to the principal living space and the ground floor, became bedrooms. It is not uncommon in certain neighborhoods to see an elevated street and houses with front doors that open onto the street. Yet on the rear side of these houses, lawns remain at the original grade 5–8 ft below street level (Cain 1972; Writers' Program 1941).

A second example of land raising took place in Galveston following the 1900 hurricane. Devastated by the storm which claimed some 8000 lives and destroyed nearly two-thirds of the city's structures, officials commissioned a report to chart a path toward protecting the city from a repeat calamity. The report recommended building a sizable seawall and raising the grade of the city behind it (Bixel and Turner 2000, 95; City of Galveston 1903). This project would be a massive undertaking, financed largely by the city, with an estimated price tag of $3.5 million. Public support was overwhelmingly in favor of authorizing the bonds to launch the project and construction of the seawall began in 1902 (Bixel and Turner 2000, 96–97). Bonds for the grade raising, which had to be funded separately, and gained approval from voters in 1903. The grade raising board received considerable authority to press forward with its mission. Home owners were required to raise their houses or authorize the contractors to do it for them for a fee. If they failed to complete the job by the time work was to begin in their neighborhood, the contractor had sweeping powers to "fill under, around, within, and over said structure" (City of Galveston 1903, para. 33). Home owners received assurances that even the humblest house owner would be able to pay for the cost of raising the structure. Sand fill for this massive undertaking came from shoals off shore and from Galveston Bay. Raising of structures included hundreds of houses and even the large Catholic church. Boardwalks were constructed for residents to continue residing in the area as the grade raising progressed. Some houses were relocated from their original sites to make way for the infrastructure required to deliver the sand fill. By 1911 the raising project was completed. Some 500 blocks had been raised, along with over 2100 structures, and they stood atop 16.3 million cubic yards of sand (Bixel and Turner 2000).

Changing the Urban Alignment

Since the earliest years of European settlement in this country, city founders have imposed a cadastral system in advance of urban development (Reps 1980). Commonly, developers file a street pattern and cadastral plan with the county/parish. The original New Orleans plan included what is now called the French Quarter. Subsequent "additions" enlarged the city's geographic territory and expanded its street network. All later development took place within this framework and the associated legal rights that come with private property. The city maintains the rights to streets and private land owners acquire certain property rights when they purchase tracks of land. A wholesale raising of the city's elevation would impinge on these rights. A review of historical precedents offers valuable insights to how other cities have coped with this challenge.

In an extensive analysis of urban rebuilding after major nineteenth century fires, Christine Rosen demonstrates the persistence of the urban cadaster, despite a clear logic to reconfigure the urban landscape. In order to assemble land for major street and land ownership realignments, property owners have had a near monopoly and proved to be highly resistant to major changes. Not only did cities face obstacles to acquiring land for alterations, but the "capital-intensive, land-extensive character" of major infrastructure projects slowed rapid transformations. The protracted process of purchasing land for new rights of way and securing public approval for bonds ran counter to the desire to rapidly restore a city's economic functions after disasters. In addition, cities were held responsible for any economic damages that resulted from condemnation of private property to make way for new alignments. In Baltimore, for example, a $6 million dollar road building project spent less than $250,000 on the actual construction, while the balance went to pay property owners for damages (Rosen 1986, 45; also Bowden 1982).

Street widening in Montreal in the late nineteenth century illustrates additional concerns. As the city sought to realign street-side structures, it met resistance from some owners, and the city incurred considerable debt reimbursing land owners from their expenses. Some owners welcomed condemnation since it provided a subsidy to rebuild, albeit a few feet back from their pre-existing positions. Perhaps more importantly, the Montreal example demonstrates that public opinion favored street widening to deal with an existing congestion problem, and not to make way for future development (Gilliland 2002). In the context of New Orleans's situation, concern with current issues rather than future ones might suggest that land owners might favor continued reliance on levees rather than a major reworking of the city's topography.

Raising the city's grade in Chicago used the existing street grid for the above-grade sewers. Despite legal challenges that sought assistance with individual house raising, the courts sided with the city and denied claims for damages. (Writers' Program 1941, 18). Galveston created a special commission and gave it exceptional powers to compel land owners to raise their properties themselves, or pay the city's contractor to do it on the commission's time table (City of Galveston 1903).

Both Chicago and Galveston used dramatic calamities to initiate actions – a series of epidemics in Illinois and a major hurricane in Texas. Major alterations to public polices guiding hazards management tend to follow disasters. However, Rosen's analysis of fires in Chicago, Boston, and Baltimore indicates resistance to cadastral realignment, despite tragic events. Yet, in the cases of raising Chicago in the mid-nineteenth century and Galveston in the early twentieth century, extraordinary legal authority, enacted in the wake of disasters, enabled reworking the topographic landscapes. In Louisiana, following the flood of 1927, public bodies received exceptional powers to take property in order to expand levees and create spillways (Reuss 1998; Camillo and Pearcy 2004). The acquisition of property for the Bonnet Carré spillway and easements in the Atchafalaya proceeded with limited opposition and largely ineffective opposition (Colten forthcoming).

Raising New Orleans

The prospects for rejuvenating De Russy's plan for land building in New Orleans is complicated by many factors. The city is much more densely settled than it was in the mid-1850s and has a vastly larger footprint. Public resistance to previous realignments of urban space after Hurricane Katrina reveals the need for effective public engagement in the earliest stages of deliberations on this prospect. Also, the river carries less sediment than it did in the 1850s. Furthermore, the costs would be far greater than in the mid-nineteenth century.

To gain some notion of the costs of this undertaking and without prescribing a location for raising to take place, researchers reviewed Google Earth images of a portion of the Fountainbleau/Broadmoor neighborhoods. A rough tabulation of an area of approximately 250 square blocks found some 7420 houses. This area contained a mixture of sizable homes and also many modest shotguns and doubles. Most of the properties in this neighborhood already are raised off the land surface. Older houses may stand 5 to 6 ft above grade. Newer and many older double shotguns are anywhere from 18 inches to 4 ft above grade in places where the land surface is generally is more than 4 ft below sea level. Estimates for raising pier houses after Katrina ranged from $40,000 to $60,000. Most houses would need to be raised to stand above sea level. Using the lower end figure for this neighborhood, rasing houses to accommodate filling would top $296 million – to say nothing of businesses, public buildings like schools, and other infrastructure.

A comparable survey of a portion of the New Orleans East neighborhood was done. This is a section of the city developed largely in the last quarter of the twentieth century and most homes are slab-on-grade. In a particularly low area containing about 180 square blocks, there are approximately 3600 houses. Using a post-Katrina estimate of $65,000 to elevate a slab home before land raising would cost on the order of $234 million without the additional costs of public and private non-residential structures and related infrastructure. Thus, the pre-land building expenses would be formidable. If the city undertook to develop five colmates at $500 million each, the total would come to $2.5 billion using somewhat dated house raising estimates.

Following the Chicago model, the land raising project might consider using existing roadways for the placement of retaining levees for the "colmates." They would then serve as the bed for elevated roadways following land raising. Obtaining suitable fill for the levees and street grades along with erecting the retaining levees would be additional costs.

To accommodate the recently unveiled New Orleans Water Plan, the land raising project might seek to consolidate existing, but thinly populated neighborhoods into raised territories and creating flood-retention basins in those territories with few existing structures. By focusing land raising on areas with the geological potential to support this type of activity and dedicating other areas to recreation and flood retention, the project could align with the Water Plan.

An additional concern would be the length of time required to complete a land raising process using river-delivered sediments. In 1922, when a crevasse in the levees below New Orleans allowed flood waters to spread across portions of St. Bernard Parish, observers reported that some areas saw accumulations of 6 to 7 ft of new ground (Saxon 1922). One of the leading authorities of the time estimated water passing through the crevasse contained sediment concentrations of 800 ppm, based on analysis by the New Orleans water purification plant, and that such volumes of sediment could raise the bed of Lake Borgne 1 foot in a year (Krebs 1923). More recent analyses of the Bonnet Carre spillway have documented that sediment accumulated as much as 5 to 10 ft between between 1936 and 1969. And this accumulation took place with only four openings of the spillway (Day et al. 2012). Obviously regular, controlled openings could yield even more rapid land building. With the potential delivery of 5 ft of sediment in 4 years, five "colmates" could be raised in sequence over 20 years. This is less time than projected for most of the state's Master Plan projects.

Residents in the areas to be raised need to be considered and this demands consideration of social equity. Public educational forums, stakeholder engagement, and planning that allows for effective public engagement should be a central component of any further deliberations and budgets will need to accommodate this process (see next section). There has been much criticism of the post-Katrina recovery which has focused development of charter schools and medical facilities in more affluent sections of the city. To avoid the somewhat unexpected "right to return" movement after Katrina, public involvement in this type of undertaking must begin with the initial public discussions (Olshanksy and Johnson 2010; Bullard and Wright 2009). Careful analysis and consideration of social issues would need to be carried out to ensure fairness in terms of both displacement during the raising and equal access to raised areas after they are completed. Resettlement projects for other environmental projects, such as dams and reservoirs, have given rise to opposition since they produce permanent displacement. Raising would enable residents to return – which was the case in Galveston. This presumes the restoration of communities and pre-existing social networks. Temporary sheltering during the displacement would be an important concern and a cost that has to be calculated – both in economic and social terms. Likewise costs of integrating new residents in areas receiving new residents must be included in budgeting.

If the city undertook to develop five colmates at $500 million each, the total would come to $2.5 billion. This investment could possibly offset the costs that might be anticipated from future floods. Recent experience has clearly illustrated the expense of flooding the city is dramatically higher. Damages from Katrina were on the order of $135 billion and recovery spending was on the order of $45 billion (Data Center 2015).

Environmental Equity and Citizen Engagement

One of the great tragedies of the post-Katrina recovery effort was the so-called "green dot" map that depicted a realigned New Orleans with sizable areas covered by large green dots representing open space that would serve as flood retention areas in the future. This cartographic travesty inspired a vigorous citizen response that came to be known as the "right to return." Fearing that their lands would be seized, their houses demolished, and their neighborhoods converted to green space while they were exiles, residents demanded a voice in the planning process and forced a more citizen-engaged recovery planning process (City of New Orleans 2007; Nelson et al. 2007; Olshanksy and Johnson 2010; Bullard and Wright 2009). This episode underscores the need to involve local citizens from the outset in any discussion of land raising. Any and all deliberations and discussions of this ambitious concept need to include a wholly inclusive public participation from the outset. Social impacts and not just biophysical and economic impacts must be accorded equal consideration. The mere suggestion of a project with social impacts can prompt social repercussions.

There has been a dubious history of environmental injustices in this country. From the forced removal of poor farmers in the Tennessee Valley to the permitting of hazardous waste facilities to the Road Home program that disadvantaged low-income home owners with poor claims to titles to their property and also renters, inequity has been a common outcome of environmental management programs (McDonald and Muldowny 1981; Green and Olshansky 2012). The literature of social impact assessment calls for community engagement before the initial announcement of projects because that is when the social impacts begin (not with the initiation of construction). This body of applied scholarship recommends multiple forms of public engagement to ensure that citizens understand the issues and that they have an opportunity to inform the planners of local issues (Gramling and Freudenburg 1992; Vanclay 2012; Walker 2010; Glucker et al. 2013; Colten and Hemmerling 2014). Social impact assessment authorities recognize that social impacts can begin with early discussions of an environmental change, well before the first shovel of dirt is turned (ICPGSIA 2003).

Conclusions

Grade raising been carried out in several historical instances with success. In both Chicago and Galveston, raising the level of the city's base elevation involved considerable disruption to normal urban life, but in the long term these efforts provided the benefits sought – drainage for sewage in Chicago and storm surge protection in Galveston. Raising the grade behind the lakefront seawall in New Orleans likewise proved successful. After the levee breaches in 2005, the elevated neighborhoods suffered far less damage than nearby homes that were in neighbors that had subsided below sea level.

By merging land raising with the Water Plan, the city and region can provide still another line of defense. Elevated neighborhoods and streets along with flood retention basins within the levees could enhance safety. Furthermore, grade raising would shift a portion of the expense to the area with the most valuable real estate rather than imposing costs on rural families outside the hurricane protection levees. Although the costs would be considerable, the tradeoff with potential flood damages in the future make the investment much more reasonable. In an age of rising sea levels, exacerbated by regional subsidence, risk of flooding from tropical cyclones will increase over time and further justify the expenses.

Central to any advancement of this line of thinking is thorough and respectful citizen participation in the planning process. Public engagement must start in the earliest stages and decisions much reflect input from concerned stakeholders.

As New Orleans and coastal Louisiana seek to build their international capacity and expertise for managing water, this would be a showpiece that could prove worthy of exporting to other coastal areas around the globe facing the challenges of sea level rise.

References

Bixel PB, Turner EH (2000) Galveston and the 1900 storm. University of Texas Press, Austin

Bowden MJ (1982) Geographical changes in cities following disaster. In: Baker ARH, Billinge M (eds) Period and place: research methods in historical geography. Cambridge University Press, Cambridge, pp 114–126

Bullard RD, Wright B (2009) Race, place, and environmental justice after Hurricane Katrina: struggles to reclaim, rebuild, and revitilize. Westview Press, Boulder

Burke and Lamantia Architects (1962) East lakefront flood protection and reclamation. Orleans Levee Board, New Orleans

Butzer KW (2002) French wetland agriculture in Atlantic Canada and its European roots: different avenues to historical diffusion. Am Assoc Geogr 92(3):451–470

Cain LP (1972) Raising and watering a city: Ellis Sylvester Chesbrough and Chicago's first sanitation system. Technol Cult 13(3):353–372

Camillo C, Pearcy M (2004) Upon their shoulders: a history of the Mississippi River Commission. Mississippi River Commission, Vicksburg

Chrzastowski MJ (1991) The building, deterioration, and proposed rebuilding of the Chicago lakefront. Shore Beach 59:2–10

City of Galveston (1903) Proposal for grade raising: Galveston, Texas. Cheesborough Papers, MSS #22–0024, Rosenberg Library, Galveston, Texas

City of New Orleans (2007) The unified New Orleans plan. City of New Orleans, New Orleans

Colten CE (1985) Industrial wastes in the Calumet area, 1869–1970: an historical geography. Illinois Hazardous Waste Research and Information Center Report RR 1, Savoy

Colten CE (1994) Chicago's waste lands: refuse disposal and urban growth. J Hist Geogr 20(2):124–142

Colten CE, Hemmerling SA (2014) Social impact assessment methodology for diversions and other Louisiana coastal master plan projects. Water Institute of the Gulf, Baton Rouge

Dam V, Petra JEM (2002) Ecological challenges, technological innovations: the modernization of sluice building in Holland, 1300–1600. Technol Cult 43(3):500–520

Danner H, Renes J, Toussaint B, Van den Ven GP, Zeiler FD (2005) Polder pioneers: the influence of Dutch engineers on water management in Europe, 1600–2000. Netherlands Geographical Studies, Utrecht, p 338

Data Center (2015) Facts for features: Katrina impact. http://www.datacenterresearch.org/data-resources/katrina/facts-for-impact/

Darby HC (1956) Draining of the Fens, 2nd edn. University Press, London

Day J, Hunter R, Keim RF, DeLaune R, Shaffer G, Evers E, Reed D et al (2012) Ecological response of forested wetlands with and without large-scale Mississippi River input: implications for management. Ecol Eng 46:57–67

de Kraker AMJ (2015) Flooding in river mouths: human caused or natural events? Five centuries of flooding events in the SW Netherlands, 1500–2000. Hydrol Earth Syst Sci Discuss 12(1):1437–1468

De Russy L (1859) Cost of draining the swamp lands bordering on Lake Pontchartrain. Taylor Printing, Baton Rouge

Edwards JD (1988) Louisiana's remarkable French vernacular architecture. Louisiana State University, Kniffen Cultural Resources Laboratory Monogram, Baton Rouge, p 1

Fenner E (1849) New Orleans. South Med Rev 1:17–19

Gilliland J (2002) The creative destruction of Montreal: street widenings and urban (re)development in the nineteenth century. Urban Hist Rev/Revue d'histoire urbaine 31(1):37–51

Glucker AN, Driessen PPJ, Kolhoff A, Runhaar HAC (2013) Public participation in environmental impact assessment: why, who, and how? Environ Impact Assess Rev 43:104–111

Gramling R, Freudenburg WR (1992) Opportunity, threat, development, and adaptation: toward a comprehensive framework for social impact assessment. Rural Sociol 57(2):216–234

Green TF, Olshansky RB (2012) Rebuilding housing in New Orleans: the road home program after the Hurricane Katrina disaster. Hous Policy Debate 22(2):75–99

Harrison RW, Kollmorgen WM (1947) Past and prospective drainage reclamation in the coastal marshlands of the Mississippi River delta. J Land Public Util Econ 23(3):297–320

Hatvany MG (2003) Marshlands: four centuries of environmental change on the shores of the St. Lawrence. Les Presses de l'Universite Laval, Sainte-Foy

Holt J (1879) History of the yellow fever epidemic in the fourth district. J.S. Rivers, New Orleans

Interorganizational Committee on Principles and Guidelines for Social Impact Assessment (2003) Principles and guidelines for social impact assessment. Impact Assess Proj Apprais 21(3):231–250

Kidder T (2000) Making the city inevitable: native Americans and the geography of New Orleans. In: Colten C (ed) Transforming New Orleans and its environs: centuries of change. University of Pittsburgh Press, Pittsburgh, pp 9–21
Krebs B (1923) Coast sea food interests favor spillway plan New Orleans Times Picayune, 6 November, 1–2
McDonald MJ, Muldowny J (1981) TVA and the dispossessed: the resettlement of population in the Norris Dam area. University of Tennessee Press, Knoxville
Merry JF (1910) Louisiana reclaimed lands make fertile farms. Illinois Central Railroad, Manchester
Nelson M, Ehrenfeuct R, Laska S (2007) Planning, plans, and people: professional expertise, local knowledge, and governmental action in Post Katrina New Olreans. Cityscape 9(3):23–52
Nesbit CM (1885) Tide marshes of the United States. Government Printing Office, Washington, DC
Okey CM (1914) Wetlands of southern Louisiana and their drainage. U.S. Department of Agriculture, Bulletin, Washington, DC, p 71
Olshansky R, Johnson L (2010) Clear as mud: planning for the rebuilding of New Orleans. American Planning Association, Chicago
Olson S (1980) Baltimore: the building of an American city. Johns Hopkins University Press, Baltimore
Pile LH (1857) Report on drainage communicated to the common council. Bulletin, New Orleans
Reps J (1980) Town planning in frontier America. University of Missouri Press, Columbia
Reuss M (1998) Designing the bayous: the control of water in the Atchafalaya Basin, 1800–1995. U.S. Army Corps of Engineers, Office of History, Alexandria
Rosen CM (1986) The limits of power: great fires and the process of city growth in America. Cambridge University Press, New York
Rubin J (2011) A negotiated landscape: the transformation of San Francisco's waterfront since 1950. Center for American Places, Chicago
Saxon L (1922) St. Bernard flood victims preparing for fresh start. New Orleans Times Picayune, 25 June, 20
Seasholes N (2003) Gaining ground: a history of landmaking in Boston. MIT Press, Boston
Steinberg T (2014) Gotham unbound. Simon and Schuster, New York
Taylor WR (2015) Architecture after Katrina: lessons from the past or designs for someone else's future. In: Taylor W, Levine MP (eds) The "Katrina Effect": on the nature of catastrophe. Bloomsbury, London, pp 303–330
Vanclay F (2012) The potential application of social impact assessment in integrated coastal zone management. Ocean Coas Manag 68:149–156
Walker G (2010) Environmental justice, impact assessment and the politics of knowledge: the implications of assessing the social distribution of environmental outcomes. Environ Impact Assess Rev 30:312–318
Wlasiuk J (2014) A company town on common waters: Standard Oil in the Calumet. Environ Hist 19(4):687–713
Writers' Program of the WPA (1941) Up from the mud: an account of how Chicago's streets and buildings were raised. Works Project Administration, Springfield

Raising Buildings: The Resilience of Elevated Structures

Jori A. Erdman, Elizabeth A. Williams, Christopher W. James, and Giovanni P. Coakley

Abstract

Living on the Lousisana coast has always been defined by the influences of the deltaic environment, climate, and the resilience of the communities and individuals who inhabit this dynamic place. Unfortunately, a variety of factors have contributed to the continuous degradation of the coast; the land upon which people depend for inhabitation and settlement is disappearing beneath the water at an accelerated pace. In addition, flooding events have become more frequent and costly. The most common response is to elevate structures above the land in order to minimize damage and reduce risk, however, living in this condition presents unique challenges for people and communities. Current building practices and materials, including slab-on-grade construction and drywall assemblies making post-flood events more damaging and costly than necessary as well as extending the disruptions of post-flood recovery. Elevating buildings is one strategy for improving the resilience of communities, individuals and structures. Living above the land means that structures must adapt through the use of vertical circulation in the form of ramps, stairs and elevators. This chapter will address the building practices and methods employed in south Louisiana as well as impacts on the individual and collective living.

Keywords

Elevated structure • Building codes • Architecture • Resilience

Introduction

> Funny Miss Twiggley lived in a tree with a dog named Puss and a color TV.... The road and the meadow were one stormy sea, and right in the middle stood the house in the tree. ...'Ahoy!' called the mayor, 'Miss Twiggley, yoo-hoo! Is there room in your house for a wet friend or two?' ...'There's room for you all,' Miss Twiggley replied. 'Climb up where it's warm; come up and be dried.' ... Now the townsfolk were glad, just as pleased as could be, that funny Miss Twiggley lived in a tree... And Miss Twiggley found out something wonderful, too: When emergencies come, you don't think about you. You help all you can. And you never ask why. – Dorothea Warren Fox, Miss Twiggley's Tree (1966).

In this children's book from 1966, the story doesn't explain what happens after the storm has passed. That's where fiction leaves off and the adult world begins. Right now, millions of people live in coastal zones globally and 2.56 million in Louisiana are faced with the prospect of increasing frequency and intensity of flooding. These threats are not due just to storms, as Miss Twiggley faced, but to sea level rise and subsidence. Given this reality we must ask ourselves: How do we build in this zone and what will happen to our cities and communities as a result of any new strategies for flood resistant settlement? It has been shown that a complete reliance on structural protection is

J.A. Erdman (✉) • C.W. James • G.P. Coakley
School of Architecture, Louisiana State University, Baton Rouge, LA, USA
e-mail: jerdman@lsu.edu; cjame32@lsu.edu; gcoakl1@lsu.edu

E.A. Williams
Foundation for Louisiana, Baton Rouge, LA, USA
e-mail: liz@lizabethannewilliams.com

J.W. Day, J.A. Erdman (eds.), *Mississippi Delta Restoration*, Estuaries of the World, https://doi.org/10.1007/978-3-319-65663-2_10

folly. For coastal settlements to thrive and prosper, new solutions based on integrated natural and manmade systems must be sought. A culture of resilience rather than resistance must be encouraged that follows the commonly understood definition of resilience as the capacity of a system, enterprise, or a person to maintain its core purpose and integrity in the face of dramatically changed circumstances. Individual buildings and collections of buildings must seek structural and material solutions that can withstand flooding events, while maintaining the social bonds of community.

Repeatedly over the past 15 years, people in south Louisiana worked to help people in flooded areas – first refugees from hurricanes Katrina and Rita (Fig. 1) and more recently from flooding in the Hammond and Baton Rouge areas (Fig. 2). Coastal communities, particularly along the Atlantic Ocean and Gulf of Mexico, are accustomed to a pattern of disaster and relief, but it is not sustainable when the rate of disasters is increasing and resources are stretched thin. Over the past 11 years in Louisiana, costs from Hurricane Katrina and 2016 floods alone are estimated at somewhere between $96 billion and $135 billion. In addition, flood disaster events have high personal costs for residents including loss of belongings not covered by insurance and loss of wages due to clean-up and recovery efforts. It is imperative that we build better and smarter if we are to break the disaster/recovery cycle.

For the over 40% of the global population living within 100 km of the coast (UNEP) and 10% subject to coast flooding (McGrahan et al 2007), the most cost effective and expedient way to accommodate flooding is to build above it. The greatest impacts from an urban or civic point of view are to relatively low density urban communities, such as New Orleans. In total, there are 2.56 million people living within the working coastal zone of Louisiana (Barnes et al. 2015). So, what happens when we build up instead of moving out? How do we begin to build above the ground or street? And further, what does it mean when we build above the ground? Our building practices in coastal zones need to be more thoroughly examined and effectively implemented. Architects and builders need to more robustly consider the natural environment and develop building strategies that can respond to changes dynamically. The 2017 Master Plan produced by the Coastal Protection and Restoration Authority for Louisiana addresses this by recommended increased enforcement of building codes that requires flood elevation, as well as

Fig. 1 Flooding in New Orleans after Hurricane Katrina, LA (Credit: Public Domain, https://commons.wikimedia.org/wiki/File:Hurricane_Katrina_Flooding.jpg)

Fig. 2 Flooding in Baton Rouge, LA, 2016 (Credit: US Department of Agriculture, CC by 2.0 https://www.flickr.com/photos/usdagov/28426228304/)

recommending the adoption of scientifically proven freeboard elevation requirements that are locally specific.

Shedding water stands as one of the originating causes of the need for shelter. The basic determinations of human physiology dictate that we find ways to protect ourselves water, whether it be from water coming directly from the sky in the form of rain, or water coming across the land in the form of flooding. All of architectural history, through buildings and cities, can be conceived of as the advancement of how to alternately resist or control water, particularly as water moves freely due to low viscosity and gravitational pull. Climate change acceleration, particularly sea level rise, in combination with more frequent stronger hurricanes and extremely heavy rainfall (see chapter "Optimum Use of Fresh Water to Restore Baldcypress – Water Tupelo Swamps and Freshwater Marshes and Protect Against Salt Water Intru sion: A Case Study of the Lake Pontchartrain Basin" on climate change) in addition to population growth, has simultaneously accelerated our consideration of water in architectural thought. In this chapter, we will consider the effects of water, as relates to structures and cities. We will examine the ways in which we elevate and how we traverse those elevations technically and architecturally. We will also examine the materials and means by which we built structures in, around, and above watery conditions. Finally, we will consider the impact of elevating structures to communities.

There are 5 basic types of flooding:

- Areal (across a land mass);
- Riverine (channel);
- Estuarine and coastal (along the coastal ocean);
- Urban flooding (related to infrastructure) and
- Catastrophic (ultimately destructive, permanently changing the environment).

An additional category that is being seen more frequently is "nuisance flooding" or "sunny day flooding" in cities such as New Orleans, Norfolk and Charleston as well as across Florida. This flooding occurs during high tides when sea water breaches storm drains and seawalls.

All of these types of floods have occurred in South Louisiana in the past several decades. Around the world, we have seen the effects of these types of flooding, sometimes alone and sometimes in combination, and in any given year, there are multiple accounts of environmental disasters caused by extreme weather and flooding, such as Hurricane

Matthew (Haiti and US) and the Baton Rouge floods of 2016. Our globe has become so populous that any significant flooding event will almost always either directly or indirectly touch the lives of millions of people.

As an attempt to escape the effects of flooding, humans originally would have retreated to higher ground, further away from rivers and other water bodies. In a delta, there isn't usually a lot of high ground, so eventually humans also began to build higher ground, which was convenient for protection from invaders as well as the natural elements. There are hundreds of examples of constructed mounds adjacent to low sloping river beds and coastal zones around the world that predate the written word. Thousands of these existed in the Louisiana coastal zone in the form of kitchen middens where native Americans worked and lived. Evidence suggests that these mounds were significantly occupied but ultimately abandoned as conditions changed across centuries. But in our post-industrial global environment, with so much building stock already in existence, the prospect of abandonment is not an easily achievable option.

The United Nations Environmental Programme issued a guidebook on climate change adaptation under the Technical Needs Assessment series titled "Technologies for Climate Change Adaptation – Coastal Erosion and Flooding." This guide book offers an overview of the global concerns relative to flooding. The authors argue that there are three approaches to flooding: protect, accommodate, and retreat. Protection measures include hard and soft infrastructures such as seawalls, dikes, beach and marsh re-nourishment among the strategies possible. Accommodation measures include regulating building development, flood proofing and hazard mapping. Retreat measures include pushing development inland and managing coastal realignment.

The Federal Emergency Management Agency (FEMA) is the leading source for information and research regarding flood plain design for architects and planners around the world. The agency provides documents as well as software that enable better planning for every day as well as emergency event. The "Technologies for Climate Change Adaptation" guidebook makes extensive use of the FEMA guidelines as they identify two methods of flood-proofing: wet and dry. They also point to the National Flood Insurance Program for further guidance on how to approach the technical details of building structures in a flood zone. The United Nations programs as well as FEMA (as informed partly by the International Building Code) provide the most up-to-date and current thinking about building practices in flood-prone conditions; albeit frequently adapted and augmented for localized situations.

Louisiana Response to Building

Louisiana has developed a robust system for studying and proposing resilience strategies across the coastal zone in the form of a Coastal Master Plan that is produced and updated every 5 years by the Coastal Protection and Restoration Authority. Experts in coastal sciences, engineering and more recently, planning, participate in the process of creating this plan each iteration. The 2017 Coastal Master Plan also includes "non-structural" recommendations that include building specific implementations as shown in the following excerpts (Fig. 3):

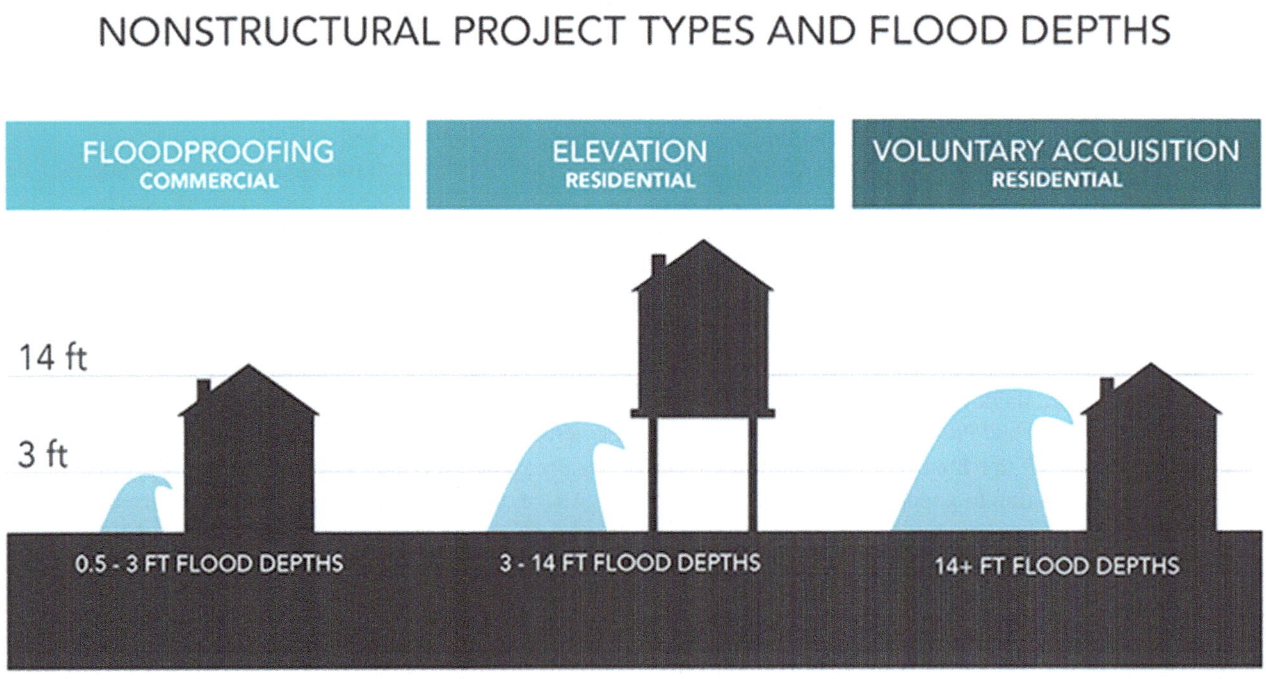

Fig. 3 Non-structural project types and flood depths (Credit: DRAFT of the 2017 Coastal Master Plan. http://coastal.la.gov/a-common-vision/2017-master-plan-update/ccrp/resilience-approach/building-codes/)

Uniform Construction Code Council:

Increase resilience of building stock by updating building standards for high risk structures in the floodplain and continuing to provide resources for local implementation and enforcement; create a state wide standard process for building code enforcement; adopt higher regulatory standards such as increased freeboard, additional levels of protection for structures behind levees, or cumulative substantial damage tracking requirements.

Parish and Municipal Governments:

Increase resilience of building stock by strengthening building standards for high risk structures in the floodplain in accordance with ASCE 24–14, and increase enforcement of these standards."

Nonstructural Projects in the 2017 Coastal Master Plan

Several types of structures were evaluated including both residential and non-residential structures. Residential buildings included single family, multi-family, and manufactured homes. Non-residential buildings included commercial, industrial, and institutional structures.

- ***Floodproofing:*** *Recommended for non-residential structures in areas with flood depths of 3 ft or less.*
- ***Elevation:*** *Recommended for residential structures in areas with flood depths between 3 and 14 ft (including two feet of freeboard).*
- ***Voluntary Acquisition:*** *Recommended for residential structures which would need to be elevated greater than 14 ft.*

CPRA's Flood Risk and Resilience Program is envisioned as a state-led, coast wide, nonstructural flood mitigation program for coastal Louisiana parishes. CPRA developed a risk reduction strategy that coordinates state resources and prioritizes areas of high risk, while parishes will play a lead role in implementing projects and selecting specific structures to be mitigated. The program is intended to take advantage of nonstructural project funding outside federal grant programs in order to maximize flexibility and speed the implementation of "shovel ready" projects that further comprehensive coastal risk reduction goals. The 2017 Coastal Master Plan recommends 32 nonstructural project areas that include the mitigation 26,569 structures at a cost of $6.06 billion over the next 50 years. It should be noted that these recommendations are general and do not address any specific structures. Feedback from our working partnerships with the Community Focus Group, Parish Floodplain Managers Group, Flood Risk and Resilience Stakeholder Group and other coastal citizens was essential in developing this effort.

Elevated Structures: Foundations

Structurally, building designers must first consider the soil on which their edifice will sit. Is it hard and water resistant such as the land found along the western coast of Italy or is it viscous and organic such as the land found at the mouth of the Mekong Delta or the Mississippi Delta? When building in delta region, one often has no solid ground on which to build, no matter how far down the structure reaches. There are two possible foundation types: open foundation or filled foundation.

Open foundation: Spatially, elevating structures using pilings, piers or walls can provide opportunities for new programmatic elements that enhance individual use and community functioning when the elevation is high enough to allow for occupation below. For instance, the space can be used as a garage for vehicles and replaces the typical attached garages found in suburban homes. The space can also function in the same way that porches operate, as a mediating space between the public space of the street and the private space of the dwelling, business or institution.

Raising a structure on pilings involves driving deep foundational shafts and placing all inhabited elements on the pilings. Pilings can be made of a variety of structural materials but are typically wooden or concrete. Wooden pilings are susceptible to degradation from pests and water exposures. Reinforced concrete pilings are more expensive and can be susceptible to structural degradation due to improper manufacturing and installation as well as salinity. The benefits of elevating a building with piers include increased spatial flexibility below the enclosed and protected space above, optimal cost, and significant structural efficiency. Pilings are also the preferred method of elevation in areas with fast moving flood waters that carry the potential of debris.

Fill foundation: Elevating a structure on fill is typically costlier and difficult when used as a form of retrofitting (Fig. 4). FEMA and the International Building Code (IBC) do not accept structures built on fill as part of the Federal Housing Insurance Program (FHIP). Fill is particularly dangerous in areas where flooding involves high velocity water that can erode fill quickly if not properly compressed and sealed. However, when appropriate, elevating on fill can provide additional opportunities for a more integrated spatial relationship with the surrounding landscape as well as the possibility of exterior and interior space that is protected from flooding.

Elevated Structures: Access

Once a structure has been elevated, there is the question of how people and their belongings will get in and out of the structure on a daily basis. There are exterior architectural elements that are employed when elevating a structure. In the modern world, there are three techniques for traversing elevated ground:

- Stepped ground – Stairs, ladders
- Sloped ground – Ramps
- Mechanical lift – Elevators, Escalators, Dumbwaiters, Lifts

The stepped ground and sloped ground techniques are workable solutions to a point but anything much beyond 50 ft of vertical elevation becomes cumbersome and exhausting to all but the ablest bodied of people on a routine basis for daily activities and work. Beyond 50 vertical feet,

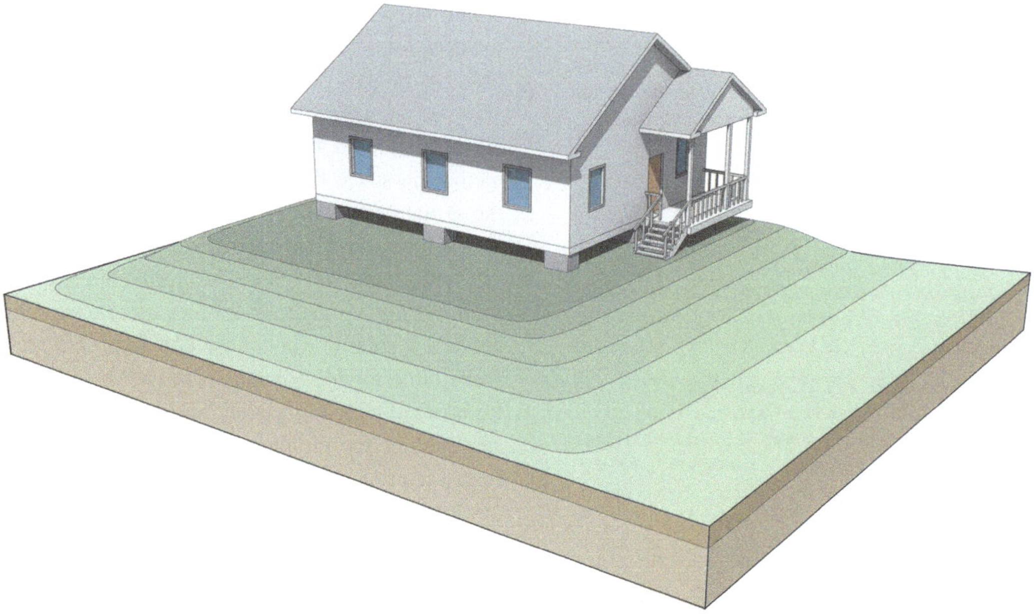

Fig. 4 Diagram of an elevated structure on fill (Credit: Giovanni Coakley, LSU School of Architecture)

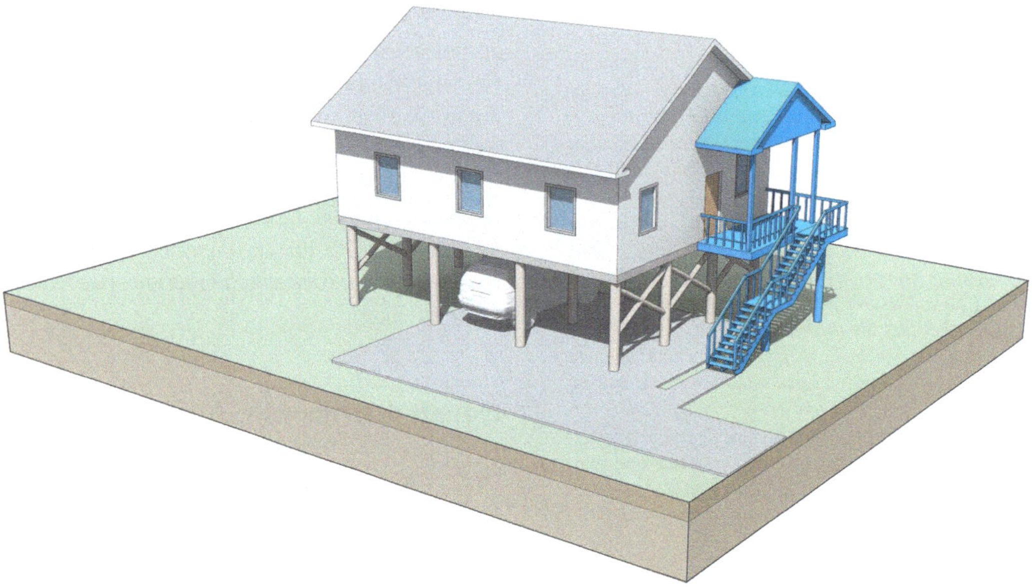

Fig. 5 Diagram of an elevated structure on piles with stairs (Credit: Giovanni Coakley, LSU School of Architecture)

the use of mechanical means is then mandatory but susceptible to degradation and malfunction. For our purposes in this chapter, we will only be focusing on shorter building structures.

Stairs

Stairs are the most typical solution to resolving elevational differences between spaces of occupation (Fig. 5); we use stairs to get us up, along with our own bodily ability to climb stairs. In the case of elevated structures, stair design becomes very important and one of the primary features of a structure. Across the coast, we see examples of stairs being added to elevated structures, creating a filigree of posts and risers in an otherwise unused space below the home. Most of these staircases are being built with little thought to any factor, other than expediency. And in most coastal environments, which are primarily flat, stairs are seen as the necessary evil for the price of living along the coast.

The ad hoc nature of retrofitted structures reveals an amazing array of solutions and combined solutions amongst the inhabitants.

Historically, we seem to have been much more gracious in our use of stairs, such as the stairs at Drayton Hall near Charleston, SC or the stairs at Laura Plantation in Louisiana (Fig. 6) These exterior stairs are expressive and integrated – their rise to run ratio allows for a slow but steady and relatively easy climb whereas many of the stairs we see attached to contemporary homes are much steeper and appear to be afterthoughts, rather than the first thought. The great American architect, Frank Lloyd Wright posited that the hearth was the first building instinct of humanity and therefore, the hearth, figural or literal, should be at the core of every home design. (Etlin, 27). In the case of coastal building, it could be possible to think of the stair as the core of every home as it provides the necessary component to come and go freely.

There are additional health benefits to the use of stairs as well. The National Academy of Health, advocates for buildings to incorporate stairs that encourage walking and climbing, instead of mechanical elevators. There are code issues with fire safety officials, who see stairs as primarily egress routes that must be fully enclosed and protected, to creating more hospitable and enjoyable stairs, but architects continue to struggle with these codes and design buildings that are safe, and navigable. The majority of stairs in southern Louisiana homes are exterior rather than interior, so fire egress codes have less impact on them, however, codes do effect institutional and commercial buildings and their use of stairs (Figs. 7 and 8).

Ramps

Sloped ground ramps have been employed since our earliest constructed forms. Sometimes designed to emulate natural hills and sloped terrains, ramps are one of the easiest ways to ascend through space. Ramps are generally take up more area than stairs but do offer ease of movement and

Fig. 6 Laura Plantation in Vacherie, LA (Credit: Jori Erdman, LSU School of Architecture)

Fig. 7 Elevated homes with shared staircase in Cocodrie, LA (Credit: LSU Coastal Sustainability Studio)

accessibility as well as the option of moving wheeled vehicles and conveyances (Fig. 9). Since 1990, the Americans with Disabilities Act (ADA) has been part of Federal law that requires all public and commercial buildings be accessible to people with a range of disabilities and increased use of ramps for building access has been one of the major changes to building construction in the US since that time. Elevated structures in coastal environments can make use of ramps to make buildings more generally accessible but ramps are particularly helpful for the aging population.

While private residences are generally not bound by the ADA, they would be well served to follow the generally guidelines when constructing ramps. Guidelines such as using a slope ratio of 1:12, maintaining a 36″ width for the run of a ramp, installing landings for every 30″ of vertical rise, and establishing workable clearances at top and bottom. Ramps are not the most prevalent option for vertical ascension in coastal Louisiana but there are many interesting interpretations and examples of ramped structures (Figs. 10 and 11).

Elevators

Another feature of newly elevated homes is the use of mechanical lifts, or elevators (Fig. 12). Assumedly these are used primarily by residents who lack the ability to climb stairs or for the convenience of delivering supplies, furniture and other heavier items to the residences above ground. There are many variations of these mechanisms, depending on how they are being used. The elevators are generally appended to an existing porch on the outside of the structure (Figs. 13 and 14).

Amphibious Structures

Amphibious structures are structures that are designed to float when flooding occurs. Elizabeth English, of The Buoyant Foundation Project at, http://www.buoyantfoundation.org, has been one of the most vocal and persistent advocates for this type of construction in the US. In principle, the

Fig. 8 Elevated home with stairs in Dulac, LA (Credit: LSU Coastal Sustainability Studio)

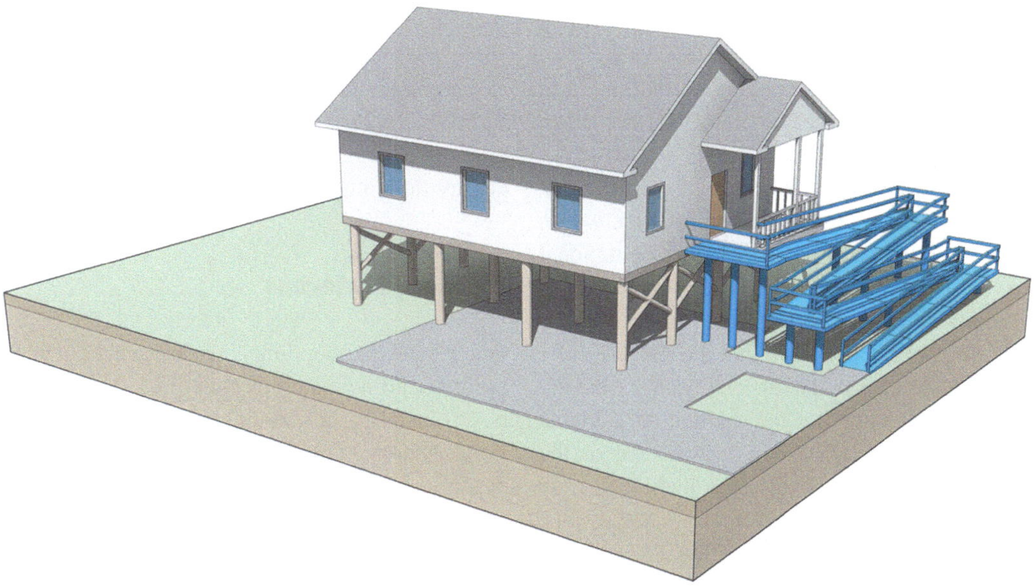

Fig. 9 Diagram of an elevated structure on piles with a ramp (Credit: Giovanni Coakley, LSU School of Architecture)

Fig. 10 Elevated home in Lafitte, LA with ramped access (Credit: LSU Coastal Sustainability Studio)

amphibious structure concept allows builders to continue using conventional construction methods, except for the connection to the foundation and some modifications under the house. These structures are built with conventional means and foundations, but are undergirded by a layer of flotation materials and moored to the foundation, rather than being rigidly attached. The undercarriage of the structure must be designed and constructed to hold together when floated, like a barge or raft.

There are certain drawbacks to this type of construction. FEMA does not recognize or recommend this method within their guidelines. It is not addressed in the International Building Code and is not allowed by building inspectors in most jurisdictions and cannot currently be covered in by FHIP. Elizabeth English has proposed that the IBC consider this method and has written code for it as well as having conduction research that supports the method as a viable alternative to elevation through other means. One of the major concerns among critics is that the method is only reasonable to consider in low or no velocity flood events, such as the Dutch example of IJburg in which 2-story floating houses, similar to houseboats, have been constructed and joined together in a community (Fig. 15). The "buoyant foundation" is not designed to withstand forceful impact with debris, nor fast moving water such as in coastal Louisiana where there is also hurricane surge and waves. It's most appropriate use is within floodplains where water rises slowly across a large area, not riverine or coastal storm surge flooding events. But these limitations also apply to piers and the perimeter wet proofing methods allowed by IBC. It is reasonable to think that in the future, amphibious structures will find application in cases where low energy flooding is possible.

Building Materials

One reason that flooding is so destructive is that our contemporary building materials either lack flood resistant qualities, or are installed in a manner that puts them within reach of flood waters. This section addresses the flood resistant capacity of historic building materials and methods and then turns a critical eye to those used in contemporary US construction and explains the shortcomings and strengths of each type with regard to flooding. The components of structures that are most likely to fail due to flooding are

Fig. 11 Elevated warehouse under construction in Dulac, LA with ramped loading dock (Credit: LSU Coastal Sustainability Studio)

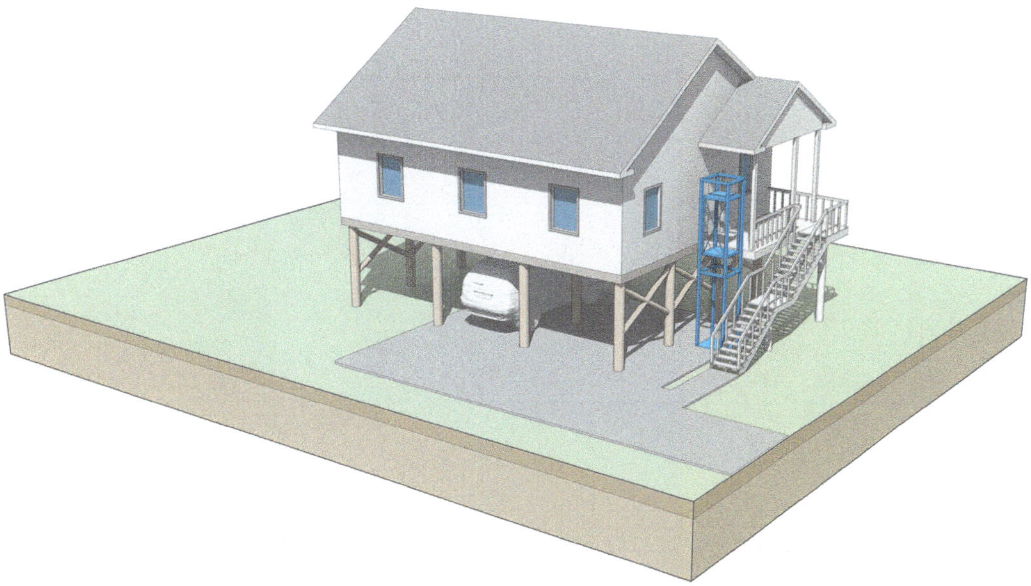

Fig. 12 Diagram of an elevated structure on piles with an elevator (Credit: Giovanni Coakley, LSU School of Architecture)

Fig. 13 Elevated home with elevator in Cocodrie, LA (Credit: LSU Coastal Sustainability Studio)

Fig. 14 Elevated home with elevator in Dulac, LA (Credit: LSU Coastal Sustainability Studio)

Fig. 15 Ijburg housing development in Amsterdam (Credit: zoetnet, CC by 2.0 https://www.flickr.com/photos/zoetnet/)

foundations, structural frames and walls. Roof types have some variation and have evolved due to more stringent code requirements, and for the most part, roofs have consistently provided protection and are not the main point of failure (other than due to lack of maintenance or poor original construction) in flood prone areas. Residential buildings comprise the majority of structures impacted by flooding and sea level rise, even in urban environments. They are also the most susceptible to damage and that is why FEMA has dedicated so many resources to the questions of construction and retrofitting of coastal residences.

Historically, in the pre-modern era, there were several techniques for dealing with flooding, particularly the seasonal flooding caused by river overflows and ocean storms. The most effective method was to raise a building on piers, however, building materials and methods of construction also contributed to the resilience of pre-industrial era structures. One factor that increased the resilience of historic buildings was the reliance on locally sourced materials and adaptive practices of construction. By using materials found in the local environment, builders could expect that materials of the environment would still be sustainable when shaped into inhabitable structures.

One the of the primary components of early coastal construction in the US was the use of heavy timber, from locally harvested woods, such as cypress in Louisiana (Fig. 16). Cypress trees are native to swamps and therefore have evolved the cellular structure that makes them resistant to damage from water and pests commonly found in the environment. These types of trees have a cellular structure that includes a large amount of sap, which hardens into an extremely durable, non-toxic and water impermeable material. When harvested and dried, the resulting lumber would harden and become virtually impermeable to flood, fire and pestilence. Today's home builder seek cypress as a building material both for the durability and beauty of the wood. Given the water-resistant characteristics of the dried wood, it was a suitable material for use in foundational pilings and structural elements as well as it's familiar use in boats of all types.

To infill heavy timber frames, early residents of Louisiana used a material known as bousillage (Fig. 17). A combination of horse hair or Spanish moss and clay, the

Fig. 16 Traditional Louisiana Cypress Timber construction in Lafayette, LA (Credit: Jori Erdman, LSU School of Architecture)

resulting material would be packed between the heavy timbers and allowed to dry before being coated with a protective lime-based plaster or stucco. The bousillage infill of heavy timbers also provided the benefit of being thermally dense, meaning that it did not transmit temperature differences between interior and exterior leaving spaces cooler in the summer and warmer in the winter. The stucco mixture of lime and sand provided an additional layer resistance to water damage and pests, but required regular maintenance to address the tendency to crack and spall over time. The labor intense process of creating and placing bousillage assemblies was replaced over time with more industrialized construction processes but there is not an assembly process available today that is as non-toxic and resilient as bousillage. When properly covered in stucco, the bousillage structure is virtually impermeable by water infiltration, and any water that does penetrate is still able to dry because it is breathable.

Plaster and lath construction is similar to the system used in bousillage construction, although not as thermally inert. Because the lath replaced the bousillage infill in terms of structure, the cavity between timbers or other structural framing would simply be air or sometimes other light weight materials. Plaster and lath was used across the US until the 1950s when plasterboard (also known as drywall and gypsum board) was introduced as a speedier method of making interior walls and ceilings. Although slower to install, once in place, plaster has certain strengths that make it more resilient in flooding and humid environments. Chemically, plaster is similar to concrete in that once hardened, it is irreducible back to its original form. This makes it resilient in humid or wet climates because it will not deform when wet and is breathable so it will not mildew as easily either. It is also more resistant to fire. The deficiencies of drywall in hot, wet climates will be addressed later in this chapter.

Fig. 17 Traditional Bousillage construction detail (Credit: Jori Erdman, LSU School of Architecture)

Brick construction would occur when a community had access to local clay deposits, or funding to import bricks from further inland. Many plantation main buildings are constructed of brick as are churches, banks and courthouses (Fig. 18). Stone construction can be found in more affluent situations.

Pre-modern institutional and commercial building such as banks, schools, churches and stores, would often be constructed with the same materials as residential buildings unless the community had more resources to building in brick or stone. Because historical structures were largely built out of natively found materials, they were more resilient in the case of flooding. In addition, because historic structures did not incorporate air handling systems, they were more able to dry out after flooding events. Finally, historic structures were built and maintaining by craftspeople who were accustomed to the construction techniques of their location and better able to repair any damage that occurred. With the advent of modern, industrialized building materials and methods shifted towards more globalized systems that created buildings that were cheaper and easier to construct, but lacked the resilience of earlier vernacular examples.

Contemporary building practices differ significantly from historical, pre-industrial methods because industrialization is predicated on "efficiency" which most often means standardization across the marketplace on a national scale. The International Building Code, which is in effect in all 50 states of the US, sets the standards of building construction. Contemporary residential construction is largely controlled by industry associations and manufacturers of home building products who have little regard for the nuances of place and climate, particularly that in on the coasts and in flood plains. Their operational strategy has consistently sought to nationally (and globally) standardize a set of building materials that often make no sense for a given climate. Much of the extensive and expensive damage suffered by coastal homes during hurricanes, and even through annual seasons, is due to this standardization. Standardization has tended to focus on the construction needs of non-coastal and northern populations, rather than the subtropical climates found in the American South.

In addition to the standardization enabled by industrialization, population expansion in the US predicated faster and more universal methods of construction. During the Post-WWII population expansion, new construction methods were developed including slab-on-grade and light weight wood frame construction. In addition, the development of artificial ventilation as well as the radical introduction of air conditioning made the homogenization of construction possible across the country, despite dramatically different climatic conditions. Most construction related problems in flood prone are areas due to: slab-on-grade foundations; particle board; drywall; and placement of core systems within the structure.

Developments in concrete construction lead to the mass market implementation of what is called slab-on-grade construction. This is an inexpensive and expedient method in which concrete is poured directly onto a prepared ground to form the floor of a building. This means that buildings with this type of foundation are more susceptible to flooding and water infiltration. In addition, this type of foundation makes it virtually impossible to raise the structure after construction. Beginning as early as the 1940s and continuing until today, this type of construction is typical for residences and commercial structures throughout the US (Fig. 19).

Another construction method developed since WWII is the assembly wall system comprised of manufactured sheet material, such as particle board and drywall, over wood stud framing. Particle board is comprised of very small pieces of wood that are suspended in a glue or resin until hardened and formed into boards. While particle board does share some of the same characteristics as wood in structural capacity, it does not generally perform well when wet or moist and tends to degrade rapidly when exposed to the environment.

Fig. 18 Brick construction of the Laurel Valley Sugar Plantation in Thibodaux, LA (Credit: Library of Congress Prints and Photographs Division Washington, DC. 20540 USA – HAER LA,29-THIB,1A—2 – http://www.loc.gov/pictures/item/la0195.photos.072549p/)

Particle board is often used in conjunction with batt insulation which also doesn't perform well once it has gotten wet. Paper-backed gypsum board, also known as plasterboard, wallboard, sheetrock or drywall, has become a ubiquitous interior building material around the world. Generally considered to be cheaper and faster than plaster and lath, in principle drywall performs the same function. However, there are important differences and climatic factors that make drywall a less than optimal material in hot, humid climates. The chemical composition of gypsum is not strong enough to withstand humidity or prolonged exposure to wet. These materials, which are typically found in suburban homes built since 1950, as well as many commercial and institutional buildings, quickly decompose as well as being highly susceptible to toxic mold and mildew. In the event of significant flooding, these materials are particularly difficult to properly dry and must be replaced in order to sustain the integrity of the interior walls and ceilings, as well as maintaining the health of the occupants (Fig. 20).

Finally, any mechanical or electrical systems that have been exposed to flood waters can be severely compromised. For that reason, homes that follow the standard practice of placing electrical and cable outlets between 12″ and 16″ off the floor can find that they have to replace all of their wiring. One of the dangerous aspects of this, is that damage to electrical systems is not visible to and systems can continue to corrode long after other repairs have been completed, leaving a strong risk of fire. The Federal Alliance for Safe Homes (www.flash.org) recommends that all electrical system components be elevated 12″ above the Base Flood Elevation although standards are difficult to monitor and enforce, especially if local building officials have not adopted this policy.

Fig. 19 Concrete slab-on-grade construction (Credit: Peter Kapitola, CC by 2.0 https://commons.wikimedia.org/wiki/File:Slab_on_grade.JPG)

The National Association of Home Builders is actively campaigning to resist former President Obama's Presidential Executive order E.O. 13690 from January 2015, which amends and strengthens the resilience of residential construction in the US. Climate appropriate building techniques can have a huge impact on the health and well-being of our structures and our communities.

We have examined that building practices and construction methods that have made flood events so costly and damaging, as well as some of the ways that we can build that would serve to minimize the damage, and one thing is clear: if folks want to continue to live in flood-prone environments, we need to start building smarter. Elevating a structure and building with water resistant materials and methods is much easier to do as you go, rather than after the building has been constructed. Now we turn our attention back to the questions asked at the beginning of the chapter about what elevation of structures means for communities and those living in them and we start with the concept of risk.

Risk

> Risk comes from not knowing what you're doing. – Warren Buffett

Risk is defined as the potential for losing something of value. But in its uncertainty, risk also allows for the possibility that value could be gained as well. All human endeavors involve some degree of risk but the question is what you are willing to lose in hopes of gaining more? And how do we mitigate the risk of loss through knowledge and better decision-making processes? And in the case of elevating structures, communities and cities, we can question the value of risk itself and the discussion surrounding risk valuations. As the essay below by Elizabeth Williams, Coastal Communities Resiliency Program Officer at The Foundation for Louisiana outlines, it is imperative that communities engage in cooperative planning in order to assure the best outcome for all.

Fig. 20 Mold damage as a result of moisture wicking up drywall (Credit: Claudette Reichel – LSU Ag Center, 2005 http://www.lsuagcenter.com/topics/family_home/home/health_safety/indoor_air_quality_mold/avoiding-mold-hazards-in-your-waterdamaged-home)

Nonstructural Approaches: Risk, Resilience and Holistic Adaptation

Across the coast and into the higher grounds, communities are in transition. Due to the effects of persistent land loss in combination with sea level rise, Louisiana's most well-resourced coastal residents are already migrating to higher, safer ground. These environmental challenges exacerbate existing socio-economic stressors and drastically reduce the region's ability to develop solutions and adapt locally. As more affluent residents relocate, leaving coastal communities to face the consequences of deteriorating tax bases, already strained communities become further stressed.

Some residents are facing the realities of relocation in places where resettlement is the only viable option. Others are seeing new stressors on existing infrastructural systems; "receiving" communities are forced to reshape and grow in both size and density often without an overarching strategy to expand without inducing risk. Receiving communities to resettlement territories create a spectrum of grey areas where economic viabilities determine the capacity to adapt, reshaping and retrofitting to confront new and changing challenges. Coastal communities across Louisiana must consider and create holistic strategies to address risk and increase resilience, considering the socioeconomic variables and influencers as they determine how to adapt to the changing landscape.

(continued)

Marsh creation, structural and nonstructural methods have been developed to reduce risks across the breadth of the changing coastal landscape. The most recent plans illustrate an understanding that the coast of coming decades must be supported to maintain the socioeconomic fabric and cultural vitality of Louisiana, but also that the coast of the future will look very different than that of the past. (See the most recent Coastal Master Plan at www.coastal.la.gov) As multiple lines of defense are restored across integrated landscape zones, the significance of coastal Louisiana to the national economy is asserted to legitimize and encourage necessary spending for this infrastructural ecosystem. The national relevance must be heartily supported and exhibited, but it cannot be genuinely referenced without an acknowledgement of the local systems at play and that play part in that assertion.

A local workforce supplies much of the power to maintain the relevant industries nationally critical within that framework, yet the risks to the local communities will challenge the ability for the adjacent necessary personnel to remain. Structural risk reduction measures, including levees, seawalls and floodgates, are investments that prove viable and worthwhile when the population and economic assets can justify the cost. Still, any barrier, protection or structural risk reduction measure should have an asterisk beside it at this point in Louisiana history. Nonstructural measures of flood proofing, elevations and voluntary acquisitions assist in the adaptation processes within coastal communities.

Measures to rebuild, sustain and even abandon parts of the coastal ecosystem influence the spectrum of communities in transition and the actions needed to adapt; however, these risks and possibilities for resilience also hinge on other, everyday socioeconomic realities. As open water increases around communities across the southern part of the state, various forms of data are indicating that many Louisiana residents have already begun to migrate. USPS Active residences signpost resettlement and migration after an acute disaster and throughout the chronic disaster that is the collapse of the Mississippi River Delta wetlands (Plyer et al.). However, upon assessing US census data as a supplement, median income levels for census tracks reveal that coastal areas with a decreasing population are also seeing falling median income levels. As the tax base for a parish or state decreases, existing resources used to sustain built and unbuilt infrastructures are depleted, and decline in shared support systems for quality of life persists. In every community, these environmental challenges exacerbate existing socioeconomic stressors and reduce the region's ability to develop solutions. For residents without access to traditional credit or a savings account, acute and chronic disasters are financially catastrophic. Existing and growing social and economic inequalities reduce the capacity for residents and communities to address risks, become more resilient, or holistically adapt to changing environmental circumstances.

Often, residents most affected by these challenges are left out of discussions regarding how to address and respond to them. Difficult conversations lie ahead, including: the need for economic and workforce development opportunities; elevating homes and forced migration; retrofitting of infrastructure; and resource shifts caused by nearby coastal restoration projects. These challenges require innovative solutions with an understanding of local and regional sensitivities to the risks as well as opportunities.

Strategic adaptations for future environments incorporate the need to reshape, retrofit and resettle the coastal zone. Strategic investments must be made that are cost effective as well as equitable; innovative with multiple benefits; and resilient across a future life cycle within a changing climate. As precipitation levels increase, development continues to sprawl into floodplains and wetlands to support exploding populations. Development trends have not considered these risks across the Louisiana landscape; high rain and water events have begun to affect more people, who are unaware that they reside in harm's way. Nonetheless, as receiving communities swell, strategic decisions to support transportation, education and healthcare infrastructures for everyday are as critical as investments in storm water management and risk reduction measures. Capable of withstanding flood depths between 3 and 14′ from future storms, "retrofit zones" bolster existing community assets and support the needs of embedded industry.

Communities facing the realities of resettlement are those with the most difficult conversations to come. In most cases, the flurry of extreme storm events within the last decade have already influenced the departure of many longtime residents. For every individual, the decision to depart from lands where one's ancestors may have resided for generations is a challenging one. However, in many cases, the ability to make that

(continued)

decision is also unaffordable. For a resident to make that complex decision and the financial ones that follow, that resident must have the means to do so. As the chronic disaster of coastal land loss continues, property values continue to deteriorate. The question is what to do when the value of an investment, coastal residence, a person or family's home, is reduced to zero. What is to be done when there are no buyers for property and the local tax base is eroding? Can residents be supported in their efforts to mitigate risks and move to higher, safer grounds without entirely depleting local revenues and catalyzing tipping points for the parish as a whole? How might creative land use transition vacant properties to productive landscapes in the context of restoration without supporting economic assets of the parish in the form of traditional development? Answers to these questions are critical in order to promote holistic adaptation for all Louisiana residents.

Parish leadership and governments are unlikely to speak plainly about the environmental challenges faced in coming decades. To openly discuss the dire conditions of environmental challenges would be to worsen existing financial problems. . .to further convince a dwindling tax base that the future within the parish might look very different than in previous generations. However, innovative incentives to furnish risk-aware communities that receive growing populations are possible with an understanding and engaged public. The retrofitting of communities on higher grounds must be supported – dense development with more permanent elevated structures might include other programs to promote social and economic activities, multiple uses and shared public spaces elevated comingling. High ground must be prioritized within each parish and new development regulated in floodplains across Louisiana watersheds. In order to promote regional resilience, coastal residents must adapt within each parish while capitalizing on the cultural, financial and environmental systems that cross multiple parishes. Decisions are made at the parish level but risk and resilience are not limited to those regulatory lines, water has no respect for political boundaries.

For Louisiana to adapt to the coastal crisis, socioeconomic conditions further strained by environmental stressors must be addressed in a holistic manner. Local governments and state agencies must collaborate across disciplines while continuously incorporating resident inputs to equitably determine how to adapt. As residents migrate to higher grounds, developments to accommodate those populations also require measures to reduce risk, accounting for the new pressures on existing infrastructural systems. From receiving communities seeing that population growth to those forced to consider relocation or planned resettlement, a spectrum of communities will need to prepare measures to adapt holistically across social, economic and environmental systems. Solutions to new and continuing coastal challenges require innovative and synergistic strategies towards a more resilient region.

Reference:

Plyer A, Bonaguro J, Hodges K (2010) Using administrative data to estimate population displacement and resettlement following a catastrophic US disaster. Popul Environ 31(1–3):150–175

Across the globe, we see evidence of risk-taking in various forms: from the communities who seek to maintain a livelihood in the coastal environment; to universities and designers who envision new forms of occupation and building; to advocacy agencies who work to mediate between the community and state and federal concerns. We know that doing nothing to change the direction of the environment in locations like Louisiana means rapid loss of land and population. . .the question is what to do and how much risk are we willing to take as we struggle to sustain inhabitation in these environments. Despite repeated destruction and rebuilding, many Louisiana communities remain committed to their lifestyle, economy and culture. They believe the risk of inhabitation balances in the favor of value gained rather than lost. . .the fruit of life for them is out on the limb of the little spit of sand they call home.

The Civics of Individual Buildings

What did it mean for Miss Twiggley to build above the land in the first place? Is it necessary for building above the land to mean divorcing oneself from place and citizenship? Jane Jacobs, in her iconic 1961 tome, The Death and Life of Great American Cities, makes the argument that the street (or ground) is where democracy and citizen truly flourish. She laments the modernist urban proposals that privilege the automobile over the pedestrian and seek to house everyone in sterile glass towers. But living above the land is a practice found throughout history.

We can see this adaptive behavior in larger cities such as New York, where the original brownstone houses were

Fig. 21 A row of Brownstone homes in Brooklyn, NY (Credit: JDH Rosewater, CC by 2.0 https://www.flickr.com/photos/regulusalpha/)

elevated to create a basement (for servants and service spaces) with a piano noble for the family. This model was also adapted for apartment buildings throughout the city. Out of this arrangement developed stoop culture, adding to the social realm of sidewalks and streets, where residents come down from their domiciles above, and participate in the life of the street, especially in the warm summer months (Fig. 21). In other urban situations, street level spaces are typically occupied by commercial ventures with residences above. This typology is found in the fabric of the old French Quarter in New Orleans with a shop on the street front and residences behind or above (Fig. 22).

In tropical and sub-tropical climates homes and other buildings were always elevated for environmental reasons including improved ventilation, decreased pestilence, as well as avoiding areal flooding and dirt from the street. The elevation was often minimal, just a few feet, but still necessitated stairs. In the American South, and particularly in New Orleans, pre-war housing (shotguns, etc), porches were and are an important feature of urban life (Fig. 23). Jacobs wrote, "Streets and their sidewalks, the main public places of a city, are its most vital organs." In the Southern city, porches also play an important social role. They are a space that is neither completely public, such as the sidewalk, nor completely private, like the interior of a house. They serve as a safe play space for children, but also provide an exterior view of the workings of the street and neighborly communication. In a hot, sunny climate, porches provide respite from the sun as well as stuffy interior spaces. People understand that porches are private, but also shared. Neighbors share information about goings on from the porch. They can observe visitors and provide surveillance of the neighborhood. And porches are places where people tend to gather and tell stories. . .without a utilitarian function, front porches are also "vital organs" of the city in Southern climates.

Larger institutions (churches, schools, banks, libraries) face different challenges to elevation and civic presence but these institutions have always navigated elevation. Churches were often elevated above the street to accentuate the spiritual experience. Banks and libraries often elevated

Fig. 22 The French Quarter of New Orleans with shops on the street front and residences above (Credit: Cazz, "French Quarter" CC by 2.0, https://www.flickr.com/photos/cazzjj/)

to increase their prominence and overall massing within the city block. While contemporary construction and norms have challenged some of these architectural typologies with the ubiquity of slab on grade construction, people are accustomed and can adjust to institutions which must elevate (Figs. 24, 25 and 26).

But what happens when houses are elevated above eye level (around 6 ft) from the street or sidewalk? In addition to changing the materials and methods with which buildings are constructed, post WWII construction has diminished street life in both urban and suburban/rural areas. Newer buildings contribute less and less amenities, such as sidewalks, stoops, and porches, meaning there is less of the traditional space for street life. But is it inherent that we will lose our vital organ that make a cities and communities social entities? When we distance ourselves from "terra," are we less connected to the earth? I would argue that this is not the case.

On the coast, residences need to be high, strong and self-sufficient during flooding events. Critics argue that elevation strategies compromise the life of the street. Examples such as the redevelopment of the Lower Ninth Ward in New Orleans reinforce their argument that elevation of structures at or above a full story of height are used to demonstrate the challenges of a less direct connection between residents and communal space of the street (Fig. 27). However, I would argue that it is not only the elevation that causes this diminishment, rather, it is a lack of density that has more of an impact. It is understandable that in the case of New Orleans, where vast amounts of the city fabric were damaged so badly that it had to be removed and repopulation is slow, that redevelopment efforts are sparse and density has not returned to previous levels. This would be true in many cases where homeowners are slow to return due to economic challenges. However, urban environments demonstrate that elevated living does not necessarily mean a negation of the

Fig. 23 Traditional shotgun home in the Marigny neighborhood, New Orleans, Louisiana. (Credit: Infrogmation, CC by 2.5 https://commons.wikimedia.org/wiki/File:Marigny14May07RampartYellowShotgun.jpg)

public space of the street. In dense cities around the world, streets are full of life and interactions amongst people despite the fact that very few people actually live at street level. Density is the factor that drives elevated living and a thriving public domain.

An argument for density in elevated structures means that society will have to reconsider issues of ownership and private property in areas where street life is a valued asset and desired condition. Use of space below and around structures that have been elevated will have to be reconsidered in light of the necessity for programming non-inhabitation at the ground level. This is a complex issue that is not attempted to be solved in this article, but one that deserves further study by designers, planners, and social scientists. Consideration for new opportunities can serve to offset the perceived loss of connection between private interior spaces and the public space of the street.

The Structures of Coastal Resilience program that developed in the wake of Hurricane Sandy is one of the latest attempts by designers to look at resilient designs for coastal environments, particularly urban and suburban conditions. Their proposals primarily focus on the types of soft infrastructure such as wetlands creation/restoration, berming, and water banking that are low impact but large scale. One team designing for Atlantic City came up with a system for elevating houses and reconsidering the public domain, notably through an examination of easements (Figs. 28 and 29). However, their study stops short of projecting the social outcomes of such efforts and the residual impacts on human activity and civics.

Fig. 24 The original slab-on-grade structure of First Baptist Church in Grand Isle, LA is now elevated on piers (Credit: LSU Coastal Sustainability Studio)

Fig. 25 Completed in 2011, the gymnasium of Grand Isle School is elevated approximately 12 ft above grade and built to withstand 150 mph winds and the harsh saltwater environment of Grand Isle, LA (Credit: LSU Coastal Sustainability Studio)

Fig. 26 The elevated community center in Grand Isle, LA (Credit: LSU Coastal Sustainability Studio)

Observing communities in south Louisiana and New Orleans that have had to elevate for flood protection, it is apparent that residents take advantage of this newly revealed space below their houses and rather than leaving it empty, have filled these spaces with the functions previously encouraged by porches. The newly released ground plane is, of course, occupied by vehicle storage, but also by barbecue grills, chairs and rockers, toys and other recreational items (Fig. 30). This reveals some of the adaptive qualities of humans, that we can appropriate spaces to meet our social needs, even in a challenging environment. The Center for Planning Excellence has conducted several major studies of coastal communities including "The View from the Coast" and "The Community Rating System: Making it Work for Louisiana." In "The View from the Coast," local officials expressed their concern about "the long-term impact of elevated houses — or of a partly or completely elevated neighborhood — on residents. As homes move away from the street and become isolated from each other, some fear that community, sociability, and neighborliness will atrophy" (CPEX, 24). Social scientists have not yet taken on this subject for study, but observationally, it seems apparent that the resilience of the Louisiana culture is already mitigating these concerns.

As noted in chapter "Introduction – Changing Conditions in the Mississippi Delta from 1700 to 2100 and Beyond: Avoiding Folly", the Changing Course competition http://changingcourse.us was one of the most comprehensive and rigorous attempts to address the current Mississippi deltaic condition and all of the three finalists projected a retreat from the lower coast. However, the 2017 Coastal Master Plan, in projecting the costs of such an endeavor, clearly indicates that complete retreat is not feasible. What is the balance between making more resilient structures and a deteriorating coast? There are no clear answers here but architects, planners, and social scientists need to be active participants in the discussions moving forward as they are the professionals and experts who can

Fig. 27 Elevated homes in the Lower Ninth Ward of New Orleans (credit: Drew from Zhrodague, CC by 2.0, https://www.flickr.com/photos/drewzhrodague/8123959293/in/photolist-dnTq4F-8LeiR-dnTw2d-dnTyEG-dnTqFZ-8Le5g-dnTuYa-dnTxpc-61hgs1-8Le5W-8Le4W-72Kkfq-dnTv8r-dnTv2x-dnTqMc-8LdUm-8LekS-dnTxVX-8Leoh-dnTr6e-8LdTi-8LdWy-dnTwLY-8Lefs-8Le4v-dnTwLD-dnTAMb-8Lemi-8LemG-dnTwo7-8LdZz-dnTwRS-dnTvVG-dnTqcF-dnTDnE-dnTDeo-8Lehp-dnTBJN-dnTsRj-dnTCmJ-8Le1T-87NxWK-6PqXkP-dnTr8k-dnTrna-dnTt93-d nTnPx-8Le4f-8Le5B-8Le6n)

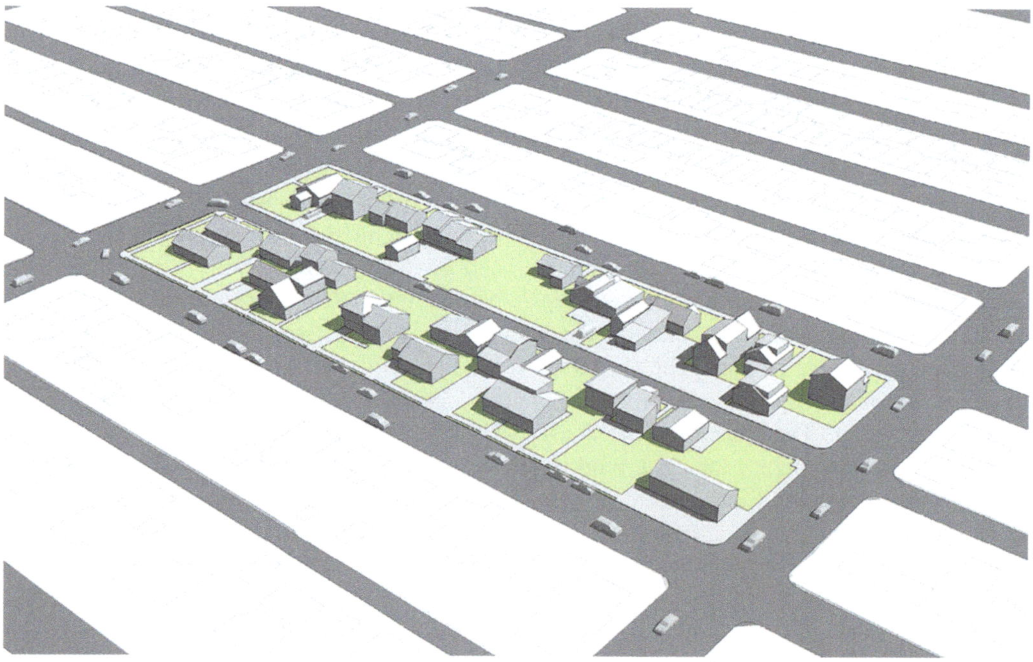

Fig. 28 Typical block of Chelsea Heights neighborhood of Atlantic City, NJ in the year 2014 (Credit: Lewis. Tsurumaki. Lewis (LTL) Architects – Structures of Coastal Resilience Atlantic City, NJ Proposal http://ltlarchitects.com/structures-of-coastal-resilience)

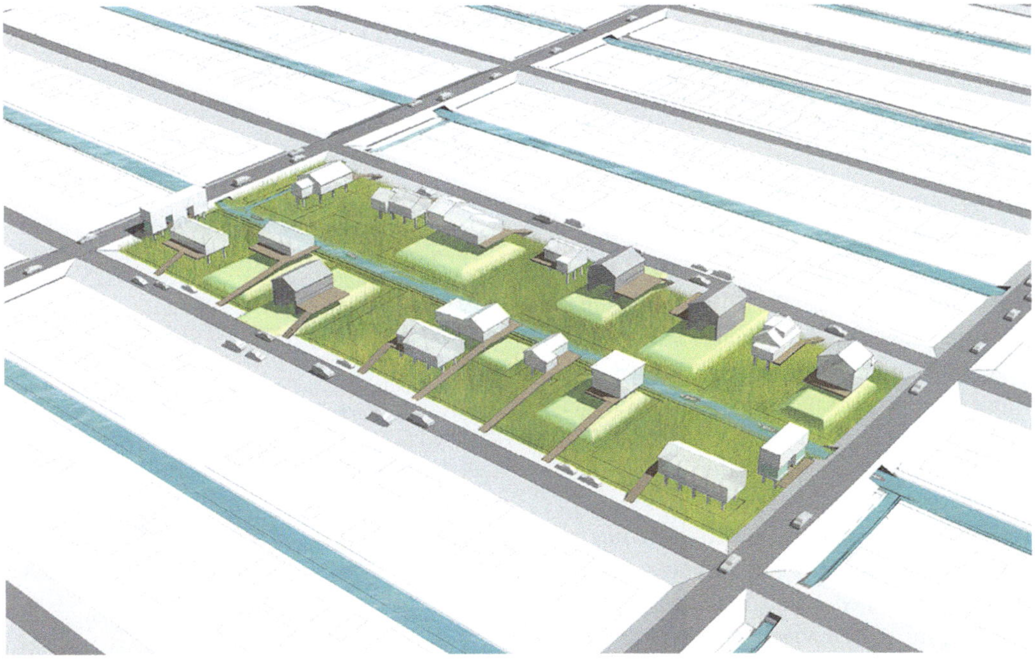

Fig. 29 Typical block of Chelsea Heights neighborhood in the year 2090 with the amphibious suburb proposal implemented (Credit: Lewis. Tsurumaki. Lewis (LTL) Architects – Structures of Coastal Resilience Atlantic City, NJ Proposal http://ltlarchitects.com/structures-of-coastal-resilience)

Fig. 30 The occupied ground plane of a neighborhood of elevated structures in Dulac, LA (Credit: LSU Coastal Sustainability Studio)

most readily address the cultural and social impacts that will result from mass elevation and/or migration.

as possible for elevating structures and communities in order to serve the needs and lives of everyday citizens.

Conclusion

The losses we suffer due to climate change are examples of what the microbiologist Garrett Hardin termed "the tragedy of the commons" in 1968 when he referred to the tendency for selfishness to takes primacy over the belief in the value of the common good in human/natural systems. However, the opposing human tendency towards decency and good will for one another can also be our salvation if we can maintain and reenergize the more optimistic practices of the commons. The possibilities of mass migration due to climate change presents real dangers to global peace and security as recognized by agencies around the world including the World Economic Forum and the CNA Military Advisory Board. The option of provided stable, resilient and flood-proof structures along the coastal zone will help to minimize the necessity for migration in many areas and reduce the stressors on the global ecosystem.

This chapter is an attempt to demonstrate how elevated structures can still provide individuals, families and communities with a quality of life and space in which they can prosper and survive flooding events with property intact. As FEMA points out, the only real option, absent hard infrastructure or building elevation changes, is relocation and for many in south Louisiana this is not only financially impossible, it is spiritually untenable. The life of most people in southern Louisiana is uniquely tied to their environment economically and culturally. Architects, builders, planners and municipal agents should advocate as strongly

References

Barnes S, Bond C, Burger N, Anania K, Strong A, Weilant S, Virgets S (2015) Economic evaluation of coastal land loss in Louisiana. Louisiana State University and the Rand Corporation. Published online. http://coastal.la.gov/economic-evaluation-of-land-loss-in-louisiana/. Accessed March 2017

Broome C, Dubinin J, Jenkins P (2015) The view from the coast: local perspectives and policy recommendations on flood-risk reduction in South Louisiana. Published online. http://www.cpex.org/resources/. Accessed March 2017

DeRussy LG (1859) Special report relative to the cost of draining the swamp lands bordering on lake Pontchartrain. J. M. Taylor, State Printer, Baton Rouge

Coastal Protection and Restoration Authority (CPRA) (2017) 2017 Louisiana coastal master plan. Published online. http://coastal.la.gov/ourplan/2017-coastal-master-plan/. Accessed April 2017

Etlin R (1994) Frank Lloyd Wright and Le Corbusier: the romantic legacy. Manchester University Press, New York

Fox DW (1966) Miss Twiggley's tree. Parents' Magazine Press, New York

http://www.unep.org/pdf/TNAhandbook_CoastalErosionFlooding.pdf

McGranahan G, Balk D, Anderson B (2007) The rising tide: assessing the risks of climate change and human settlements in low elevation coastal zones. Environ Urban 19(1):17–37

Napier HE (1847) Florentine History, from the Earliest Authentic Records to the Accession of Ferdinand III. Grand Duc of Tuscany. Vol. 5. Moxon:1847

Nordrum A (2014) Architects place priority on public health for 2015. Scientific American, December 15:2014. https://www.scientificamerican.com/article/architects-place-priority-on-public-health-for-2015/

World Health Organization (1991) Surface water drainage for low-income communities. Published online. https://www.ircwash.org/resources/surface-water-drainage-low-income-communities. Accessed March 2017

Raising New Orleans: The Marais Design Strategy

Jori A. Erdman, Christopher W. James, Giovanni P. Coakley, and Elizabeth A. Williams

Abstract

New Orleans, Louisiana, is an iconic American city that is located in one of the most dynamic deltaic environments in the world. Called "The Accidental City" by author Lawrence Powell, New Orleans provides a case study for all of south Louisiana, as well as cities around the world that are increasingly threatened by sea level rise as to how settlement can continue in such a precarious location. In this chapter, we show a design proposal for elevating the city of New Orleans as an adaptive course of action. Our two-part strategy begins by reinforcing the lake front edge of New Orleans, along Lake Ponchartrain using infill to extend the higher, buildable ground. The higher ground would be fronted by a new cypress swamp and urban edge. The second part of the strategy aims to build a series of levee-like structures called polders by the Dutch, although we use the French term, "marais," across the city by following existing infrastructure. The design proposal further develops edge and fill tactics to complete an elevation of the city, in whole or part.

Keywords

New Orleans • Land-building • Urban design • Architecture • Urban infrastructure

Introduction

> Upon commencing the necessary reconnaissances for the execution of the work entrusted to us, in June last, we proposed to suggest the adoption of the system known as the 'System of Colmates,' for filling and elevating, by river deposits, the large district of country in rear of the city of New Orleans, and bordering the Lake Pontchartrain, instead of the establishment of regular and permanent drainage;... – excerpt from Special Report Relative to the Cost of Draining the Swamp Lands Bordering on Lake Pontchartrain by Lewis G. DeRussy, Baton Rouge, 1859

J.A. Erdman (✉) • C.W. James • G.P. Coakley
School of Architecture, Louisiana State University, Baton Rouge, LA, USA
e-mail: jerdman@lsu.edu; cjame32@lsu.edu; gcoakl1@lsu.edu

E.A. Williams
Foundation for Louisiana, Baton Rouge, LA, USA
e-mail: liz@lizabethannewilliams.com

J.W. Day, J.A. Erdman (eds.), *Mississippi Delta Restoration*, Estuaries of the World, https://doi.org/10.1007/978-3-319-65663-2_11

Lewis De Russy was a man ahead of his time. When he proposed using a system of "colmates" (Italian for little mountain but also referring to a natural system of land building through flooding and sediment deposition), the Board of Swamp Land Commissioners resisted. He, and others before him, believed that the long-term benefits of elevating the swamp land surrounding the center city of New Orleans outweighed the alternative land reclamation strategy of drainage. The city fathers, thinking only of more immediate concerns, opposed the long-term view and insisted on a system of drains and pumps that form the basis of New Orleans city water management. By contrast, in the same time period Chicago also addressed flooding but determined to elevate the entire city (Fig. 1). The results are that Chicago no longer struggles with surface flooding while New Orleans remains tied to an aging infrastructure and the laws of gravity that mean continuous flooding with every rain event and catastrophic failure in the face of a major weather event as well as climate change.

Fig. 1 A photograph along the Chicago River illustrating the multiple levels of Wacker Drive with the buildings of downtown on the elevated plinth beyond (Credit: Christopher James)

Fig. 2 New Orleans is constantly threatened by the combined flooding effects of subsidence and sea level rise (Credit: Christopher James)

New Orleans is perhaps the most well-known example of a US city that is eminently threatened by flooding due to subsidence and sea level rise (Fig. 2). Of course New Orleans is not the only example, even in Louisiana; in fact, all cities and communities in large river deltas around the world are facing the same double threat. Flooding events create loss of assets, productivity and life with increasing regularity. This condition is not sustainable and has become the subject of much study. Termed "resilience" in popular culture, attempts to respond to the effects of climate change before emergency situations happen have lead to proposals and infrastructure projects that range in scale from comprehensive planning to discrete interventions. However, with every passing year and new climate change analysis that is released, communities are realizing that flooding is happening more frequently and with greater ferocity than current infrastructure can accommodate.

Flooding itself is not a new issue for settled communities. Over the centuries and particularly since the dawn of the industrial age, humans have endeavored to address the effects of flooding that have been exaggerated by occupying fertile and strategic locations along rivers, lakes and oceans. Mitigation attempts including seawalls and drainage have been augmented with mechanical means such as pumps and movable barriers. Operating, maintaining and servicing such methods is both expensive and resource intensive and may ultimately be unsustainable as long-term solutions for communities.

Pre-industrial cultures developed strategies such as mounding as a means of controlling water. Levees, known to the Dutch as dikes, are natural or man-made mounds, adjacent to rivers or water bodies, that are generally constructed to prevent flooding. As a form of control, the levees are effective and can even allow for operational openings when properly designed. The Mississippi River relies on this form of water control to maintain navigation and commerce along the river, as well as protection of the lands behind the levee.

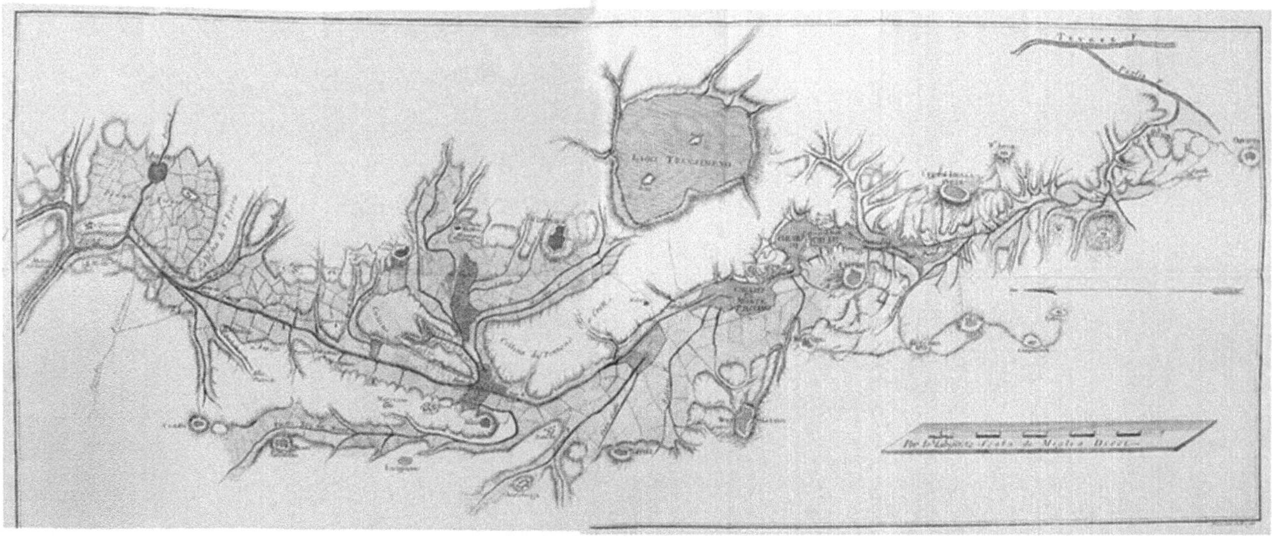

Fig. 3 Map of Fossombroni's land reclamation proposal in the Val-de Chiana region of Tuscany (Credit: Illemorie Idraulico atoriche copra la Val di Chiana, 1789)

Another, less used, means of protection has been the elevating of significant areas in order to maintain habitable environments. For example, in the 1780s an Italian engineer named Vittorio Fossombroni proposed the elevation, rather than drainage of the Val-de Chiana region in Tuscany. It is the Fossombroni proposal that DeRussy referenced in his 1858 Report to the Commissioners of Louisiana, he outlined a process of building "colmata," or cumulative land building by allowing an area to flood with sediment laden water, then quarantined until the sediment settles and clear water is drained away (Fig. 3). The process in repeated until the desired elevation change is achieved. Based on research documented by Henry Edward Napier in his 1847 multi-volume Florentine History, the Fossombroni plan was enacted and successful in changing the elevation and entire floodplain drainage direction of the valley.

More contemporaneously, as outlined by Professor Craig Colten in chapter "Raising Urban Land: Historical Perspectives on Adaptation," both Chicago, IL, and Galveston, TX, successfully endeavored to elevate their entire city area as a response to flooding threats. Colten also points to DeRussy's proposed use of the colmate system to address flooding in New Orleans as a counter to the drainage plan in the study for the city of New Orleans. City officials declined to consider the elevational system as proposed and, instead, continued the trajectory of drainage methods already implemented and soon to be augmented by mechanical pumps. . .a high energy cost endeavor that relies on continuous and un-interrupted energy and maintenance. This decision, made more than 150 years ago, has lead to problems that continue to plague the city today including rain induced nuisance flooding, dependence on energy intensive pumping stations and ongoing subsidence of major parts of the city.

In 2016 professors and designers from Louisiana State University came together to design a new plan for addressing the challenges of subsidence and sea level rise in the Mississippi River delta of Louisiana. When approaching design strategies for resilient settlements, the LSU design team decided to take on New Orleans as the basis for a case study. Our design team of ecologists, architects, engineers, and social scientists addressed a range of settlement related issues. The group came together through the idea that we could work to propose an alternative to the current reliance on energy and artificial drainage (Fig. 4). We drew on the scientific understanding of natural processes to inform our proposal through an understanding of the way that deltas naturally build land through seasonal flooding and deposition of sediment from upstream. This process has produced the fertile land and vibrant economies that are associated with delta environments around the world.

The Marais Design Strategy

DESIGN TEAM: Giovanni Coakley, Craig Colten, John Day, Jori Erdman, Christopher James, Robert Lane, Jeff Rutherford, Adrian Weigman, Jim Wilkins, Elizabeth Williams, and Clint Willson.

Our team proposes elevational and protective design strategies that allow New Orleans to remain the intact but with substantial change to the way the city functions relative to water and infrastructure. Our design process and conversations are based on a stated framework of fundamental beliefs and goals:

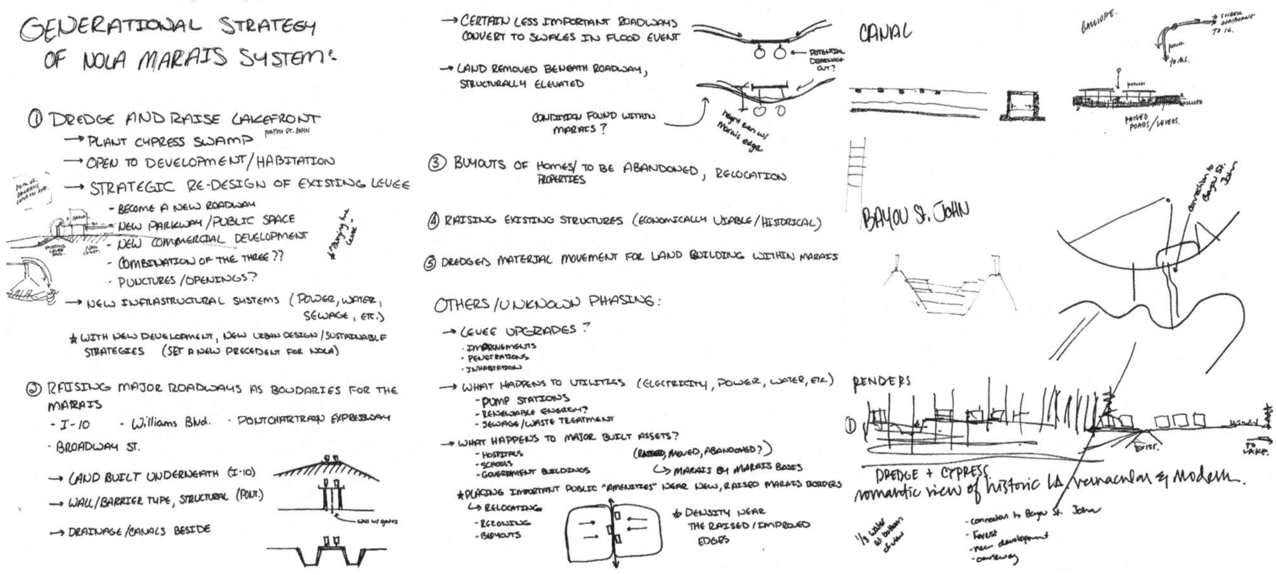

Fig. 4 Sketches and notes from the design charrette held on January 10th, 2016 (Credit: Christopher James)

	Drainage	Soils	Subsidence	Reliance on Infrastructure	Population/development	Socio-economic status	Insurance rates
High Ground	Gravity	Stable	<1mm/year	Minimal	Growing	Affluent	Lower
Sustainable but at-risk	Gravity/Pumps	Firm	<1mm/year	Moderate	Growing/high risk	Mixed	Mixed
Greater risk	Pumps/Gravity	Loam/clay	>1mm/year	High	Low density	Vulnerable	Highest
Exposed	None	Organic	>1mm/year	None	Shrinking	Vulnerable	No insurance
Must fix	Pumps	Variable	>1mm/year	Full	Dense residential	Industrial	Variable

Fig. 5 Flooding Risk Rubric. These conditions are found across the entire Louisiana coast, as well as within the city of New Orleans (Credit: Jori Erdman)

1. We are seeking options that integrate NATURAL and BUILT SYSTEMS.
2. We recognize the following CONSTRAINTS on the design options: Energy scarcity, Climate change, Degrading natural systems in the coastal zone, Change resistant legal; socio-economic and political systems
3. We seek options that MAINTAIN or HEIGHTEN the existing socio-economic and cultural conditions

As we worked we also recognized that within the broad context, there are variable conditions that determine different levels of flooding risk. In reviewing the flooding risk of settlement patterns in differing conditions, we formulated a rubric (assessing drainage, soils, subsidence, population, socio-economic status, and insurance rates) in order of increasing difficulty of adaptation (Fig. 5).

Parts of New Orleans fall into all five of these categories. We assume that some parts of the city cannot or will not be able to sustain occupation in the future due to increased threats from hurricanes as well as the expense of pumping and infrastructure. However, our strategies could provide much larger margin of possibility (lower risk) for continued inhabitation.

Currently, the city of New Orleans relies on a series of levees and sea walls that surround the city for the majority of protection from external flooding (Figs. 6, 7, 8 and 9).

Fig. 6 The levee along the shore of Lake Pontchartrain in Kenner, a northwestern suburb of New Orleans (Credit: Christopher James)

Fig. 7 The flood gate along the shore of Lake Pontchartrain in Kenner, a northwestern suburb of New Orleans (Credit: Christopher James)

Fig. 8 The levee along 17th Street Outfall Canal (Credit: Christopher James)

Fig. 9 The levee along Orleans Avenue Outfall Canal (Credit: Giovanni Coakley)

Fig. 10 Pump station in Metarie (Credit: Giovanni Coakley)

While primarily passive in nature, these artificial devices do require constant maintenance and surveillance and are susceptible to degradation and obsolescence due to the effects of increasing sea level rise. For areal flooding within the protective walls and levees, the city relies on a series of energy intensive, high-volume pumping stations (Figs. 10 and 11). Elevated highways cross the city and keep interstate traffic moving but ground level streets are often blocked due to water infiltration and lack of drainage. As seen most vividly during Hurricane Katrina, but also observable during any modest rain event, these measures cannot keep pace with the impact of water on the citizens of New Orleans.

The Marais Design Strategy is aimed at adaptation rather than mitigation. For example, one primary design strategy uses the existing infrastructure as a foundation for interventions (Fig. 12). Tactically, we employed the use of marais, which is the use of levees to enclose low-lying areas of land (Fig. 13). The leveed marais edges follow existing major canals and avenues as well as the elevated interstate highway. As the edges are built up, using sediment from the Mississippi River (Figs. 14 and 15), the internal measures of elevating individual structures will increase individual building resilience. Over time, the city will be elevated to a height that is less risk-prone and more sustainable. Using natural systems thinking, we were able to make a proposal that maintains the character and structure of the city, while providing a way forward that protects the people and property of the city.

Tactics

1. Address the most vulnerable area, along the coast of Lake Ponchartrain by using dredge taken from the lake bed and creating an 800-yard-deep cypress forest along the edge. Directly adjacent to the forest, on the land side, densely developed urban edge, architecturally structured to sustain hurricane events, this natural edge will greatly reduce the risks associated with storm surge by dissipating the energy of hurricane induced water inundation.
2. Identify a large, unpopulated area within the city to be used as a sediment catchment area.

Fig. 11 17th Street Canal Pump Station (Credit: Giovanni Coakley)

Fig. 12 New Orleans regional map with existing flood infrastructures, land elevation, and proposed marais extents; created after the initial design charrette (Credit: Elizabeth Williams)

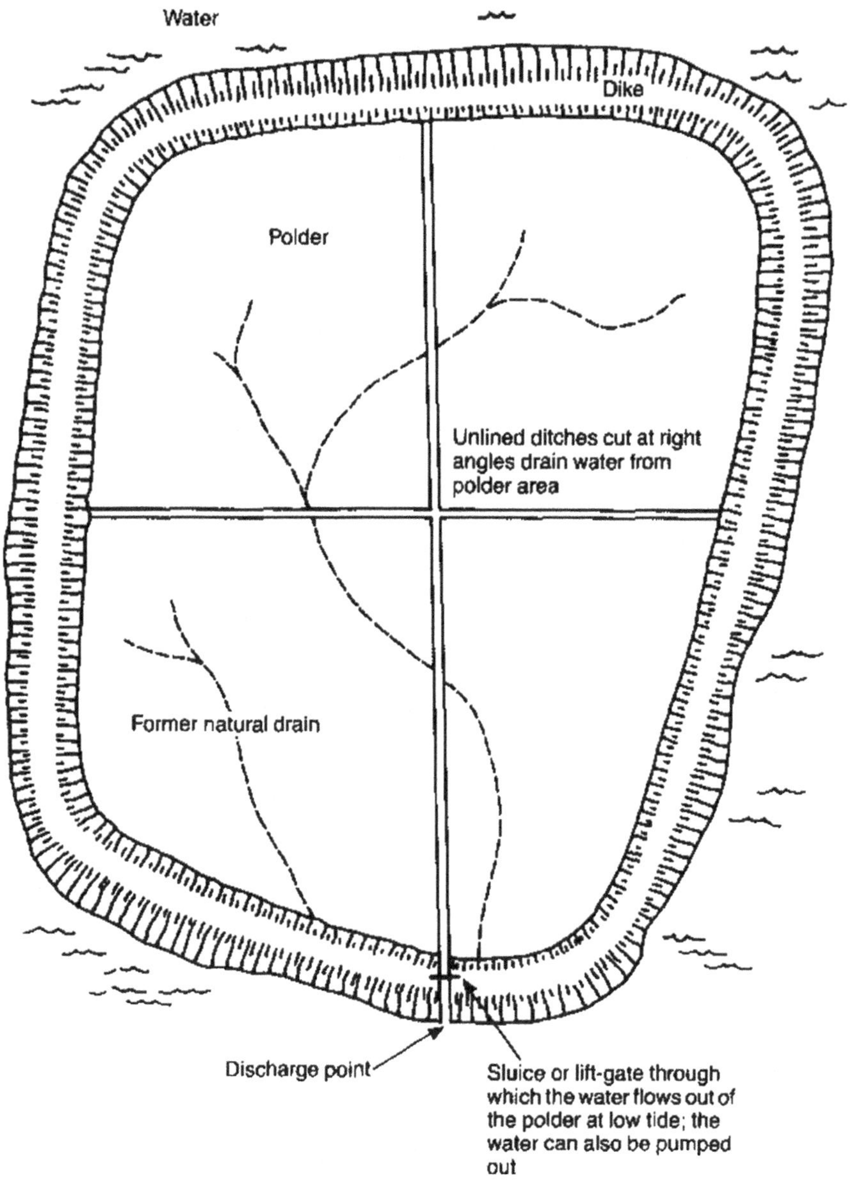

Fig. 13 A schematic diagram of a traditional Dutch Polder (Credit: Surface water drainage for low-income communities. Publication of the World Health Organization, 1991. Published online. https://www.ircwash.org/resources/surface-water-drainage-low-income-communities. Accessed March 2017)

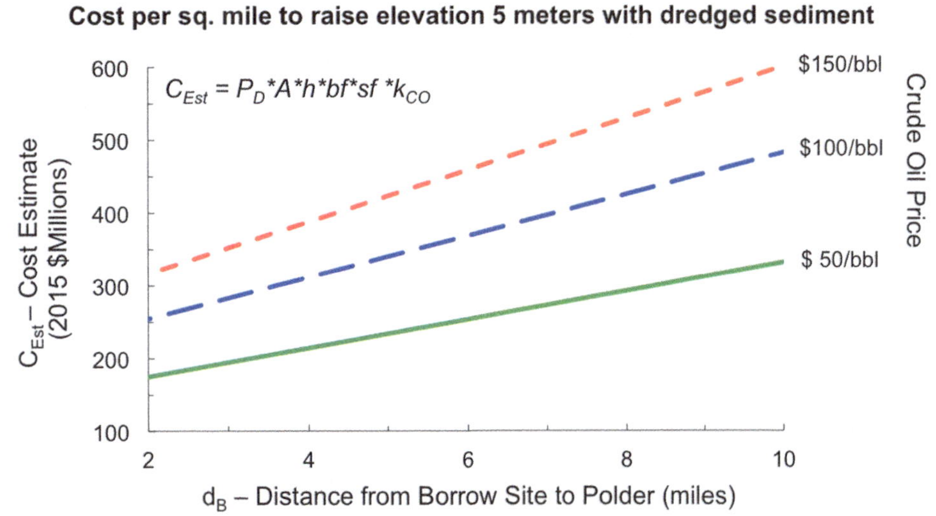

	Notes on Cost Estimates ($C_{Est} = P_D*A*bf*sf*k_{CO}$):
d_B	– pumping distance, distance from sediment "borrow site" to fill area
P_D	– price of dredging (2015 $ per cubic yard) at a given distance
estimated by the	
	following equation $P_D = 0.58*d_B + 3.99$ (from Figure 7 in Chapter 7)
A	– area, 1 (miles^2) = 3,098,000 (yards^2)
H	– height, 5 (meters) = 5.47 (yards)
bf	– bulking/loss factor, 1.5 (33% loss of borrow volume)
sf	– settlement factor, 1.5 (33% compaction)
k_{CO}	– factor for the price of crude oil (derived from Figure 5 in Chapter 7)

Fig. 14 Cost per square mile to pump 5 m of dredged sediment from the Mississippi river into a polder. Cost estimates are given for various sediment pumping distances and energy prices, calculations are from models given in Figs. 5 and 7 in chapter "The Costs and Sustainability of Ongoing Efforts to Restore and Protect Louisiana's Coast" (Credit: A.R.H. Wiegman)

Fig. 15 Sediment from the Mississippi River delivered to marsh-building projects by the Coastal Wetlands Planning, Protection, and Restoration Act (Credit: CWPPRA – https://lacoast.gov/new/About/)

Fig. 16 Locals socializing on the neutral ground of Broad Street in the Treme neighborhood of New Orleans (Credit: Giovanni Coakley)

3. Construct pipelines from the Mississippi River to the catchment area following current infrastructure such as highways.
4. Using current technology, direct dredge to the catchment area along the pipeline.
5. Once sediment is deposited, clear water will be released to Lake Pontchartrain.
6. Collected sediment will be used to build marais edges (similar to levees).
7. Continuous collection will allow areas of town to be infilled within the marais and elevate large or small areas to be determined through further study.
8. Resulting geospatial characteristics are a more topographical diverse urban environment with areas of higher and lower altitude (hills and valleys).

The defining characteristics of New Orleans urban fabric are a vibrant street life denoted by porches, activated neutral ground (medians), and frequent impromptu parades (Figs. 16 and 17). Citizens of the city identify strongly with place as defined by neighborhoods. Over time, infrastructure projects including canals and highways have further defined and distinguished neighborhoods from one another (Fig. 18). Our project design seeks to both recognize the unique qualities of New Orleans city life as well as build connectivity through infrastructure projects that are inhabited, lively and gracious. Note the "Activated Flood Walls" found in Marais Edge Type 4 that provides space for markets and other traditional neutral ground activities (Fig. 18).

Fig. 17 An impromptu parade about to begin on N. Rampart Street in the French Quarter of New Orleans (Credit: Giovanni Coakley)

Fig. 18 The numerous drainage canals throughout the suburb of Metarie are a fact of life in these quiet residential neighborhoods (Credit: Christopher James)

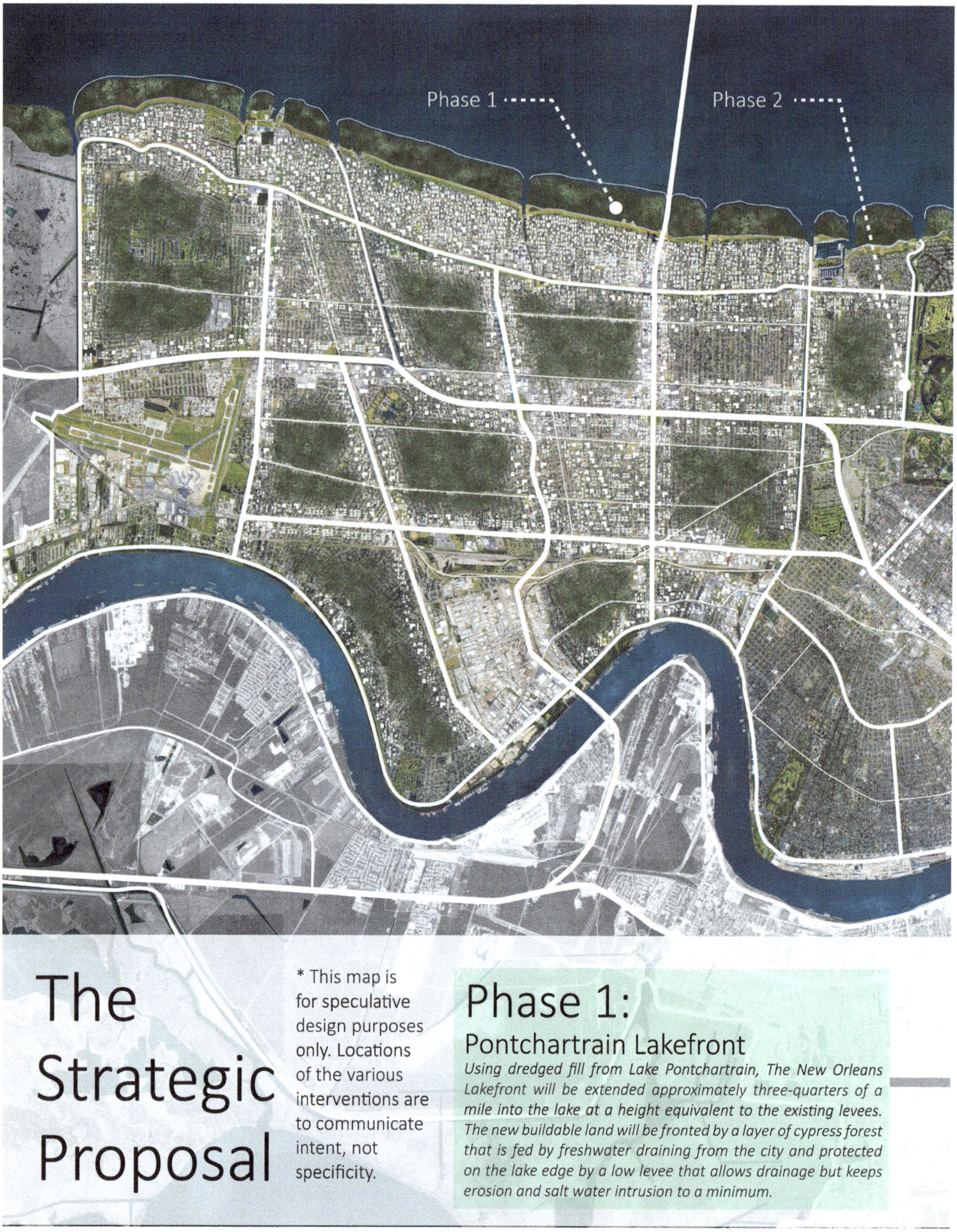

Design Spread 1 The strategic proposal

Design Spread 1 (continued)

PHASE 1: *New Pontchartrain Lakefront*

The first phase of the design calls for a re-imagined lakefront for the city of New Orleans. Instead of a hard, leveed edge prone to storm surge, the new Pontchartrain Lakefront consists of layers that break down those threats- the first as a new cypress forest that stretches the entire length of the city, and the second as reclaimed land that is elevated to existing levee height, running parallel to the Lake Vista Neighborhood which was built in a similar way.

This new high ground will be an invaluable asset to the city that encourages denser, more sustainable settlement on the high ground, and the re-introduction of the cypress forest habitat which is vital to the Louisiana coastal ecosystem. The Lakefront will also accommodate the relocation and temporary housing for populations displaced by the land-building and retrofitting efforts proposed by the later phases of the design.

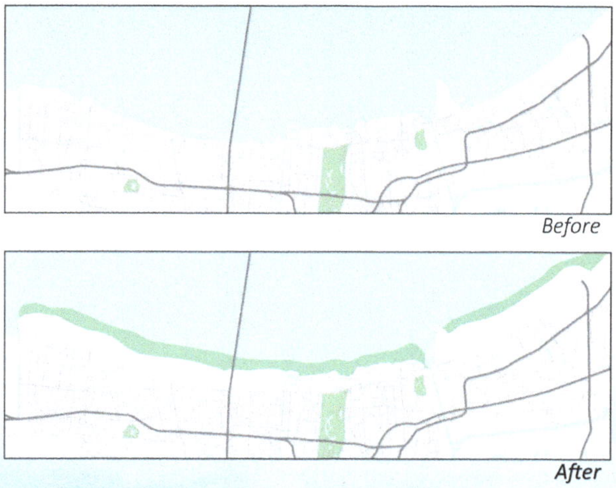

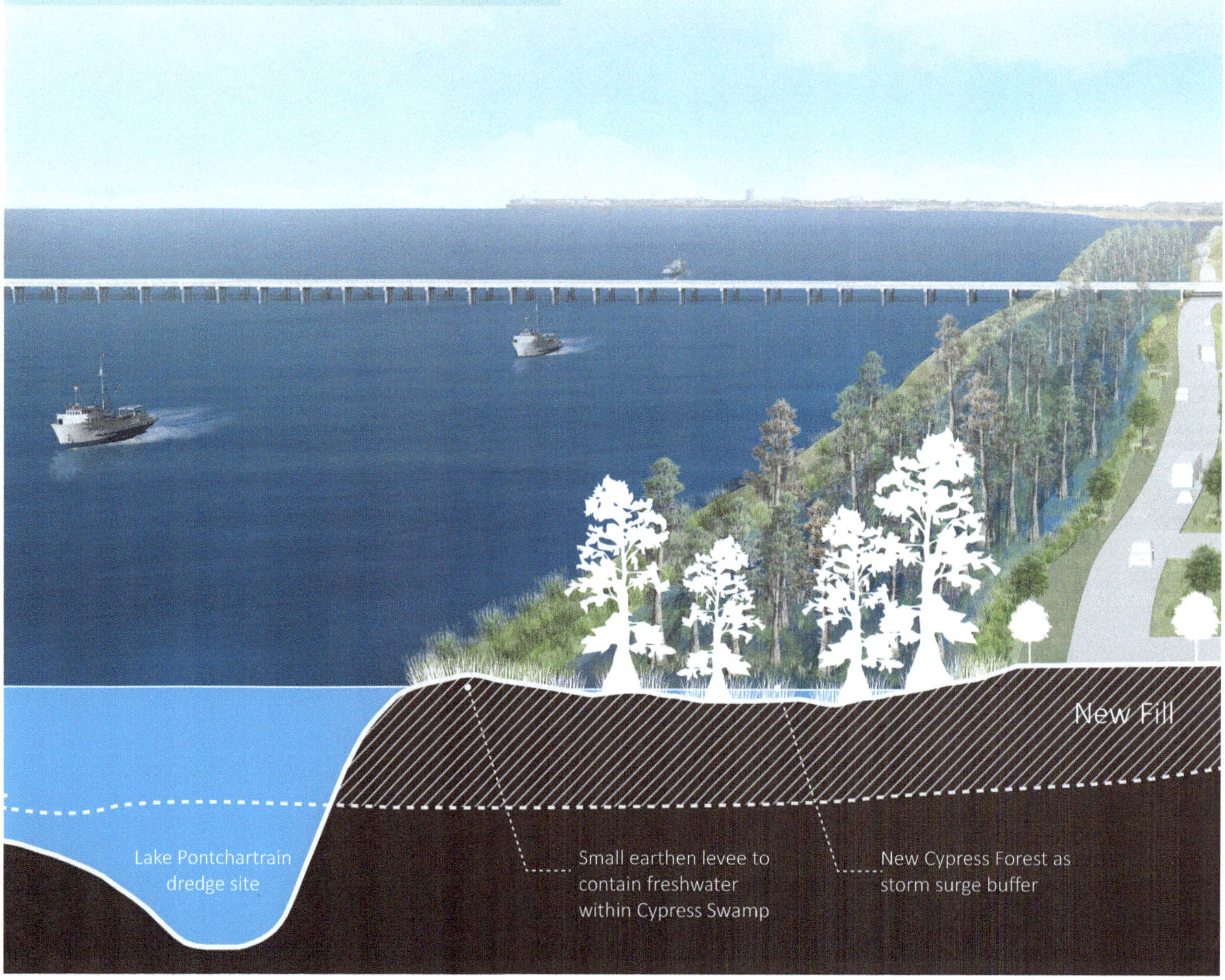

Design Spread 2 Phase 1 – New Pontchartrain Lakefront

Case Study: Lakefront Development

The Lakeshore and Lake Vista subdivisions were created along the shore of Lake Pontchartrain as part of a land reclamation strategy by the Orleans Levee Board in the early 1900's. Previously a swamp with unbuildable land, the project was initiated as a way to accommodate the growing city as well as experiment with the Garden City ideas that were popular in England and the US. The work began with a seawall and then continued as a levee project under the WPA. Although the elevation of the neighborhood is right at sea level, the land drains into lower lying areas inland from the constructed land. Amazingly, the Lakefront area did not flood during the Post-Katrina disaster.

Sediment Deposition Toolbox

Area of New Land: *9.4 square miles*

Area of New Cypress Forest: *12.2 square miles*

Height to Be Raised: *approx 20 feet*

Sediment Volume Required: *444 million cubic yards* *(approx 99 Louisiana Superdomes)*

Source of Sediment: *Lake Pontchartrain Dredge material*

Approximate Cost (per sq. mile): $$$

Notes: **Louisiana Superdome volume: 4.5 million cubic yards*
- Height of new developed land raised to existing levee height.
- Cypress forest raised to just below sea level.

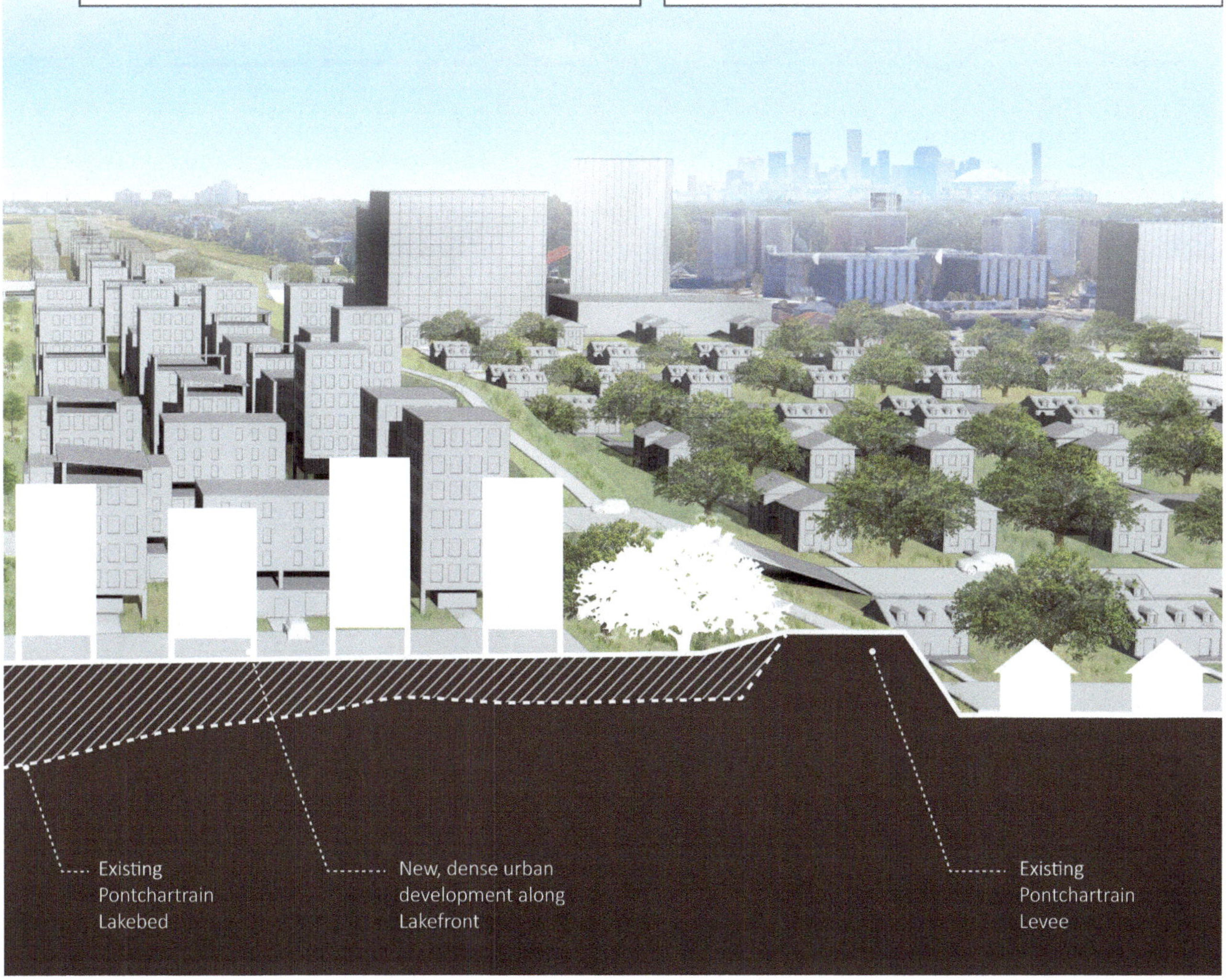

Design Spread 2 (continued)

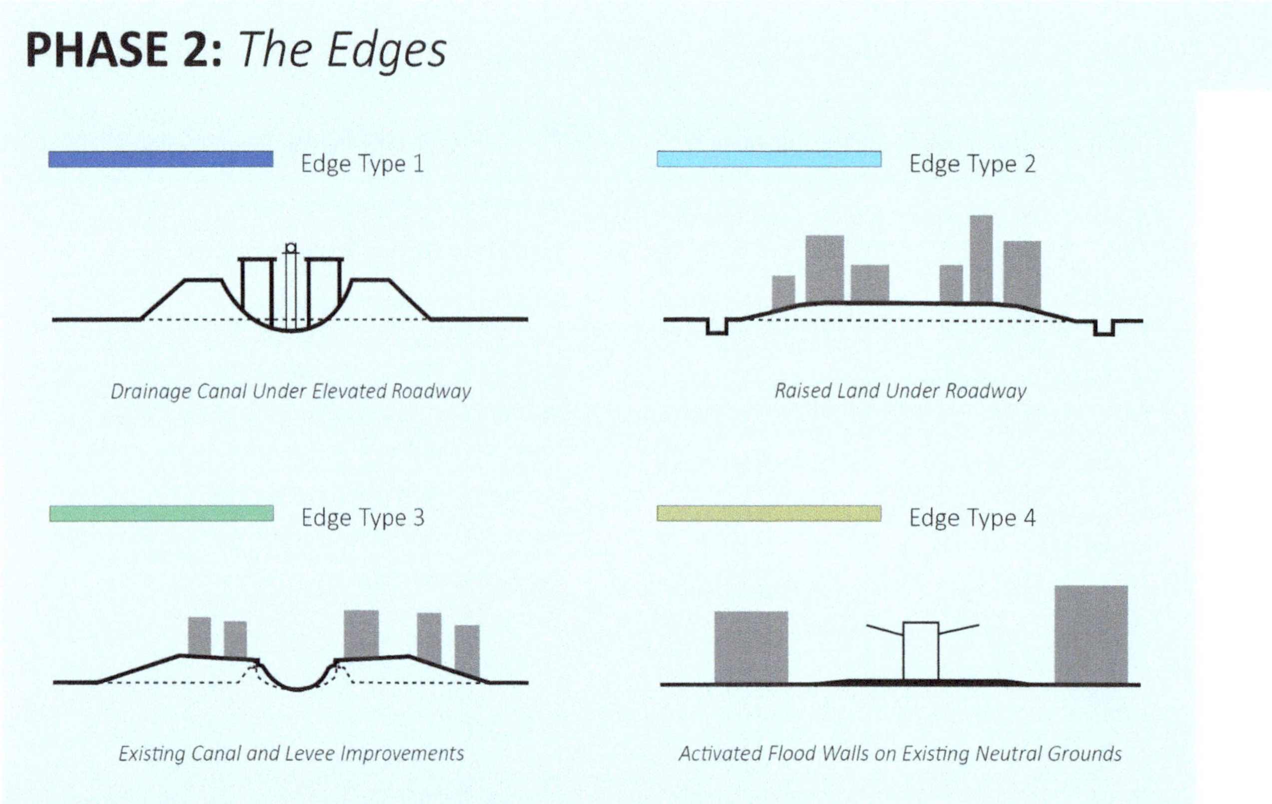

Design Spread 3 Phase 2 – Edge typologies, edge construction

Edge Construction

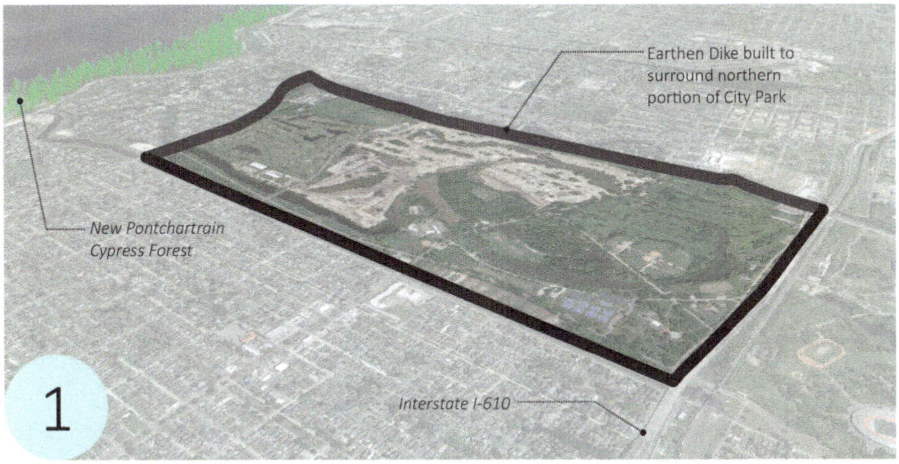

Build:

In order to construct the raised edges throughout New Orleans, a large amount of fill material is required to elevate the land. Using City Park as an example, various locations within the city that are sparsely developed will be designated as "sediment catch basins". In preparation to receive sediment pumped from the Mississippi River, levees will be built around the perimeter to contain the dredge material and demarcate its extents.

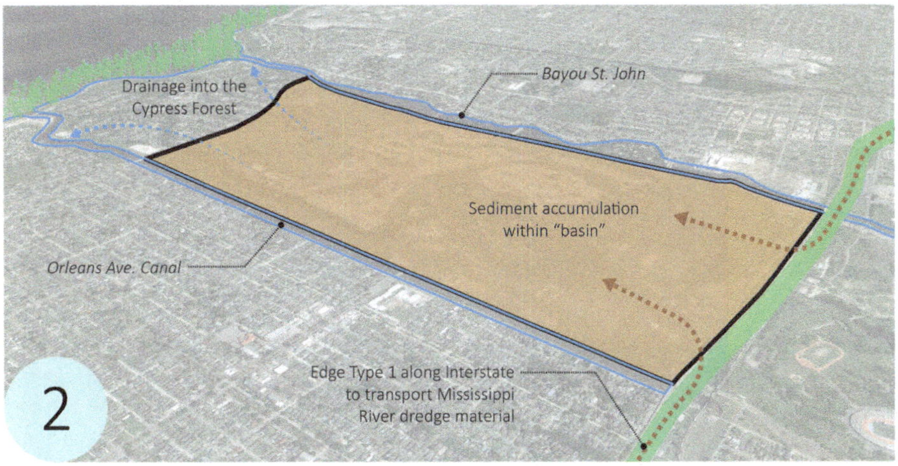

Fill:

After the levees are built, sediment will be pumped from the Mississippi River along Edge Type 1 into the northern portion of City Park. Over time, sediment will accumulate within the "basin" while the excess water will drain into the adjacent bodies of water (Bayou St. John and The Orleans Ave. Canal). This influx of freshwater from the River will then flow into the new cypress forest along the Pontchartrain lakefront.

Distribute:

As sediment accumulates in the basin over time from the continuous pumping of dredge from the river, the material can now be distributed via truck throughout the city to the various edges that are to be raised. This elaborate process of acquiring fill material essentially replicates the natural process of deltaic sediment deposition in a highly engineered way, harnessing the vast amounts of sediment in the river to combat flooding within New Orleans.

Design Spread 3 (continued)

Edge Type 1 *Drainage Canal Under Elevated Roadway*

The first elevated edge type becomes the means of transport of dredge material from the Mississippi River to land-building projects within the city. A large pipe carrying the sediment is attached to the median space between the lanes of traffic of the elevated interstate highways in New Orleans. The sediment is first transported to the deposition area located in the city such as the northern portion of City Park. Once deposited, the fill material is used to build the other edge types as well as elevating the access roads on either side of the interstate. After the "dikes" are built, sediment carried along this first edge type is then pumped to the interior marais of the city, building land over time.

Underneath the elevated roadway and sediment transport system is a drainage canal that aids in collecting water in low-lying areas of the city as well as creating a new, aqueous recreational space previously uninhabitable for the city.

Sediment Deposition Toolbox
Length of Edge Type (total): 37.8 miles
Width of Edge Type (typ.): 300 feet
Height to be Raised (typ.): 20 feet
Sediment Volume Required: 1.1 million cubic yards per mile 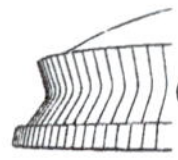 *(approx 0.25 Louisiana Superdomes per mile)*
Source of Sediment: sediment deposition areas
Approximate Cost (per sq. mile): $$$
*Notes: *Louisiana Superdome volume: 4.5 million cubic yards - Sediment utilized to raise access roads on either side of the interstate, containing the drainage canal within*

Design Spread 4 Phase 2 – Edge type 1, edge type 2

Edge Type 2 *Raised Land Under Roadway*

The second edge type, which would occur primarily in low-density suburban areas, raises large arterial roadways on top of a gradually sloping ridge, approximately thirty feet above existing grade. These new linear tracts of high ground then become focused areas of residential and commercial development. Also, construction of public infrastructures such as transportation and utilities will occur along these edges out of the reach of flooding. This is the most sediment-intensive condition where elevated ground is being created in low lying areas

This method of creating linear ridges of high ground within the city will improve resilience by encouraging continued habitation and settlement in areas above flood levels. Edge Type 2 maintains the urban morphology of the city with commercial corridors occurring along broad avenues and residential development within.

Sediment Deposition Toolbox
Length of Edge Type (total): *41.5 miles*
Width of Edge Type (typ.): *1200 feet*
Height to be Raised (typ.): *30 feet*
Sediment Volume Required: *7 million cubic yards per mile* *(approx 1.5 Louisiana Superdomes per mile)*
Source of Sediment: *sediment deposition areas*
Approximate Cost (per sq. mile): $$$$
Notes: **Louisiana Superdome volume: 4.5 million cubic yards* - *Exact height and width of edge depends on the elevation of the existing grade and the proposed land use of the new, elevated land.*

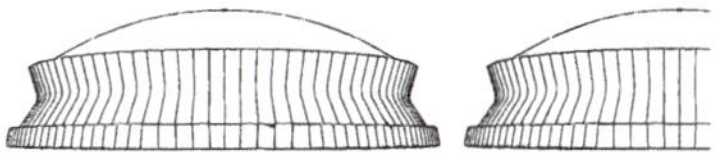

Design Spread 4 (continued)

Edge Type 3 *Existing Canal and Levee Improvements*

Along existing levees and outfall canals within the city, dredge material from the Mississippi River will be used to build up the ground behind the levee, creating an elevated tract of land that gradually slopes back to existing grade. The land will be raised to existing flood control height, essentially becoming a wide, inhabitable levee. Like the second edge type, this new high ground will present the opportunity for dense, sustainable development.

This edge type also proposes a new relationship with waterways both within and surrounding New Orleans. Currently, the city's relationship to water is agnostic, given the very real and pressing threat of flooding. By raising the land adjacent to the canal, water can be eliminated through gravity drainage, and people can access the canals without penetrating a wall. The opportunities for creating valuable urban space along the water are promising.

Sediment Deposition Toolbox
Length of Edge Type (total): 114 miles
Width of Edge Type (typ.): 600 feet
Height to be Raised (typ.): 20 feet
Sediment Volume Required: 2.3 million cubic yards per mile 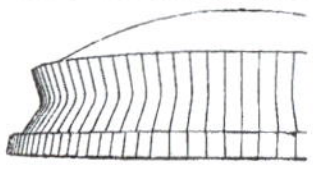 *(approx 0.5 Louisiana Superdomes per mile)*
Source of Sediment: sediment deposition areas
Approximate Cost (per sq. mile): $$
*Notes: *Louisiana Superdome volume: 4.5 million cubic yards - Existing levees will need to be analyzed and updated prior to the addition of fill on their back face.*

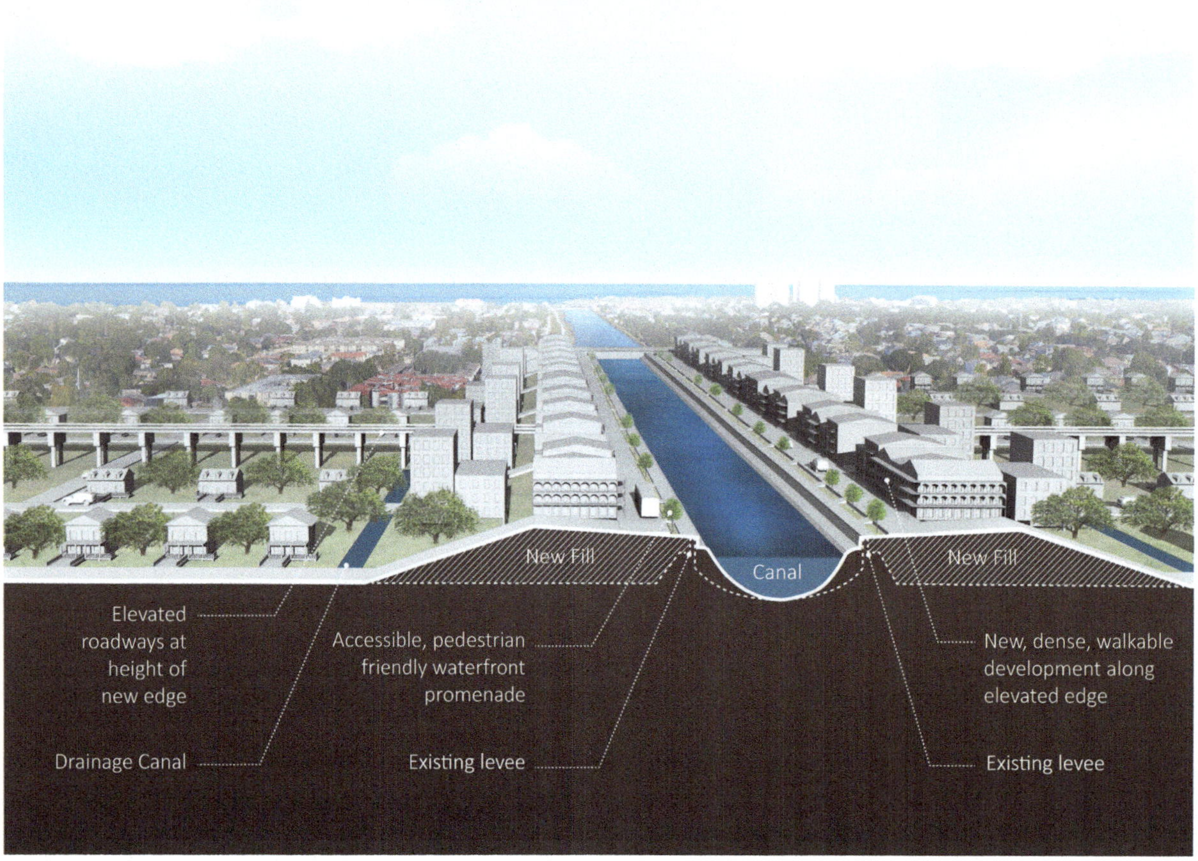

Design Spread 5 Phase 2 – Edge type 3, edge type 4

Edge Type 4 *Activated Flood Walls on Existing Neutral Grounds*

In sensitive historic districts where large-scale land-building is prohibitive, an architectural flood wall will be constructed to ensure the integrity of the marais that it is framing. The "neutral ground" will be elevated on flood walls. "Neutral grounds" are the medians that run along the many boulevards through the city and have strong social importance for New Orleanians who frequently use the neutral grounds to celebrate, parade and gather. Edge Type 4 maintains the social importance of the neutral ground, while elevating the roadways above flooding.

The design of the wall will vary along its length, ranging from inhabitable trails and green space on the elevated neutral ground, to permeable, movable doors and gates along the wall at ground level. The architecture and landscaping of the wall will become a new public amenity for the city while simultaneously compartmentalizing flooding risk.

Compared to the other edge types, this strategy is the least sediment intensive, since much of the wall will be constructed primarily of concrete. This type disrupts the least amount of total land area during its construction.

Sediment Deposition Toolbox
Length of Edge Type (total): 10 miles
Width of Edge Type (typ.): 60 feet
Height to be Raised (typ.): 20 feet
Sediment Volume Required: 250,000 cubic yards per mile  *(approx 0.05 Louisiana Superdomes per mile)*
Source of Sediment: sediment deposition areas
Approximate Cost (per sq. mile): $
*Notes: *Louisiana Superdome volume: 4.5 million cubic yards - Sediment will be utilized strategically along the length of the walls to create berms and landscaping for public use.*

Design Spread 5 (continued)

Marais Type 1: *Filled*

This marais type is suitable for sparsely populated districts where relocation is manageable and feasible. After all residents have moved out of the area, similar to a controlled spillway, Mississippi River sediment will freely flow into the enclosing marais, accumulating land over time. Once the ground has reached the height of the surrounding elevated edges, it may be resettled.

Land Development Considerations

Marais Type 1 requires complete agreement from the residents that they will relocate to the edges of the infill area in order for the sediment to completely fill the interior of the site. This type must involve strong negotiation with residents. Legally binding agreements with the state and the city must protect the rights of the residents to return to their now elevated property following the fill time frame.

1

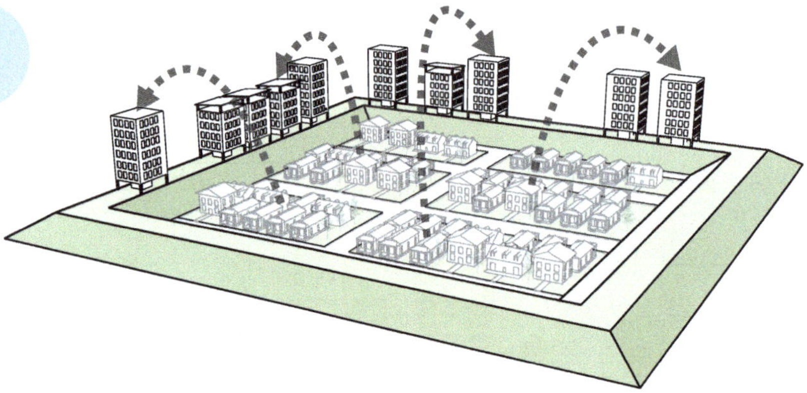

Relocation:

Over time, residents of the selected marais will relocate to new residential developments either along the elevated edges or the new Pontchartrain Lakefront. If possible, homes may also be physically transported to higher ground. Community engaged planning and funding mechanisms must be in place to facilitate the move.

2

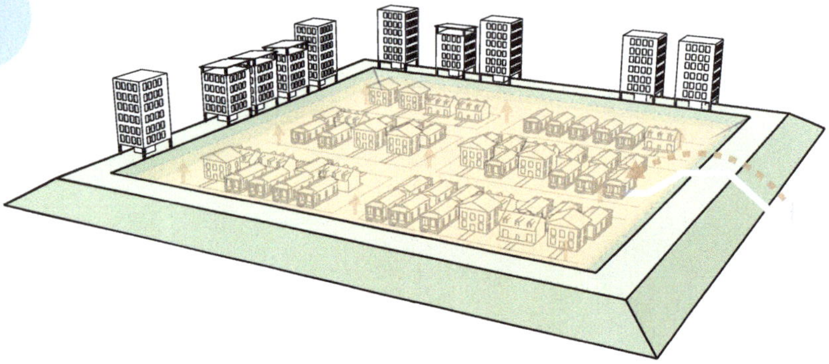

Fill:

Mississippi River dredge material will be pumped through the pipes along edge type 1 to the marais being retrofitted. This allows sediment to accumulate in the low ground over time, eventually raising the entire portion of the city above sea level, even with the surrounding edges. Of all marais types, this strategy is most sediment intensive.

3

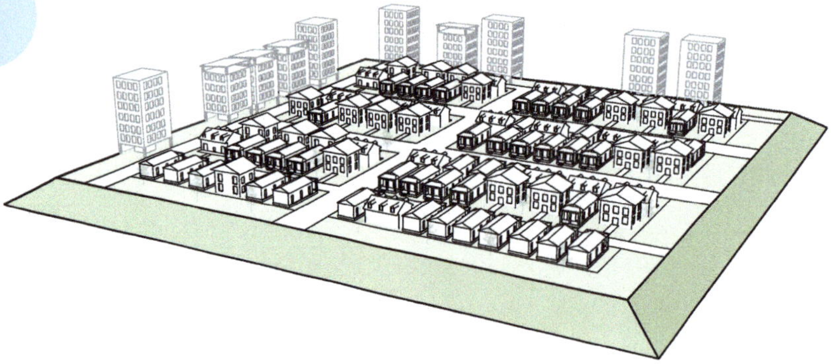

Resettlement:

After the sediment has settled and the water receded, the marais is now ready for re-population. People who moved away will have first rights to the land and assurances that they will not be priced out of the market, given the new land's increased value since it is now above sea level.

Design Spread 6 Phase 3 – Marais type 1, marais type 2

Marais Type 2: *Transition or Preservation*

The second type of marais essentially leaves the existing conditions as is. In some cases, the interior area will be best suited to returning to a natural state - reverting to the marsh and swamp that existed before human settlement. However, some conditions in areas like the French Quarter and the Central Business District, are too valuable or too expensive to retrofit or relocate, so those areas of the city will be preserved without many changes.

Land Development Considerations

By allowing land to continue subsiding, there is inevitable deterioration for inhabitants in less historically significant areas of the city. These residents should be provided with viable and affordable options for relocation. Historic areas and the Central Business District are more stable and less susceptible to subsidence in addition to providing economic assets for the city, thus justifying their protection.

A

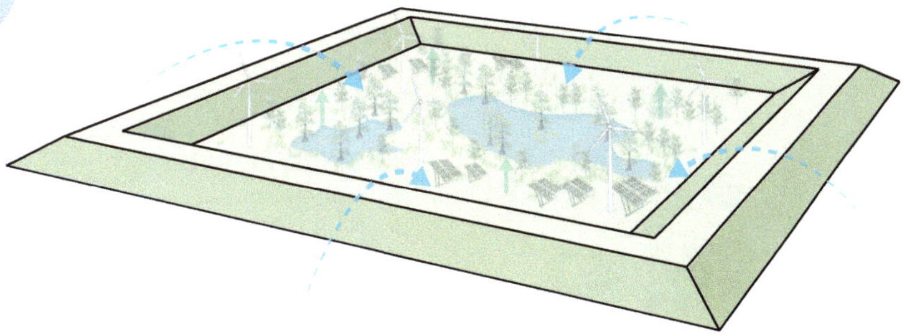

Return to Natural State:

Shrinking the overall inhabitable footprint of the city, these marais will bring greenspace into the urban fabric. They can serve multiple functions, including habitat creation and preservation, water drainage and storage from other populated marais, and the opportunity for new industries such as forestry, aquaculture, and renewable energy creation through solar and wind power.

OR

B

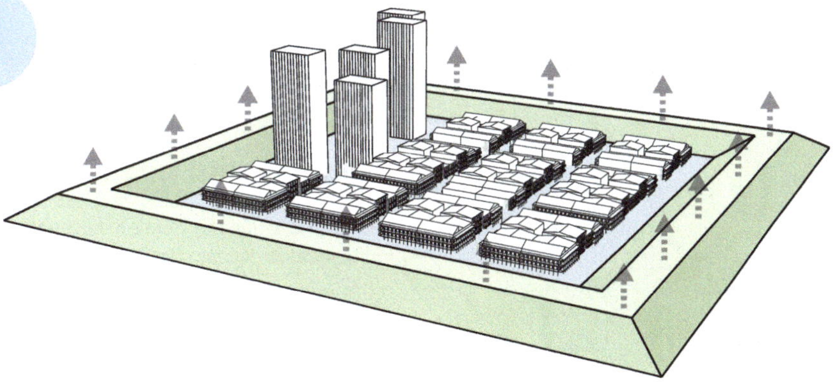

Preserve Existing:

Certain portions of the city of New Orleans are already on higher ground, including historically sensitive areas such as the French Quarter. Retrofitting this marais type will consist of improving/maintaining existing flood management systems to preserve the existing built environment within.

Design Spread 6 (continued)

Marais Type 3: *Phased Retrofit*

The most common strategy for the marais is partial fill and partial retrofit. Existing buildings will be elevated to a height consistent with the surrounding elevated edges and connected through a series of elevated roadways or better maintained ground level roadways. Other portions of the marais will be elevated with Mississippi River dredge material in a similar fashion to the edges. Lastly, remaining low-lying land will become areas for water collection, drainage, and reforestation.

Land Development Considerations

In this hybrid strategy, a mixture of elevation and relocation will largely be driven by individual property owners and neighborhood associations. In this typology, infrastructure improvements will assist in allowing residents to continue inhabitation of their neighborhood and property.

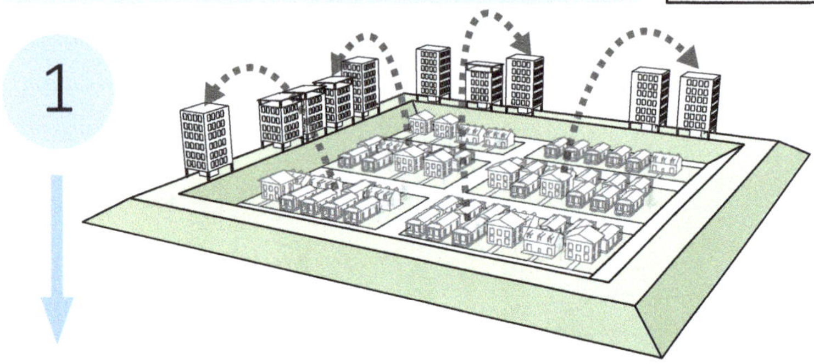

Relocation:

Through community involvement, residents of a particular marais can choose this final marais type as the retrofit-strategy of choice. The first step is for those who are willing and able to relocate their homes and businesses to the high ground of the surrounding edges or the new Pontchartrain Lakefront. Funding mechanisms must be in place to facilitate the move.

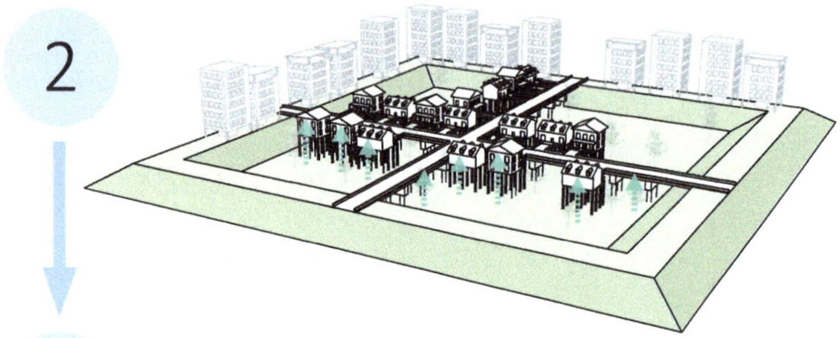

Retrofit/Elevate:

For those who decide to not relocate from their property, they will be encouraged to elevate their homes and buildings above flood levels. Public infrastructures such as roadways will also be elevated, connecting these newly elevated properties above the flood waters. Architecturally, a new, constructed ground plane is created, making elevated housing more spatially humane.

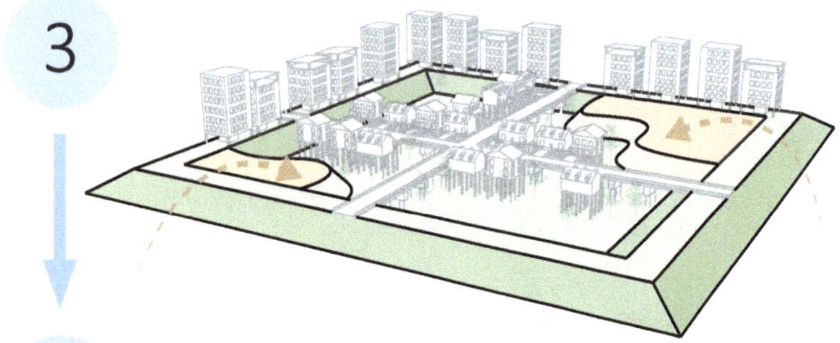

Selective Land-Building/Islands:

After residents have either relocated to the edges or elevated their homes, portions of the marais can be elevated in a similar way to Marais Type 1 - utilizing Mississippi River dredge material transported along Edge Type 1. This will in turn create a series of islands within the marais for people to resettle, with preference given to valuable civic and public institutions.

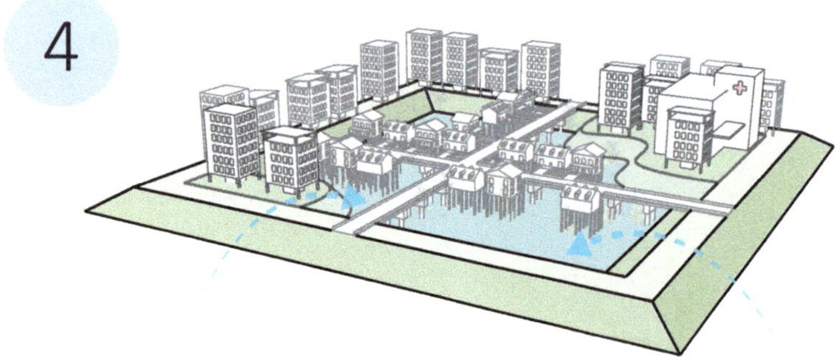

Resettlement/Flood Management:

The combination of land building and elevating will create within the marais a patchwork of flood mitigation strategies. The portions of the land that remain low-lying can become areas for water drainage and greenspace while the the denser, more populated areas will be concentrated on the high ground of the edges and the newly built "islands".

Design Spread 7 Phase 3 – Marais type 3; construction timeline

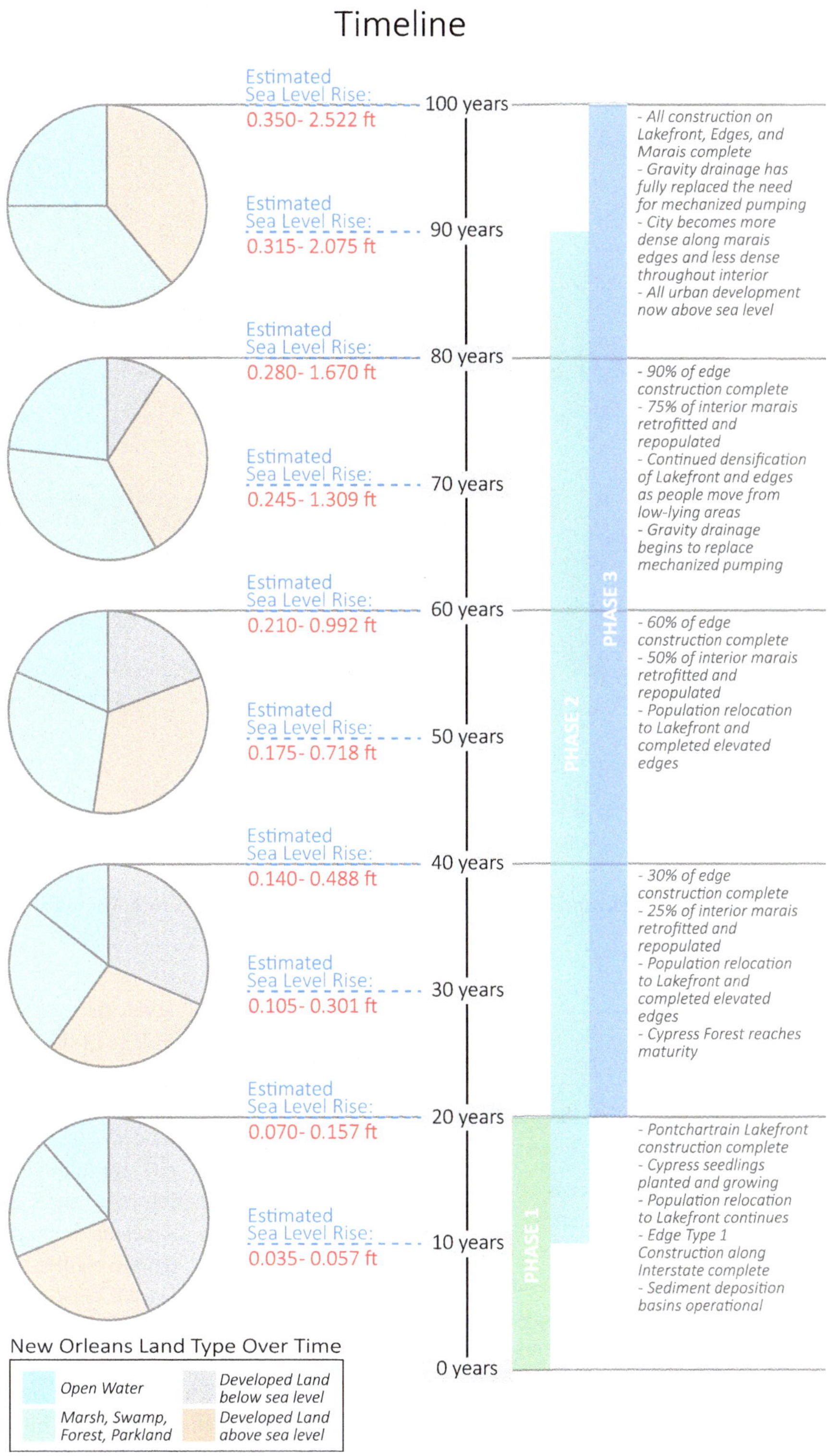

Design Spread 7 (continued)

Fig. 19 A rendering of the new Lake Pontchartrain Lakefront based on the design proposal (Credit: Giovanni Coakley)

Conclusions

Why not go out on a limb? That's where the fruit is. – Will Rogers

By working through a design process that included disciplines of engineering, social sciences and coastal sciences, as well as designers, we were able to produce a proposal that is more holistic and well-informed. The strategies and tactics we developed for New Orleans are useful provocations for other deltaic cities around the world. Only by having a group with such varied expertise, could we have developed a proposal that considers the natural, human and infrastructural systems as codependent and mutually reinforcing. One of the less tangible but significant outcomes of this project, is the demonstrable value of engaging interdisciplinary teams to address issues of climate change and sea level rise as we move into the future. Through the critical lens of the architectural design process, observations became structures, forms and strategies for adaption and risk reduction for coastal inhabitants.

This project provided an exciting opportunity for to provide innovative and speculative ideas for continued and evolving inhabitation of a fragile environment. The dynamic landscape of the Louisiana delta, particularly areas on the far reaches of the deltaic outlet, demand an architectural endeavor that defies traditional architectural design processes. While some aspects of the project may seem to be unrealistic, it is also unrealistic to think that New Orleans will continue to be a viable city without such large-scale interventions.

From the perspective of architecture, we strongly believe that the risk of engaging in novel projects has the potential for far greater value gain than the prospect of maintaining a more conventional path. Through this project, our understanding of

Fig. 19 (continued)

how architecture performs in the natural environment has evolved. Architects can no longer assume that our role is limited to space planning and basic enclosure within a static and relatively unchanging environment. Buildings and cities are part of the ecosystem and many disciplines need to work together if we want to continue our trajectory towards a more just and sustainable global environment.

Our project, Raising New Orleans: The Marais Design Strategy employs New Orleans as a model to propose dramatically rethinking approaches to sea level rise in deltaic conditions (Fig. 19). This design proposal may seem radical but we believe that this is the best strategy for continued inhabitation of New Orleans. In a pre-peak oil world, we may still have the energy stores necessary to elevate a city such as New Orleans. A ready supply of sediment combined with a retooling of current infrastructure can be applied to assist in adapting to sea level rise and the impact of flooding. If action is not taken soon, the residents of New Orleans and other coastal areas will continue to fight a battle with water that is unwinnable because walls and levees require constant monitoring and inevitably fail. Retreat will be expensive and difficult to accomplish. Our studies demonstrate an alternative that will also be difficult and expensive, but in the long run, offers a way for the city to survive, mostly intact, and preserves much of what is valued about the culture and atmosphere of New Orleans.

Coastal settlements around the world are being challenged by climate change and sea level rise. Ultimately, many will opt to retreat from current settlements and relocate further inland. Urban settlements pose a particular problem because the magnitude of population, economy and infrastructure that are effected would be nearly impossible to replicate in a wholesale fashion. By bringing together interdisciplinary teams to address these environments and issues, we will arrive at more robust, complex and ultimately, more successful adaptive solutions to the challenges of a warming planet.

References

DeRussy LG (1859) 1796–1864: Special report relative to the cost of draining the swampy lands bordering on Lake Pontchartrain. J.M. Taylor, State Printer, 1859, Baton Rouge. Also by Louisiana. Board of Swamp Land Commissioners (page images at HathiTrust) https://catalog.hathitrust.org/Record/100410097

Napier HE (1847) Florentine history, from the earliest authentic records to the accession of Ferdinand III. Grand Duc of Tuscany, vol 5. Moxon, 1847

Publication of the World Health Organization (1991) Surface water drainage for low-income communities. Published online. https://www.ircwash.org/resources/surface-water-drainage-low-income-communities. Accessed March 2017

Eroding Communities and Diverting Populations: Historical Population Dynamics in Coastal Louisiana

Scott A. Hemmerling

Abstract

The traditional notion of inherent resilience as a continual functioning of community during "non-crisis" periods does not apply perfectly to coastal Louisiana. This research suggests that operating in a non-crisis mode in the face of the omnipresent threat of coastal land-loss is counterproductive and that inherent resilience must necessarily incorporate adaptive resilience factors if coastal communities are to persist and thrive in a changing coastal environment. Resilience includes both the system's capacity to return to the state that existed before a disturbance and the system's ability to advance the state through learning and adaptation. Put simply, a community that persists is stable, while one that persists and adapts is resilient. A community that has withstood many storm events over several years and maintained its population has shown itself to be stable. But with each successive shock, that community may begin to see a coincident erosion of community resilience. This could ultimately cumulate in a situation where a historically stable population that has withstood numerous large storm events begins to decline precipitously when impacted by smaller shocks or events. The key to successfully navigating coastal change is to understand the multiscale historical processes that have been driving that change and assuring that any regime shifts result in positive outcomes for residents and the ecosystems that they rely upon in both the short term and long term. New Orleans and its surrounding region is a prime example of a population that may have been deemed stable in the years following the hurricanes of the 1960s while at the same time becoming less resilient. This dichotomy highlights the importance of a deeper understanding of community resilience, an understanding that acknowledges resilience as an historical process, one with a preexisting trajectory that will influence the ways in which communities respond to disasters and slow moving environmental change.

Keywords

Inherent resilience • Historical resilience • Coastal communities • Population change

Introduction

Residents of coastal Louisiana have often been described as being a resilient people who have endured numerous natural and technological hazards and yet persist, firmly anchored to the landscape by long held social networks and the resource-based economies upon which they depend for their livelihoods (Colten et al. 2017). Another narrative holds

S.A. Hemmerling (✉)
The Water Institute of the Gulf, Baton Rouge, LA, USA
e-mail: shemmerling@thewaterinstitute.org

J.W. Day, J.A. Erdman (eds.), *Mississippi Delta Restoration*,
Estuaries of the World, https://doi.org/10.1007/978-3-319-65663-2_12

that residents of Louisiana are far less resilient than suggested, with many residing in areas highly vulnerable to both coastal and riverine flooding. In many cases, residents have foregone the traditional elevated housing types for lower cost slab on-grade ranch housing that became increasingly common in American suburbs in the post-World War II era. At the same time, social processes have segregated racial and ethnic groups from one another, particularly in the urban centers such as New Orleans, but also in smaller rural communities throughout the region. This dichotomy highlights the importance of a deeper understanding of community resilience, an understanding that acknowledges resilience as an historical process, one with a preexisting trajectory that will influence the ways in which communities respond to disasters and slow moving environmental change. All too often, researchers treat resilience as though it were a temporally static phenomenon, paying scant attention to the historical processes that have built and sustained individual communities through time. Most measures of resilience do not get to the heart of what makes one community more resilient than another. They do not capture the details of why certain communities thrive and flourish, while others, statistically similar in many ways, fail.

Population change has long been a key component of community resilience as residents adapt to changing environmental, economic, and social conditions by either remaining rooted in place or relocating to new places (Bailey et al. 2014). The loss of population often begins long before people migrate away from a region. Communities without the means to effectively adapt to changing conditions begin to lose their sense of wellbeing and security and often their sense of place itself (Adger et al. 2002). Ultimately, this loss of community can trigger migration away from a place. Historically, when population migration has occurred, it has operated at multiple scales (Dalbom et al. 2014; Darlington and Woodell 2006). People have relocated to different neighborhoods within the same town or city, to different communities within the same region, or to entirely new regions, sometimes breaking long-held connections to place and community. While traditional ideas of resilience tend to view migration away from a place as a loss of resilience, such movement could potentially enhance overall regional resilience. In practice, however, the overall impacts of migration and movement of coastal residents have had varying effects on community resilience, with the set and setting of this migration often determining the ultimate outcome.

In Louisiana, population change and migration and community relocation have become increasingly important topics as people and places within the state's coastal zone are increasingly forced to contend with the cumulative impacts of decades or more of environmental degradation and the increased likelihood that climate change and sea level rise will place more and more residents at risk (Bailey et al. 2014). Residents who choose or are forced to remain in increasingly hazardous locales must find innovative ways to adapt. In coastal Louisiana, efforts to reduce levels of impervious surface, for example, through the development of greenways and raingardens have allowed some communities to reduce the impacts of some extreme weather and rainfall events (Hemmerling 2017). The degree to which communities are able to adapt or relocate, however, is often dependent on pre-existing socioeconomic conditions. Lack of income, lack of transport, age, gender, and minority status are all factors which may contribute to the ability of a community to relocate out of potentially hazardous environments. The effect of these demographic factors on migration is most likely to be seen through interaction with other drivers, particularly economic (Dalbom et al. 2014). Indeed, in many cases, the interaction of these factors have resulted in situations where the most vulnerable members of society have moved into locations of higher risk, compounding their vulnerability.

Defining and Measuring Resilience as a Factor in Population Change

There are inherent factors, developed through time that become built in to the "fabric" of a place, which allow certain communities to persist, despite the detrimental impacts of centuries of natural and technological change. Inherent resilience factors persist in the face of numerous threats and risks to community well-being. These factors may be economic. They may be social. They may be physical. Taken together, it is these factors which, when lost, indicate a potential loss of resilience and the decline of community. Resilience, as defined by Susan Cutter, is the ability of a community to respond to and recover from environmental and economic disturbances and includes those inherent conditions that allow the system to absorb impacts and cope with an event, as well as post-event, adaptive processes that facilitate the ability of the social system to re-organize, change, and learn in response to a threat (Cutter et al. 2008). Resilience determines the persistence of relationships within a system and is a measure of the ability of these systems to absorb changes of state variables, driving variables, and parameters, and still persist. Further, resilience can be thought of as a system's capacity to absorb disturbance and re-organize into a fully functioning system. It includes not only a system's capacity to return to the state (or multiple states) that existed before the disturbance, but also to advance the state through learning and adaptation (Holling 1973).

Cutter and others have suggested that community resilience has two qualities: inherent and adaptive (Cutter et al.

2008). Inherent resilience allows communities to function well during non-crisis periods while adaptive resilience is a measure of a community's flexibility in response to disasters. Both inherent and adaptive resilience can be applied to infrastructure, institutions, organizations, social systems, and economic systems. These systems, taken in their entirety, are components of what we can broadly think of as "community." The artificial bifurcation between inherent and adaptive operates on the assumption that disasters are short-term shock events. Certainly in coastal Louisiana, the increasing threat of extreme weather events, tropical and otherwise, are ever-present. Events such as hurricanes, extreme rainfall events, and river flooding are constant threats to communities living along the coast. Yet, this is only part of the story of community risk in coastal Louisiana. Underlying the annual and seasonal hazards to coastal communities is a more long-term, almost imperceptible threat. Coastal land loss is a crisis that has been occurring over several decades and, with continued relative sea level rise and loss of the states coastal wetlands, one that is expected to continue or worsen in the coming decades. This type of slow-burn crisis presents almost an existential threat to coastal communities. Residents are aware of this threat, but their ability to adapt may be hampered by uncertainty and a lack of immediacy, sometimes even a cultural heritage that ties them to these places. The traditional notion of inherent resilience as a continual functioning of community during "non-crisis" periods does not apply perfectly to coastal Louisiana. Indeed, it could be suggested that operating in a non-crisis mode in the face of the omnipresent threat of coastal land-loss is counterproductive, indeed self-destructive. Inherent resilience must necessarily incorporate adaptive resilience factors if coastal communities are to persist and thrive in the coastal environment.

It is important to realize that the degree of resilience, as well as physical risk and social vulnerability, possessed by communities in Louisiana's coastal zone is, in large part, a function a number of diverse but interrelated factors. First, the physical geography of coastal Louisiana serves as a limiting factor on settlement and land use. The highest land on Louisiana's deltaic plain lies atop the natural levees that border the Mississippi River and its distributaries. On the backslope of the levees, the soil elevation drops, the soil texture becomes finer, the amount of water and organic matter in the soil body increases, and the water table rises ever closer to the surface (Campanella 2010; Saucier 1962, 1994) Much of the population of the region resides atop the limited high ground on narrow threads of land reaching down towards the coast. In some cases, levees and flood control structures surround these communities, often resulting in development on the backslope of the natural levees.

Second, the economy and lifeways of coastal residents are often intricately tied to this inherently hazardous landscape. Louisiana possesses a wealth of natural resources, both renewable and nonrenewable, that requires a local workforce to extract and process. In fact, Louisiana has the greatest proportion of coastal employment of the five Gulf of Mexico states, with nearly 34% of the state's workers residing in coastal parishes (Adams et al. 2004). A large portion of this employment centers on specific coastal economy sectors, including oil and gas production, commercial and recreational fisheries, marine construction, ship and boatbuilding, tourism and recreation, and marine transportation (Kildow et al. 2014). These immovable industries form the backbone of Louisiana's coastal economy.

Importance of Understanding Historical Resilience and Population Dynamics

In many ways, communities in coastal Louisiana can be considered prototypical social-ecological systems, with a large proportion of coastal residents highly dependent upon both renewable and nonrenewable industries for their economic wellbeing. As a result, many of these communities face increased levels of vulnerability due to environmental variability, global economic pressures, and state interventions (Adger 2000). Such social-ecological systems are therefore highly dynamic, with internal vulnerabilities constantly shifting through time, being impacted by any number of discrete exogenous shocks and slow moving crisis events. Understanding how historical changes and processes have shaped present day social and ecological landscapes is vital to planning and managing efforts designed to enhance community and regional resilience.

Social-ecological systems exhibit a dynamic complexity that is not easily captured in either the engineering or ecological resilience models. Structural protection projects, for example, are engineered and designed to promote stability and to protect the status quo. But a social-ecological systems model rejects the notion of a consistent, stable state. Just as a community can never reach a state of perfect resilience, it can never be perfectly stable. Just as attempts to build a utopian social system are bound to fail, attempts by scientists and planners to build a perfectly resilient "utopian physical environment" are also bound to fail because, at the most basic level, they are attempting to achieve a stable state in the face of constant change. Communities exist in a state of flux, constantly striving for stability and attempting to build resilience, but given the dynamic complexity of socioeconomic systems, always subject to innumerable endogenous and exogenous shocks to the system.

Attempts to constantly return to some type of "optimal stable state" that is presumed to have existed prior to a major shock effect, may succeed in the short term. The community may appear to have rebounded to a point where it resembles the pre-event community on the surface. But, without adaptation, that optimal stable state can become less stable and may lose resilience with each successive shock. These shock effects, which may occur on many temporal or geographical scales, begin to compound to the point where a community may fail to rebound. A community that has withstood numerous large exogenous shocks through the years may fail when faced with a single small endogenous shock to the system. Similarly, a community that has coped with a long term, slow exogenous shock, such as coastal land loss, but not adapted to the changing conditions may begin to lose its inherent resilience in a slow but inevitable way. This community may be impacted by several large shocks from which it physically recovers, but without adaptation, that community is continually being eroded along with the ground it was built upon.

The specifics of community response to systems shocks, whether acute or chronic, can only be determined by the socioeconomic context of specific communities and the patterns of vulnerability generated by this context (Black et al. 2011). For example, while environmental change may result in the displacement and mobility of local populations, preexisting socioeconomic conditions and historical community resilience trajectories can determine which populations have the resources to relocate, where they relocate to, and the timing of that relocation. Conversely, preexisting conditions can also determine which residents have the resources to stay. Typically then, the wealthiest and the poorest residents often remain in communities impacted by acute and chronic shocks, although with significantly different levels of adaptation.

The question then becomes, how do we effectively incorporate temporal change into the existing models of community resilience? This is further complicated, as noted above, by the fact that coastal communities, particularly in Louisiana but most certainly in other subsiding deltaic regions as well (e.g., Day et al. 2016), face two main categories of disturbance. The first category of disturbance consists of those type of acute exogenous discrete events, or system shocks, such as extreme natural events and technological hazards, that have a clearly delineated beginning and ending, in terms of the event itself. System shocks cover a multitude of phenomena, from hurricanes, through ecosystem collapses, to abrupt climate changes. In coastal Louisiana, the most visible system shock events have been hurricanes. Coastal Louisiana has been the epicenter of hurricane strikes along the Gulf of Mexico for as long as hurricanes have been tracked. While the impacts of individual storm events are variable in their spatial impacts, over time, certain regions emerge as hotspots of risk (Fig. 1). Similarly, certain locations have historically experienced a higher rate of storm surge than other coastal locations (Fig. 2). For acute events like these to be experienced as a system shock, it must have some impact on human wellbeing. The impacts of these events can be direct via loss of life or property, or indirect via the social-ecological systems that people depend upon, such as agriculture or fisheries (Lenton 2013). While the system shock itself may have an endpoint, the impacts on community resilience cannot be so clearly delineated. The impact of a single shock on resilience, for example, could be measured in a single decade while it might take a generation to learn how a region changes as a result of repeated shocks (Pendall et al. 2007).

The second category of disturbance consists of endogenous slow-moving challenges or "slow burns." The best example of this in coastal Louisiana is coastal land loss, a continual, on-going threat to many coastal communities. According to the U.S. Geological Survey, coastal Louisiana lost over 4000 acres of land in last half of the twentieth century (Couvillion et al. 2011). The impacts of coastal land loss and other slow burn events, such as sea level rise or saltwater intrusion, are difficult to measure in the short term as change occurs almost imperceptibly, but the cumulative environmental impacts represent a significant deterioration of the environment. Because slow burn events are not discrete, with a clearly defined start and end point, they are much more likely to slowly corrode community resilience. With this type of event, more than a generation of observations might be necessary to understand the resilience response of communities to slow burn challenges (Pendall et al. 2007).

Often, with the quantification of resilience factors, the notion that community resilience is based on an understanding that the natural state of human systems, like any other biogeophysical system, is one of change rather than one of equilibrium is neglected (Nelson et al. 2007). The cumulative impacts of repeated system shock events, along with coastal land loss and a myriad of other environmental, social, demographic, and economic changes that occur over time, foreground the fact that resilience is not a static event. Rather, resilience must be considered a continuously evolving process rather than a static quality of communities. The level of community resilience present today represents but one point on a temporal spectrum of change. The level of resilience in communities is constantly changing based upon any number of endogenous or exogenous factors, sometime rapidly and sometimes slowly and imperceptibly. The type and magnitude of change is not always predictable, but change will occur. For this reason, human systems need to be managed for flexibility rather than for maintaining stability (Nelson et al. 2007).

Ultimately, if environmental management programs designed to protect coastal communities are to succeed, it

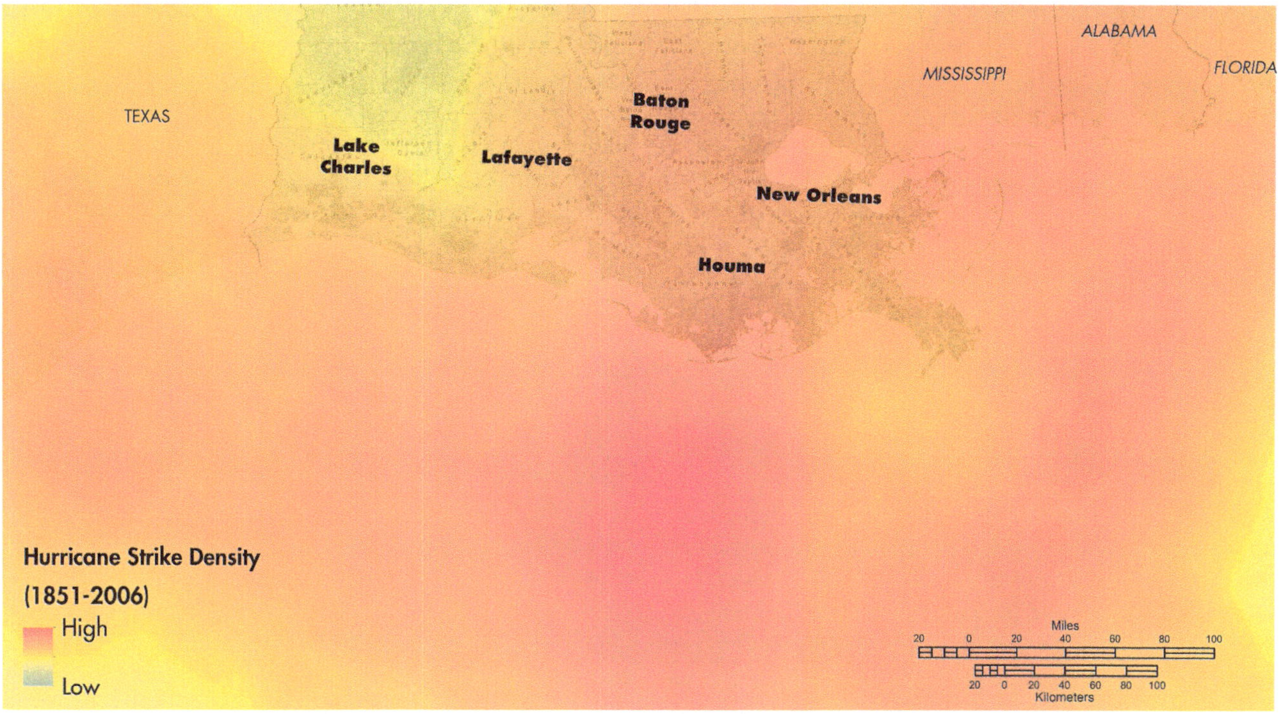

Fig. 1 Historical hurricane strike density in the Northern Gulf of Mexico (1851–2006)

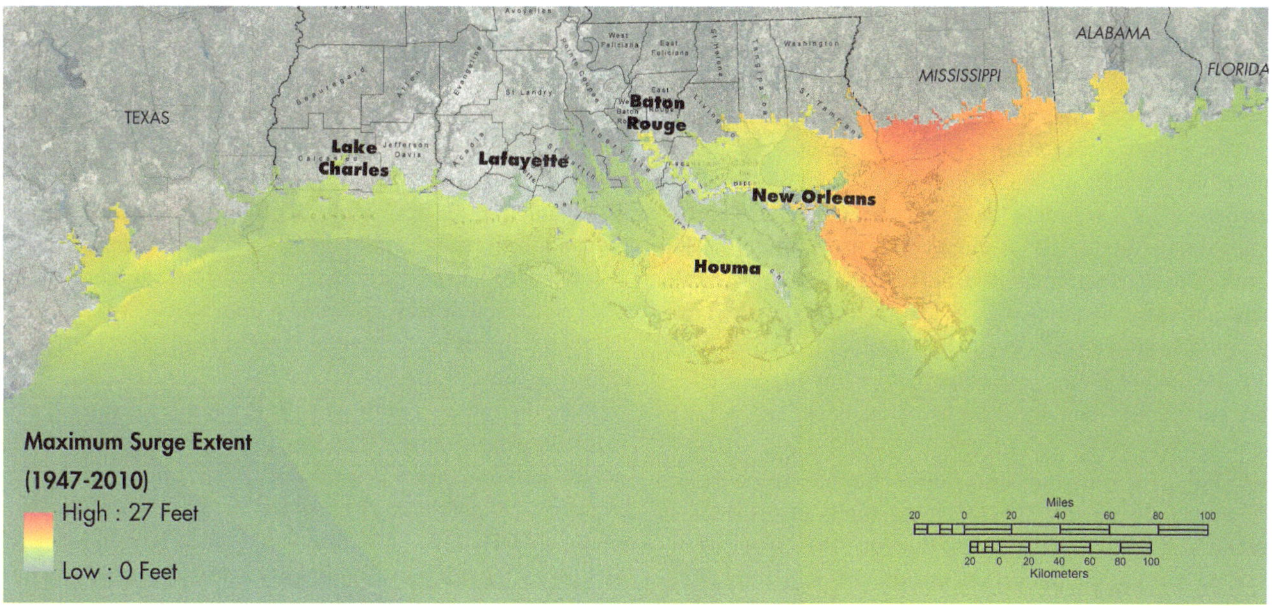

Fig. 2 Historical maximum recorded storm surge depth in the Northern Gulf of Mexico (1947–2010)

must be understood that resilience is, ultimately, an historical process. There is a momentum to community resilience, a trajectory that will either need to be enhanced or overcome for coastal management plans to succeed. Without understanding that resilience is an historical process, communities cannot fully anticipate multi-hazard threats, respond to hazards events, recover from disasters, or reduce vulnerability to hazard threats (Colten et al. 2008). Without accounting for preexisting environmental, social, demographic, and economic trends, communities may continue to become less resilient, despite any amount of development and protection provided to those communities.

Resilience and Recovery in Coastal Louisiana

Coastal restoration scientists and community planners are constantly faced with the dilemma of how best to balance community protection and preservation with the need to restore and protect the coastal landscape at large. For coastal restoration and protection plans to succeed, it is vital that planners understand not only the physical processes that helped form the present-day landscape, but the social processes as well. A clear understanding of these processes and the inherent community resilience factors that persist through time, despite numerous shocks and perturbations to the system, is vital to assuring that Louisiana's working coast remains socially and economically significant.

Many studies contend that the resilience of a community can be measured in the ability of that community to rebound from large scale disturbances. Often times, rebound is defined by the return of the population to the community. There is an inherent flaw in this notion. Even though population might return to pre-storm levels, the community may not necessarily function as a more resilient community. Repeated shocks to the system may erode the inherent resilience of a community. While the population may return to the community, that community may be weakened, making it less resilient.

According to Holling, there is a clear differentiation between the stability of a community and that community's level of resilience. Stability represents the ability of a system to return to an equilibrium state after a temporary disturbance. Resilience, on the other hand, is a measure of the persistence of systems and their ability to absorb disturbances and still maintain the same relationship between system variables and populations (Holling 1973). Conceptualized in this way, the return of population to a community after a storm event or other disaster is not necessarily an indicator that the community is resilient. Rather, that community can be thought of as being temporarily stable.

Clearly, these two measures of community response are not mutually exclusive. Indeed, a community that is stable is much more likely to be resilient than one that is not. However, we must be cautious when assuming that there is a one-to-one relationship between stability and resilience. A community that has withstood many storm events over several years and maintained its population has shown itself to be stable. But with each successive shock, that community may begin to see a coincident erosion of community resilience. This could ultimately cumulate in a situation where a historically stable population that has withstood numerous large storm events begins to decline precipitously when impacted by smaller shocks or events.

The key to building and maintaining community resilience then is the adaptive capacity of that community. Resilience includes both the system's capacity to return to the state that existed before a disturbance, but the system's ability to advance the state through learning and adaptation. In other words, resilience is the capacity of a community to respond and recover from disasters and includes those inherent characteristics that allow the system to absorb impacts and cope with an event, as well as those adaptive processes that facilitate the ability of the social system to re-organize, change, and learn in response to a threat (Cutter et al. 2008). Put simply, a community that persists is stable, while one that persists and adapts is resilient.

New Orleans and its surrounding region is a prime example of a population that may have been deemed stable in the years following the hurricanes of the 1960s while at the same time becoming less resilient. Based on outward appearances, it could be argued that Southeast Louisiana had effectively recovered from both Hurricane Betsy and Hurricane Camille in short time. Indeed, by 1970, the population of the region had largely bounced back, with most coastal parishes experiencing high levels of growth between 1960 and 1970 (Fig. 3). Most of this growth centered on the circum-Pontchartrain region, those parishes encircling Lake Pontchartrain, as well as the suburban parishes neighboring New Orleans. New Orleans itself, it should be noted, experienced a slight loss of population over this same time period, although the population continued to expand outward into areas such as New Orleans East, a broad expanse of reclaimed marshland transformed into suburban-style residences and subdivisions, located to the east of the historic center of New Orleans. In addition to an expanding population, total unemployment was at or above the state average across the region (Fig. 4) and home vacancy rates were among the lowest in the state (Fig. 5). By all measures, the region appeared to have recovered from the devastating storms of the 1960s.

Indeed, when contrasted with the recovery from Hurricanes Katrina and Rita (as well as the myriad smaller storms striking the region in the 2000s), the growth and recovery of southeast Louisiana in the 1960s appears remarkable. By 2012, when the U.S. Census Bureau released its five-year American Community Survey data on population and housing, the region appears almost as a reverse image of 1970 (Fig. 6). Orleans Parish and its immediate neighboring parishes on the south shore of Lake Pontchartrain (Jefferson, Plaquemines, and St. Bernard) all experienced significant sustained population loss. Also of note is that Cameron Parish in southwest Louisiana, which bore the brunt of Hurricane Rita, also experienced tremendous population loss over this time period. Unlike in 1970, several parishes in south Louisiana, including Orleans and

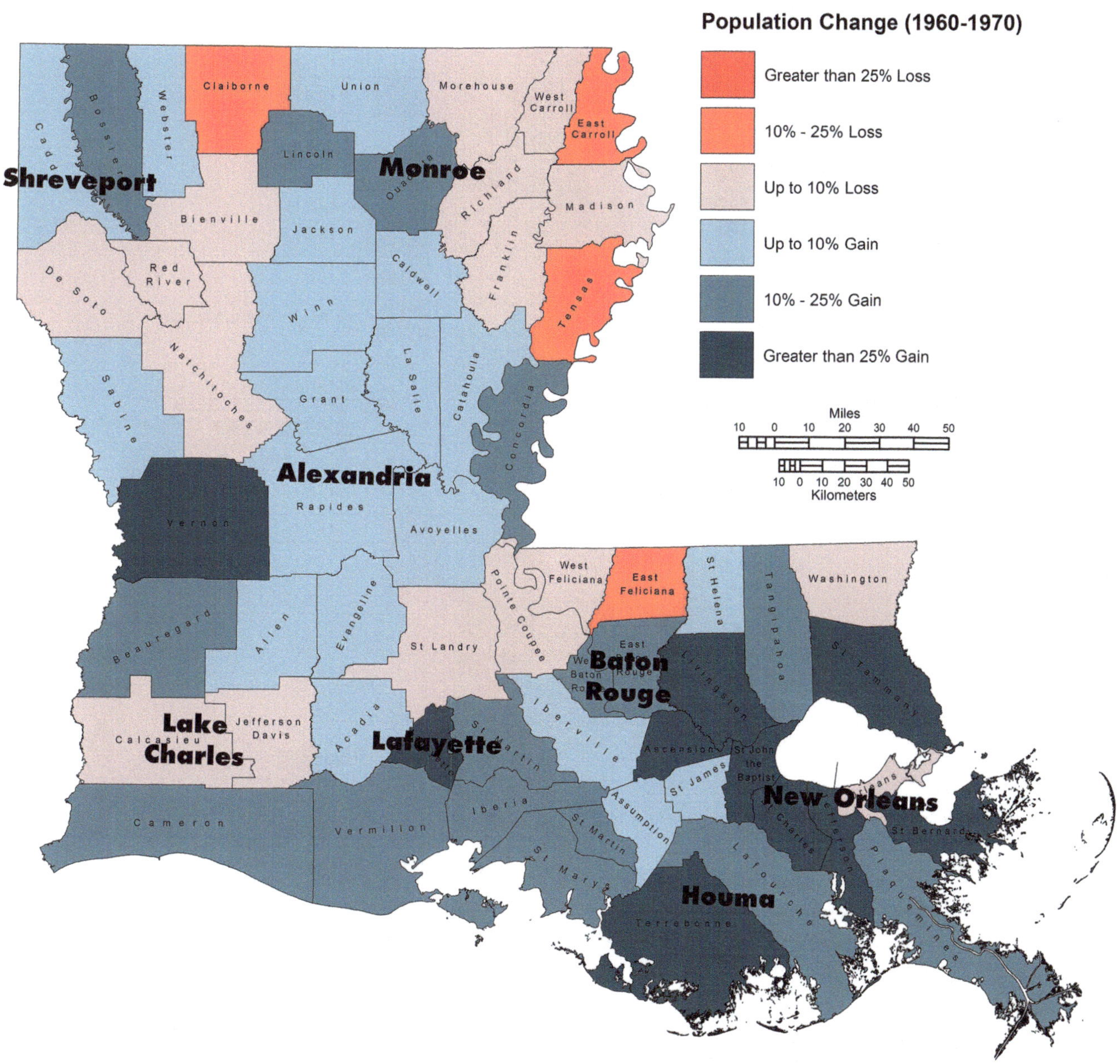

Fig. 3 Percent population change in Louisiana by county (1960–1970)

St. Bernard had levels of unemployment significantly higher than the state average (Fig. 7). Orleans and St. Bernard also had significantly high levels of vacant homes according the 2008–2012 American Community Survey (Fig. 8). Based on this comparison of just a handful of indicators of recovery, it becomes readily apparent that there were significant qualitative differences in the recovery of southeast Louisiana from Hurricanes Katrina and Rita compared to Hurricanes Betsy and Camille. While many of these differences can be ascribed to the impacts and physical characteristics of the storms themselves and the failure of the protection systems, the role of community resilience cannot be understated.

According to a report published by the National Academies' Committee on Increasing National Resilience to Hazards and Disasters, New Orleans had indeed suffered from a loss of resilience in the years between Hurricane Betsy in 1965 and Hurricane Katrina in 2005 (Olson 2011). One of the key elements involved in this loss of resilience was the construction of new housing in areas below sea level, much of which had previously flooded (Campanella 2010; Olson 2011). This expansion was driven in part by the construction of hurricane protection levees around the city which excluded the entire city from the 100-year floodplain (Olson 2011). The construction of the levee system was a

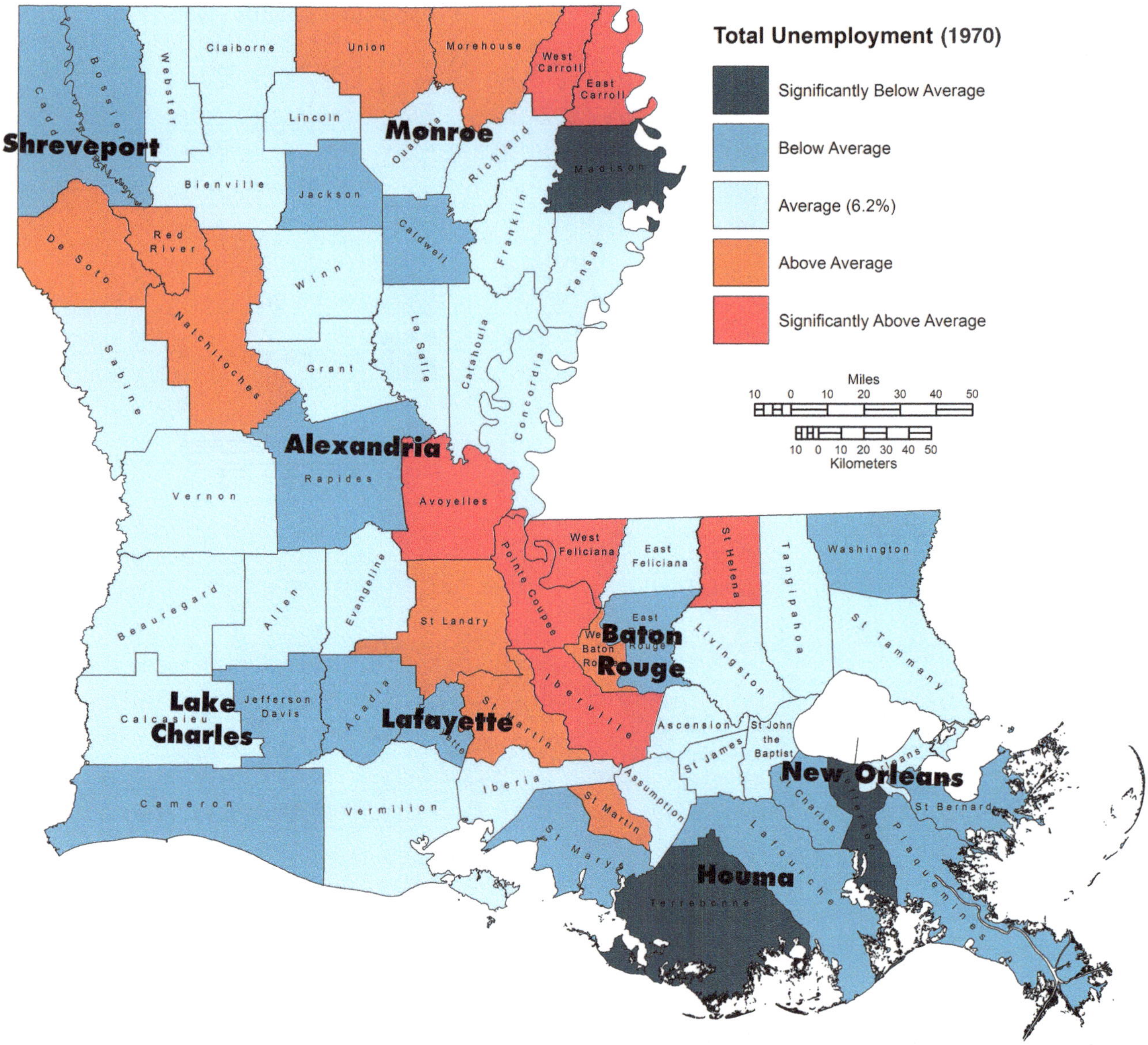

Fig. 4 Unemployment in Louisiana by county (1970)

structural adaption that should have increased the resilience of the city of New Orleans and other communities within the system. Yet, by removing much of the enclosed area from the "official" floodplain, residents were encouraged to develop land and construct dwellings in flood-prone regions, including areas flooded by Hurricane Betsy. The risk to these residents was compounded by the increased levels of subsidence that occurred within the New Orleans levee system since their construction (Burkett et al. 2002; Dixon et al. 2006). Much of the newly developed land within the levees was constructed atop either drained and urbanized wetlands or land augmented by engineered fill (Dixon et al. 2006). Over the course of several decades, compaction of soils combined with municipal water drainage resulted in a situation where the much of the land within the system subsided below sea level (Campanella 2010; Dixon et al. 2006) (Fig. 9). In New Orleans East, for example, subsidence rates have historically been the highest in south Louisiana with land sinking 3 to 5 m below sea level.

While levee construction and municipal drainage was quite effective at reducing the impacts of both river flooding and rainfall-induced flooding, the unanticipated consequences of these actions would have deleterious effects on much of the city during Hurricane Katrina. This is one example of a maladaptation that ultimately played a significant role in reducing the resilience of the region. Such maladaptations are generally driven by non-climatic forces and serve short-term human goals, but often with unanticipated costs to individuals, communities, and society (Smithers and Smit 1997). These costs might not be

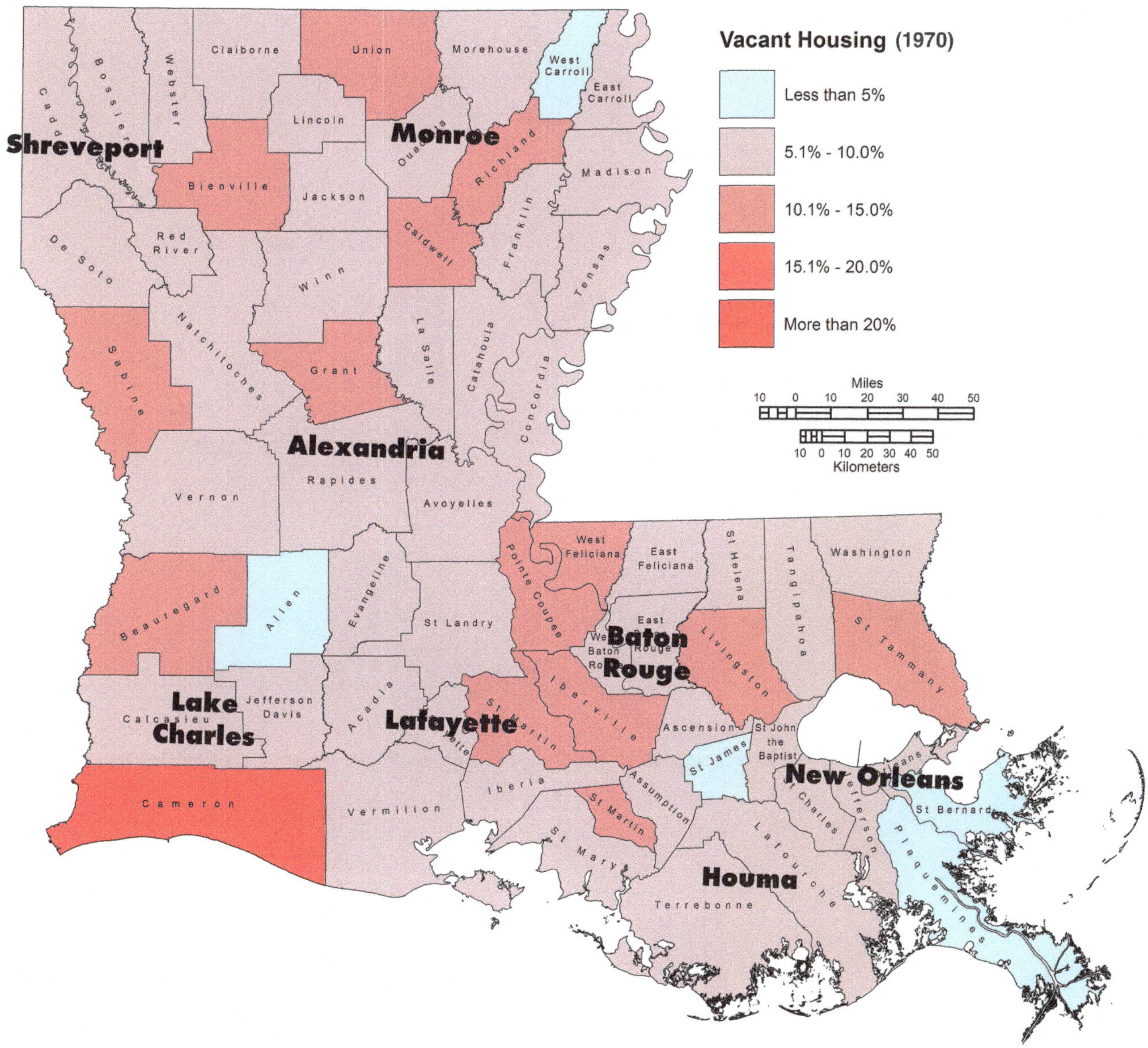

Fig. 5 Vacant housing in Louisiana by county (1970)

immediately apparent, but over time, especially when compounded with other factors, they can be significant in terms of future vulnerability.

Another compounding factor that contributed the historical loss of resilience in coastal Louisiana, particularly in the greater New Orleans region, was the continuation and expansion of the construction of slab on-grade housing in high risk areas. Despite being impacted by both Hurricane Betsy in 1965 and Hurricane Camille in 1969, the New Orleans metropolitan region was experiencing what could be considered a housing boom, with home vacancy rates in 1970 among the lowest in the state. Yet, much of this boom consisted of low cost slab on-grade housing constructed in areas that had previously flooded (Fig. 10). Prior to World War II, houses within the city limits were elevated above the floodplain to provide some protection against floods. However, most of the approximately 75,000 new houses constructed within the city and the surrounding region after Hurricane Betsy were built on concrete slabs raised just a few inches above ground level (Campanella 2010). These factors laid the groundwork for the disastrous consequence of Hurricanes Katrina and Rita, as well as many other flood events in the state, up to and included the spring floods of 2016 in the northwestern city of Shreveport and the August 2016 floods in the Baton Rouge metropolitan area.

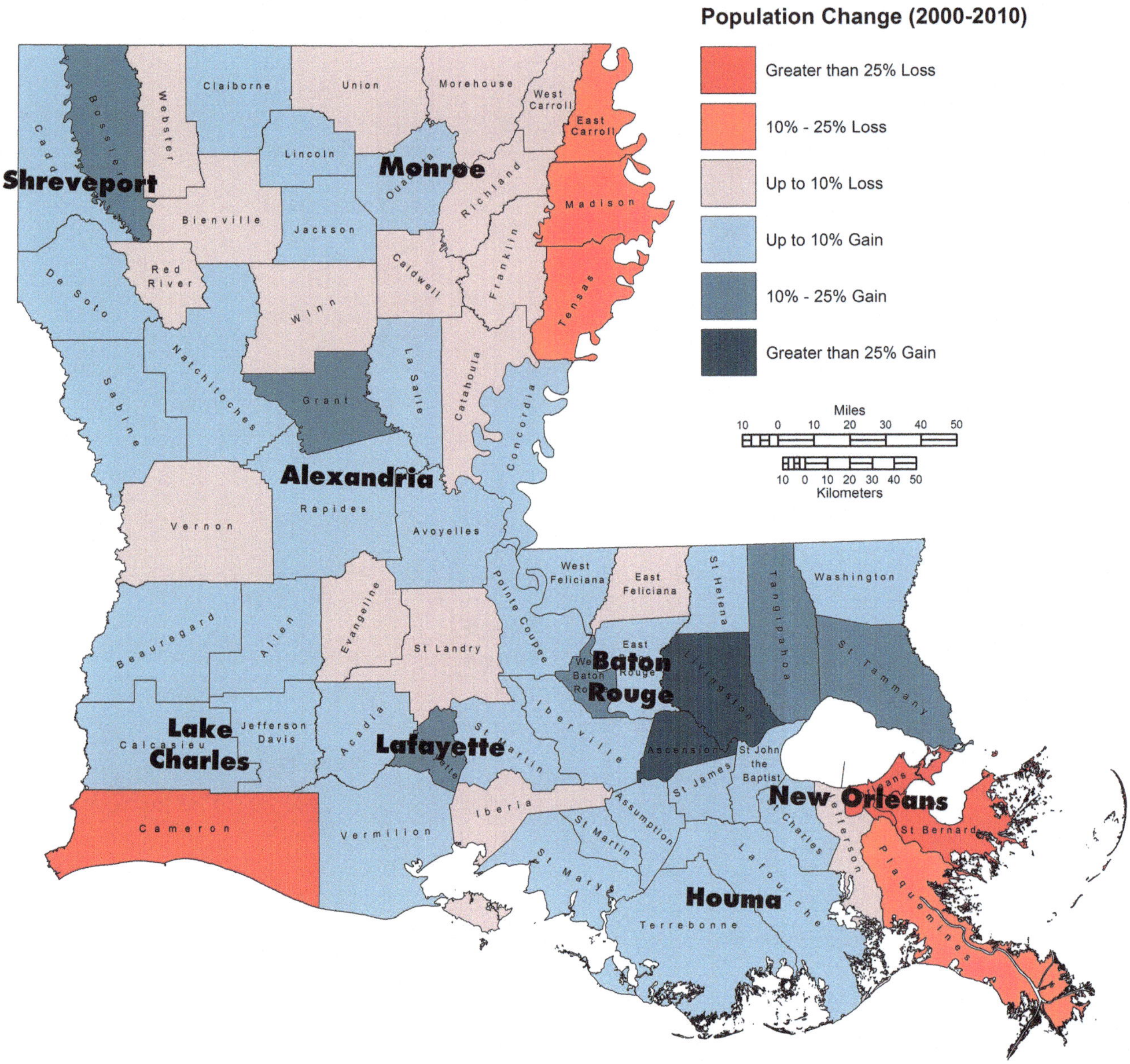

Fig. 6 Percent population change in Louisiana by county (2000–2010)

Historical Changes in Social Vulnerability

To reiterate, resilience is the ability of a system to respond and recover. It includes those inherent conditions that allow the system to absorb impacts and cope with a shock, as well as post-event adaptive processes that facilitate the ability of the system to reorganize, change, and learn in response to the shock. According to Cutter et al. (2008), vulnerability is one of those inherent conditions of a system that creates the potential for a differential ability to recover following a shock. Vulnerability is a function of exposure and the sensitivity of the system (the degree to which people and places can be harmed). It is a function of local socioeconomic conditions and the nature of the hazard to which the human population is exposed (Adger et al. 2004). While overall vulnerability is dependent upon exposure to specific hazards, social vulnerability represents the inherent characteristics of a community or population group that influence how it is able to respond to and recover from any number of theoretical hazards events. According to the Intergovernmental Panel on Climate Change (IPCC 2012), social vulnerability refers to the propensity of exposed elements such as human beings, their livelihoods, and assets to suffer adverse effects when impacted by hazard events. Social vulnerability is related to predisposition, susceptibilities, fragilities,

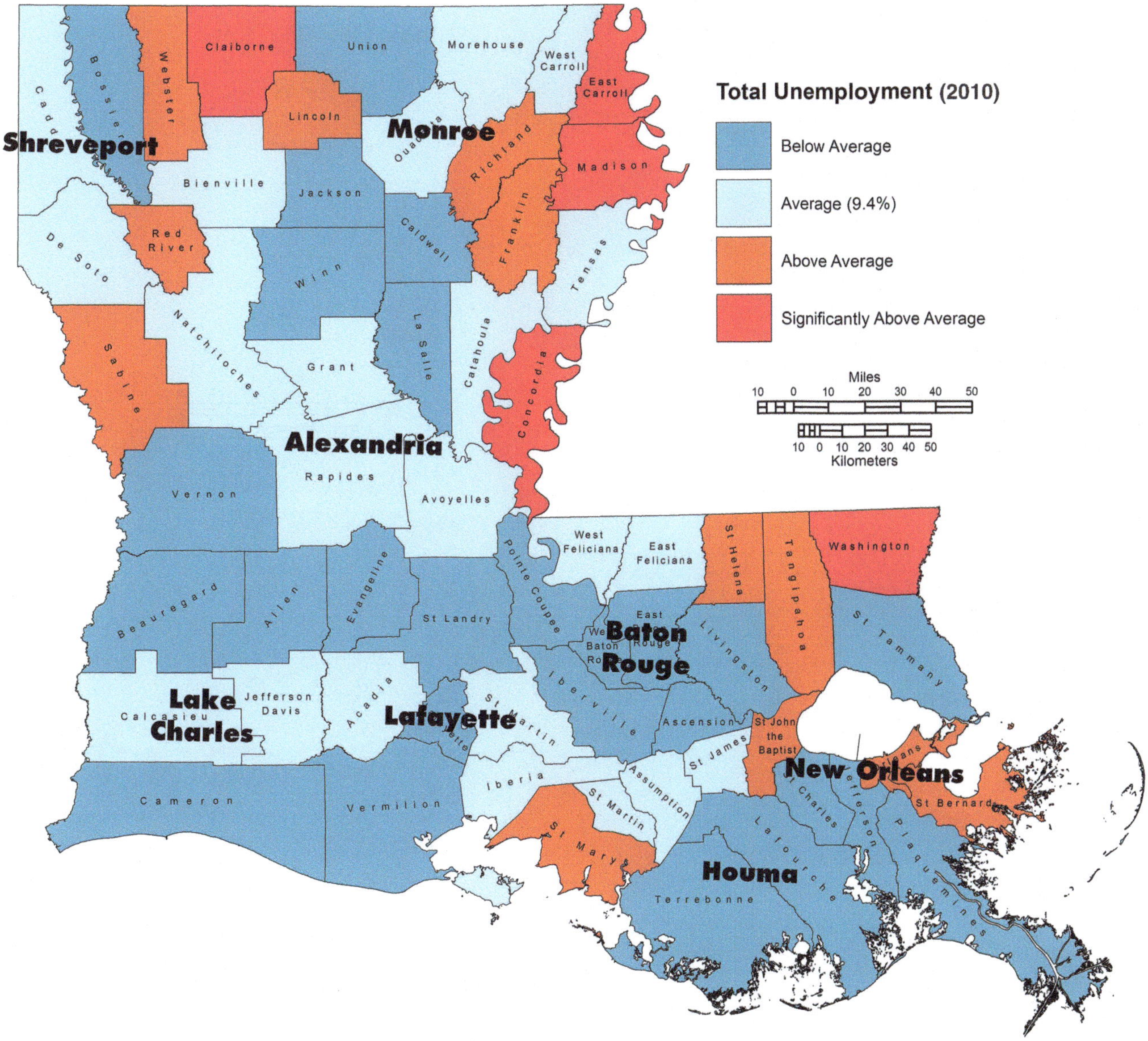

Fig. 7 Unemployment in Louisiana by county (2010)

weaknesses, deficiencies, or lack of capacities that favor adverse effects on the exposed elements (IPCC 2012). Many factors contribute to the ability of communities to respond adaptively to changing conditions and these factors can be represented by any number of indicator variables. Indicator variables are either quantitative or qualitative measures derived from observed facts that simplify the reality of complex situations (Cutter et al. 2010).

While certain factors such as poverty, minority status, and age are determinants of vulnerability across a wide spectrum of different hazards, other factors make communities more vulnerable to certain types of hazards. In resource dependent communities, for example, disruption of livelihoods can result from the loss of land and animals for farmers, or boats and nets for fishers (Wisner et al. 2004). As a result, high levels of natural resource employment are an important determinant of a coastal community's social vulnerability to the impacts of land loss, sea level rise, and tropical storm events (Hemmerling and Hijuelos 2016).

As with physical vulnerability, social vulnerability is far from stable, particularly in states that are economically dependent upon renewable and nonrenewable natural resources. Social and economic conditions are constantly in flux. This is particularly notable at smaller scales, both spatially and temporally. Changes in social vulnerability within individual communities, for example, can often change from once census decadal period to the next. When clusters of communities are examined at a larger regional

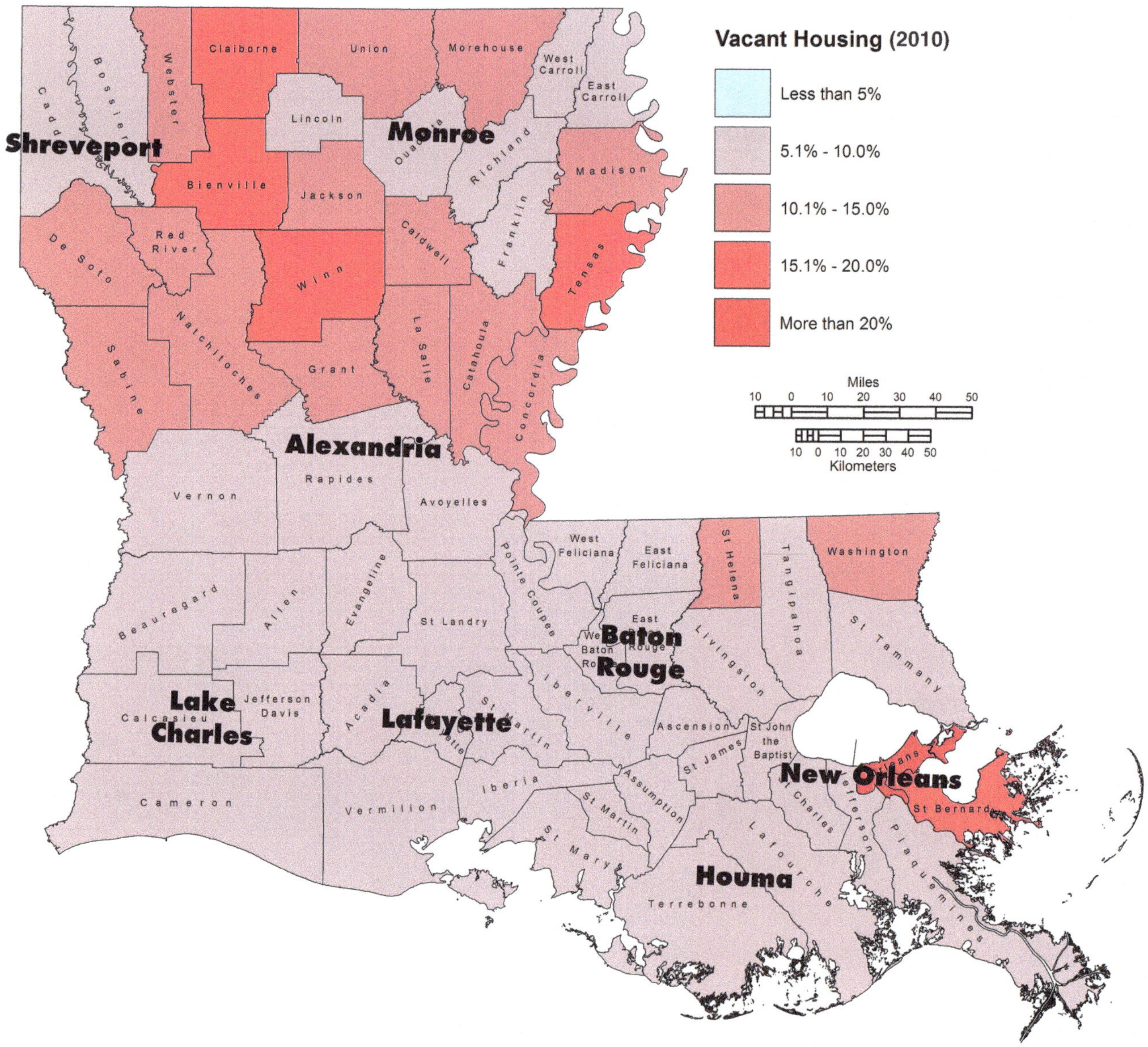

Fig. 8 Vacant housing in Louisiana by county (2010)

scale, however, general patterns that persist through time begin to appear (Fig. 11). The development of the interconnected Baton Rouge-Houma-New Orleans region has led to increasing economic wellbeing in many communities and an overall reduction in social vulnerability in many more.

Other areas become more vulnerable through time. Regions of persistent poverty, for example, may develop and last through several decades (Fig. 12). This is particularly true in areas that have historically been heavily dependent on agriculture as in the region stretching between the cities of Lafayette and Alexandria. With the growth of industrial agriculture, rural communities that historically relied on farming were forced to adapt. Some areas, such as Lafourche and Terrebonne parishes, were able to effectively transition to an oil and gas-based economy. Other areas in the center of the state, where oil and gas deposits are scarce, were unable to transition in this way, leading to higher levels of poverty and vulnerability and out-migration. In some cases, as in lower Plaquemines Parish, on the bird's foot delta of the Mississippi River, fisheries-dependent communities persisted not only on the edge of the land, but on the edge socioeconomically. The combination of Hurricane Katrina in 2005 and the Deepwater Horizon Oil Spill in 2010 tipped this area's demographic and economic composition such that it is now more vulnerable than at any time in recent decades (Bailey et al. 2014). In general, communities that are heavily reliant on

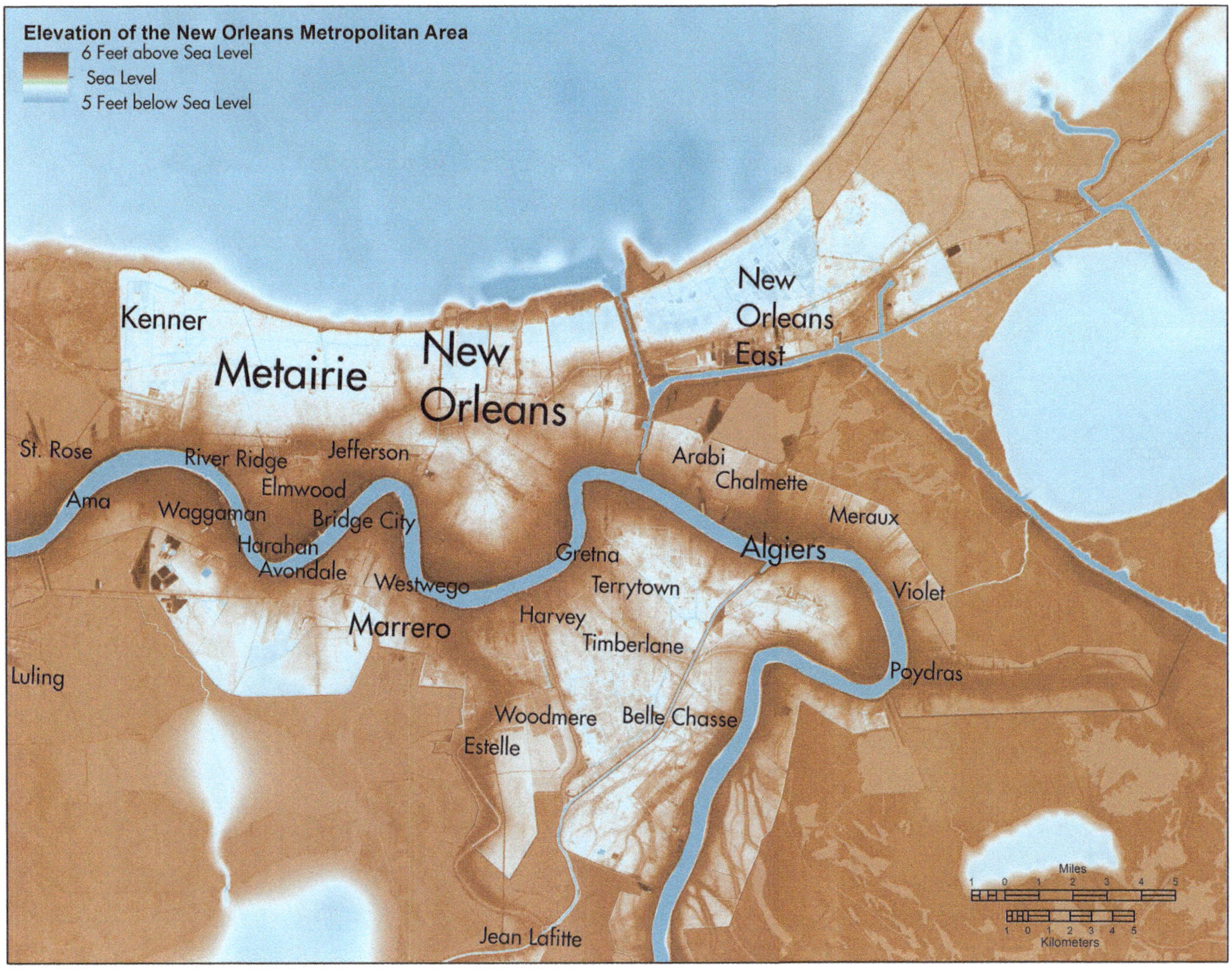

Fig. 9 Elevation of the New Orleans Metropolitan Area. Historical high ground is located on top of the natural river levee and on old river crevasses and distributary channels. The lake Ponchartrain shoreline north of New Orleans has been dredged to higher elevation

natural resources tend to be more vulnerable to changing social and economic conditions.

It should be noted that what is true for the community in general may not necessarily hold for individual residents or neighborhoods within that community. This is especially true in large metropolitan areas like New Orleans, where there are very clear structural divisions between the wealthy and the poor. The greater the levels of inequality in a community, the more vulnerable that community is. Inequality affects vulnerability directly by constraining the options of households and individuals when faced with an external shock (Adger 1999). The idea that individuals within a particular community many not necessarily share the same level of vulnerability or resilience of the community in general holds true for communities at the regional level. Overall regional resilience may increase even if some communities within that region do not experience any change in resilience. In New Orleans, for example, the flight of the white middle class to the suburban parishes, triggered in large part by the integration of schools in the 1960s, set in motion changes that led to a decline of population. This out-migration was racially selective, and after 1980 the city of New Orleans had a black majority, although surrounding suburban parishes, including those on the north shore of Lake Pontchartrain, did not (Fussell 2007). By 2005, when Hurricane Katrina struck the city, this racial differentiation had created a class of residents who were exceptionally vulnerable to the catastrophic impacts of the storm. The median income of the city's residents was two-thirds of the national average and over a quarter of the city's population lived in poverty. Consequently, race and class conditioned not only how many residents of New Orleans were able to evacuate and how and whether they would return and rebuild (Fussell 2007).

When examined at the broader coastal scale, different patterns begin to appear. Despite a clearly delineated regionalization of social vulnerability throughout the state, when communities are analyzed based upon their coastal status,

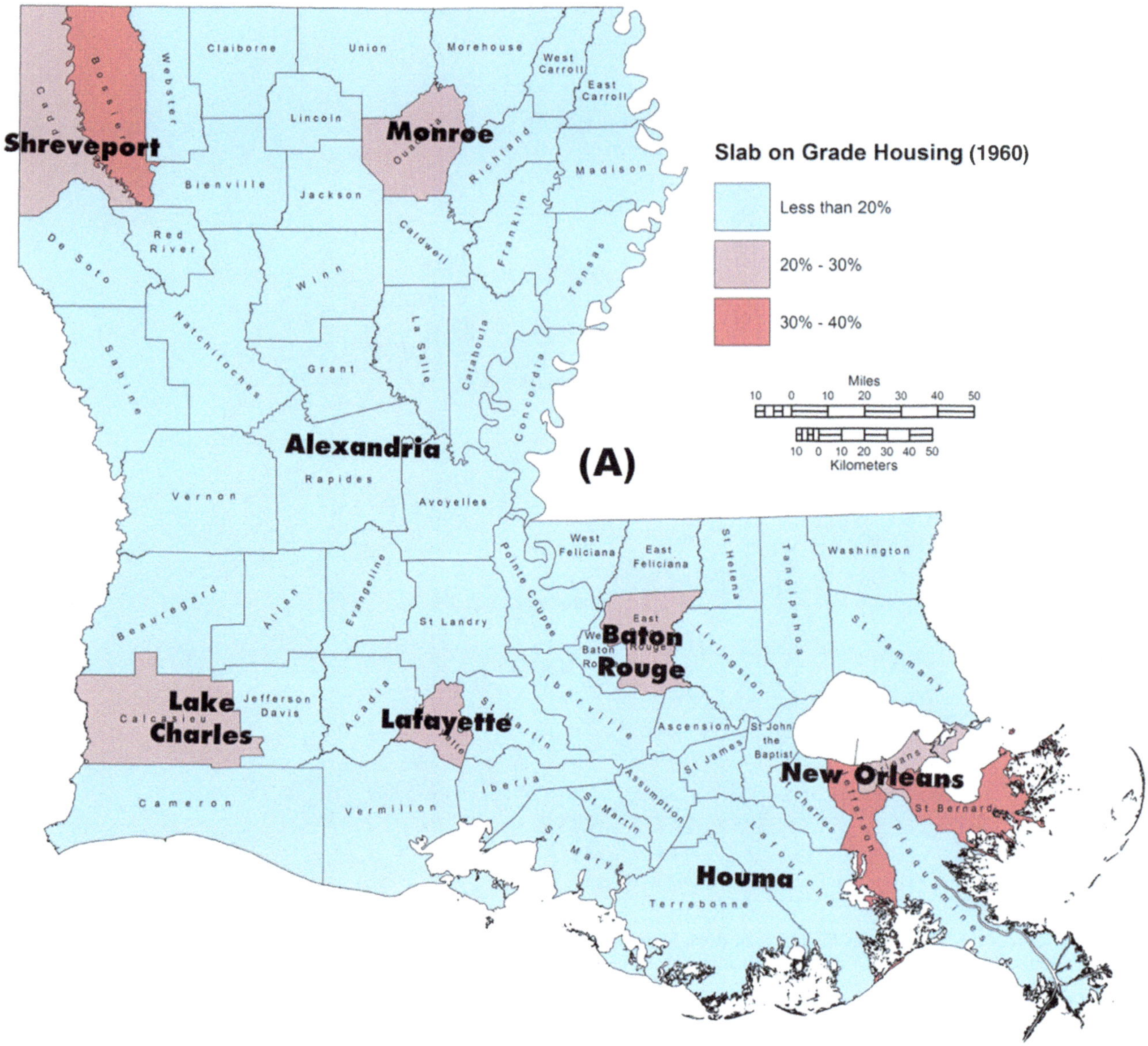

Fig. 10 Slab on grade housing in Louisiana in 1960 (**a**) and 1970 (**b**)

either coastal shoreline or coastal watershed, different temporal patterns emerge (Fig. 13). Coastal shoreline counties have been defined in the NOAA Coastal Assessment Framework as those parishes directly adjacent to the open ocean and major estuaries. These parishes host the majority of the economic production associated with coastal resources but also bear much of the direct effects of coastal hazards. This includes all of those parishes adjacent to the Gulf of Mexico as well as those bordering Lake Pontchartrain and Calcasieu Lake. Coastal watershed counties, on the other hand, are those parishes that have a substantial watershed-based impact on coastal and ocean resources but are not generally involved in coastal economic activities. The coastal watershed parishes extend up through the central portion of Louisiana, as far north as Rapides Parish, and includes the urban centers of Baton Rouge and Lafayette. Throughout the latter decades of the twentieth century, communities within the coastal watershed parishes maintained fairly stable levels

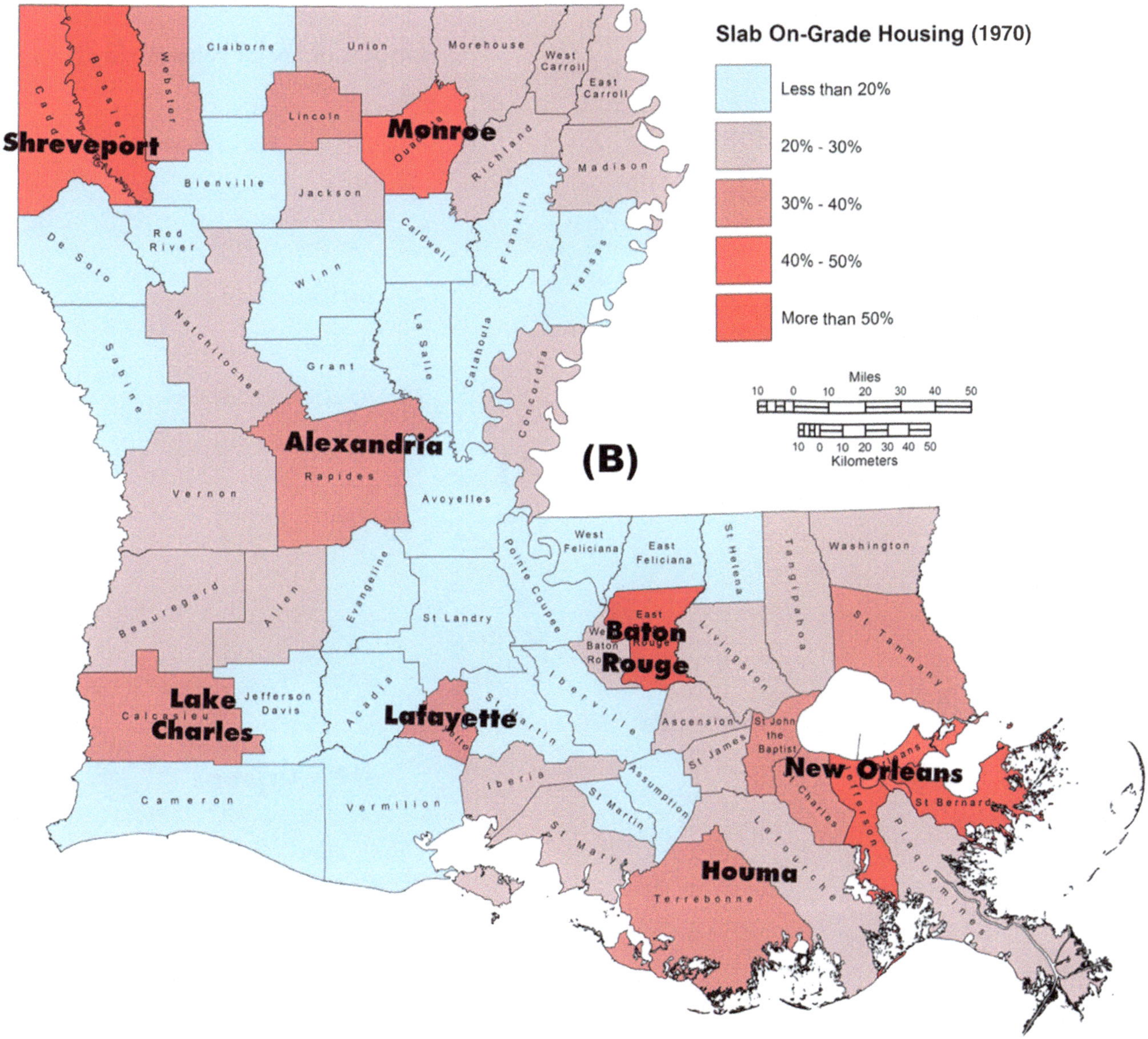

Fig. 10 (continued)

of vulnerability, until 1990, when the levels of social vulnerability in these communities began, on average, began to decline relative to other communities in the state. The coastal shoreline parishes have consistently had levels of social vulnerability below the state average while the noncoastal parishes have generally had higher levels. Since 1990, there has been a general trend of declining relative social vulnerability in the coastal watershed parishes and increasing social vulnerability in the coastal shoreline and noncoastal parishes. It is important to note that the baseline here is the state average. So, even with increasing social vulnerability in the coastal shoreline communities, overall social vulnerability of communities in this region is still lower than the state average.

Historical Population Change and Transitions

One measure of the inherent resilience of a community is the rate of natural increase, or net reproduction per individual, a value equal to the difference between the crude birth rate and the crude death rate. This is essentially a measure of the

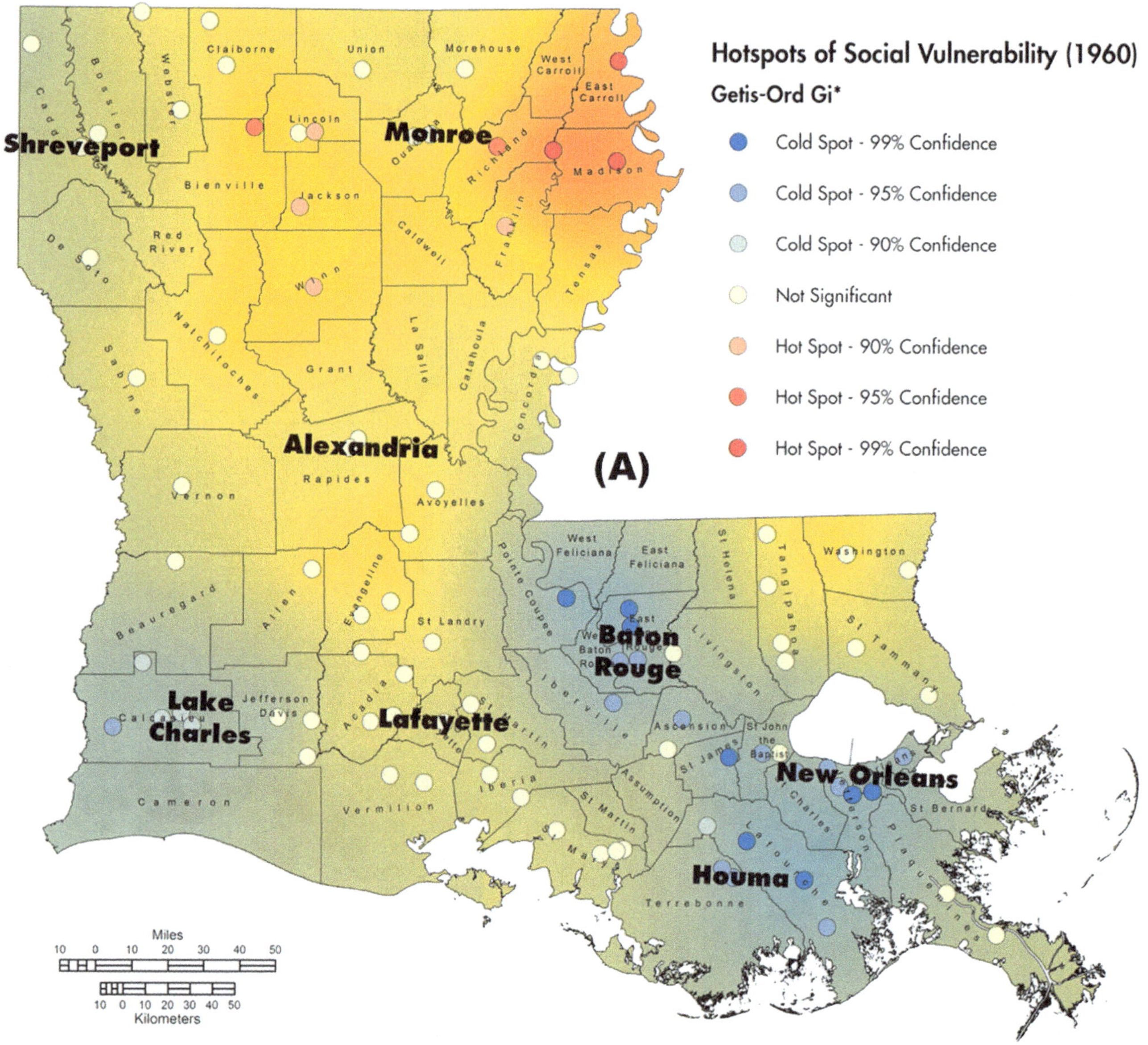

Fig. 11 Social vulnerability index by decade in Louisiana. Shown here: 1960 (**a**), 1970 (**b**), 1980 (**c**), 1990 (**d**), 2000 (**e**), 2010 (**f**)

population change exclusive of the effects of migration. The rate of natural increase is independent of the size of the community, allowing for a direct comparison among populations of differing sizes. A low or declining rate of natural increase is often correlated to several social and economic factors, including an aging population and slow economic growth and vice versa. In Louisiana, one of the primary aspects of resilience tying residents to the landscape is family and kin networks. A high rate of natural increase in a community is to some degree a measure of the growth and expansion of kin networks.

In the ecological literature, the rate of natural increase of populations is indicative of the replacement rate of the population, the time required for recovery, and the ability of a population to persist in the face of demographic stochasticity (Denney et al. 2002). These characteristics are clearly analogous to the concept of community resilience in human systems. As in ecological systems, human communities exhibiting high rates of natural increase are often more resilient to collapse (Demeny 2011).

While the correlation between the rate of natural increase and community resilience, wellbeing, and vulnerability has not been explored in human systems as extensively as ecological systems, an initial examination of parish-wide values reveal patterns that are consistent with several other measures of community resilience (Fig. 14). While overall rates of natural increase have broadly declined over time, regional pockets of high and low growth have persisted.

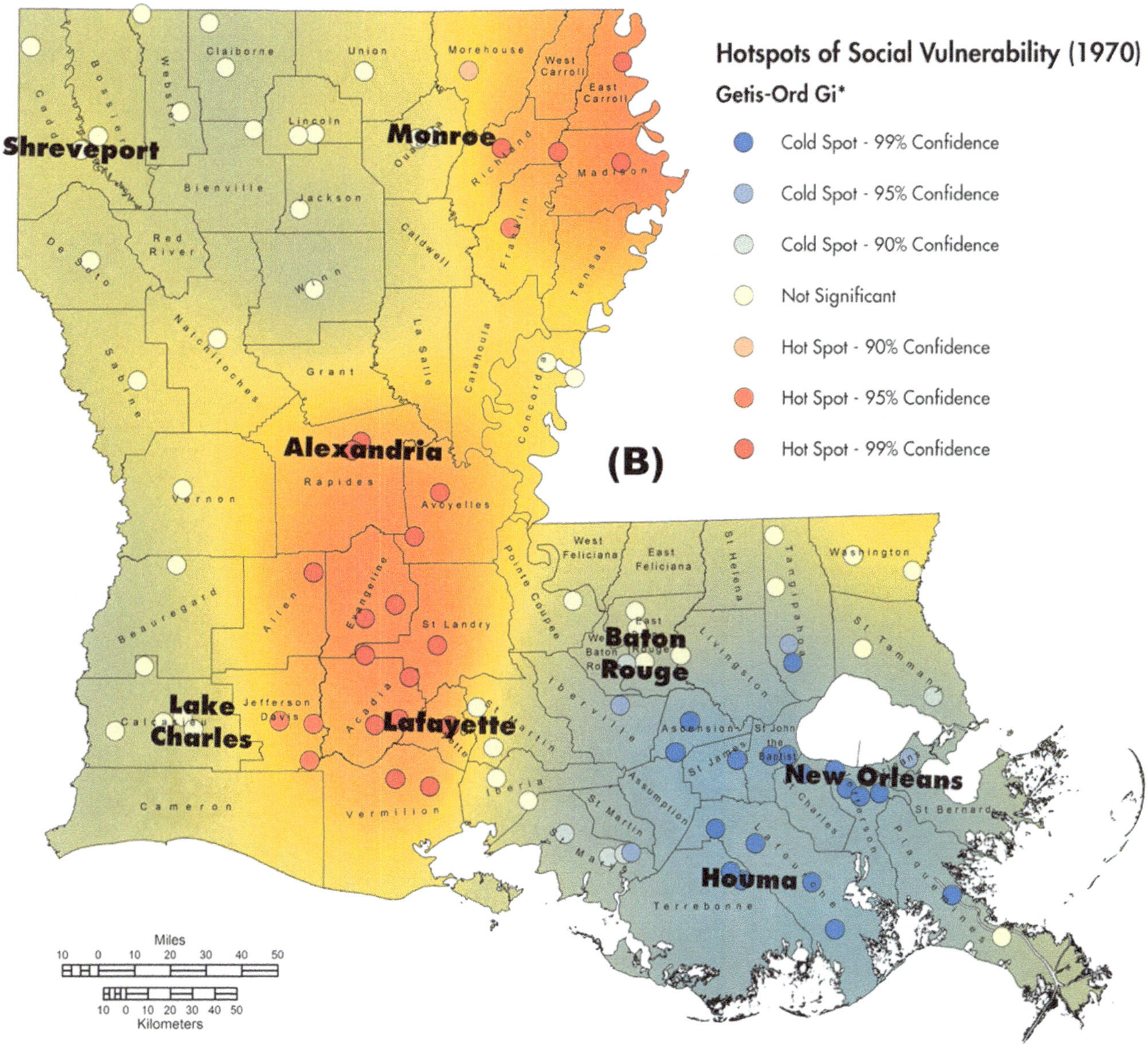

Fig. 11 (continued)

Many of the rural-agricultural regions of the state, particularly the core area of Acadiana stretching up through northeastern Louisiana, have maintained rates of natural increase generally higher than other regions in the state.

In coastal Louisiana, there is a clear divide between the southwestern and southeastern regions of the state. Rates of natural increase have been consistently lower in the southwest, with Cameron Parish being the first to drop below a rate of 0.5. The coastal parishes of Southeast Louisiana, particularly those rural-agricultural parishes where sugarcane is grown, from Lafourche Parish up to and including Pointe Coupee Parish, have continued to have relatively high rates of natural increase. Only one coastal parish has experienced a negative rate of natural increase. Terrebonne Parish saw a fluctuation in the rate of natural increase over a several decade timespan, ultimately seeing the death rate exceed the birth rate in 2010. Whether this represents a short term adjustment following the storms of the 2000s or the

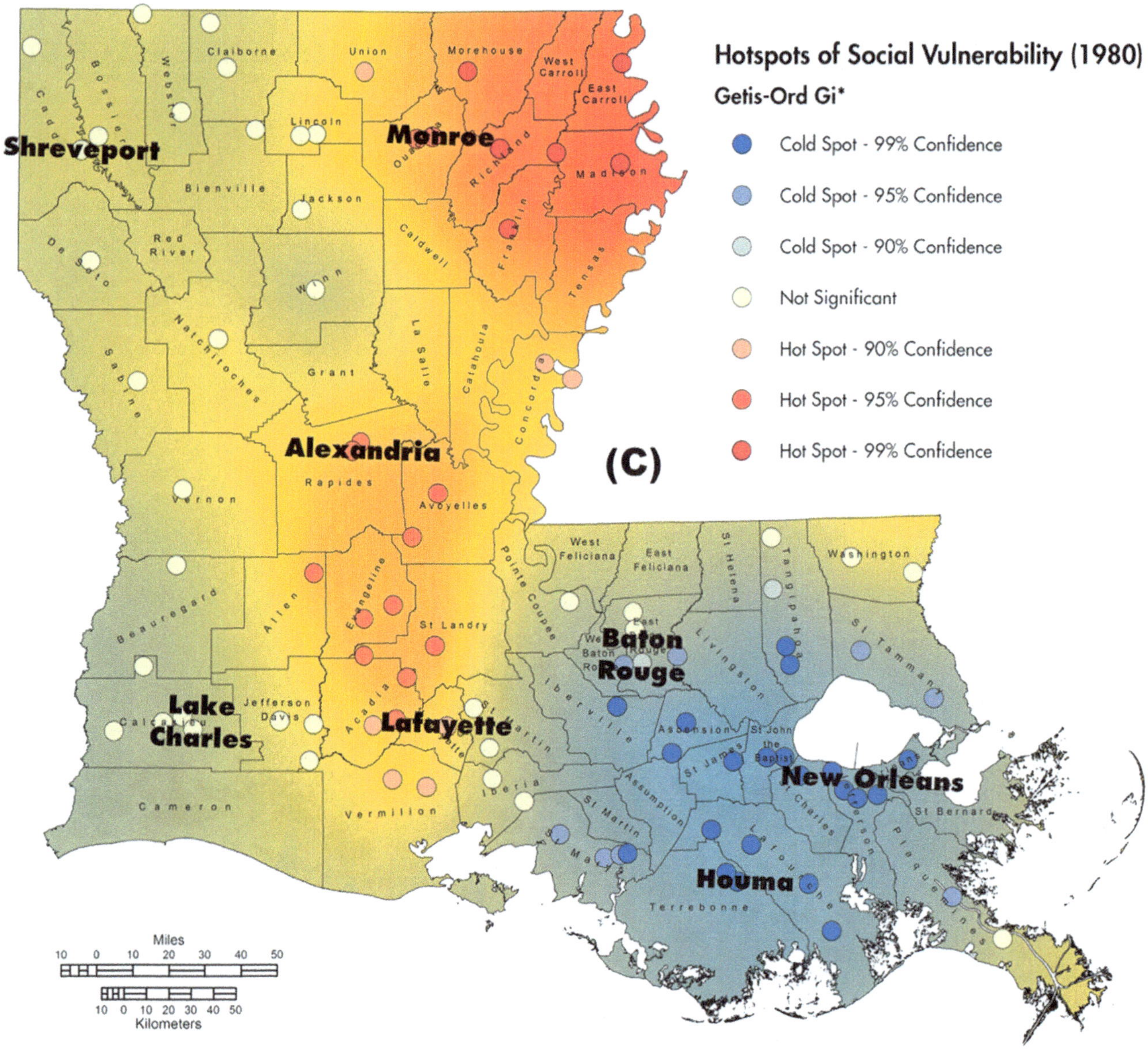

Fig. 11 (continued)

Deepwater Horizon oil spill remains to be seen. It is worth noting however, that other coastal parishes in southeast Louisiana such as Lafourche and even Plaquemines, which experienced a tremendous loss of population in the immediate aftermath of Hurricane Katrina, experienced increases in the rate of natural increase. The causes for the instability seen in Terrebonne Parish cannot be determined, but this parish is one that has experienced extremely high levels of coastal land loss in the latter half of the twentieth century, much of this occurring proximate to communities outside of structurally protected areas.

Increasing Adaptive Capacity

The impacts of extreme natural events such as Hurricanes Katrina and Rita, as well as technological tragedies like the Deepwater Horizon oil spill, have had and will continue to

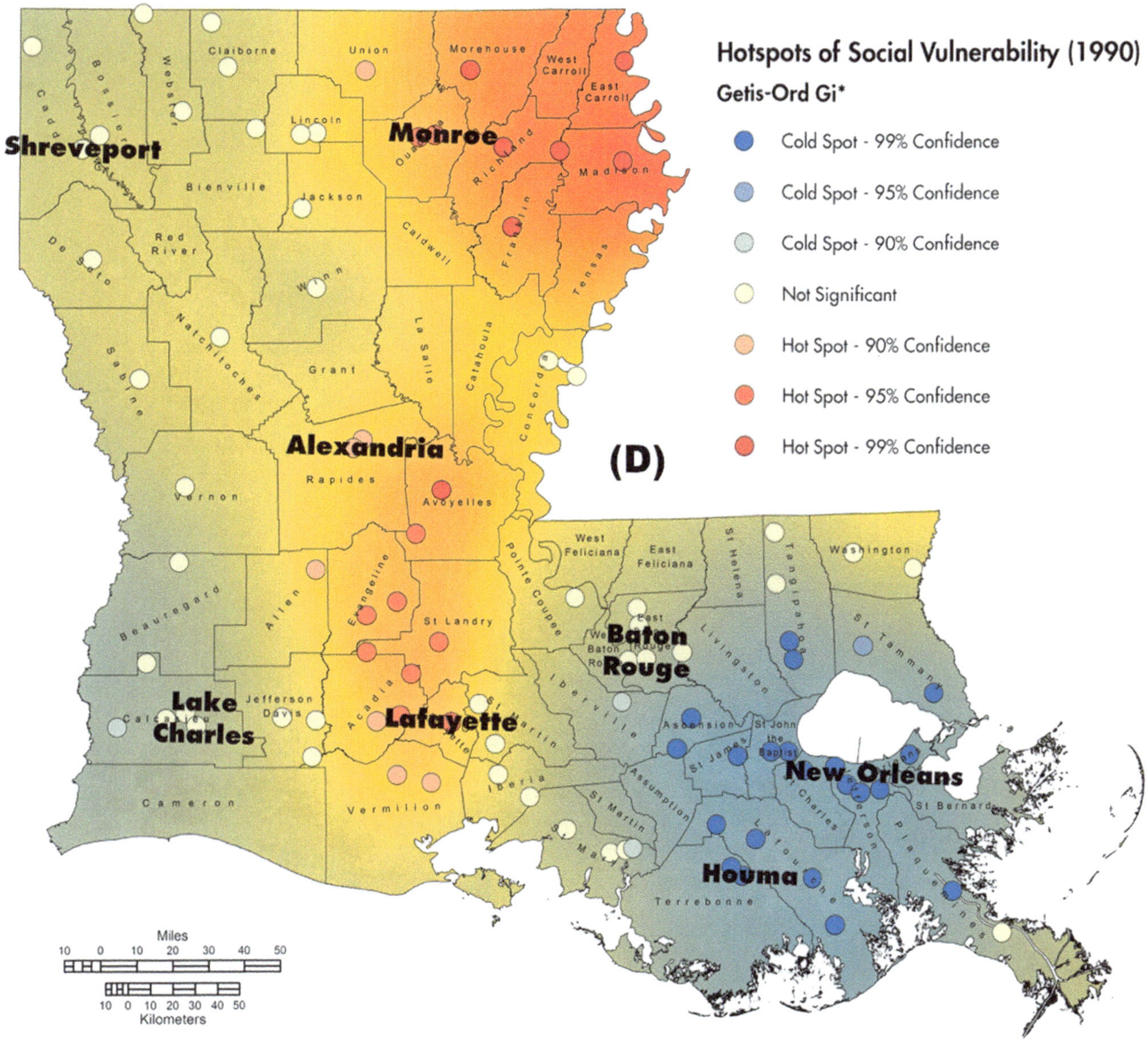

Fig. 11 (continued)

have profound and lasting effects on the communities of coastal Louisiana. For better or for worse, these disasters have the potential to trigger a regime shift in a way that previous disasters did not. While the full impact of these events remains to be seen, evidence suggests that Louisiana may have learned from the maladaptive mistakes of the past and has begun to integrate change-oriented adaptation strategies into many of the actions of state and local government. Such adaptation strategies are directed toward a deliberate transformation of the status quo and attempt to alter the nature of human activity in order to achieve better integration with local and regional environmental conditions (Smithers and Smit 1997).

The development of resilience-based planning strategies will reduce the types of maladaptive development that have, historically, placed Louisiana's coastal populations in

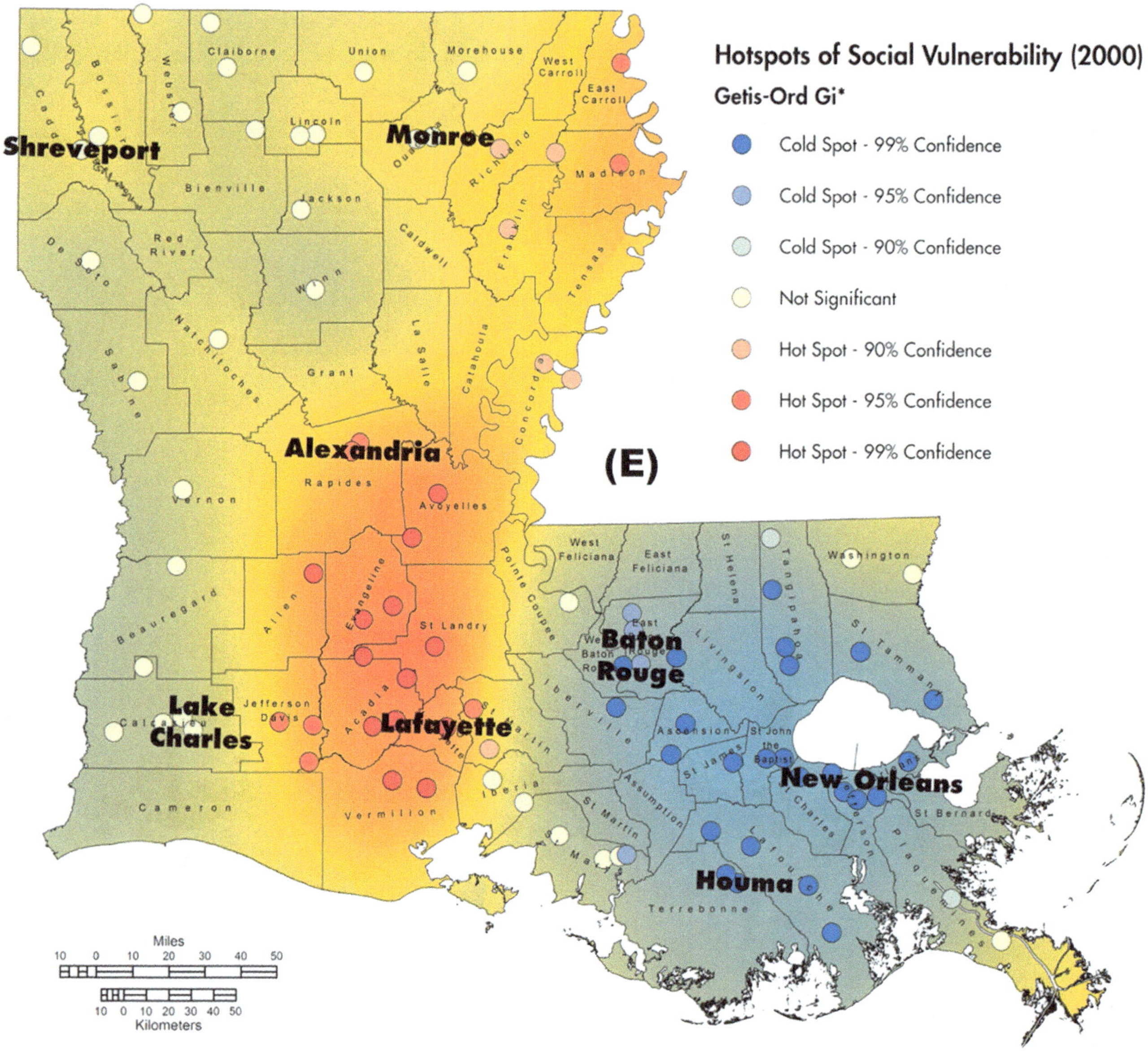

Fig. 11 (continued)

harm's way. These strategies must necessarily operate on a multiscale level. Locally, residents must begin to incorporate nonstructural elements in to their homes and neighborhoods. At a minimum, homes need to be elevated above base flood elevation to immediately reduce risk to residents. At a broader regional level, an integrative resilience-based strategy should include policy-related adaptations, including targeted land development, coordinated transportation investments, stronger building codes, and a strategic approach to land regulation (CPEX 2012). Finally, if coastal land loss is going to be halted and coastal communities are going to be protected from future tropical storm events, a number of large-scale structural and nonstructural measures will need to be undertaken by governmental and nongovernmental organizations. The key to this effort must be a fully integrated coastal protection and

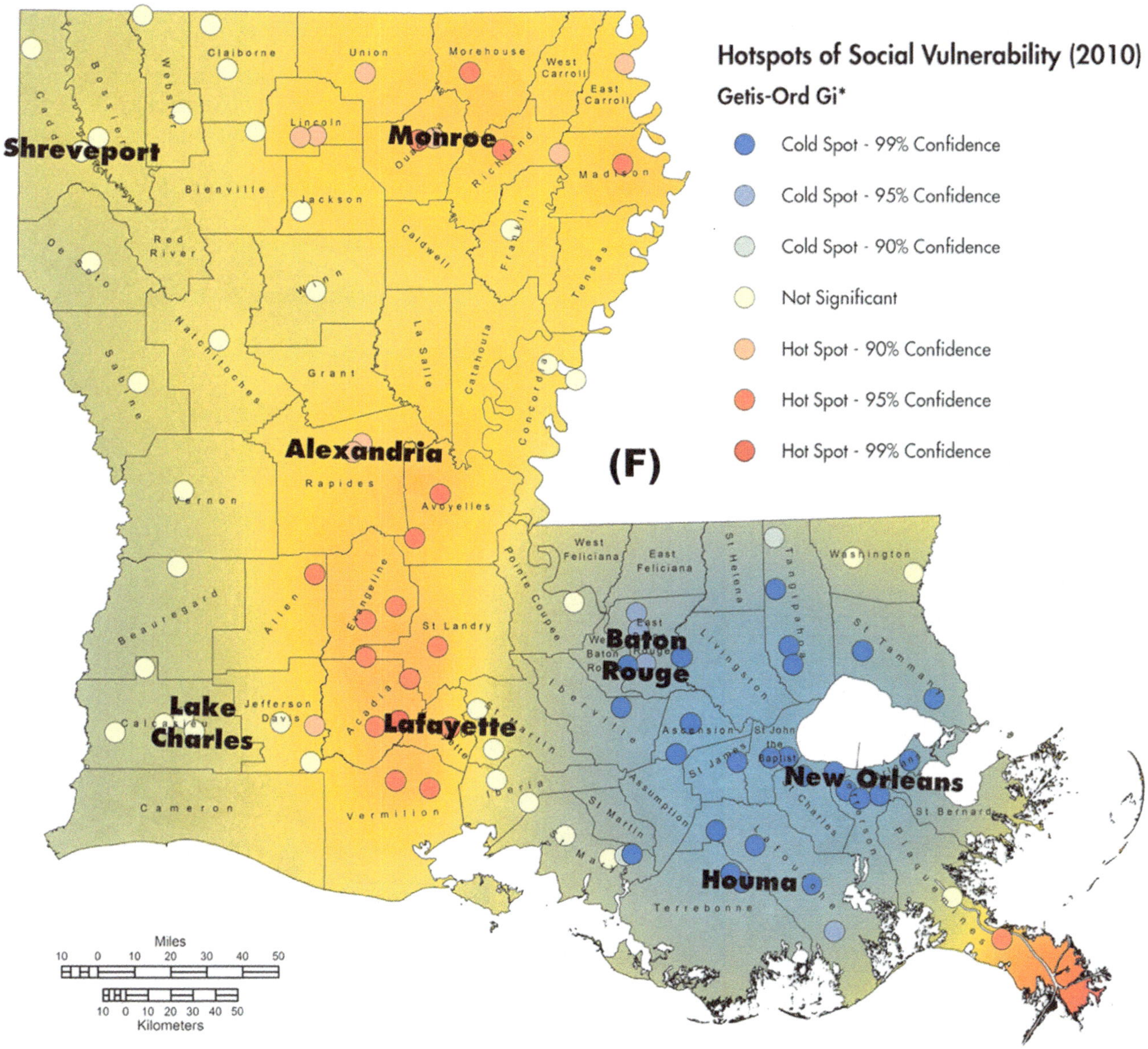

Fig. 11 (continued)

restoration plan, one that incorporates aspects of the physical, biological, and human environments.

Creating community resilience is not an easy process, however. On some levels, these adaptation strategies have forced officials to make trade-offs and decisions that past officials were either unwilling or unable to make. As Colten et al. (2008) note, creating community resilience is an extended process that can take years, even decades, to accomplish. This process is complicated by the fact that the decisions that need to be made to proactively improve the elements of resilience often involve difficult, sometimes unpopular, decisions. Following disasters, impacted residents often feel an innate desire for their lives to return to normal as quickly as possible. During the disaster

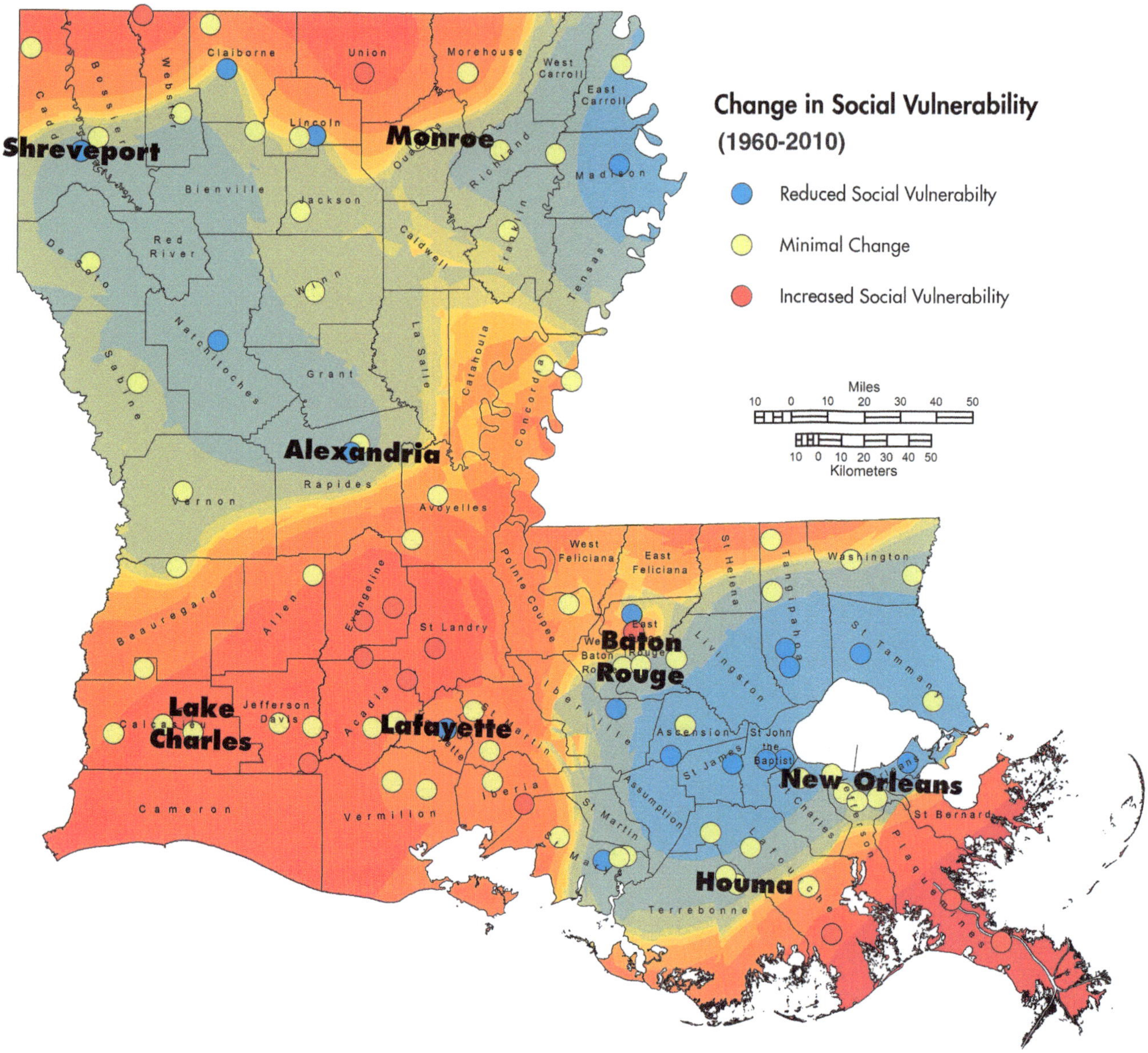

Fig. 12 Change in social vulnerability index (1960–2010)

recovery period, public officials and civic leaders often need to balance the conflicting policy goals of rapid recovery, safety, betterment, and equity (Kates et al. 2006).

In the immediate aftermath of Hurricane Katrina, for example, despite the clear need to protect individual homes and structures by elevating above base flood elevation, many New Orleans residents were reluctant to elevate damaged buildings due to the cost and effort involved. However, many damaged residences received city building permits before FEMA issued recovery guidance stating that every substantially damaged structure needed to be elevated at least 3 ft above the Highest Existing Adjacent Grade. As a result, they were exempt from complying with the FEMA requirements and allowed to rebuild with little to no elevation. Many homes with less than 50% damage restored

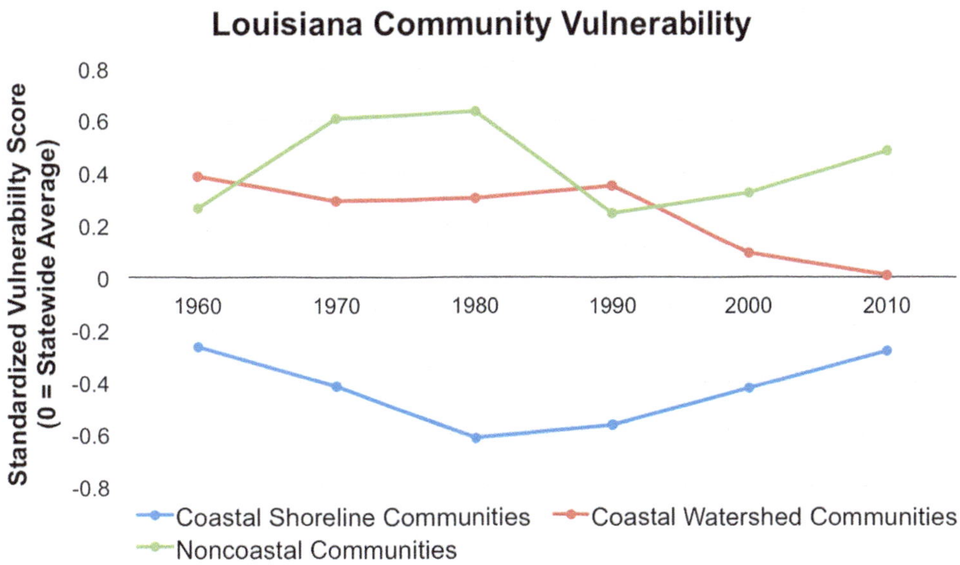

Fig. 13 Social Vulnerability Index (SOVI) scores for 94 Louisiana communities, averaged across three broad regions. These data were derived from 30 socioeconomic variables that were continuously available for each of these communities

without elevating – only about 3000 houses in New Orleans were raised after Katrina.

This example highlights the difficulties in trying to change the status quo. Recent history of building and development in New Orleans and the surrounding region emphasized rapid recovery at the expense of safety in the decades following Hurricane Betsy. This paradigm dominated the initial stages of recovery in New Orleans after Hurricane Katrina, despite calls from planners and others to rebuild in a more resilient fashion. In addition, the question of where redevelopment should be allowed to occur raised issues of equity, as many of the residents of low lying flooded areas in the city were African American and poor.

Later, the state of Louisiana passed a law requiring mandatory adoption and enforcement of building codes. Louisiana now enforces the 2012 International Residential Code, which requires the elevation of the ground floor of buildings within established flood zones and the use of flood resistant materials for the portion of any structure located below base flood elevation. This law in many ways may have extended the length of time needed for communities to rebuild and recover and made it difficult for many residents to rebuild at all. But, the adoption of building codes designed to prevent the proliferation of slab on-grade housing that exploded in the years following Hurricane Betsy in 1965 will ultimately serve to increase the resilience of coastal communities in the long term.

As previously noted, these individual level adaptations represent only one part of a multiscale resilience strategy. Changes at the broader municipal and regional levels have the potential to provide many additional layers of protection unavailable at the individual level. Catastrophic disasters provide an opportunity for communities to rebuild in a smarter, more resilient way, essentially righting the wrongs of the past (Johnson and Olshansky 2010). In the years following Hurricane Katrina, for example, the city undertook several extensive planning efforts to guide the recovery and rebuilding of the city. These efforts culminated in the Resilient New Orleans strategy, a comprehensive plan to address community resilience in a systematic fashion, addressing issues ranging from developing sustainable infrastructure to supporting coastal restoration and protection to fostering improved economic and environmental equity within the city (City of New Orleans 2015). Coincident with the implementation of this plan, the city is creating New Orleans' first comprehensive resilience district in Gentilly, a neighborhood of approximately 11,000 homes, mostly low and moderate income. The neighborhood, which struggled to rebuild after Hurricane Katrina, will be redesigned with medians excavated and turned into "blue corridors," and others will become "green corridors" with swales, tree canopies and water permeable sidewalks (LaRose 2016). This type of redevelopment is designed to reduce flood risk and allowing for the absorption of groundwater.

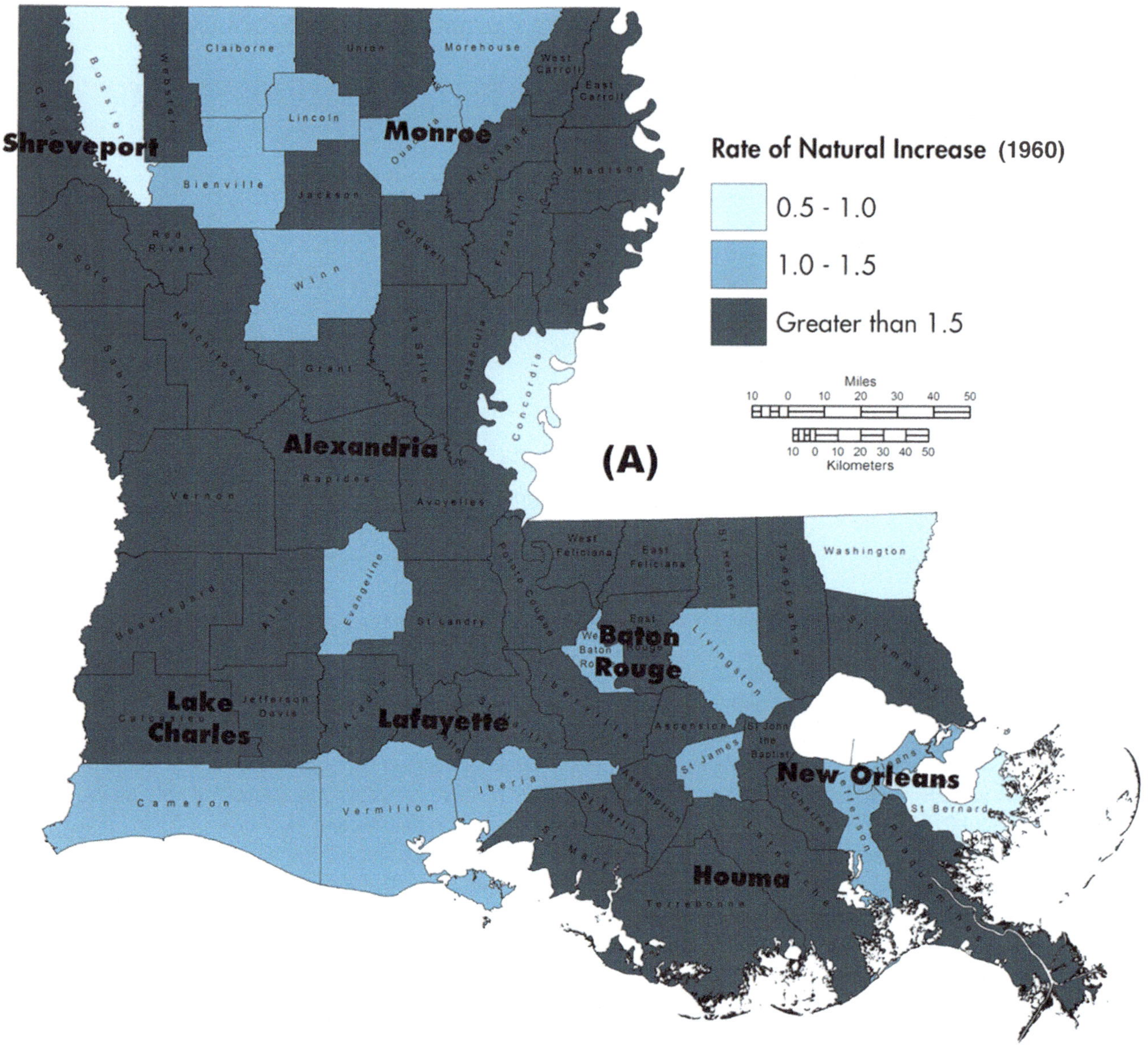

Fig. 14 The rate of natural increase in Louisiana by decade. Shown here: 1960 (**a**), 1970 (**b**), 1980 (**c**), 1990 (**d**), 2000 (**e**), 2010 (**f**). The rate of natural increase is another term for the net reproduction per individual (see Historical Population Change and Transitions in text)

Finally, at the broader coastwide scale, efforts are underway to slow the rate of coastal land loss and in some cases restore land previously lost to erosion, subsidence, and sea level rise. The state of Louisiana developed its Comprehensive Master Plan for Sustainable Coast in cooperation with a broad network of state, federal, private, and academic researchers. This effort incorporates a broad suite of coastal restoration and protection projects to develop a plan that would maximize both protection of communities and the restoration of the environment (CPRA 2012). For this plan to be fully implemented policy makers will, again, need to make some difficult decisions. Similar to recovery from acute hazards events, efforts to slow down or reduce land loss, must balance rapid recovery, safety, betterment,

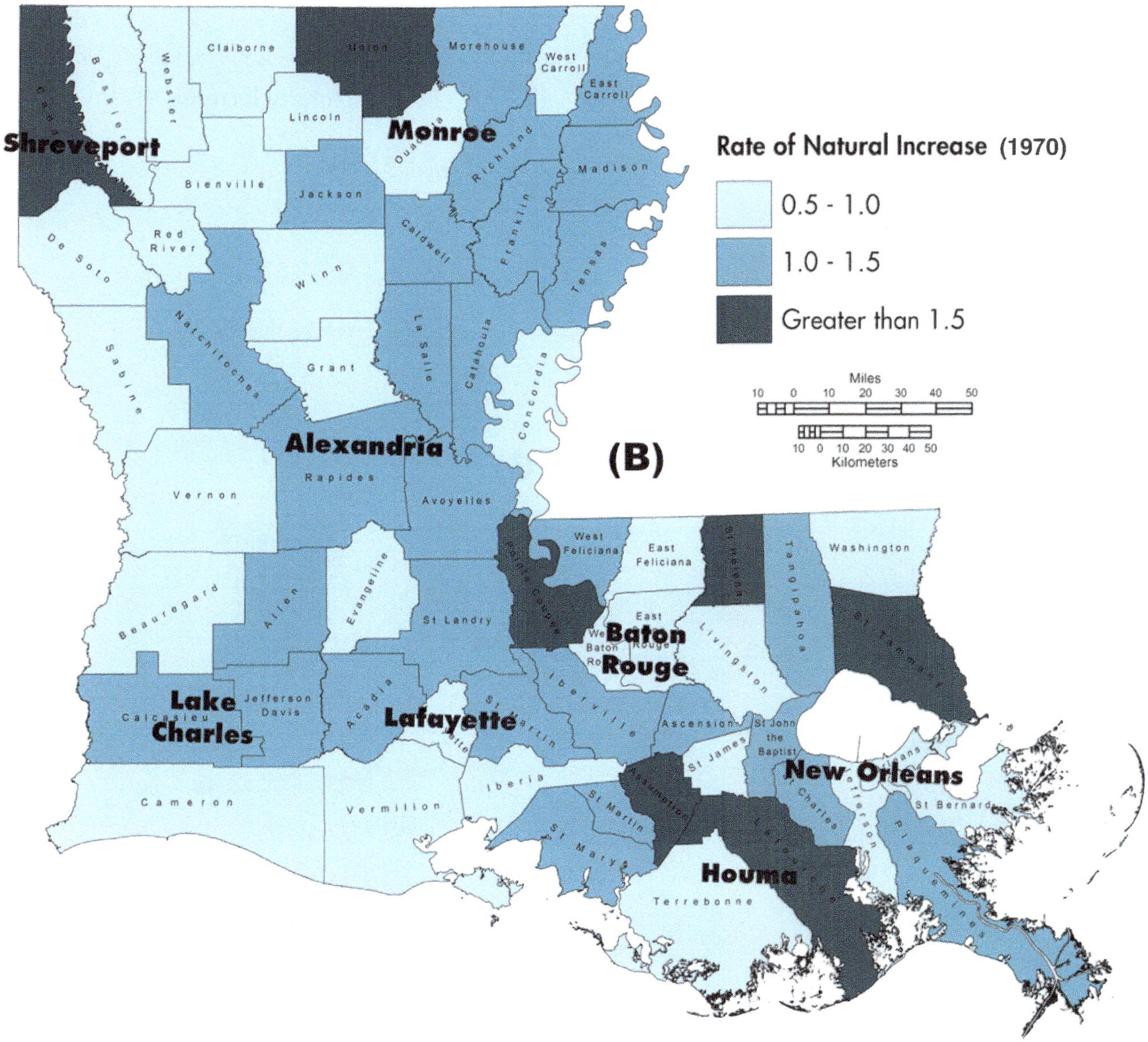

Fig. 14 (continued)

and equity factors in the planning process. Tradeoffs will need to be made for this plan to be successful. Some of these decisions will be purely economic, such as finding the funding to construct the projects. Other decisions will need to be made related to balancing equity issues with the idea of the broader public good, including betterment and safety. In some cases, this involves acknowledging that all areas of the coast cannot be restored or protected and providing alternatives for residents unable to be protected. In other cases, it means acknowledging that many of the

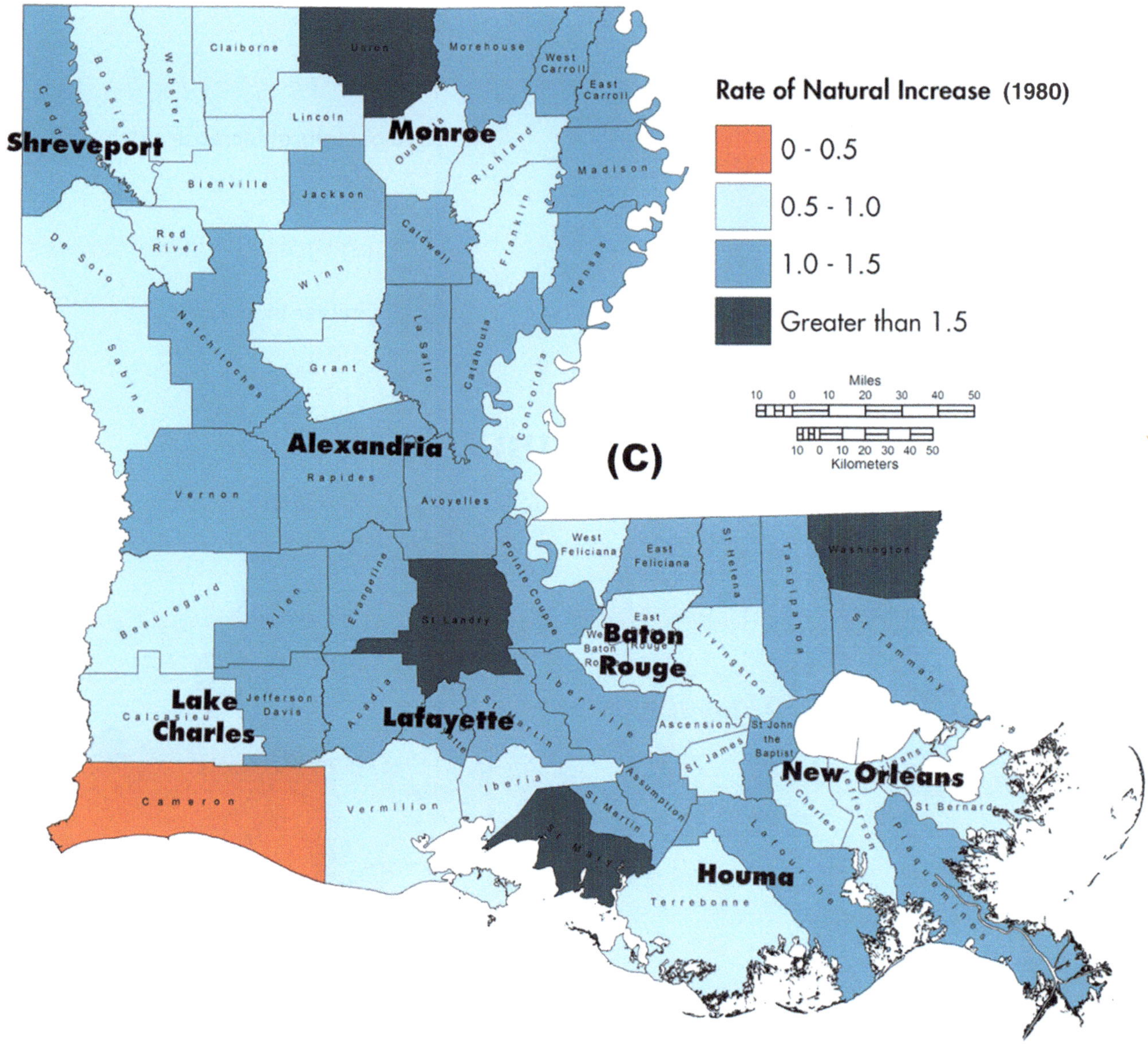

Fig. 14 (continued)

projects that are designed to build land will necessarily impact the livelihoods of coastal resident such as commercial fisherfolk.

The challenges around building more resilient communities in coastal Louisiana and other locations around the world revolve largely around adaptively managing change. Attempts to control change and build a perfectly resilient, stable community are bound to fail, simply because, in reality, a perfectly resilient, stable system cannot exist. This is particularly true in social-ecological systems

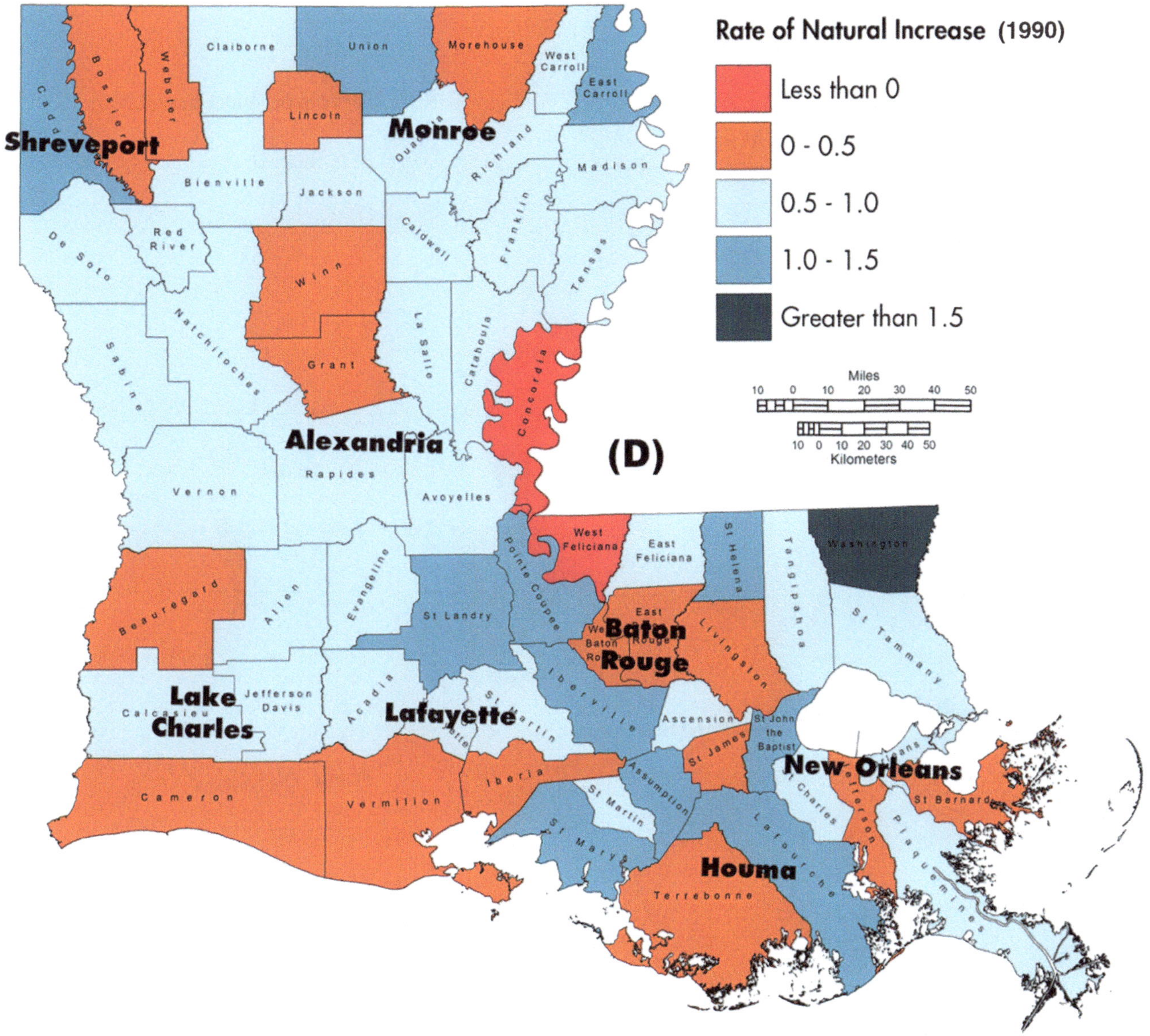

Fig. 14 (continued)

where any number of internal and external perturbations force coastal residents to adapt and adjust. Stability is a temporary state that exists in the midst of change and upheaval. Seemingly stable states can suddenly shift and become "something new, with internal controls and aggregate characteristics that are profoundly different from those of the original" (Kinzig et al. 2006). Although these regime shifts often occur because of external perturbations, they also occur because of complex interactions within the system that operate across spatial and temporal scales. These

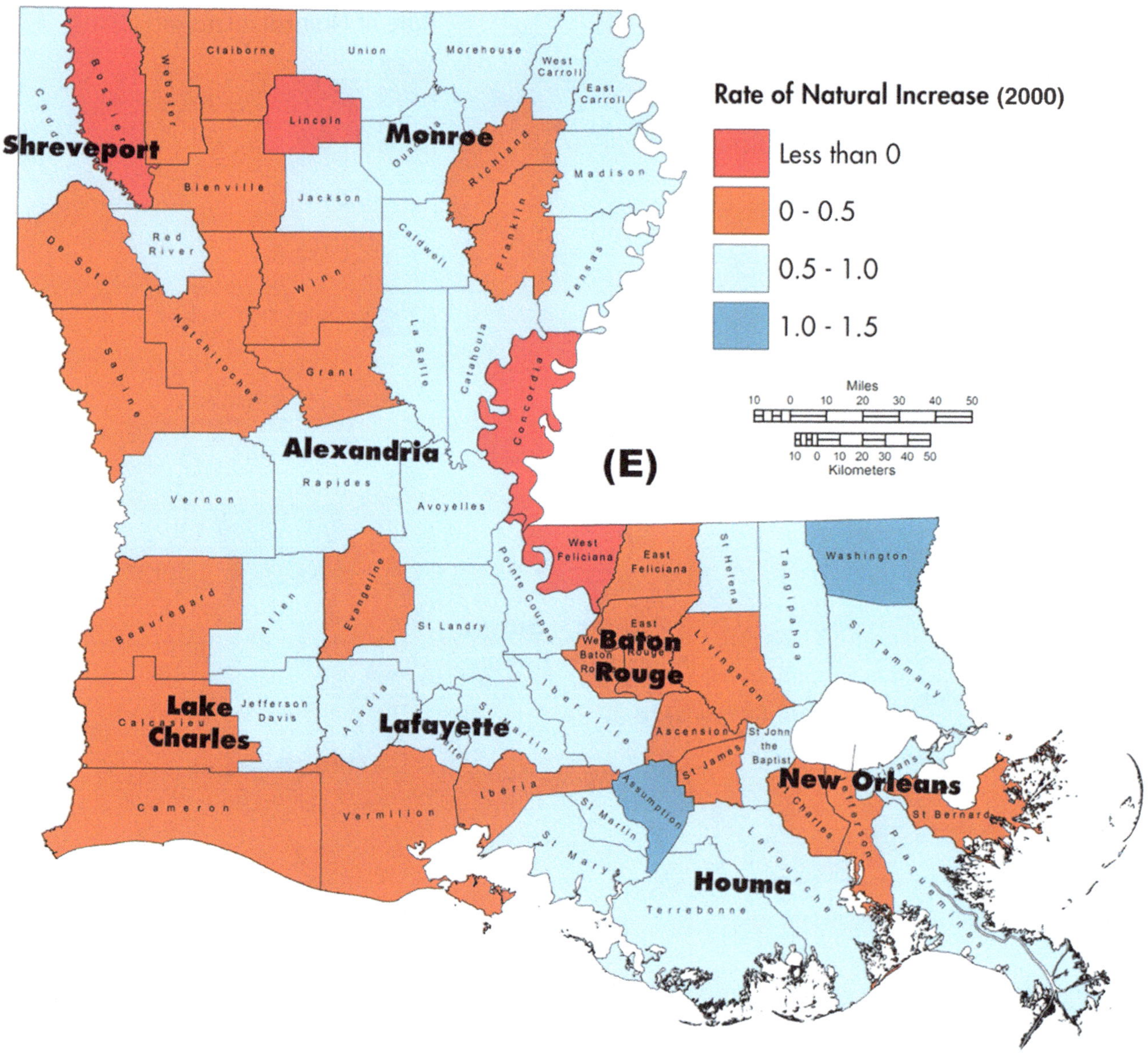

Fig. 14 (continued)

regime shifts often involve changes in the inherent resilience of communities. The key to successfully navigating coastal change is to understand the multiscale historical processes that have been driving that change and assuring that any regime shifts result in positive outcomes for residents and the ecosystems that they rely up in both the short term and long term.

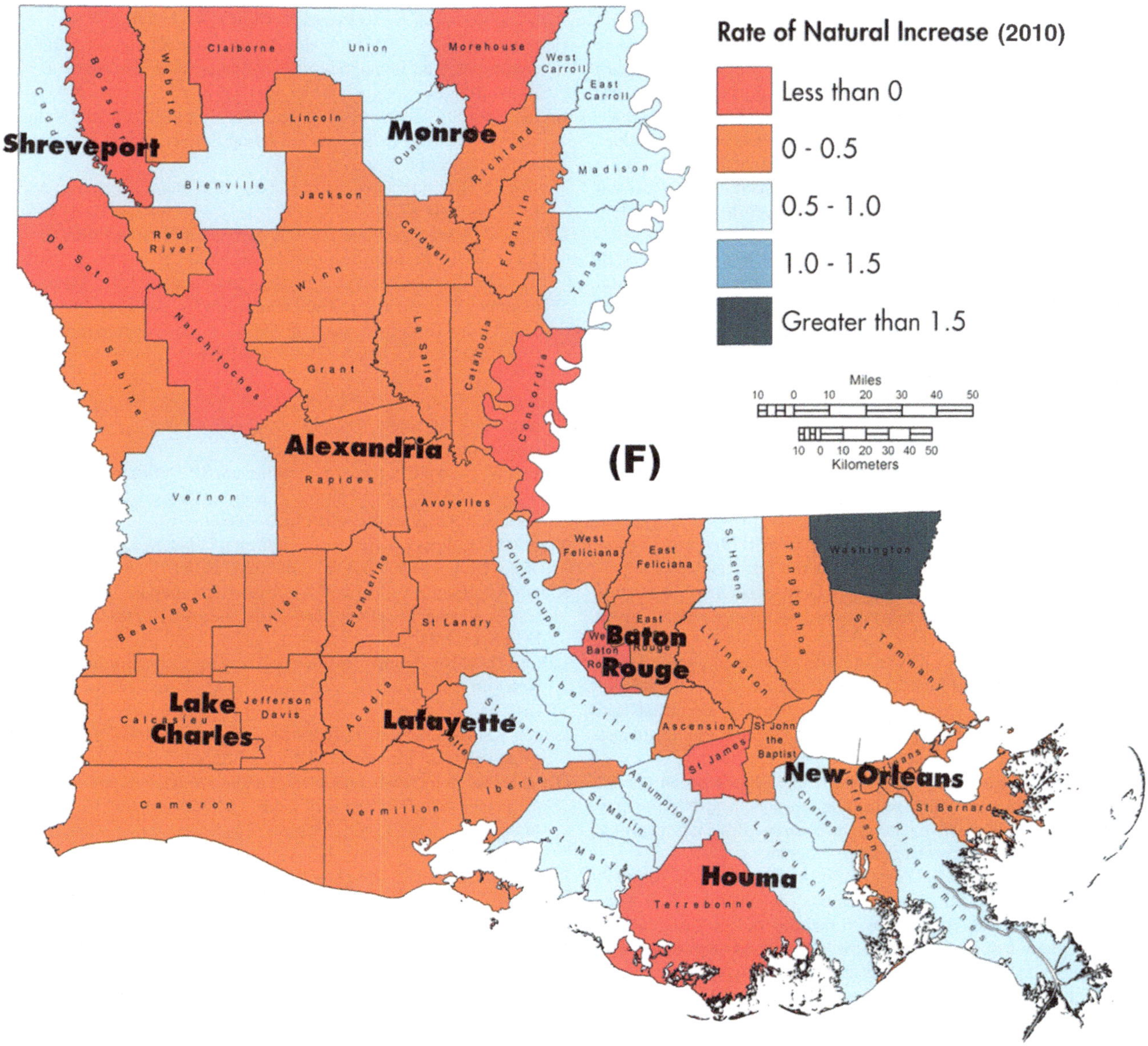

Fig. 14 (continued)

References

Adams CM, Hernandez E, Cato JC (2004) The economic significance of the Gulf of Mexico related to population, income, employment, minerals, fisheries and shipping. Ocean Coast Manag 47 (11–12):565–580

Adger WN (1999) Social vulnerability to climate change and extremes in coastal Vietnam. World Dev 27(2):249–269

Adger WN (2000) Social and ecological resilience: are they related? Prog Hum Geogr 24(3):347–364

Adger WN, Kelly PM, Winkels A, Huy LQ, Locke C (2002) Migration, remittances, livelihood trajectories, and social resilience. AMBIO J Hum Environ 31(4):358–366

Adger WN, Brooks N, Bentham G, Agnew M, Eriksen S (2004) New indicators of vulnerability and adaptive capacity, vol 122. Tyndall Centre for Climate Change Research Norwich

Bailey C, Gramling R, Laska SB (2014) Complexities of resilience: adaptation and change within human communities of coastal Louisiana. In: Perspectives on the Restoration of the Mississippi Delta. Springer, Dordrecht, pp 125–140

Black R, Adger WN, Arnell NW, Dercon S, Geddes A, Thomas D (2011) The effect of environmental change on human migration. Glob Environ Chang 21:S3–S11

Burkett VR, Zilkoski DB, Hart DA (2002) Sea-level rise and subsidence: implications for flooding in New Orleans, Louisiana. In: US Geological Survey Subsidence Interest Group Conference: proceedings of the technical meeting, Galveston, Texas, 27–29 November 2001, pp 63–71

Campanella R (2010) Delta urbanism: New Orleans. American Planning Association, Chicago
City of New Orleans (2015) Resilient New Orleans strategy: strategic actions to shape our future city. City of New Orleans, New Orleans
Colten CE, Kates RW, Laska SB (2008) Community resilience: lessons from New Orleans and Hurricane Katrina. Oak Ridge National Laboratory, Community and Regional Resilience Initiative, Oak Ridge
Colten CE, Grimsmore AA, Simms JRZ, Hemmerling SA (2017) Social justice and mobility in coastal Louisiana, USA. Regional Environmental Change. doi:https://doi.org/10.1007/s10113-017-1115-7
Couvillion BR, Barras JA, Steyer GD, Sleavin W, Fischer M, Beck H, Trahan N, Griffin B, Heckman D (2011) Land area change in coastal Louisiana (1932 to 2010) (Scientific Investigations Map No. 3164). US Department of the Interior, US Geological Survey, Reston
CPEX (Center for Planning Excellence) (2012) Best Practices Manual for Development in Coastal Louisiana. CPEX (Center for Planning Excellence), Baton Rouge, p 120
CPRA (Coastal Protection and Restoration Authority) (2012) Louisiana's comprehensive master plan for a sustainable coast. CPRA (Coastal Protection and Restoration Authority), Baton Rouge, p 188
Cutter S, Barnes L, Berry M, Burton C, Evans E, Tate E, Webb J (2008) Community and regional resilience: perspectives from hazards, disasters, and emergency management CARRI Research Report No. 1. Community and Regional Resilience Institute, Oak Ridge
Cutter SL, Burton CG, Emrich CT (2010) Disaster resilience indicators for benchmarking baseline conditions. Journal of Homel Secur Emerg Manag 7(1)
Dalbom C, Hemmerling SA, Lewis JA (2014) Community resettlement prospects in Southeast Louisiana: a multidisciplinary exploration of legal, cultural, and demographic aspects of moving individuals and communities (Issue Paper). Tulane Institute on Water Resources Law & Policy, New Orleans, p 49
Darlington JD, Woodell G (2006) The relationship between coastal restoration and community relocation: an annotated bibliography and analysis of alternative relocation scenarios. Research report for Governor's Applied Coastal Science Program
Day JW, Agboola J, Chen Z, D'Elia C, Forbes DL, Giosan L, Kemp P, Kuenzer C, Lane RR, Ramachandran R, Syvitski J, Yañez-Arancibia A (2016) Approaches to defining deltaic sustainability in the twenty-first century. Estuarine, Coastal and Shelf Science
Demeny P (2011) Population policy and the demographic transition: performance, prospects, and options. Popul Dev Rev 37(s1):249–274
Denney NH, Jennings S, Reynolds JD (2002) Life-history correlates of maximum population growth rates in marine fishes. Proc R Soc B Biol Sci 269(1506):2229–2237
Dixon TH, Amelung F, Ferretti A, Novali F, Rocca F, Dokka R, Sella G, Kim S-W, Wdowinski S, Whitman D (2006) Space geodesy: subsidence and flooding in New Orleans. Nature 441(7093):587–588
Fussell E (2007) Constructing New Orleans, constructing race: a population history of New Orleans. J Am Hist 94(3):846–855
Hemmerling SA (2017) A Louisiana coastal atlas: resources, economies, and demographics. Louisiana State University Press, Baton Rouge
Hemmerling SA, Hijuelos AC (2016) Coastal Louisiana Social Vulnerability Index (SVI). Version I. Coastal Protection and Restoration Authority, Baton Rouge, p 27
Holling CS (1973) Resilience and stability of ecological systems. Annu Rev Ecol Syst:1–23
IPCC (2012) Managing the risks of extreme events and disasters to advance climate change adaptation. In: Field CB, Barros V, Stocker TF, Qin D, Dokken DJ, Ebi KL, Mastrandrea MD, Mach KJ, Plattner GK, Allen SK, Tignor M, Midgley PM (eds) A special report of Working Groups I and II of the Intergovernmental Panel on Climate Change. Cambridge University Press, Cambridge, p 582
Johnson L, Olshansky R (2010) Clear as mud: planning for the rebuilding of New Orleans. American Planning Association. Planners Press, Chicago
Kates RW, Colten CE, Laska S, Leatherman SP (2006) Reconstruction of New Orleans after Hurricane Katrina: a research perspective. Proc Natl Acad Sci 103(40):14653–14660
Kildow JT, Colgan CS, Scorse JD, Johnston P, Nichols M (2014) State of the US ocean and coastal economies 2014
Kinzig AP, Ryan PA, Etienne M, Allison HE, Elmqvist T, Walker BH (2006) Resilience and regime shifts: assessing cascading effects. Ecol Soc 11(1)
LaRose G (2016) Gentilly Resiliency District goes against flow of how New Orleans handles stormwater. The Times-Picayune, New Orleans
Lenton TM (2013) What early warning systems are there for environmental shocks? Environ Sci Policy 27(Suppl 1):S60–S75
Olson S (2011) Increasing national resilience to hazards and disasters: the perspective from the Gulf Coast of Louisiana and Mississippi: summary of a workshop. National Academies Press, Washington, DC
Pendall R, Foster KA, Cowell M (2007) Resilience and regions: building understanding of the metaphor. Institute of Urban and Regional Development, Berkeley
Saucier RT (1962) Recent geomorphic history of the Pontchartrain Basin, Louisiana. DTIC Document
Saucier RT (1994) Geomorphology and quaternary geologic history of the lower Mississippi Valley, vol 1. US Army Corps of Engineers, Vicksburg
Smithers J, Smit B (1997) Human adaptation to climatic variability and change. Glob Environ Chang 7(2):129–146
Wisner B, Blaikie P, Cannon T, Davis I (2004) At risk, 2nd edn. Routledge, London

Developing Legal Strategies for a Sustainable Coast

James G. Wilkins

Abstract

Legal mechanisms exist for moving people involuntarily from high risk areas but property rights issues and political costs may make such actions counterproductive. Takings law at the state level has narrowed the range of allowable public purposes for eminent domain. Eminent domain, however unpopular it is as a government function, may in some cases be the only option. Relocation assistance such as the federal Uniform Relocation Assistance and Real Property Acquisition Act can help lessen resistance to forced relocation but there are also a number of programs administered by FEMA and HUD that provide assistance for voluntary relocation. The limiting factor for any type of relocation is, almost invariably, money which depends on the political will of the governing bodies who appropriate funds. Large federal programs currently in existence, GOMSEA and RESTORE may be flexible enough to provide funding for hazard mitigation. Market forces, such as the escalating cost and availability of insurance, will at some point be the major driver of relocation and elevation in hazardous coastal areas. Proposed revisions to the National Flood Insurance Program have caused a great deal of anxiety and political backlash in flood prone communities but seem inevitable especially in light of climate change. The escalating cost of federal disaster relief will also be a factor in relocation and land use planning in high hazard areas. Taxpayers will tire of subsidizing repeated irresponsible decisions of individuals and local governments and the diminished availability of disaster relief will result in relocation. Local governments will be increasingly brought to task for poor land use planning decisions. There are several legal and policy changes that could affect the actions proposed in this book such as how structural protection is factored into the NFIP, making flood insurance mandatory; restricting disaster relief; increasing NFIP requirements on local governments; requiring developers to post bonds; requiring full disclosure of risk by the real estate industry; controlling and coordinating land use planning for hazard mitigation at the state level. Third party standards such as certified-flood-resistant or climate-smart-investment and private regulation such as real estate and lending institutions requiring flood insurance based on the extent of historic floods could be powerful market forces.

J.G. Wilkins (✉)
Louisiana Sea Grant Law and Policy Program, Center for Natural Resource Economics and Policy, Louisiana State University, Baton Rouge, LA, USA
e-mail: jwilkins@lsu.edu

J.W. Day, J.A. Erdman (eds.), *Mississippi Delta Restoration*,
Estuaries of the World, https://doi.org/10.1007/978-3-319-65663-2_13

Keywords
Eminent domain • Relocation and property acquisition • National flood insurance program • Land use planning • FEMA • Disaster relief

Introduction

The water is coming, what to do? The actions humans take in response to sea level rise and other aspects of climate change will have profound effects on the environment, institutions and culture.[1] The ideas proposed in this book offer some possible solutions that may seem radical at first but could be much cheaper and more permanent than the standard measures currently employed. Much can be accomplished with enough capital and political will. Political will involves overcoming public resistance and having the authority to act. The first impulse is always to resist any significant change in the status quo. It will take a great deal of effort to influence cultural and individual psychology enough to allow effective adaptation, or it may just take one or two more disastrous events, we just don't know. The question this chapter asks is whether legal systems can assist the process or at least not hinder it too much. Legal issues affecting adaptation are usually focused on property rights, civil rights, and enabling of risky behavior by creating entitlements to government programs.[2] The extent to which legal systems can evolve with changing conditions has always been a difficult discussion especially when Constitutional issues are involved. Adaptation strategies that attempt to affect property rights are the foremost example[3] but other types of "rights" can arise, some created by statute. Flood insurance is an example of a statutorily created system that, judging by reaction to proposed changes to it, has come to be considered an entitlement.[4] Changes to the National Flood Insurance Program (NFIP) seem inevitable[5] and could be a vehicle for effective adaptation. In addition to the NFIP, there are other legal mechanisms that affect adaptation. These include eminent domain which is the governmental power to take private property for a public purpose as long as the private property owner is fairly compensated; mineral rights issues involving loss of the right to acquire subsurface minerals when surface rights are lost; the Uniform Relocation Assistance and Real Property Acquisition Policies Act (URA) which provides assistance to people displaced federal projects; the Gulf of Mexico Energy Security Act (GOMESA) that allows funds to be used for hurricane protection and infrastructure and the Resources and Ecosystems Sustainability, Tourist Opportunities, and Revived Economies of the Gulf Coast States Act (RESTORE Act) which has provisions for coastal flood protection and related infrastructure.[6] In addition there are market forces that can be affected by law and policy and could have significant effects on risk adaptation strategies. These include financial lending practices and rules, real estate practices and insurance industry practices and reforms.

National Flood Insurance Program

Whose responsibility is it to prevent flooding? The U.S. Army Corps of Engineers was authorized to get into the flood control business in the late 1800s, as long as it

[1] James E.M. Watson, *Human Responses to Climate Change will Seriously Impact Biodiversity Conservation: It's Time We Start Planning for Them*, Conservation Letters: A journal of the Society for Conservation Biology, Jan.- Feb. 2014, at 1, 1; Robert J. Nicholls & Richard S. J. Tol, *Impacts and responses to sea-level rise: a global Analysis of the SRES scenarios over the twenty-first century,* Philosophical Transactions of the Royal Society a: Mathematical, Physical and Engineering Sciences, Apr. 2016, at 1073, 1073; Susanne C. Moser, *Impact assessments and policy responses to sea-level rise in three U.S. states: An exploration of human-dimension uncertainties,*15 Global Environmental Change, 353, 353 (2005); R.J. Nicholls, K.C. Dennis & C.R. Volonté, *Impacts and Responses to Sea-Level Rise: Qualitative and Quantitative Assessments,* Journal of coastal Research, Spring 1995, at 26, 26.

[2] Robert Meltz, Congressional Research Service, Climate Change and Existing Law: A Survey of Legal Issues Past, Present, and Future (2014); Wilkins J. *Is Sea Level Rise Foreseeable? Does It Matter?* 26 Journal of Land Use & Environmental Law 437 (2011).

[3] Lara D. Guercio, *Climate Change Adaptation and Coastal Property Rights: A Massachusetts Case Study*, 40 B.C. Envtl. Aff. L. Rev. 349 (2013), http://lawdigitalcommons.bc.edu/ealr/vol40/iss2/2; Bill Higgins, Institute for Local Government, Regulatory Takings and Land Use Regulation: A Primer for Public Agency Staff (2006), *available at* http://mrsc.org/getmedia/FB8A8201-E2CC-453B-BE00-AECEFA196562/m58takings.aspx

[4] Barry Zalma, *Why is a National Flood Insurance Program Policy Not Insurance?*, LexisNexis Legal Newsroom, September 23, 2015, at 1.

[5] *Changes to the National Flood Insurance Program (NFIP)*, FEMA (May 23, 2016 12:34 PM), https://www.fema.gov/changes-national-flood-insurance-program; *National Flood Insurance Program (NFIP),* Nat'l Ass'n of Ins. Comm'rs & The Ctr. for Ins. Policy and Research (June 21, 2016), http://www.naic.org/cipr_topics/topic_nfip.htm

[6] Holly Doremus, *Climate Change and the Evolution of Property Rights,* 1 U.C. Irvine L.Rev. 1091 (2011); Center for Climate and Energy Solutions, Climate Change Adaptation: What Federal Agencies Are Doing 7 (2012); Kelly Klima & Alessandra Jerolleman, *Integrating adaptation in hazard mitigation planning efforts,* Natural Hazard Mitigation Association (May 1, 2013), https://www.epa.gov/sites/production/files/2016-03/documents/part_6_-_integrating_adaptation_in_hazard_mitigation_planning_efforts.pdf; Letter from John Bel Edwards, Governor, State of Louisiana, to Barack Obama, President, U.S. Gov't (Feb. 12, 2016) (on file with the author).

related to navigation.[7] At the same time the Mississippi River Commission adopted their "levees only" policy and the Corps got busy building levees (see chapter "Levees and the Making of a Dysfunctional Floodplain").[8] From that time, it seems that individual responsibility for flood prevention was largely replaced by government efforts. Time has proven the folly of depending solely on public works projects to protect personal property from flooding. From the Great Flood of 1927 to the devastation of Hurricane Katrina there have been a series of failures of the levees-only policy.[9] The great south Louisiana flood of 2016 was caused by intense rainfall from a near stationary system fueled by an overheated Gulf that dumped up to 2 ft (more than 60 cm) of rain in 3 days. This clearly demonstrated that catastrophic flooding in the coastal area is not solely related to hurricane or Mississippi River flooding.[10]

Hurricane Katrina, resulted in a change in Corps terminology for its flood defense structures from "protection" to "risk reduction systems," in recognition of the reality that a levee is only the first line of defense and a necessarily imperfect one.[11] Congress acknowledged the inadequacy of structural flood protection when it enacted the National Flood Insurance Program in 1968.[12] The NFIP was intended to provide affordable flood insurance to those vulnerable to flooding and to reduce flood risk through better land use planning. However, it was designed as a quid pro quo agreement with states and local governments: affordable flood insurance in exchange for land use planning and construction practices that directed future development out of the flood risk areas, either through elevation or restricting (even preventing) development in floodplains.[13] Over time, it seems that the second half of the bargain has been neglected and the first half is now considered a sacred entitlement that individuals and local economies cannot do without.[14] The NFIP has been in the red since at least Hurricane Katrina and the deficit was exacerbated by Hurricane Sandy and will soon greatly increase from hurricanes Harvey, Irma and Maria.[15] In 2011 Congress attempted to put the program on sounder financial footing by revising the premium structure to be more in line with true actuarial costs rather subsidized rates.[16] The Biggert-Waters Act caused a firestorm of protest from home-owners and their political representatives because of the drastic increases in premiums that would result from the revision.[17] Parts of the law were temporarily suspended to further study the effects of the revisions and possible options.[18] The issue will arise again and, in light of predicted sea level rise,[19] more intense hurricanes, and more erratic and intense weather in general (witness the 2016 flood), will be more controversial and more important than ever.

The flood insurance controversy presents an opportunity for the kind of radical, forward thinking ideas suggested herein (see chapter "Eroding Communities and Diverting Populations: Historical Population Dynamics in Coastal Louisiana"). Since flood insurance has become such a necessary component of coastal economies there is a great deal of leverage that could be employed to establish a new bargain between the federal government and local governments. Elevating portions of ground within polders in New Orleans could be a more permanent solution than levees and might

---

[7] PBS, *The Mississippi River Commission and the Army Corps of Engineers,* AMERICAN EXPERIENCE, http://www.pbs.org/wgbh/american-experience/features/general-article/flood-control/.

[8] CHARLES A. CAMILLO & MATTHEW T. PEARCY, U.S. MISSISSIPPI RIVER COMMISSION, UPON THEIR SHOULDERS: A HISTORY OF THE MISSISSIPPI RIVER COMMISSION FROM ITS INCEPTION THROUGH THE ADVENT OF THE MODERN MISSISSIPPI RIVER AND TRIBUTARIES PROJECT 28 (2nd Ed. 2006).

[9] *Id.* at 112, 150, 153, 157, 180, 194.

[10] Chris Dolce, *Why the Louisiana Flood Happened, And 4 Other Things to Know,* THE WEATHER CHANNEL (Aug. 15, 2016 10:00 PM), https://weather.com/storms/severe/news/louisiana-flooding-why-it-happened-things-to-know

[11] *Risk Reduction Plan,* US ARMY CORPS OF ENGINEERS, http://www.mvn.usace.army.mil/Missions/HSDRRS/Risk-Reduction-Plan/; NATIONAL RESEARCH COUNCIL OF THE NATIONAL ACADEMIES, LEVEES AND THE NATIONAL FLOOD INSURANCE PROGRAM: IMPROVING POLICIES AND PRACTICES 33–48 (2013), http://planning.usace.army.mil/toolbox/library/PCoP/S3_Moving%20from%20Damage%20Reduction%20to%20Risk%20Managment_HalpinCocchieri.pdf

[12] P.L. 90–448 Sec 1302(a)(2) 1968: "The Congress finds that. . .despite the installation of preventive and protective works and the adoption of other public programs designed to reduce losses caused by flood damage, these methods have not been sufficient to protect adequately against growing exposure to future flood losses."; 42 U.S.C. § 4011 (West 2012).

[13] *Supra* note 5.

[14] Zalma, *supra* note 4.

[15] CAROLYN KOUSKY & ERWANN MICHEL-KERJAN, HURRICANE SANDY, STORM SURGE AND THE NATIONAL FLOOD INSURANCE PROGRAM: A PRIMER ON NEW YORK AND NEW JERSEY 8 (2012), http://www.rff.org/files/sharepoint/WorkImages/Download/RFF-IB-12-08.pdf; Meghan Milloy, *The NFIP Is Due For Some Major Reforms,* AMERICAN ACTION FORUM (March 9, 2016), https://www.americanactionforum.org/research/the-nfip-is-due-for-some-major-reforms/.; DOUGLAS HOLTZ-EAKIN, AMERICAN ACTION FORUM, IMPLEMENTATION OF THE BIGGERT-WATERS FLOOD INSURANCE ACT OF 2012: PROTECTING TAXPAYERS AND HOMEOWNERS 4 (2013), *available at* http://financialservices.house.gov/uploadedfiles/hhrg-113-ba04-wstate-dholtzeakin-20131119.pdf; Milloy, *supra* note 15.

[16] Biggert-Waters Flood Insurance Reform Act of 2012, 126 Stat. 916 (codified as amended at 42 U.S.C. §§ 4001–4131 (2012)).

[17] NAT'L ASS'N OF INS. COMM'RS & THE CTR. FOR INS. POLICY AND RESEARCH, *supra* note 5.

[18] Andrew G. Simpson, *House Passes Flood Insurance Bill; Key Senators Sign On,* INSURANCE JOURNAL (March 4, 2014), http://www.insurancejournal.com/news/national/2014/03/04/322194.htm; Bjorn Philip Beer *Underwater,* EARTH ISLAND JOURNAL, WINTER 2015, http://www.earthisland.org/journal/index.php/eij/article/underwater/

[19] Rebecca Lindsey, NOAA, *Climate Change: Global Sea Level,* CLIMATE.GOV (June 20, 2016), https://www.climate.gov/news-features/understanding-climate/climate-change-global-sea-level

well cost less in the long run (see chapter "Eroding Communities and Diverting Populations: Historical Population Dynamics in Coastal Louisiana"). There are procedural and logistical challenges as well as financing issues, but elevating polders could be the most effective and permanent solution to flooding risks absent wholesale retreat. It would be in line with the goal of the NFIP to direct development out of flood risk zones, in this case a vertical retreat. The question is how to convince people to give up living in the areas to be raised, at least temporarily, and how to finance the projects.

On the issue of convincing people to move, at least temporarily out of areas to be elevated, the first strategy will always be voluntary buyouts. The Department of Homeland Security and FEMA have provisions for buyouts of various classifications of property[20] including the Hazard Mitigation Grant Program for property which is damaged in a Presidentially declared major disaster[21] and the Flood Mitigation Assistance Grant Program for repetitive loss properties covered by NFIP policies.[22] Both of these are available only for post disaster properties and the land must be relegated to permanent open or green space.[23] However, it might be possible to amend the law to forgo the open space requirement if the land level is raised significantly above the local base flood elevation. FEMA also has the Pre-Disaster Mitigation Program that provides funding for acquisition of properties at risk of flooding and like the post-disaster programs the acquired land may only be used for open space.[24] The FEMA acquisition programs do not have a permanent source of funding but rely on congressional appropriations on an annual basis or appropriations for individual disasters, and usually require non-federal cost sharing.[25] The details of these programs are beyond the scope of this chapter but FEMA has produced extensive guidance for anyone interested in learning more.[26]

Actuarial studies could find that preemptive buyouts will be much more economically efficient than those done post disaster. Figuring in the cost of construction and maintenance of structural flood protection over time and the cost of recovery after a flood disaster versus the more "permanent" solutions might show a significant long-term advantage to preemptive action. Some commentators have suggested that the NFIP be amended to allow pre-disaster agreements for buyouts, that would take place post disaster, and pre-disaster buyouts of flood prone properties in exchange for flood insurance premium discounts.[27] The strategy would probably work best in low income areas that have a high flood risk and therefore high flood insurance premiums. People in those areas might be more willing to accept the deal.

An issue looming over the discussion is the solvency of the NFIP. The program has been billions of dollars in debt to the federal Treasury for over a decade.[28] If sea level rises as predicted it seems unlikely that the NFIP can survive in its current configuration.[29] The NFIP is due for reauthorization in 2017 and there are bound to be serious debates concerning its very existence or, at least, concerning its structural soundness. Strong political support from coastal areas will probably prevent repeal of the NFIP but it is unlikely to come out of the process doing business as usual. There could be significant increases of premiums for individual homeowners, probably not to the extent that Biggert-Waters would have imposed, but enough to affect individual decisions about where to live.[30] There also could be some reforms on the other end of the quid pro quo bargain that the NFIP was supposed to be. Lawmakers could for instance require much stricter local flood ordinances that would significantly tighten land use restrictions.[31] One idea would be to completely remove the existence of structural protection from the calculation of elevation requirements for new construction and substantially damaged properties. That would make the failure of levees and other structural protection much less catastrophic because stricter requirements for

[20] Roy E. Wright, FEMA, Hazard Mitigation Assistance Guidance (2015), *available at:* https://www.fema.gov/media-library-data/1424983165449-38f5dfc69c0bd4ea8a161e8bb7b79553/HMA_Guidance_022715_508.pdf

[21] *Hazard Mitigation Grant Program,* FEMA (Sept. 28, 2016 9:05), https://www.fema.gov/hazard-mitigation-grant-program

[22] *Flood Mitigation Assistance Grant Program,* FEMA (August 30, 2016, 12:34 AM), https://www.fema.gov/flood-mitigation-assistance-grant-program.

[23] Homeland Security Grants, Severe repetitve Loss (SRL) Grant Program, *available at:* http://www.homelandsecuritygrants.info/GrantDetails.aspx?gid=21795

[24] *Pre-Disaster Mitigation Grant Program,* FEMA (Nov. 09, 2016, 10:38), https://www.fema.gov/pre-disaster-mitigation-grant-program; Fed. Ins. & Mitigation Admin. Fact Sheet, FY 2016 Pre-Disaster Mitigation (PDM) Grant Program Fact Sheet (2016), *available at:* https://www.fema.gov/media-library-data/1455711373912-17d561db31cc299667dc5c608111165d1/FY16_PDM_Fact_Sheet.pdf

[25] Wright, S*upra* note 20.

[26] *Id.*

[27] Rob Moore, *Flood, Rebuild, Repeat: The Need for Flood Insurance Reforms,* NRDC (Aug. 11, 2016), https://www.nrdc.org/experts/rob-moore/flood-rebuild-repeat-need-flood-insurance-reforms.

[28] Biggert-Waters, *supra* note 16.

[29] Rachel Cleetus, Union of Concerned Scientists, Overwhelming Risk: Rethinking Flood Insurance in a World of Rising Seas 1–8 (Trudy E. Bell ed., 2013), *available at* http://www.ucsusa.org/sites/default/files/legacy/assets/documents/global_warming/Overwhelming-Risk-Full-Report.pdf

[30] Nat'l Ass'n of Ins. Comm'rs & The Ctr. for Ins. Policy and Research, *supra* note 5.

[31] FEMA, Floodplain Management Requirements 6–16 (2015), *available at* https://www.fema.gov/pdf/floodplain/nfip_sg_unit_6.pdf; Jennifer B. Wriggins, *In Deep: Dilemmas of Federal Flood Insurance Reform,* 5 U.C. Irvine L. Rev. 1443, 1450 (2015).

structures would presumably make them better prepared to withstand flooding. It would also help prevent the "if you build it they will come" effect of levees as can be seen in the area of the Morganza to the Gulf levee that is currently being constructed by the Terrebonne Parish and other local governments.[32] In New Orleans such a policy would result in "uneven" neighborhoods with older slab on grade houses next to those 20 ft in the air. This effect is already being seen in some areas of New Orleans as people voluntarily raise their houses and in New Jersey where levee protection is absent.[33] The ideas proposed in this book to raise the ground level above the flood zone would eliminate that effect and allow more convenient building practices. The cost savings of reduced flood insurance premiums could be used to leverage the cost of polder projects.

One commentator has suggested that flood insurance should be structured in a similar manner as life insurance, with the understanding that the risk of disaster will rise over time as the threat of sea level rise becomes more imminent.[34] Under this model premiums would rise with the risk until those who could not afford the high cost of insurance would leave the coast. One might say that a market based approach such as the proposed Biggert-Waters reforms would accomplish the same end but some argue that the NFIP operates based on extreme events and does not account for long term trends and therefore is not actuarially sound. They contend that flood insurance is intended to deflect the true risks of living in coastal areas and is essentially a social welfare program that seeks to maintain the status quo. Thus, it does not communicate the actual risk of coastal development especially in light of sea level rise.[35]

Eminent Domain

Governments always have the power of eminent domain (the authority to take private property for a public use) at their disposal to carry out valid public purposes as an exercise of their police power.[36] When government does not expressly take private property by eminent domain but restricts the use of property the action may still be considered a "regulatory taking" of private property requiring compensation.[37] The actions proposed in chapters "Raising Buildings: The Resil ience of Elevated Structures" and "Eroding Communities and Diverting Populations: Historical Population Dynamics in Coastal Louisiana" would almost certainly require explicit use of eminent domain requiring just compensation. A distinction has arisen in federal constitutional takings law between government actions that are merely for a public purpose and those that are taken to prevent an imminent threat to public safety under their police power.[38] In the latter situation, government taking of property might not require compensation under the U.S. Constitution's Fifth Amendment Takings Clause as interpreted in some court decisions but the threat would have to be dire.[39] One Louisiana Supreme Court case has suggested that coastal protection and restoration could justify such an extreme use of police power.[40] On the other hand there has been pushback at the state level to another line of U.S. Supreme Court cases holding that eminent domain can be used to bestow economic benefits to private entities.[41]

---

[32] Jacob Batte, *Morganza levee system nears milestone as gap is closed*, HOUMATODAY.COM (June 19, 2016 5:40 PM), http://www.houmatoday.com/news/20160619/morganza-levee-system-nears-milestone-as-gap-is-closed; Len Bahr, *Morganza-to-the Gulf project could become a coastal development bonanza,* LACOASTPOST (Jan. 28, 2015), http://lacoastpost.com/blog/?p=47758; Karl Gommel, *Morganza build progressing: Flood protection system aimed at keeping Terrebonne dry,* HOUMA TIMES, June 8, 2016, *available at* http://www.houmatimes.com/business/morganza-build-progressing-flood-protection-system-aimed-at-keeping-terrebonne/article_942f8bdc-2c1f-11e6-b950-8b6ded0d8e56; ROBERT TWILLEY ET AL., MORGANZA-TO-THE-GULF TECHNICAL PANEL REVIEW (December 5, 2008), *available at* https://www.researchgate.net/publication/255507660

[33] Jeff Adelson, *After Hurricane Katrina, a look at New Orleans' uneven recovery among its neighborhoods,* THE ADVOCATE, Aug. 23, 2015, http://www.theadvocate.com/baton_rouge/news/article_9906ab1b-a1e7-5eb7-8e52-824adb52d6a3.html; *Can the Lower Ninth Ward ever recover from Katrina?,* PBS (Dec. 28, 2013, 12:00 AM), http://www.pbs.org/newshour/bb/nation-july-dec13-lower9th_12-28/

[34] Edward P. Richards, *Applying Life Insurance Principles to Coastal Property Insurance to Incentivize Adaptation to Climate Change*, 43 B. C. ENVTL. AFF. L. REV. 427, 427 (2016), http://lawdigitalcommons.bc.edu/cgi/viewcontent.cgi?article=2198&context=ealr

[35] *Id* at 450.

[36] The Editors of Encyclopedia Britannica, *Police power,* ENCYCLOPEDIA BRITANNICA (July 20, 1998), https://www.britannica.com/topic/police-power (Generally defined as "the permissible scope of federal or state legislation so far as it may affect the rights of an individual when those rights conflict with the promotion and maintenance of the health, safety, morals, and general welfare of the public."); Santiago Legarre, *The Historical Background of the Police Power*, 9 U. PA. J. CONST. L. 745 (2007). Available at: http://scholarship.law.upenn.edu/jcl/vol9/iss3/3

[37] "The Constitution of the United States," Amendment 5; *Pennsylvania Coal* Co. v. *Mahon*, 260 U.S. 393 (1922).

[38] Jan G. Laitos, *The Public Use Paradox and the Takings Clause*, 13 J. ENERGY NAT. RESOURCES & ENVTL. L. 9 (1993).

[39] *Keystone Bituminous Coal Ass'n v. DeBenedictis*, 480 U.S. 470, 107 S. Ct. 1232, 94 L. Ed. 2d 472 (1987); Margaret Teresa Harris, *Compensable "Takings'-and Why Not? An Analysis of the Fifth Amendment Just Compensation Clause and Police Power Regulatory "Takings'*, 10 AM. J. TRIAL ADVOC. 365 (1986).

Jan G. Laitos, *The Public Use Paradox and the Takings Clause*, 13 J. ENERGY NAT. RESOURCES & ENVTL. L. 9 (1993).

[40] *Avenal v. State*, 2003–3521 (La. 10/19/04), 886 So. 2d 1085.

[41] La. Const. Ann. art. I, § 4; *Kelo v. City of New London, Connecticut* 545 U.S. 469, 490 (2005); John J. Costonis, *Avenal v. State: Takings and Damagings in Louisiana*, 65 LA. L. REV. 1015 (2005).

State Powers – "Kelo Issues"

In 2005 the U.S. Supreme Court decided the case of Kelo v. New London, a Connecticut town engaged in urban renewal projects. New London sought to condemn private property and allow it to be redeveloped by other private enterprises under the premise that increasing real estate values is a public purpose. The property owners contended that taking private property to give to other private parties purely for commercial purposes was not a valid use of takings power under the U.S. Constitution because it did not serve a "public use" as required by the Constitution.[42] The U.S. Supreme Court held that the taking was constitutional.[43] The decision shocked private property advocates and generated a wave of state anti-takings legislation that limited state government takings authority for purposes of private economic development.[44] Louisiana joined the majority of states by enacting legislation that in the opinion of some legal scholars severely hampers the state's ability to use eminent domain to conduct land use reforms for coastal protection and restoration purposes if such actions are seen to benefit private economic interests.[45] On the other hand, Louisiana may have wide eminent domain authority based on its police power for the protection of public safety.[46] Land use reforms that have the primary goal of reducing flood risk with the indirect effect of increasing the value of private property would likely survive scrutiny under the state constitutional prohibitions. The topic needs more extensive discussion and examination, already started by others, but is beyond the scope of this chapter. Suffice to say that any land use changes affecting private property by the use of eminent domain, or takings resulting from regulatory restrictions, or damage from government action, should have as the primary purpose of the action the protection of public safety with any economic benefit to private parties as an incidental but added justification.

There are at least two other legal issues that can affect Federal Emergency Management Agency (FEMA) buyouts and eminent domain. The first involves clouded titles to immovable (real) property. Some cultural subsets in Louisiana eschew formal (legal) transfer of property when the owners die. In Louisiana that would entail going through the succession process in the courts. Without that process an heir who has merely taken possession of the real property of a deceased relative cannot "legally" sell the property even if they have been in possession of it for years. FEMA will not buy out properties unless the owners (possessors) have a clear title.[47] This issue of clouded titles could also affect eminent domain. Even though a government may condemn property that has a clouded title and deposit the money with a court until the ownership is sorted out, disputes with the presumptive owner can cause significant delays.[48] At least one non-profit organization, Louisiana Appleseed, offers people without clear title to their property assistance to obtain such title.[49]

Another issue unique to Louisiana involves the issue of subsurface minerals. Under Louisiana law the various ownership rights in land that have been separated and vested in more than one person usually returns complete or perfect ownership rights in one person by default.[50] Thus, with a few exceptions, there is no permanent ownership of subsurface minerals in Louisiana apart from ownership of the surface as there are in other states like Texas.[51] The ownership of subsurface minerals eventually returns to the owner of the surface.[52] When land is appropriated via eminent domain it may be possible to reserve mineral rights in the seller for a temporary period.[53] However, if the mineral rights are not exercised within the time prescribed by law, usually 10 years, they revert to the surface owner, the state or another party to whom the state has transferred ownership.[54] For many in coastal Louisiana losing mineral rights is an

[42] *Kelo v. City of New London, Connecticut* 545 U.S. 469, 490 (2005).

[43] *Id.*

[44] Donald E. Sanders & Patricia Pattison, *The Aftermath of Kelo*, 34 Real Est. L.J. 157 (2005); Ilya Somin, *The Judicial Reaction to Kelo*, 4 Alb. Gov't L. Rev. 1 (2011); Scott P. Ledet, *The Kelo Effect: Eminent Domain and Property Rights in Louisiana*, 67 La. L. Rev. 171 (2006).

[45] La. Const. Ann. art. I, § 4; Scott P. Ledet, *The Kelo Effect: Eminent Domain and Property Rights in Louisiana*, 67 La. L. Rev. 171 (2006); John J. Costonis, *Avenal v. State: Takings and Damagings in Louisiana*, 65 La. L. Rev. 1015 (2005).

[46] *Avenal v. State*, 2003–3521 (La. 10/19/04), 886 So. 2d 1085.

[47] *For Communities Plagued by Repeated Flooding, Property Acquisition May Be the Answer*, FEMA (May 28, 2014, 12:25), https://www.fema.gov/news-release/2014/05/28/communities-plagued-repeated-flooding-property-acquisition-may-be-answer

[48] *Cadeville Gas Storage, LLC v. 30.68 Acres of Land In Ouachita Par., La.*, No. CIV.A. 12–2837, 2013 WL 6712975, at *1 (W.D. La. Dec. 20, 2013).

[49] *Louisiana Appleseed is dedicated to solving our state's toughest problems at the root cause*, Louisiana Appleseed (2017), http://louisianaappleseed.org/; *Southeast Louisiana Legal Services*, SLLS (2017), https://slls.org/

[50] La. Civ. Code Ann. art. 530; A. N. Yiannopoulos, *Usufruct: General Principles - Louisiana and Comparative Law*, 27 La. L. Rev. 369, 369 (1967) *available at* http://digitalcommons.law.lsu.edu/cgi/viewcontent.cgi?article=3465&context=lalrev

[51] Guy E. Wall, *Imprescriptible Mineral Interests In Louisiana*, 42 La. L. Rev. 123, 123 (1981) *available at:* http://digitalcommons.law.lsu.edu/cgi/viewcontent.cgi?article=4607&context=lalrev

[52] *Id.*

[53] Robert A. Dunkelman, *Consideration of Mineral Rights in Eminent Domain Proceedings*, 46 La. L. Rev. 827, 829 (1986).

[54] E.g., *Frost-Johnson Lumber Co. v. Salling's Heirs*, 150 La. 756, 91 So. 207 (1922).

important negative effect of moving. There is already a provision in the Louisiana Constitution (Article 9 Section 4) allowing the state to negotiate the ownership of mineral rights when coastal restoration projects build land in navigable water bottoms that the adjacent property owner, who previously lost land to erosion, could have rebuilt and become the owner of under the Louisiana Constitution and revised statutes.[55] An amendment to that provision or a similar statute could allow reservation of mineral rights in situations where land is condemned and rebuilt for resiliency purposes. Perhaps mineral rights would follow property owners to relocation areas and back to the original property. Or mineral rights could be pooled with each of the former inhabitants having an equal share similar to the unitization statutes for common mineral reservoirs. There are probably a number of creative ways to maintain mineral rights during dislocation caused by the polder projects, which would help alleviate apprehension of the property owners affected.

Federal Programs That Affect Relocation

When the federal government is involved in a project that results in the displacement of human settlement, there are federal laws that can be brought to bear to help resolve the financing of property and relocation issues. I have already discussed the "Takings clause in the 5th Amendment of the U.S. Constitution and the power of eminent domain. At present there is no single state or federal agency tasked with the responsibility of overseeing or coordinating the relocation of human populations affected by state or federal public works projects. A recent survey of federal government agencies and programs involved in relocation was conducted in 2014.[56] From that list the following could apply to the proposals suggested herein.

FEMA Programs[57]

- FEMA – Hazard Mitigation Grant Program – provides grants after a major disaster declaration to state and local governments for implementation of long-term hazard mitigation measures. Funds can be used to acquire and demolish property and relocate displaced residents.
- FEMA – Flood Mitigation Assistance Grant Program–provides for property acquisition, demolition, and relocation of repetitive loss properties that are insured under the National Flood Insurance Program to reduce or eliminate risk of flood damage.
- FEMA – Pre-Disaster Mitigation Program – provides annually distributed funds for property acquisition and relocation as part of hazard mitigation planning.

HUD Programs

- U.S. Department of Housing and Urban Development (HUD) – Community Block Grant Entitlement Program – provides grants to local governments for development in urban areas. The emphasis is on low- and moderate-income residents and grants can be used to acquire real property for relocation or demolition.[58]
- HUD – Community Development Block Grant Disaster Recovery Program – applies to Presidentially declared disasters for the purchase of damaged properties in the flood plain and relocation of displaced residents to safer areas, including payments for people and businesses displaced by disaster.[59]
- HUD-HOME Investment Partnership Program – provides formula grants to states and local communities to develop affordable housing for low-income individuals. The funds may be used for direct rental assistance or to develop housing stock for rent or ownership; for building, buying, and/or rehabilitating buildings or for site acquisition and improvement, including demolition of dilapidated housing to make way for HOME-assisted development and[60] to cover some relocation costs.[61]

The programs listed above all have specific requirements that must be met before they are available to individuals or governments. The general theme of the programs is to improve resiliency.

There are HUD funded efforts underway to relocate communities affected by climate change such as the

[55] La. Const. Ann. art. IX, § 4; La. Const. Ann. art. IX, § 3, La. R.S. 41:1702–1714.

[56] Melissa Daigle, Research Associate & James Wilkins, Director, La. Sea Grant Law and Policy Program The Water is Coming- What to Do? 16–21 (2014); U.S. General Services Administration, Government wide Relocation Advisory Board Findings and Recommendations 2 (2005) *available at* www.gsa.gov/.../Findings_and_recommendations_R2-yMA0_0Z5RDZ-i34K-pR.doc

[57] *Supra* notes 20–26.

[58] *Community Development Block Grant Entitlement Program,* HUD Exchange (2016), https://www.hudexchange.info/programs/cdbg-entitlement/

[59] *Community Development Block Grant Disaster Recovery Program,* HUD Exchange (2016), https://www.hudexchange.info/programs/cdbg-dr/

[60] *HOME Investment Partnerships Program,* U.S. Department of Housing and Urban Development (2016), http://portal.hud.gov/hudportal/HUD?src=/program_offices/comm_planning/affordablehousing/programs/home/

[61] *Id.*

community of Isle De Jean Charles in coastal Louisiana's Terrebonne Parish.[62]

Uniform Relocation Assistance and Real Property Acquisition Policies Act (URA)

The URA was enacted in 1970[63] to provide consistent and equitable treatment of people displaced by acquisition of property for federal projects and federally assisted programs including state and local programs. It calls for "the unique circumstances" of displaced persons in "essentially similar situations" to be considered uniformly so they can be afforded similar treatment under the benefits of the law when they are subject to governmental relocation plans or mandates. The program's benefits include moving expenses, temporary housing and payments for the costs of purchasing or renting comparable replacement housing rather than just fair market value. Compensation can also be provided for moving expenses, closing costs and security deposits.

The URA is administered by the Federal Highway Administration ("FHA"). URA benefits apply to persons displaced as a "*direct* result of the acquisition, rehabilitation, or demolition of real property[64] by "a direct Federal program or project" and "Programs and projects receiving Federal financial assistance."[65] The URA establishes a right to "expeditious acquisition of the property, advance notice of the agency's intent to acquire property, offers of "just compensation," and certain litigation expenses.[66] For the Road Home[67] program the URA applied to relocation funded by federal disaster assistance. The methods and participants in the proposed polder projects will determine if the necessary triggering events for the application of the URA are present.

Gulf of Mexico Energy Security Act (GOMESA), and the Resources and Ecosystems Sustainability, Tourist Opportunities, and Revived Economies of the Gulf Coast States Act (RESTORE)

GOMESA and RESTORE offer two possible sources of future funding for nonstructural flood protection measures. GOMESA allows for revenue sharing of outer continental shelf oil and gas leasing activities with Alabama, Louisiana, Mississippi, and Texas.[68] RESTORE establishes a fund and sets out the distribution plan for 80% of the civil penalties levied in the case of the Deepwater Horizon Oil Spill.[69] Under RESTORE, 30% of the amount made available to Louisiana must go directly to coastal zone parishes, with the amount assigned based on a formula that takes into consideration the miles of shoreline oiled, the population, and the land mass of the parish. There are some conditions to this requirement. Specifically, "as a condition of receiving amounts allocated under [RESTORE], the chief executive of the eligible parish shall certify to the Governor of the State that the parish has completed a comprehensive land use plan."[70] In a recent study conducted by the Louisiana Sea Grant Law & Policy Program[71] one interviewee mentioned that the parish where he lived had submitted a plan for approval to CPRA, but no one really knew what the comprehensive land use plan is supposed to contain. That parish decided to focus on issues raised in the 2012 Master Plan, including projects such as bank stabilization and marsh creation. Their plan also used vague terminology regarding proposed levees that they hoped would eventually tie into the neighboring parish's proposed levees. As part of a comprehensive plan structural protection projects may be effective but by themselves they in no way substitute for more permanent measures such as avoiding development in flood hazard areas either by elevating or moving back from flood plains. Parishes could benefit greatly from taking the RESTORE requirement to develop comprehensive land use plans seriously and receiving direction on what exactly should be included in the plan but it seems that many

[62] Coral Davenport & Campbell Robertson, *Resettling the First American 'Climate Refugees'*, N. Y. Times (May 2, 2016), https://www.nytimes.com/2016/05/03/us/resettling-the-first-american-climate-refugees.html?_r=0; Carolyn Van Houten, *The First Official Climate Refugees in the U.S. Race Against Time*, Nat'l Geographic (May 25, 2016), http://news.nationalgeographic.com/2016/05/160525-isle-de-jean-charles-louisiana-sinking-climate-change-refugees/; Bob Marshall, *The people of Isle de Jean Charles aren't the country's first climate refugees*, The Lens (Dec. 6, 2016, 3:22 PM), http://thelensnola.org/2016/12/06/the-people-of-isle-de-jean-charles-arent-the-countrys-first-climate-refugees/

[63] Uniform Relocation Assistance and Real Property Acquisition Policies for Federal and Federally Assisted Programs, 42 U.S.C. §§ 4601–4655 (West 2016); Uniform Relocation Assistance and Real Property Acquisition Policies for Federal and Federally Assisted Programs, 49 C.F.R. §§ 24.1–24.603 (West 2016).

[64] 49 C.F.R. §24.2(9) (West 2016).

[65] 49 C.F.R. §24.101(a), (b) (West 2016).

[66] 49 C.F.R. §24.102(a), (b), (d)–(g) (West 2016).

[67] *The Road Home Program Overview*, The Road Home (2017), https://www.road2la.org/; *Road Home Program- Grant Imbursements As Of Monday, March 27, 2017*, The Road Home (2017), https://www.road2la.org/HAP/

[68] Gulf of Mexico Energy Securities Act, 43 U.S.C. §1337 (West 2016).

[69] Resources and Ecosystems Sustainability, Tourist Opportunities, and Revived Economics of the Gulf Coast States Act (RESTORE Act), Subtitle F of Pub. L. 112–141 (2012).

[70] *Id.*

[71] Melissa Daigle, Research Associate & James Wilkins, Director, La. Sea Grant Law and Policy Program The Water is Coming- What to Do? 9 (2014).

parishes are attempting to make do with what is already in place, which is often the bare minimum.

Both RESTORE and GOMESA allow funds to be used for flood protection projects, but neither have a set amount that must be used for nonstructural or even mention nonstructural measures. The use of RESTORE funds is limited to a list of eligible activities related to ecosystem rehabilitation and direct human benefits. In that list is a category that could apply to the projects proposed in chapters "Raising Buildings: The Resilience of Elevated Structures" and "Erod ing Communities and Diverting Populations: Historical Pop ulation Dynamics in Coastal Louisiana":

> "(B) Use of funds.
>
> (i) Eligible activities in the Gulf Coast region. Subject to clause (iii), amounts provided to the Gulf Coast States under this subsection may only be used to carry out 1 or more of the following activities in the Gulf Coast region:
>
> **(VII) Coastal flood protection and related infrastructure.**"[72]

Under GOMESA, in the list of authorized uses is this provision:

> "(1) In general. Subject to paragraph (2), each Gulf producing State and coastal political subdivision shall use all amounts received under subsection (b) in accordance with all applicable Federal and State laws, only for 1 or more of the following purposes:
>
> (A) Projects and activities for the purposes of coastal protection, including conservation, coastal restoration, **hurricane protection, and infrastructure** directly affected by coastal wetland losses.[73]

It appears that the provisions of both GOMESA AND RESTORE reference to "coastal flood protection and related infrastructure" and hurricane protection and infrastructure" could be interpreted to allow the use of those funding streams for non-structural flood protection measures such as land use plans that include resiliency measures restricting development in high flood risk areas and/or requiring elevating neighborhoods either by structural elevation of houses or building up the grade in polders as described in chapter "Eroding Communities and Diverting Populations: Historical Population Dynamics in Coastal Louisiana".

Discussion

In order to understand the feasibility of the proposals in chapters "Raising Buildings: The Resilience of Elevated Structures" and "Eroding Communities and Diverting Populations: Historical Population Dynamics in Coastal Louisiana" it is necessary to look at the wider efforts of flood risk mitigation in the U.S. because those efforts will affect these measures. The current legal regime for dealing with disasters is becoming obsolete in the face of changing environmental conditions, increased human activity in floodplains and deltas and advancing technology that can afford a better understanding (and prediction) of risk.[74] Likewise, social and economic factors are beginning to argue for more responsible decision making by individuals and government in accepting and allocating risk.[75] The admirable humanitarian concept of helping people who are injured by the unpredictable whims of nature is being eroded by a system that fails to demand responsibility from individuals and government, especially when risks are reasonably foreseeable, and, indeed, seems to encourage and enable the pseudo-acceptance of risk with tacit assurances that other people's money will come to the rescue, thereby enabling the cycle to repeat ad infinitum.[76]

In most jurisdictions land use planning is under the sole authority of local governments subject to restrictions from other statutes and regulations that control specific activities and/or protect specific resources. Examples of legal regimes that could affect local government land use decisions would be regulation of development in wetlands, development affecting scenic streams and rivers, protection of endangered species habitat, coastal zone management regulations for areas within the statutorily designated coastal zone and local flood ordinances overseen by the National Flood Insurance Program.[77] All these have their limitations and some

[72] *Id.* Emphasis added.

[73] *Id.*

[74] David Passeri et al., *Tidal Hydrodynamics Under Future Sea Level Rise and Coastal Morphology in the Northern Gulf of Mexico*, 4 AGU Publications Earth's Future 159 (2016); Matthew V. Bilskie et al., *Dynamic Simulation and Numerical Analysis of Hurricane Storm Surge Under Sea Level Rise With Geomorphologic Changes Along the Northern Gulf of Mexico*, 4 AGU Publications Earth's Future 177 (2016); M. V. Bilskie et al., *Data and Numerical Analysis of Astronomic Tides, Wind-Waves, and Hurricane Storm Surge Along the Northern Gulf of Mexico*, 121 Agu Publications Journal of Geophysical Research: Oceans 3625 (2016); Donald T. Resio & Joannes J. Westerink, *Modeling the Physics of Storm Surges*, Physics Today 33 (September 2008)

[75] Richards, *supra* note 34.

[76] Amy Wold, *Flood, rebuild, repeat: Louisiana top state in repetitive flood losses; experts cite 2 key reasons*, Baton Rouge Advocate (Oct. 8, 2016 2:17 PM), http://www.theadvocate.com/baton_rouge/news/environment/article_8721adde-8742-11e6-aaa9-aba6cfaa21ba.html; Rob Moore, *Flood, Rebuild, Repeat: The Need for Flood Insurance Reforms*, Nat'l Res. Def. Council (Aug. 11, 2016), https://www.nrdc.org/experts/rob-moore/flood-rebuild-repeat-need-flood-insurance-reforms.

[77] Clean Water Act, 33 U.S.C. § 1344 (1987); Wild and Scenic Rivers Act, 16 U.S.C. §§1271–1287 (2014); Endangered Species Act, 16 U.S.C. §§ 1531–1544 (2014); Coastal Zone Management Act, 16 U.S.C. §§ 1451–1465 (2009); National Flood Insurance Act, 42 U.S.C. §§ 4001–4005.

seem to be routinely circumvented.[78] Most states in the Gulf of Mexico region have resisted establishing state planning authorities with the only exception being Florida, which has state hazard mitigation requirements for development in hazardous coastal areas.[79] As developed areas have expanded it is obvious that patchwork, weak-willed local government planning for development is insufficient to address regional hazards. Development in watersheds and coastal areas can, and often does, adversely affect neighboring communities.[80] The increase in impervious surfaces, the displacement of flood waters by structures and the effect of levees and flood walls has caused much adverse impact both within communities and to neighboring ones.[81]

Disaster relief from federal and state governments seems to always be available to assist with recovery and rebuilding thereby pushing the cost off on taxpayers and unsuspecting homeowners who always seem to shoulder the costs of poor land use planning. Local governments and developers are not usually held responsible for their poor planning decisions. However, the Association of State Floodplain Managers has done extensive work in developing legal strategies to deal with unwise development in flood prone areas and holding governments accountable when development is allowed that causes existing developed areas to flood. Their work should be thoroughly examined by anyone seeking better solutions to flooding hazards.[82] Local governments have historically not been held liable for allowing development in flood risk areas unless the new development causes other development to flood.[83] That may be changing. As technology becomes increasingly more accurate in predicting risk it will be much harder for governments to hide behind uncertainty and policy making discretion as defenses to liability.[84]

The failure to acknowledge the need to leave enough space for water has caused untold misery that is usually blamed solely on "extreme" natural events ("that has never happened before and will never happen again") and lack of flood control projects ("if we had only built that structure we would have been saved").[85] The denial of reality is often the cause of our misery.

An interesting example of this disconnect are the flood insurance rate maps (FIRMs) produced by FEMA that are used to estimate flood risk and establish flood insurance premium rates. FIRMs are updated based on new information and changing conditions, and may be redrawn after a disaster.[86] In that process, based on the new data from the most recent flooding event, the special flood hazard areas (SFHAs)are typically, but not always, expanded, making more land subject to the local flood ordinance.[87] The flood ordinance imposes more stringent restrictions on development within the SFHAs and that is seen by many as a hindrance to economic growth. Quite often local governments will contest the expansion of the SFHAs and try to get FEMA to redraw the maps to reduce the size of the

[78] *News roundup: Planning Commission approves Willows at Bayou Fountain subdivision... Subsidizing developers rebuilding in vulnerable areas is 'social welfare,' FEMA chief says... AMC Entertainment wins US antitrust approval to buy Carmike Cinemas,* Greater Baton Rouge Business Report (Dec. 20, 2016), https://www.businessreport.com/article/news-roundup-planning-commission-approves-willows-bayou-fountain-subdivision-subsidizing-developers-rebuilding-vulnerable-areas-social-welfare-fema-chief-says; Alexandria Burris, *Baton Rouge Sierra Club has beef with plans for Burbank subdivision,* Greater Baton Rouge Business Report (Dec. 19, 2016), https://www.businessreport.com/article/local-sierra-club-raises-flags-proposed-burbank-subdivision; Matthew Tresaugue, *Review: Developers failing to follow wetlands mandate,* Houston Chronicle (Aug. 2, 2015, 8:58 PM), http://www.houstonchronicle.com/news/houston-texas/houston/article/Review-Developers-failing-to-follow-wetlands-6417918.php; Chris Kardish, *Southern Louisiana Picks a Fight With Big Oil to Save the Wetlands,* Governing the States and Localities (Aug. 25, 2015), http://www.governing.com/topics/transportation-infrastructure/gov-louisiana-wetlands-lawsuits.html.

[79] Fla. Stat. Ann. § 163.3178 (West).

[80] *No Adverse Impact,* Association of State Floodplain Managers (2017), http://www.floods.org/index.asp?menuID=460

[81] Shiqiang Du et al., *Quantifying the impact of impervious surface location on flood peak discharge in urban areas,* 76 J. of the Int'l Soc'y for the Prevention and Mitigation of Natural Hazards 1457, 1457–71 (2015) (Discussing Natural Hazards); Mark Riebau et al., No Adverse Impact Floodplain Mgmt., Community Case Studies (2004), http://www.floods.org/PDF/NAI_Case_Studies.pdf

[82] *No Adverse Impact Legal Issues,* Ass'n of State Floodplain Managers (2009), http://www.floods.org/index.asp?menuID=352&firstlevelmenuID=187&siteID=1

[83] James Wilkins, *Is Sea Level Rise "Foreseeable"? Does it Matter?,* 26 Fla. St. U. J. Land Use & Envtl. L. 437 (2011).

[84] *Id.*; Institute for Research on Public Policy & The Canadian Academy of Engineering, Round Table Report: Making Better use of Science and Technology in Policy-Making 1–3 (2016), *available at* http://irpp.org/wp-content/uploads/2016/03/roundtable-report-2016-03-08.pdf; Jon Kusler, Government Liability and Climate Change: Selected Issues For Wetland and Floodplain Managers 1–4, 19–20 (Sharon Weaver & Marla Stelk eds., 2016), *available at* http://www.aswm.org/pdf_lib/government_liability_and_climate_cahnge_kusler_0416.pdf

[85] Steve Hardy, *In the mire: Meeting environmental rules for long delayed Comite Diversion Canal is a bureaucratic slog,* The Advocate, Sept. 3, 2016, http://www.theadvocate.com/baton_rouge/news/article_7d9c877a-6fc8-11e6-8053-f39613910654.html; David J. Mitchell, *FEMA won't change maps after floods; some worry of consequences,* The Advocate, Sept. 24, 2016.

[86] *Id.* David J. Mitchell; Becky Hayat & Robert Moore, *Long-Term Resiliency Through the National Flood Insurance Program,* 45 Envtl. L. Rep. 10338, 10341 (2015); Toledo News Now, *Flood plan maps redrawn change residents affected,* Fox 19 now (2012), http://www.fox19.com/story/18213961/flood-plain-maps-redrawn-change-residents-affected; https://msc.fema.gov/portal/resources/faq

[87] Hayat & Moore, *Id.*

area designated as SFHAs.[88] In the flooding that devastated south Louisiana in August of 2016 a different tune was heard. FEMA, some complained, had not been expansive enough in determining the SFHAs and thus failed to warn enough people of their risk and the inadvisability of living where they lived, or the prudence of buying flood insurance.[89] In fact, just a short time before the flood at least one affected parish had been successful in lobbying FEMA to redraw the flood maps in the parish to shrink the size of the high risk flood areas on the FIRMS. Some people who flooded had cancelled their flood insurance policies based on the change.[90] This disingenuous outrage illustrates perfectly the self-contradiction that often accompanies our thinking and discussions about risk and what it means to truly accept it. Could we use the August 2016 Baton Rouge flood event and the inevitable next one wherever it may happen to finally have an honest dialog about risk and truly accepting the consequences of our actions?

Moving Forward

As it stands now the only direct methods of relocating people out of high risk areas are eminent domain or voluntary buyouts. Given enough political will, other incentives, disincentives and market based approaches could be effective in accomplishing the goals expressed in chapter "Erod ing Communities and Diverting Populations: Historical Population Dynamics in Coastal Louisiana" as well as wider flood hazard mitigation goals. The following measures should be considered. Some measures are independent and some can work in conjunction with other measures. The ideas are merely proposals for discussion purposes and without many details. All are aimed toward a greater acknowledgment of risk and a fairer allocation of accountability. As such all will be very unpopular and politically difficult.

- Remove structural protection from consideration when determining flood insurance requirements, premium rates and elevation requirements. Levees always fail at some point. If buildings are elevated and insured, recovery will be easier and more effective. Such a measure would also make the idea of raising land more attractive because flood insurance would be cheaper or maybe not required on raised land and buildings would not have to be elevated.
- It is becoming apparent that for the NFIP to remain viable flood insurance should be mandatory in the 1000 year flood plains and coastal areas irrespective of whether the property has a mortgage.[91] One frustrated mayor actually suggested just that after the Baton Rouge flood.[92] This is especially important in light of sea level rise that will make many areas much riskier than the current flood maps depict.[93] The insurance would continue to be prorated based on relative risk and ability to pay, and with many more in the pool it could still be subsidized for existing properties. The premiums could be partially used for moving development out of the flood plain.
- In some other states insurance companies are taking the initiative by requiring people seeking to buy homeowners' insurance to also carry flood insurance.[94] This market based pressure may alleviate some of the problems with flood insurance but will not completely substitute for mandatory requirements.
- Alternatively, if people are allowed to decline flood insurance and are not required to build above the 1000 year base flood elevation, they should be required to permanently and irrevocably opt out of receiving any form of government assistance for damage associated with flooding before receiving a permit to build.
- The concept of 100 or 500 year, or longer, event should be replaced with a risk assessment based around the

[88] Rebekah Allen, *Many Central homes dropped from 'high risk' zones, allowed to lose insurance weeks before flood*, The Advocate (Sept. 3, 2016), http://www.theadvocate.com/baton_rouge/news/article_c6278d8e-6576-11e6-a301-173b12fac153.html

[89] Steve Hardy, *Insurance concerns: Half the flooded East Baton Rouge homes not in 'high-risk' areas*, The Advocate (Aug. 25, 2016), http://www.theadvocate.com/louisiana_flood_2016/article_58886b9a-6b02-11e6-9557-cfdf7ce00cf7.html

[90] Allen, *supra* note 88.

[91] National Academies Keck Center, Reducing Flood Losses: Is the 1% Chance (100-Year) Flood Standard Sufficient? 4 (2004), *available at* https://www.nrcs.usda.gov/Internet/FSE_DOCUMENTS/16/nrcs143_009401.pdf; U.S. Geological Survey, 100-Year Flood- It's All About Chance 3 (2010), *available at* https://pubs.usgs.gov/gip/106/pdf/100-year-flood-handout-042610.pdf; *NAT'L FLOOD INS. PROGRAM, Continued Progress Needed to Fully Address Prior GAO Recommendations on Rate-Setting Method,* U.S. Gov't Accountability Office Highlights of GAO-16-59, a report to the Ranking Member, Comm. on Financial Services, House of Representatives March 2016, http://www.gao.gov/assets/680/675855.pdf

[92] Mitchell, *supra* note 85.

[93] Jonathan Gregory, Projections of sea level rise 11 (IPCC AR5 Working Group I 2013), *available at* https://www.ipcc.ch/pdf/unfccc/cop19/3_gregory13sbsta.pdf; *2017 Coastal Master Plan,* Coastal Protection and Restoration Authority, http://coastal.la.gov/2017-coastal-master-plan/

[94] *Flood Insurance: Do I Really Need It?*, Allstate Ins. Co. (Sept. 2016), https://www.allstate.com/tools-and-resources/home-insurance/do-i-really-need-flood-insurance.aspx; *Does My Homeowners Insurance Cover Flooding?*, Ins. Info. Inst., http://www.iii.org/article/does-my-homeowners-insurance-cover-flooding; *Give your home more protection, so you have peace-of-mind*, Progressive Specialty Ins. Agency, Inc., https://www.progressive.com/flood-insurance/Personal communication with Ms.Tammy Hatch of Eastern Insurance Group in Marshfield, Massachusetts on March 24, 2017.

inundation of record with the assumption that such an event can and probably will reoccur and that a more severe inundation is entirely possible and even likely. The method for determining the 100 year flood risk areas seems to have been a political compromise to achieve some level of risk reduction while not making those areas too expansive in the interest of protecting economic development.[95] The process should be revamped because flooding events keep demonstrating that the FIRMS are too conservative in that they underestimate risk.[96]

- The local government regulation part of the bargain also needs to be reworked to be much more stringent regarding development in floodplains. This will require federal legislation but given the current laxity in the NFIP requirements for local flood ordinances it is a crucial part of the solution to repeated disasters.
- Developers also bear responsibility for flooding disasters. They routinely lobby governments to allow development in marginally safe areas where the land is cheaper because, of course, it is only marginally safe.[97] Developers should be required to post a bond on every development within at least the record inundation area for the life of the structures. Their insurers would have the incentive to more accurately assess the flood risk of an area. The bond should be sufficient to pay for relocation or elevation of damaged structures.
- Real estate brokers should be required to fully disclose flood hazard based on the best available information not limited to FIRMs. This would include all known historical floods. Informed consent is an important issue in disaster events. After the August flood in south Louisiana many people were in disbelief that they were living in areas that could be flooded. This is despite the fact that they live in a river basin that had experienced a similar, but not quite as extensive, flood 33 years earlier. Companies usually have better access to information than average citizens and should be required to provide the best available data to prospective home buyers.
- Hazard mitigation planning should be regulated at the state level for whole watersheds/drainage basins. Piecemeal development and structural flood defense will only move the problems from one place to another and cause an overall increase in flooding events a severity. It is obvious that the lack of coordinated development in watersheds cannot continue, the taxpayers can't afford it.
- Public resistance to the polder project could be blunted somewhat by offering the displaced people development rights on the created higher ground.

Third Party Standards/Private Regulation

- As consumers become more quality conscious and more aware of social and environmental costs of economic activity, businesses have begun to accommodate demands for higher quality, greener and more socially appropriate goods and services. One method to accomplish this end has been the establishment of standards by industry and non-governmental organizations (NGOs) that are outside governmental regulatory processes but fill a vacuum in the absence of government regulation or complement existing regulations. An example of this would be certified-flood-resistant or climate-smart-investment real estate. Properties that carry very low risk, particularly in areas not flooded by the inundation of record could be identified in some way that would be attractive to certain discriminating buyers and command higher prices.[98]
- Lending institutions could require flood insurance based not solely on FEMA's FIRMs but also on the extent of historic floods. That requirement would more accurately convey risk and levy a cost of accepting that risk as well as protecting lending institutions.

[95] National Academies, *supra* note 91.

[96] National Academies, *supra* note 91; *Louisiana Flood of 2016 made worse by growth-focused policies*, The Times-Picayune, New Orleans (Sept. 23, 2016, 9:00 AM), http://www.nola.com/news/baton-rouge/index.ssf/2016/09/louisiana_flood_of_2016_develo.html; UC Davis Center for Watershed Science, *New Baton Rouge flood map show limits of current risk and planning methods*, California WaterBlog (August 28, 2016), https://californiawaterblog.com/2016/08/28/new-baton-rouge-flood-map-show-limits-of-current-risk-and-planning-methods/

[97] Alexandria Burris, *Baton Rouge Sierra Club has beef with plans for Burbank subdivision*, Greater Baton Rouge Business Report (Dec. 19, 2016), https://www.businessreport.com/article/local-sierra-club-raises-flags-proposed-burbank-subdivision; *News roundup: Planning Commission approves Willows at Bayou Fountain subdivision… Subsidizing developers rebuilding in vulnerable areas is 'social welfare,' FEMA chief says… AMC Entertainment wins US antitrust approval to buy Carmike Cinemas*, Greater Baton Rouge Business Report (Dec. 20, 2016), https://www.businessreport.com/article/news-roundup-planning-commission-approves-willows-bayou-fountain-sub division-subsidizing-developers-rebuilding-vulnerable-areas-social-welfare-fema-chief-says

[98] Yesim Yilmaz, Cato Inst., Private Regulation: A Real Alternative for Regulatory Reform (6th ed. 1998); Timothy D. Lytton, *Competitive Third-Party Regulation: How Private Certification Can Overcome Constraints That Frustrate Government Regulation*, 15 Theoretical Inquiries in Law 539, (2014), https://object.cato.org/sites/cato.org/files/pubs/pdf/pa-303.pdf; Lesley K. McAllister, *Harnessing Private Regulation*, 3 Mich. J. Envtl. & Admin. L. 291 (2014), Available at: http://repository.law.umich.edu/mjeal/vol3/iss2/6

Table of Laws

Area of law	Basic definition	Application	Potential issues
Federal Takings Clause	The 5th Amendment of the U.S. Constitution prohibits the government from taking private property for public use without just compensation. This is commonly referred to as the "takings clause."	Takings may be effectuated by physical acquisition or occupation of private property called "physical" takings or by limitations on the use of private property called "regulatory" takings. When a government takes private property by a physical or regulatory taking it must pay the private property owner just compensation for the loss of economic value suffered as a result of the government action.	Whether a claimant can obtain compensation as a result of government action often depends on a number of factors that will be discussed below.
Physical takings A. Eminent domain	A. The power of a government or its agents to expropriate private property so long as the owner is paid fair compensation and the property is taken for public purpose or use. Also referred to as "comdemnation" this power of eminent domain is inherent in a sovereign government. The takings power can be legislatively delegated to private parties or corporations upon authorization by the legislature. The right of eminent domain is subject to the Federal Takings Clause, the 5th amendment of the U.S. constitution, which prohibits the government from taking private property for public use without just compensation.	A. The authority can be delegated to government agencies and private parties, with some exceptions. "Public purpose" has a broad definition that can include economic development. The most common use of eminent domain is for roads, government buildings, and other facilities.	A. The broad nature of "public purpose" leaves a lot of discretion to the government to determine what private property may be expropriated. However, if a court finds that the government action was not for a public purpose the action can be denied or the action may be determined to constitute negligence and therefore subject to legal action for recovery of damages. Whether the government action is classified as eminent domain or a negligent action (tort) that takes private property is an important distinction because the former implies a constitutional right whereas the latter is often blocked by sovereign immunity (prohibition against suing government) or other procedural hurdles In response to the U.S. Supreme Court's holding that economic development can justify a government's exercise of eminent domain, many states passed legislation that limits the ability to use eminent domain for economic development at the state level. Louisiana forbids the consideration of economic development or tax revenues when determining what constitutes a public purpose.
B. De-facto eminent domain-physical takings without compensation	B. Occur when the government physically takes possession, either permanently or temporarily, of an interest in property for some public purpose absent a declaration of eminent domain. A government action that imposes a physical invasion of private property is a taking, even if the action achieves an important public purpose and causes minimal economic impact.	B. Examples include government occupation or take over of a leasehold interest for its own purposes or storing government equipment on private land without the property owner's permission. In one case, the government required an apartment building owner to allow a portion of the rooftop to be used for a TV cable company's equipment to service residents within. In this case a regulation resulted in a physical occupation.	B. Although a minor physical invasion may result in a taking, the overall impact will be taken into account when determining "just compensation." In other words, a minor invasion may entitle the owner to a small amount of compensation.

(continued)

Table of Laws (continued)

Area of law	Basic definition	Application	Potential issues
Regulatory takings	Involve government rules or regulations that affect the use of private property. These rules are sometimes referred to as "inverse condemnation" because the government action is a de-facto use of eminent domain without declaring so or offering compensation leaving the property owner to pursue an action Regulatory takings may be partial, categorical, or total, depending on the particular effect the government regulations have on the private land at issue.	Regulatory takings claims are evaluated on a case-by-case basis using a 3-part balancing test. The three prongs are 1. The character of the government action 2. The economic impact on the claimant and 3. The extent to which the government action has interfered with distinct, reasonable investment-backed expectations.	In partial takings cases the relative loss of economic value is an important factual finding. In cases where the loss is relatively minor (in a contextual sense) courts often find that no taking has occurred. However, if ALL economic value in the land is destroyed (total taking), this will automatically constitute a taking unless background principles of law would have allowed the state to take the action. The prohibition against any building has been determined by the U.S. Supreme Court to be a total taking.
How dual sovereignty under U.S. federalism affects eminent domain and regulatory taking	Under the principle of U.S. federalism, a dual sovereignty exists among the state and federal governments. Both the state and federal sovereigns may exercise the power of eminent domain. While the federal government is only bound by the limitations of the U.S. constitution, a state is bound both by its own constitution and the U.S. constitution. The U.S. constitution establishes a floor of individual rights, including the right to property, which may not be abridged by the federal or state governments. A state may protect the right to property to a greater extent than the Federal Constitution, but it may not provide less protection.	In response to the U.S. Supreme Court's *Kelo v. City of New London* decision that broadened a sovereign's ability to exercise eminent domain for economic development purposes, forty-seven states proposed amendments to their constitutions in order to limit the eminent domain power.	Dual sovereignty can lead to drastic differences in each state's definition of "public purpose" used to justify the exercise of eminent domain. In effect, these varying definitions will create differences in when compensation should be received as well as the amount an individual should receive depending on the state.
Louisiana Eminent Domain and Takings Law	The Louisiana Constitution's takings clause says: "property shall not be taken or damaged by the state or its political subdivisions except for public purposes and with just compensation paid…" Louisiana has placed limitations on when the state may exercise its eminent domain authority. These limitations include the following situations: for purposes of transfer of ownership to any private person or entity; for economic development purposes only; to enhance tax revenues. Exemptions include: appropriation for levee construction and levee drainage purposes; and for coastal wetlands conservation, management, preservation, enhancement, creation, or restoration activities.	Under Louisiana law, in general, there is a three-pronged analysis to determine whether a claimant is entitled to compensation for government interference with the use of private property: 1. Whether a person's legal right with respect to a thing or object has been affected; 2. Whether the property right or thing itself has been taken or damaged in a constitutional sense; 3.Whether the taking or damaging is for a public purpose. If all three of these inquiries are answered in the affirmative, then a taking has most likely occurred. The Louisiana courts do not consider a property owner's reasonable investment-backed expectations when deciding whether state government action has effectuated a taking under the state's constitutional protection of private property rights	The "taken or damaged" phrase has led to confusion between government actions that may be classified as negligent actions (torts) and takings. Prohibition against consideration of economic development in determining "public purpose" can hamper the state's ability to use eminent domain to conduct land use reforms for coastal protection and restoration purposes if the actions are seen to primarily benefit private economic interest. Land use reforms that have a primary goal of reducing flood risk with only the indirect effect of increasing value of private property have a greater chance to survive judicial review.

(continued)

Table of Laws (continued)

Area of law	Basic definition	Application	Potential issues
		Louisiana law generally allows greater compensation for the taking, loss, or damage to property rights than the "just compensation" standard that is required by the fifth amendment of the U.S. constitution. However, Louisiana law limits compensation to the fifth amendment standard for certain takings. For example, property affected by hurricane protection projects and coastal wetlands conservation or restoration. This could create issues in determining government action classification and the proper amount of compensation. Private entities authorized to use eminent domain by the state cannot do so until a judicial determination has been made that the action is for a "public and necessary purpose and with just compensation paid to the owner." The taking of property by the City of Monroe, for predominant use by a private developer, and a primary public purpose of economic development, is not a valid exercise of eminent domain authority by a municipality. Op. Atty.Gen., no. 07–0157 (Oct. 15, 2009), 2009 WL 3785540.	
Police power exception to takings challenges	Power of state and local governments to adopt and enforce laws that promote public health, safety, morals, and general welfare. Authority for this power arises from English and European common law traditions. Power is broadly interpreted so long as the statue is "appropriate and necessary" to accomplish a public purpose. Generally, a state's valid exercise of its police power does not entitle a claimant to compensation.	Specific to the National Flood Insurance Program, communities can use their police power authority to enact regulations concerning development in floodplains and other high risk areas. The Louisiana supreme court has stated the construction of a diversion project for the benefit of coastal marshlands to be a valid exercise of the state's police power.	Controversy over the exercise of state police power can arise when it conflicts with individual rights. In contrast to the doctrine of eminent domain, exercise of a state's police power does not require just compensation. Purported exercise of police power by a state that is found to be outside the scope of such power may be recognized as a "taking" or "damaging" that might require the state to pay just compensation to the individuals affected.
Mineral Estate Ownership Rights in Louisiana	Unlike many other states, Louisiana law favors maintaining complete (surface and subsurface) land ownership rights in one person rather than severing the two estates. However, separate mineral servitudes as to separate depts., levels or horizons are permitted. La. Stat. Ann. § 31:21 When land is appropriated via eminent domain, mineral rights	When land is appropriated via eminent domain, mineral rights may be reserved with the seller for a temporary period of time. However, if these mineral rights are not exercised within the prescription period (usually 10 years), the rights revert to the surface owner, which would be the state or another party to whom the state has transferred ownership.	If exercised, mineral rights may be able to secure a substantial amount of income for property owners via royalties. Therefore, the potential to lose these rights is a factor that weighs against moving for many individuals living in coastal and flood-prone areas.

(continued)

Table of Laws (continued)

Area of law	Basic definition	Application	Potential issues
	may be reserved with the seller for a temporary period of time.	In *State Through Sabine River Authority v. Miller,* a Louisiana court of appeal held that a landowner was entitled to severance damages to reserved mineral rights located on land condemned by a river authority under its eminent domain power.	
100-year flood zone or "A" zone	B. Defined as a flood event that has a 1% probability of occurring in any given year. This does not mean that a 100-year flood will likely occur only once in a 100 years. In fact, there is a 63.4% chance that a 100-year flood event will happen once or more during any given 100-year period. Put another way, there is a 26% chance of a 100-year flood occurring during a 30-year home mortgage	B. FEMA uses the 100-year flood calculations for the purpose of mapping flood hazard risk areas. These maps are then used to determine flood insurance requirements. The area that would be inundated by a 100-year flood is called the special flood hazard area in which flood insurance is required for property with a federally backed mortgage. The maps are not real-time indicators of actual flood risk.	B. FEMA flood map calculations of 100-year floods do not always represent true risk and can confuse in mislead the public. The methodology reflects a political compromise to achieve some level of risk reduction while not making those areas too expansive in the interest of protecting economic development. In addition, maps may be outdated in some regions.
Uniform relocation assistance and real property acquisition policies act (URA) 1970	Establishes minimum standards for federally funded programs and projects that require the acquisition of real (immovable) property (real estate) or displacement of persons from their homes, businesses, or farms. URA's protections and assistance apply to the acquisition, rehabilitation, or demolition of real property for federal and federally funded projects.	Qualified displaced persons can receive a myriad of benefits or compensation. For example, instead of merely receiving "market value," many property owners may obtain "replacement value" for the property subject to eminent domain.	Whenever real (immovable) property is acquired by a state agency and furnished as a required contribution incident to a federal project, the federal agency with grant authority may not accept such property unless the state agency has provided assurances that fair and reasonable relocation payments, services, assistance, and comparable replacement dwellings shall be provided to or for the displaced persons within a reasonable period of time prior to displacement. 42 U.S.C.A. § 4630 (west) however, ambiguity in the language of the URA has led to a large amount of litigation and is often the subject of much debate and criticism. 42 U.S.C.A. § 4627 (west).
Gulf of Mexico energy security act of 2006 (GOMESA)	An act that significantly enhances outer continental shelf oil and gas activities as well as revenue-sharing in the Gulf of Mexico. Allows leasing revenues to be shared with gulf-producing states and the Land & Water Conservation Fund for coastal restoration projects Bans oil and gas leases within 125 miles of a portion of Florida's eastern and central planning areas until 2022	Provided for the four gulf-producing states and their eligible coastal political subdivisions to share 37.5% of "qualified revenues" from OCS oil and gas leases issued since December 20, 2006. Revenues from a subset of leases were shared beginning with the 2009 disbursements. Revenues from a more substantial set of leases started being shared in 2017. This was done by expanding the definition of "qualified revenues" and capping the revenue-sharing between the states at $500 million per year until 2055.	Funded projects should address coastal conservation, restoration, and hurricane protection. However, some projects have been criticized for failing to meet this standard. One example is Alabama's proposal that $850 million of GOMESA shared revenue should be used for interstate expansion in order to ensure hurricane protection. This has been criticized for its potential to take away from other environmental projects.

(continued)

Area of law	Basic definition	Application	Potential issues
	Allows companies to exchange certain existing leases in moratorium areas for bonus and royalty credits to be used on other gulf leases.	12.5% of revenues were allocated to the land and water conservation fund.	Would the proposals in chapters "Raising Buildings: The Resilience of Elevated Structures" & "Eroding Communities and Diverting Populations: Historical Population Dynamics in Coastal Louisiana" qualify for funding as "coastal flood protection and related infrastructure"?
Resources and ecosystems sustainability, tourist opportunities, and revived economies of the Gulf coast states act of 2012(RESTORE act)	In response to the *Deepwater Horizon* oil spill that dumped nearly 5 million barrels of oil into the Gulf of Mexico, this was enacted in an effort to restore the affected gulf coast states. The act allocates 80% of the funding (created by civil penalties from clean water act violations related to the spill) to the Gulf coast restoration trust fund (GCRTF). The GCRTF then allocates portions of this money to the Gulf states for restoration projects.	Allocation by the GCRTF is as follows: 35% of the funding will be divided equally among the five gulf states for restoration activities. 30% will be managed by Gulf Coast ecosystem restoration council for ecosystem restoration. 30% will go to the five states, divided based on impacts. 2.5% will go towards monitoring. 2.5% will fund centers of excellence.	It has been less than 5 years since its enactment, but many are skeptical of the chances the RESTORE act has to fulfill its goal of restoring the economy and ecosystem of the Gulf coast region. Would the proposals in chapters "Raising Buildings: The Resilience of Elevated Structures" & "Eroding Communities and Diverting Populations: Historical Population Dynamics in Coastal Louisiana" qualify as hurricane protection, and infrastructure"?
National Flood Insurance act of 1968	A. Created by the National Flood Insurance act of 1968, the NFIP aims to reduce the impact of flooding by providing affordable insurance to property owners in exchange for communities' adopting and enforcing floodplain management regulations. The program promotes the purchase of flood insurance in an effort to reduce socio-economic impacts of flood disasters.	A. Government-backed flood insurance is not available in areas that do not participate in the NFIP. In communities participating in the NFIP properties with federally backed mortgages are required to carry flood insurance for the life of their mortgage. For other property purchase of flood insurance is voluntary. Insurance rates are subsidized, meaning that premiums do not reflect actuarial rates (higher). Properties built before the implementation of the NFIP are "grandfathered" out of higher premium rates established under the program.	A. NFIP rates are not set by the market risk valuation. It is less expensive than a private insurance company rate would be, which is accomplished either by the program running a deficit or subsidies from the federal government. NFIP's deficit has been steadily increasing and most likely will continue in the future without reform due to the increase in frequency of flooding events resulting from climate change. As of January 2017, FEMA owed the U.S. Treasury $23 billion.
Biggert-Waters Flood Insurance Reform Act of 2012 (Biggert-Waters Act)	Enacted in 2012, designed to extend the NFIP for a five-year period and strengthen its financial solvency. At the time the bill passed, the NFIP cumulative debt was $17 billion. The law intended to change NFIP premium rates to match actuarial risk-based premium rates that more accurately reflected the expected losses and real risk of flooding. The Biggert waters act was extensively amended in 2013 to remove some of the more draconian impacts to policy holders	Provisions in the act required significant reform of the program in an effort to progressively reduce its deficit. These reforms included phasing out almost all discounted insurance premiums, such as subsidized premiums. Additionally, the changes included removing discounts to many policies that were priced below actuarial risk targets. The practice of "grandfathering," (allows a property owner to pay a payment rate established at a time prior to current codes regardless of compliance) were also phased out. If a community were to adopt a new, updated flood insurance rate map (FIRM), grandfathered rates would be increased by 20% per year for 5 years.	The extent to which the Biggert-waters act would have reduced financial exposure of NFIP remains unclear. Upon implementation of the reforms, ratepayers experienced heavily increased rates and their agitation caused considerable political push back. Constituent fear of unaffordability led to the passing of the Homeowner Flood Insurance Affordability Act of 2013.

(continued)

Table of Laws (continued)

Area of law	Basic definition	Application	Potential issues
Homeowner Flood Insurance Affordability Act of 2013 (HFIA)	Repeals and modifies certain provisions of the Biggert-Waters act of 2012 that affected the NFIP. Makes additional changes to other aspects of the NFIP. In effect, certain subsidies that were repealed, thereby increasing premium rates under the Biggert-Waters act were reinstated and premium increases were slowed under HFIA.	The bill delays the flood insurance premium increases created under the Biggert-waters act for 4 years while FEMA came up with a proposal to make the premiums cheaper and reassess flood hazard mapping. In addition to delaying the onset of higher premiums, the bill allows homeowners selling their homes to pass the lower flood insurance premiums on to the next homeowner	May increase the NFIP's long-term financial burden on taxpayers. During the time the bill passed, the NFIP owed the U.S. Treasury increased to $24 billion. However, the act was widely viewed as necessary for economic reasons. It is believed that without these delays in rate increases and subsidy cuts, policyholders (business owners, homeowners, etc.) would not have been able to afford coverage. However, it leaves the issue of NFIP solvency unresolved.
Federal Emergency Management Act (FEMA)	An agency within the Department of Homeland Security responsible for coordinating federal disaster response. FEMA also works with state and local governments to develop plans for disaster response. Authority derived from the Stafford act.	Provides a range of federal support for disaster response, recovery, and mitigation. This can include technical support, personnel, supplies, and funding for relief efforts.	FEMA has been criticized for its slow initial responses to disasters such as hurricane Katrina. FEMA's resources have at times been overwhelmed by large scale disasters. The disaster relief functions of the Stafford act are intended to be one half of a quid pro quo between the federal government and recipients of aid, whereby the state and local government recipients take steps to reduce the impacts (and costs) of future disasters. Generally, the required disaster mitigation plans developed by local jurisdictions have not been sufficient to accomplish that goal.
Department of Housing and Urban Development (HUD)	An executive cabinet department that is responsible for programs directed towards housing needs, fair housing opportunities, and improving & developing U.S. communities. Includes Federal Housing Administration and Federal Housing Finance Agency.	HUD directs many different housing and development programs, including the community planning and development, housing, public and Indian housing, Office of Fair Housing and Equal Opportunity, policy development and research, and the government National Mortgage Associationl. These programs provide grants that are delegated to state, county, and local governments as well as eligible non-profit organizations that have housing programs. Low income is required to receive most HUD grants but some are awarded to moderate-income recipients.	HUD has been the subject of criticism for the department's alleged scandals and inefficiencies costing taxpayer dollars. Can the HUD relocation programs be applied to the proposals in chapters "Raising Buildings: The Resilience of Elevated Structures" & "Eroding Communities and Diverting Populations: Historical Population Dynamics in Coastal Louisiana".

Summary and Conclusions

John W. Day

Abstract

When European explorers first arrived in what is now south Louisiana in the late seventeenth century, the Mississippi delta was a vibrant, healthy, dynamic regional coastal ecosystem. Now however, the Mississippi delta is profoundly changed and unsustainable. Given the growing constraints imposed by climate change and resource scarcity and the projections for loss of most coastal wetlands even with the current proposed coastal master plan, it may be that a dramatically new approach will be required. The goal of this book was to provide a framework of what a new approach might look like. This chapter summarizes the book's key findings. The authors concluded that restoration should focus on activities that are more sustainable over the long-term without large energy inputs, river diversions are an example of this. Other recommendations include a dramatic intervention involving raising parts of New Orleans, municipalities focusing on maximizing freshwater input to wetlands and controlling nutria populations, land-use planners prioritizing building structures 15 ft above sea level. Regardless of the approach, expensive, energy-intensive projects that are long lasting, more sustainable, and convey long-term benefits should be done as early as possible.

Keywords

Mississippi delta • Coastal restoration • Climate change • Land-use planning • Wetland ecosystems

When European explorers first arrived in what is now south Louisiana in the late seventeenth century, the Mississippi delta was a vibrant, healthy, dynamic regional coastal ecosystem. For over two centuries, the delta provided a wealth of natural resources, what are now called ecosystem goods and services, to the growing population and economy based in the coastal area. These included abundant fish and wildlife resources, vast old growth forests, rich soils, and the Mississippi River that served as a source to freshwater and as important, an avenue for trade. The vast wetlands were not only a habitat for fish and wildlife but also protected New Orleans and other areas from hurricanes.

Now however, the Mississippi delta is profoundly changed and unsustainable. Several environmental and economic forcings will affect the delta and its restoration in the twenty-first century. Climate change will lead to a 1–2 m sea-level rise by 2100 and 3–5 m coming centuries (Hoyos et al. 2006; Knutson et al. 2010). There will also be more, larger hurricanes and river floods (Day et al. 2014; Melillo et al. 2014). Energy will become more costly and scarce (Murphy and Hall 2011; Maggio and Cacciola 2012). Current coastal restoration methods are very energy intensive. In the future, restoration will become more expensive, and perhaps prohibitively so (e.g., Tessler et al. 2015). David Muth writes: "Many are engaged in a futile effort to hold

J.W. Day (✉)
Department of Oceanography and Coastal Sciences, Louisiana State University, Baton Rouge, LA 70803, USA
e-mail: johnday@lsu.edu

J.W. Day, J.A. Erdman (eds.), *Mississippi Delta Restoration*,
Estuaries of the World, https://doi.org/10.1007/978-3-319-65663-2_14

onto what is doomed or put back what is already lost, rather than allow what could be: a vibrant new river management system that reignites the process that built the delta and its vast productivity in the first place (2014). There is great potential for restoration if the river is used, but our coastal society is mired in a system with no living memory of a naturally functioning delta."

The goals of this book were to develop conceptual plans to (1) formalize a new framework to promote safe and sustainable development in the Mississippi delta, (2) reinitiate a functioning delta by more fully utilizing the potential of the river, (3), redesign a **new** New Orleans by incrementally raising large portions of the city using river sediments, promoting gravity drainage for most of the city, and planting a healthy cypress forest to protect against hurricanes that would be nourished and protected from saltwater intrusion by river water, (4) raising and building structures that are storm resilient, energy-efficient and use renewable energy, and (5), investigate social and legal constraints and opportunities related to delta restoration. The book is highly multi disciplinary with contributions about law, architecture and landscape architecture, coastal and environmental sciences, engineering, sociology, public policy, art and literature, and others. The book is based on an analysis carried out under the auspices of the Coastal Sustainability Studio at Louisiana State University to take a provocative new look about how the Mississippi delta can be managed and restored in a sustainable manner in a climate-challenged, energy-scarce future. In addition to policy analysis, the book provides scientific information and imagery of the ecological, economic, and social history and future of coastal Louisiana. These illustrate the sobering realities and drawbacks of relying on energy-intensive industrial methods to restore the coast, and the great potential ecological and cultural benefits if we invest in restoring natural deltaic processes.

This book draws on published and unpublished work from studies done at LSU and elsewhere to present a conceptual as well as practical overview of a new coastal management strategy based largely on natural deltaic functioning. A central challenge of the book was to conceptualize these ideas in a way that could form the basis for policy. Faculty with expertise in geology, engineering, coastal and environmental sciences, law, geography, architecture and landscape architecture, and sociology interacted to write the book. Here we summarize the findings of the book.

In chapter "A Brief History of Delta Formation and Deterioration", Lane et al. describe the development and deterioration of the delta. The drainage basin of the Mississippi covers 3.2 million km^2 making it one of the largest watersheds in the world. It covers about 40% of the area of the lower 48 U.S. states, and accounts for about 90% of river discharge to the Gulf of Mexico. After sea level stabilized approximately 7000 years ago, the River formed a vast deltaic wetland complex encompassing about 25,000 km^2 of wetlands, low relief uplands, and shallow water bodies. Overbank flooding and crevasse formation deposited large amounts of sediments and nutrients in the Mississippi River delta, allowing the delta to build into the Gulf of Mexico in a series of overlapping deltaic lobes. Hundreds of crevasses overlapped to form a more or less continuous band of sand-rich deposits essential to the formation and maintenance of both natural levees and coastal wetlands. Active and abandoned distributary channels and barrier island ridges formed a network of interconnected natural levee ridges that sheltered wetlands from wave erosion and modulated the interaction of riverine and marine processes, limiting salinity intrusion.

In the eighteenth and nineteenth centuries and greatly accelerating through the twentieth century, the Delta was impacted by a variety of human activities including a reduction of sediment input by the river, levee construction and closure of distributaries that largely isolated the deltaic plain from riverine input, massive hydrological alternation, impoundments, and barrier island loss. These activities led to salt water intrusion, deterioration of the skeletal framework, increased subsidence, and massive wetland loss. These stresses are mostly the result of management in which navigation and flood control as well as tax revenues from the extraction of minerals and petroleum prevailed over the ecological sustainability of the delta. Climate change in the form of acceleration of sea-level rise, more frequent strong hurricanes, and more frequent droughts will exacerbate delta deterioration.

The only place along the Louisiana coast where there has been significant wetland gain is at the mouth of the Atchafalaya River. The Atchafalaya and Wax Lake delta complex is part of the beginning stages of a major new deltaic lobe development fed by the Atchafalaya River distributary, which carries about a third of the combined flow of the lower Mississippi River.

Colten (chapter "Levees and the Making of a Dysfunctional Floodplain") builds on the proceeding chapter to provide a more detailed and nuanced view of the isolation of the delta plain from the river. In the earliest years of settlement, French colonists recognized that the Mississippi River built the delta floodplain. Iberville noted flood marks on trees as he journeyed up the river in the spring of 1699, and residents of the incipient New Orleans discovered cypress trees buried in the alluvial sediment as they dug water wells. The fact that the river regularly deposited soil and elevated the land surface did not override their concern with the threat of flooding. An overwhelming desire to build a barrier between the emerging settlements and the river drove early levee building efforts. Construction initially focused on the colonial capital. As agricultural activity began spreading up and downstream, policy dictated that landowners had to build levees to extend the earthen bulwarks across their river-front properties.

Imperfect, at best, the levees rose only a few feet above the top of the high ground along the river and gradually extended as far upstream as Baton Rouge by the end of the colonial period. These levees were subject to many breaches, known locally as crevasses, and floods continued to deposit sediment across the wetlands regularly but episodically. The successful cultivation of rice demanded deliberate connections to the river, and planters built sluices in the levees that allowed spring floods to cover their fields. This maintained limited connections across large portions of the lower river and sometimes led to large levee breeches.

The adoption of sugar as the most promising commodity in the late eighteenth century prompted a renewed effort to fully blockade spring floods. Levees grew in size, but crevasses remained a chronic problem. State officials by the mid-nineteenth century observed that higher levees contributed to rising flood stages. Yet, after the Civil War and the restoration of the sugar economy, policies emphasized the primacy of levees as flood control devices. With the creation of the Mississippi River Commission in 1879, a "levees only" policy dominated and federal funds helped build more substantial and systematic barriers even though a minority called for regular outlets to relieve flood stress.

The failure of this massive levee system in 1927 called into question the prevailing policy, and the Corps of Engineers adopted a "levees and outlets" approach. The federal authorities adopted a plan to build two major outlets in Louisiana – the Bonnet Carré Spillway and the Atchafalaya Floodway. Never intended to restore pre-colonial connections between the river and the vast wetlands, they were to protect urban populations, agricultural activities, and transportation interests. One state official at the time noted the importance of floods in maintaining important wetland economic activities and the wildlife that lived there, but flood protection continued to dominate design and construction concerns.

These outlets have generally performed as intended and have prevented serious flooding of New Orleans along with floodplain agriculture and industry. Yet, in the planning stages, officials disregarded concerns of fishing communities, and they have disrupted fisheries when opened. More importantly, they continue to preclude the pre-historic connections between the river and its floodplain. Their design and construction, while slow moving on the human scale, have imposed relatively rapid change on the natural maintenance of the lower river delta when considered in geologic time. A continuing commitment to flood protection will perpetuate the dysfunctional floodplain created to serve floodplain development.

Sasser et al. (chapter "The Nutria in Louisiana: A Current and Historical Perspective") discuss the potential impact of nutria (*Myocastor coypus*) on sustainable coastal restoration. Nutria is an exotic aquatic rodent that was introduced to Louisiana wetlands during the early 1930s and can make coastal restoration more challenging. From 1960 to 1990, greater than 36-million nutria were taken when the fur market was lucrative. By 2000, the market had collapsed and nutria populations increased. Nutria reach sexual maturity at 4–8 months, may have 2–3 litters per year, and average 13.1 young per female per year. Nutria generally have a small home range in marsh habitats. Nutria occur primarily in fresh and intermediate marsh. The organic nature of these soils makes them particularly susceptible to destruction with grazing. Nutria are opportunistic feeders, with a broad diet comprising more than 60 plant species in Louisiana and they are attracted to wetlands that contain a reliable source of nutrient-rich fresh water, such as river diversions and assimilation wetlands. It is imperative that restoration projects that increase input of nutrient-rich fresh water into wetlands have a nutria control program in place. Nutria can consume large amounts of marsh biomass and in certain cases can cause the collapse of marsh locally. Scientific studies investigating effects of nutria on marsh habitats consistently conclude that nutria grazing is damaging to marsh and young forest vegetation. It is generally accepted that nutria damage—in addition to larger scale subsidence, sea level rise, and salinity intrusion—can create an accelerated deterioration of wetlands. Nutria grazing on baldcypress and water tupelo seedlings is extensive and remains a major factor in the inability of baldcypress-water tupelo forests to regenerate. Projects designed to restore coastal swamp forests should include a nutria control component and suitable protection of transplants should be used to minimize mortality from grazing. Eruptions of populations of nutria can cause severe wetland damage and loss. Some areas of the coast have persistent populations creating "Hot Spots" of severe damage, especially in freshwater-intermediate salinity areas of Terrebonne, Barataria, and Breton Sound basins. Nutria densities are relatively low in the Chenier Plain currently compared to historic observations and harvest records. The Coastwide Nutria Control Program (CNCP) was implemented in 2002–2003 by the LDWF, and since then there has been a reduction in 70,000 acres of marsh damage from 2003 to 2010. Approximately 446,000 nutria were harvested in 2010 in the CNCP, primarily in the deltaic plain. When considering the costs of creating new wetlands (approximately $50,000–$70,000 per acre), the CNCP is a successful wetland conservation program that has produced measureable reduction in marsh damage. Since 2002–2003, 2,571,480 nutria have been harvested. This program is a success and, from a resource management perspective, should be continued with improvement and expansion if possible. Nutrient enrichment of coastal landscapes may cause nutria population growth and habitat damage. Coastal restoration projects with areas receiving nutrient enrichment should include nutria control to ensure plant productivity, establishment, and expansion. The effect of predators on nutria population control is unclear.

Shaffer et al. (chapter "Optimum Use of Fresh Water to Restore Baldcypress – Water Tupelo Swamps and Freshwater Marshes and Protect Against Salt Water Intrusion: A Case Study of the Lake Pontchartrain Basin") discuss the wise use of freshwater resources in the Mississippi delta to help maintain freshwater wetland ecosystems. Freshwater wetlands are important in the Mississippi Delta for habitat, water quality improvement, fisheries, carbon sequestration, and as a buffer against hurricane storm surge and waves. Forested wetlands are particularly important as hurricane buffers because of their 3-dimensional structure and their resistance to blow down during hurricanes. Fresh wetlands will be severely threatened by accelerated sea-level rise, more frequent stronger hurricanes, and intense drought that will lead to progressive inundation and saltwater intrusion. The lake Pontchartrain basin contains the largest area of tidal freshwater wetlands in the delta. In order to ensure sustainability of fresh wetlands, a consistent source of freshwater is needed to counter increasing salinity levels. Sustainable restoration of baldcypress-water tupelo swamps in the Pontchartrain Basin can only be achieved through wise use of point and non-point sources of fresh water. In this paper, we identify potential sources of fresh water in the Pontchartrain Basin and determine the feasibility of engineering these to maximize sheet flow to enhance freshwater wetland health. Sources of freshwater include coastal plain rivers, Mississippi River diversions, non-point source runoff, direct rainfall on wetlands, storm water pumps, and treated municipal effluent. The latter is important because it is available even during drought periods.

Rutherford et al. (chapter "Energy and Climate – Global Trends and Their Implications for Delta Restoration") make the point that climate change and energy cost and scarcity will limit options and increase costs of coastal restoration in coming decades. Mounting evidence indicates that human activity is leading to dramatic changes in the climate system. The concentration of CO_2 in the atmosphere has increased from about 280 ppm in 1880 to over 400 ppm today. This is the most important factor causing the Earth's average temperature to increase by 0.85 °C since 1880. Future warming will depend on human economic activity, energy sources, and population growth. CO_2 levels in the atmosphere are currently increasing at rates consistent with the highest published IPCC scenarios.

Climate change is leading to an acceleration of eustatic sea-level rise due to melting of land based ice masses and ocean thermal expansion. There is a strong consensus among climate scientists that sea-level rise will continue to accelerate throughout this century by one to two meters. It is also likely that extreme weather events such as heavy precipitation, category 4 and 5 hurricanes, droughts, and large floods on the Mississippi River will become more common. The heightened level of risk from climate change was demonstrated by hurricane Katrina and a "1000-year flood" that struck Baton Rouge, Louisiana in August 2016. Such disasters are likely to become the new normal with climate change.

The functioning of modern industrial economics is closely tied to energy consumption and fossil fuels supply more than 80% of world energy use. Recent studies suggest a peak in total fossil fuel production between 2020 and 2050, with conventional fossil fuels peaking sooner. In addition, analysis of the net energy yield or energy return on investment (EROI) indicates that energy is becoming increasingly expensive and difficult to produce and that the alternatives to conventional fossil fuels have a lower net energy yield. Many advocate renewable energy to meet future needs while controlling emissions to mitigate climate impacts, but renewables have several serious limitations including intermittency, low energy density, storage, substitutability, slow market penetration, low EROI, and material constraints.

Louisiana's economy is fundamentally dependent on fossil fuels; much revenue is derived from navigation, tourism, agriculture and fisheries, and oil and gas and petrochemical industries, all of which are energy intensive. Coastal management and restoration are very energy intensive and thus vulnerable to energy availability and cost. Dredging of navigation channels, building and maintaining levees, pumping dredged sediments in pipelines, building and maintaining large water control structures, and maintaining flood protection systems for areas near or below sea level are highly energy intensive. Projected trends for climate change and energy scarcity threaten the sustainability of the MRD and its restoration. Given future climate and energy trends, Rutherford et al. conclude that management based on the natural functioning of the delta will be more sustainable.

In chapter "The Costs and Sustainability of Ongoing Efforts to Restore and Protect Louisiana's Coast", Wiegman et al. review the ongoing efforts under the Coastal Master Plan to restore the Mississippi delta and discuss how energy and financial constraints will influence the sustainability of restoration plans. The CMP has to fulfill several competing objectives, most notably creating sustainable land in the delta, while supporting economic benefits from ecosystem services and a working coast. Due to uncertainties about the delta's ecosystems and the level of specificity required to manage tradeoffs between the plans objectives, the CMP has coevolved with cutting edged delta science and numerical modeling.

An example of the difficulty meeting the CMPs dual objectives of building sustainable land and promoting economic benefits in the Mississippi Delta is the battle between diversions proponents and coastal stakeholders, especially fishers (see chapter "Large Infrequently Operated River Diversions for Mississippi Delta Restoration"). But 2012 and 2017 CMP models indicate that long-term restoration and land building is not possible without large river diversions and that reversing land loss on the delta scale is not likely. While it is clear that over the long term, river diversions will be a necessary and vital aspect of coastal restoration and sustainable ecosystems, it is possible that in the short term diversions could be a detriment to the economy, as fishers struggle to adapt to changes in the Breton Sound and Barataria estuaries, as benefits from land building from diversions are more slowly realized. CMP model projections suggest that most saline and brackish marshes will disappear even with an aggressive restoration effort (See Fig. 2 in chapter "Introduction – Changing Conditions in the Mississippi Delta from 1700 to 2100 and Beyond: Avoiding Folly").

Uncertainties regarding benefits from diversions and the need to protect coastal infrastructure now, support the necessity of hydraulic dredging for marsh creation and beach and dune restoration projects. While marsh and dune creation provide immediate results, they require re-nourishment every several decades. With increasing energy prices and rates of sea-level rise, the costs these systems may prove unsustainable (see chapter "Energy and Climate – Global Trends and Their Implications for Delta Restoration").

Another problem facing coastal restoration is lack of funding. The estimated total cost of the 50 year plan is likely in excess of $90 billion, much of which has not been identified. Increasing energy costs and climate challenges will increase the cost of coastal restoration and protection even further (see chapter "Energy and Climate – Global Trends and Their Implications for Delta Restoration"). For example, the cost sustain an area of coastal marsh with marsh creation and nourishment over from 2017 to 2100 could increase by a factor of 12 under the worst case energy and sea level forecasts (see chapter "Large Infrequently Operated River Diversions for Mississippi Delta Restoration"). Unlike marsh creation, once diversions are built they can continue to function, likely for over a century, at very little additional cost. This suggests that ecological engineering, which works with and uses the energies of nature, is a more sustainable strategy to choose restoration strategies that minimize use of energy from fossil fuels and maximize use of energy from natural processes.

Day et al. (2016, reprinted in chapter "Large Infrequently Operated River Diversions for Mississippi Delta Restoration") discuss the potential of using large infrequently operated diversions for coastal restoration. They based their analysis on a historical survey of large freshwater inputs to the coastal zone as well as some recent modeling efforts. Diversions are a central component of the State Coastal Master Plan (see chapter "The Costs and Sustainability of Ongoing Efforts to Restore and Protect Louisiana's Coast"). However, opposition to diversions has developed based on a number of perceived threats. These include over-freshening of coastal estuaries, displacement of fisheries, perceived water quality problems, and assertions that nutrients in river water leads to wetland deterioration. In addition, growing climate impacts and increasing scarcity and cost of energy will make coastal restoration more challenging and limit restoration options. They address these issues in the context of an analysis of natural and artificial diversions, crevasse splays, and small sub-delta lobes and suggest that episodic large diversions and crevasses (>5000 $m^3\ s^{-1}$) can build land quickly while having transient impacts on the estuarine system. Small diversions (<200 $m^3\ s^{-1}$) that are more or less continuously operated build land slowly and can lead to over-freshening and water level stress. The analysis of land building rates for different sized diversions and impacts of large periodic inputs of river water to coastal systems in the Mississippi delta indicates that high discharge diversions operated episodically will lead to rapid coastal restoration and alleviate concerns about diversions. Single diversion events have deposited sediments up to 40 cm in depth over areas up to 130–180 km^2. Some crevasses have operated intermittently for over century with accretion rates as high as 1–4 cm/year. This approach should have broad applicability to deltas globally.

Additional work was carried out to further investigate the impacts of diversion. To demonstrate the land building potential for a large, intermittent diversion, an existing 2-dimentional delta progradation model was used to conduct a rapid assessment of a large (7079 m^3/s), intermittent river diversion into a forested wetland system bordering Lake Maurepas. Results demonstrate that impressive land building is possible. In the central SLR scenario (SLR-3, 1.0 m by 2100), there were 123 km^2 of predicted land gain by 2050. However, these results are highly variable based depending on the rate of sea-level rise.

A marsh eco-geomorphology model was developed to investigate the influence of energy prices, sea level rise and suspended sediment concentrations on the costs and benefits of marsh creation in the Mississippi Delta. The model simulates how the introduction of sediment from a river diversion would affect the lifespan of Marsh Creation (MC) projects and the cost of sustaining the created marsh. Results indicate that diversions increase the benefits (marsh lifespan) relative to costs of sustaining coastal marsh with hydraulic dredging. Putting MC projects near the sediment plume of a diversion should be prioritized. But regardless of river input, sustaining marsh elevation into future will require ever increasing effort. Accelerating sea-level rise will require more frequent re-nourishment resulting in increased unit costs per restoration effort due to rising energy prices. The combined effects of forecasted increases in energy prices and sea-level rise resulted in large increases in the cost to sustain coastal marsh; up to eight times the cost for a no change scenario in SLR. The results indicate that it will not be possible to sustain areas of land in outer portions of the delta without diversions. Diversions and marsh creation projects will be more effective if placed in upper basins where subsidence is low and where the potential for river diversion is high.

One solution to flooding threats is to move up. Colten (chapter "Raising Urban Land: Historical Perspectives on Adaptation") reviews the history of how this has been done in the past. Around the globe, inhabitants of coastal wetlands have found creative solutions to escaping waterlogged settings. Prehistoric residents of the lower Mississippi River delta and other coastal areas, created extensive midden mounds that served as elevated community sites. French colonists built houses raised on piers to avoid inundation, and levees augmented raised houses. In early New Orleans, land owners used refuse to fill low-lying sites. Elsewhere, cities have reclaimed waterfronts by building dikes and filling behind them. As early as the mid-nineteenth century, a New Orleans engineer recommended a process to actually raise the level of existing land in this floodplain city. His plan would have required the construction of numerous ring levees and feeder canals that would direct floodwaters into the "colmates." Once impounded, the sediment would settle out inside these holding basins and raise the land level. Is such a procedure possible in a modern urban setting? This is addressed in chapters "Raising Buildings: The Resilience of Elevated Structures" and "Raising New Orleans: The Marais Design Strategy".

Two case studies offer insights into how Chicago and Galveston implemented significant programs to raise the level of urban land. Chicago, following epidemic disease outbreaks in the mid-nineteenth century, sought to improve municipal sanitation. With little topographic gradient to work with, the city laid sewers on top of existing streets and then mounded the road beds above the sewers. Businesses and residents then raised structures to accommodate the new human-made elevations. In Galveston, hurricane damage in 1900 drove a program to elevate much of the city. Relying on local financing the city pumped sand into the urban core of the city and required businesses and residents to raise their own structures to meet the new grade. In both instances, courts backed up the land rasing projects and the imposition of costs on land owners.

To carry out a similar plan in New Orleans would require substantial sums of public and private investment. The cost, however, based on very preliminary estimates, might be less than recovering from a storm with the magnitude of a Hurricane Katrina. To advance toward such an audacious plan, thorough community participation is essential throughout the entire process. These issues are addressed in chapter "Raising New Orleans: The Marais Design Strategy".

In chapter "Raising Buildings: The Resilience of Elevated Structures", Erdman et al. document the methods and materials used in elevated architectural projects in New Orleans and coastal Louisiana in general. They summarize FEMA policies and guidelines that control building construction and retrofitting within flood zones areas. It goes further than just documentation by asking the question about what it means to build above the land. . .something that society and architectural practice has not addressed directly except in special historical conditions such as Venice, Italy.

In chapter "Raising New Orleans: The Marais Design Strategy", Erdman et al. develop a design proposal for elevating much of the city of New Orleans. In a pre-peak oil world, there is still the energy necessary to elevate a city such as New Orleans. There are examples of this type of massive land reconstruction strategy in history as far back as the Renaissance in Italy but more contemporary examples of Chicago and Galveston as discussed in chapter "Raising Urban Land: Historical Perspectives on Adaptation" are also part of the inspiration.

The proposal begins by proposing that the Lake Pontchartrain waterfront of New Orleans be the first to be raised. The proposal involves dredging lake sediments and constructing a higher, buildable ground as was done in the Lakeview section of the city in the 1930s. The higher ground would be fronted by a cypress swamp. Low to fresh conditions would be maintained by river water, storm water pumps, or perhaps treated effluent. The swamp and higher ground form a significant protective boundary that would diminish impact of sea level rise and hurricane storm surge.

The second aspect of the proposal aims to build a series of polders/marees in a crisscross pattern across the city. These are conceived of as levee-like structures that elevate roadways and other existing infrastructure and create a series of contained areas within their boundaries. The ground used to create the elevated boundaries is from dredged sediment from the Mississippi River and piped to a polder, to drain naturally and become workable ground. Once dry enough to construct land with, the soil would be trucked around the city to build the enclosing high ground and connective ramps. Alternatively, polders could be filled in.

As the elevated roadways are completed, citizens would make decisions about their individual houses or work as collectives to make larger scale decisions. As individuals, houses could be raised through the most typical method of pilings or a combination of methods as designated by FEMA and building code standards. As collectives, entire areas could be elevated with sediment and ultimately, the safest method would be to fill an entire area of a maree.

Institutions, which are much more difficult to elevate due to their size, could be relocated to perimeter elevated roadways over time. Many institutions in New Orleans have already recognized their precarious situation with regard to flooding and have taken measures to elevate, move or otherwise rehabilitate their facilities.

This design proposal may seem radical but it may be the best strategy for continued inhabitation of New Orleans. Many policy and municipal changes would be necessary to make such a project possible and it would require great leadership and community buy-in to make this a realizable goal. The alternative scenario for a future New Orleans of doing nothing to address sea level rise and subsidence is to wait for another massive disaster, at the scale of Hurricane Katrina or Sandy, to completely devastate the city so badly that there is no return.

Hemmerling (chapter "Eroding Communities and Diverting Populations: Historical Population Dynamics in Coastal Louisiana") notes that the traditional notion of inherent resilience as a continual functioning of community during "non-crisis" periods does not apply perfectly to coastal Louisiana. Indeed, it could be suggested that operating in a non-crisis mode in the face of the omnipresent threat of coastal land-loss and climate change is counterproductive, indeed self-destructive. Inherent resilience must necessarily incorporate adaptive resilience factors if coastal communities are to persist and thrive in the coastal environment.

Clearly, a community that is stable is much more likely to be resilient than one that is not but there is not a direct relationship between the two. Repeated shocks reduce community resilience and may lead to precipitous decline when impacted by smaller shocks or events. The key to building and maintaining community resilience is the adaptive capacity of that community. Resilience includes both the system's capacity to return to the state that existed before a disturbance and the system's ability to advance the state through learning and adaptation. Put simply, a community that persists is stable, while one that persists and adapts is resilient.

New Orleans and its surrounding region is a prime example of a population that may have been deemed stable in the years following the hurricanes of the 1960s while at the same time becoming less resilient. By all measures, the region appeared to have recovered from the devastating storms of the 1960s. Yet, recovery from Hurricanes Katrina and Rita in 2005 appears almost as a reverse image of that seen following Hurricanes Betsy and Camille. Orleans Parish, and its immediate neighboring parishes on the south shore of Lake Pontchartrain (Jefferson, Plaquemines, and St. Bernard), as well as Camerion Parish, all experienced significant sustained population loss. Social vulnerability also plays a key role in community resilience and the ability of communities to recover from hazards events. Factors such as poverty, minority status, and age are determinants of vulnerability across a wide spectrum of different hazards.

The level of natural resource employment is also an important determinant of a coastal community's social vulnerability to the impacts of land loss, sea level rise, and tropical storm events. For example, the combination of Hurricane Katrina and the Deepwater Horizon Oil Spill tipped southeast Louisiana's demographic and economic composition such that it is now more vulnerable than at any time in recent decades.

Residents who choose to remain in increasingly hazardous locales must find innovative ways to adapt. The development of resilience-based planning strategies will reduce the types of maladaptive development that have, historically, placed Louisiana's coastal populations in harm's way. The key to this effort must be a fully integrated coastal protection and restoration plan, one that incorporates aspects of the physical, biological, and human environments. But creating such community resilience through adaptation is not an easy process.

The challenges around building more resilient communities in coastal Louisiana and other locations around the world revolve largely around adaptively managing change. Stability is a temporary state that exists in the midst of change and upheaval. Seemingly stable states can suddenly shift and become something new. The key to successfully navigating coastal change is to understand the multiscale historical processes that have been driving that change and assuring to the greatest extent possible that such shifts result in positive outcomes for residents and the ecosystems that they rely on in both the short term and long term.

In chapter "Developing Legal Strategies for a Sustainable Coast", Wilkins asks whether legal systems can assist the process or at least not hinder too much the ideas proposed in this book for a sustainable coast. Legal issues affecting adaptation are usually focused on property rights, civil rights, and enabling of risky behavior by creating entitlements to government programs. The first impulse to suggestions for change is always to resist any significant change in the status quo. It will take a great deal of effort to influence cultural and individual psychology enough to allow effective adaptation.

Adaptation strategies that attempt to affect constitutionally protected property rights are often a barrier to the type of actions proposed herein. Governments always have the power of eminent domain at their disposal to carry out valid public purposes as an exercise of their police power. Recent cases decided by the U.S. Supreme court generated a wave of state anti-takings legislation that limited state government takings authority for purposes of private economic development. Louisiana joined the majority of states by enacting legislation that in the opinion of some legal scholars severely hampers the state's ability to use eminent domain to conduct land use reforms for coastal protection and restoration purposes if the actions are seen to benefit private economic interests. Land use changes affecting private property by the use of eminent domain, or involving takings resulting from regulatory restrictions or damage from government action, should have as the primary purpose of the action the protection of public safety with any economic benefit to private parties as an incidental but added justification.

Eminent domain acquisitions and FEMA buyouts can also be affected by title issues when an heir who has merely taken possession of the property of a deceased relative cannot "legally" sell the property even if they have been in possession of it for years. Another issue unique to Louisiana involves the issue of subsurface minerals. For many in coastal Louisiana losing mineral rights is an important negative effect of moving. An amendment to that provision or a similar statute could allow reservation of mineral rights in situations where land is condemned and rebuilt for resiliency purposes.

Several federal programs can affect the financing of relocation. HUD's Community Block Grant Entitlement Program provides grants to local governments for development in urban areas. HUD's Community Development Block Grant Disaster Recovery Program applies to Presidentially declared disasters for the purchase of damaged properties in the flood plain and relocation of displaced residents to safer areas, including payments for people and businesses displaced by disaster. FEMA's Pre-Disaster Mitigation Program provides annually distributed funds for property acquisition and relocation as part of hazard mitigation planning. FEMA's Flood Mitigation Assistance provides pre-disaster funds to reduce or eliminate risk of flood damage through property acquisition, demolition, and relocation. It applies to buildings that are insured under the National Flood Insurance Program. FEMA's Hazard Mitigation Grant Program provides grants after a major disaster declaration to state and local governments for implementation of long-term hazard mitigation measures. Funds can be used to acquire and demolish property and relocate displaced residents. HUD-HOME's Investment Partnership Program provides formula grants to states and local communities to develop affordable housing for low-income individuals. The funds may be used for direct rental assistance or to develop housing stock for rent or ownership; for building, buying, and/or rehabilitating buildings or for site acquisition and improvement, including demolition of dilapidated housing to make way for HOME-assisted development and to cover some relocation costs.

A number of other programs can aid in legal issues related to restoration. The Uniform Relocation Assistance and Real Property Acquisition Policies Act (URA) was enacted to provide consistent and equitable treatment of people displaced by acquisition of property for federal projects and federally assisted programs including state and local programs.

The Gulf of Mexico Energy Security Act (GOMESA), and the Resources and Ecosystems Sustainability, Tourist Opportunities, and Revived Economies of the Gulf Coast States Act (RESTORE Act) do not not specifically list non-structural flood protection measures but it appears that their reference to "coastal flood protection and related infrastructure" and hurricane protection and infrastructure" could be interpreted to allow the use of funds for non-structural flood protection measures such as building up the grade in polders as described in chapter "Raising New Orleans: The Marais Design Strategy".

In most jurisdictions land use planning is under the sole authority of local governments and most states in the Gulf of Mexico region have resisted establishing state planning authorities. Disaster relief from federal and state governments seems to always be available to assist with recovery and rebuilding thereby pushing the cost of poor decisions off on others. Governments can be held accountable when development is allowed that causes existing developed areas to flood but historically have not been held liable for allowing development in flood risk areas unless the new development causes other development to flood. That may be changing. As technology becomes increasingly more accurate in predicting risk it will be much harder for governments to hide behind uncertainty and policy making discretion as defenses to liability. As it stands now the only direct methods of relocating people out of high risk areas are eminent domain or voluntary buyouts. Given enough political will, other incentives, disincentives and market based approaches could be effective in accomplishing the goals expressed in chapter "Eroding Communities and Diverting Populations: Historical Population Dynamics in Coastal Louisiana" and wider hazard mitigation advances.

As consumers have become more quality conscious and more aware of social and environmental costs of economic activity, business has begun to accommodate demands for higher quality, greener and more socially appropriate goods and services. One method to accomplish this end has been the establishment of standards by industry and non-governmental organizations (NGOs) that are outside governmental regulatory processes but fill a vacuum in the absence of government regulation or complement existing regulations.

Final Summary

The implications of the information presented in this book are that the timing of decisions made about Mississippi delta restoration will have a significant impact on the success of restoration. Over the 50-year timeframe of the State Master Plan, climate change will grow more challenging and resource scarcity will become more severe. By 2100, it is projected that sea level will increase by 1 to 2 m, category 4 and 5 hurricanes will increase in frequency, extreme weather such as very heavy precipitation events and droughts will become more common, and peak Mississippi River discharge will increase significantly. Resources, especially energy, will become more scarce and expensive. This means that restoration will become more challenging and expensive and options will become limited. Indeed, model projections of the 2017 Coastal Master Plan show that most coastal wetlands will be lost within 50 years even with aggressive restoration.

Restoration Should Focus on Activities That Are More Sustainable Over the Long-Term Without Large Energy Inputs Care should be taken about spending large amounts of resources on efforts that are likely not sustainable over the long-term. An example is marsh creation in areas of high relative sea-level rise that cannot be nourished with river diversions. For much of the area proposed for marsh creation (e.g., parts of the Chenier Plain), nourishment with diversions is not practicable. Relative sea-level rise of 1–2 m may overwhelm the ability of created marshes to maintain elevation.

How much of the flood protection system using levees is sustainable given projections for relative sea-level rise and increasing cost of energy? There is a need to do careful and realistic optimization analyses for these situations and to balance moving people and infrastructure up and inland versus protection in place. The relocation of the population of Isle de Jean Charles serves as an example of this.

Analyses of leveed areas that currently drain by gravity turning into pumped polders are needed. Given a sea-level rise of one to two meters, much of the areas behind hurricane protection levees will have to become pumped over coming decades; an outcome that is perhaps unaffordable. Using river sediment to create relatively small areas of high ground may be an option. Rising parts of New Orleans as described in chapter "Raising New Orleans: The Marais Design Strategy" is an example of this. There is a need to face the question of whether areas below sea level are sustainable in a climate-challenged, energy-scarce future (e.g., Day et al. 2016). New Orleans is the prime example of this, but given sea level rise projections, most of the areas protected by levees will be below sea level by the second half of this century.

What Should Be Done First?

Expensive, energy-intensive projects that are long lasting, more sustainable, and convey long-term benefits should be done as early as possible. This includes especially large effective diversions. The Maurepas diversion proposed in chapter "Large Infrequently Operated River Diversions for Mississippi Delta Restoration" is an example of this. If built within a decade, these diversions will still be relatively inexpensive but will last well into the next century. An example is the Bonnet Carré Spillway that will be a century old in 16 years and is likely to last for several decades or more after that. Once constructed, such large diversion would be cheap to operate and convey multiple benefits such as flood protection, fisheries enhancement, wetlands restoration, carbon sequestration, and others.

How Much of the Coast Can Actually Be Saved This is a critical and central question for coastal restoration but one that is not easy to answer. It is accepted by many, and a projection of CMP modeling, that significant portions of the lower coast are not sustainable given what is known about projections for climate change and energy scarcity for the next 100 years. Care should be taken not to spend large amounts of funding and resources on projects that are not sustainable. There is a need to focus on efforts that can be largely sustained by natural processes. This is ecological engineering.

In summary, given the growing constraints imposed by climate change and resource scarcity and the projections for loss of most coastal wetlands even with the current proposed plan, it may be that a dramatically new approach will be required. It is the goal of this book to provide a framework of what a new approach might look like.

References

Day J, Kemp GP, Freeman A, Muth DP (eds) (2014) Perspectives on the restoration of the Mississippi Delta: the once and future delta. Springer

Day J, Agboola J, Chen Z, D'Elia C, Forbes D, Giosan L, Kemp P, Kuenzer C, Lane R, Ramachandran R, Syvitski J, Yanez A (2016) Approaches to defining deltaic sustainability in the 21st century. Estuar Coast Shelf Sci, doi.org/10.1016/j.ecss.2016.06.018

Hoyos CD, Agudelo PA, Webster PJ, Curry JA (2006) Deconvolution of the factors contributing to the increase in global hurricane intensity. Science 312(5770):94–97

Knutson TR, McBride JL, Chan J, Emanuel K, Holland G, Landsea C, Held I, Kossin JP, Srivastava AK, Sugi M (2010) Tropical cyclones and climate change. Nat Geosci 3(3):157

Maggio G, Cacciola G (2012) When will oil, natural gas, and coal peak? Fuel 98:111–123

Melillo JM, Richmond TT, Yohe G (2014) Climate change impacts in the United States. Third National Climate Assessment

Murphy DJ, Hall CA (2011) Energy return on investment, peak oil, and the end of economic growth. Ann N Y Acad Sci 1219(1):52–72

Tessler ZD, Vörösmarty CJ, Grossberg M, Gladkova I, Aizenman H, Syvitski JPM, Foufoula-Georgiou E (2015) Profiling risk and sustainability in coastal deltas of the world. Science 349 (6248):638–643

Index

J.W. Day, J.A. Erdman (eds.), *Mississippi Delta Restoration*,
Estuaries of the World, https://doi.org/10.1007/978-3-319-65663-2

Printed by Printforce, the Netherlands